Bioactive Phytochemicals in By-products from Bulb, Flower and Fruit Vegetables

Mohamed Fawzy Ramadan
Editor

Bioactive Phytochemicals in By-products from Bulb, Flower and Fruit Vegetables

Springer

Editor
Mohamed Fawzy Ramadan
Faculty of Applied Medical Science, Clinical Nutrition Department
Umm Al-Qura University
Makkah, Saudi Arabia

ISBN 978-3-031-77398-3 ISBN 978-3-031-77399-0 (eBook)
https://doi.org/10.1007/978-3-031-77399-0

This Springer imprint is published by the registered company Springer Nature Switzerland AG
The registered company address is: Gewerbestrasse 11, 6330 Cham, Switzerland

Preface

The development of waste management is one of the cornerstones to ensuring the industry's economic, social, and environmental sustainability. Many bulb, flower, and fruity vegetables are cultivated, wherein the bio-waste produced during vegetable processing and manufacturing is millions of tons annually. By-products from the bulb, flower, and fruity vegetable processing are rich sources of bioactive phytochemicals, including carbohydrates, phenolic acids, phyto-sterols, volatile oils, carotenoids, fixed oils, fatty acids, and amino acids, which have nutritional and positive impacts on human health. Recovery of bioactive compounds involves their removal by manufacturing systems that do not affect their chemical composition and physicochemical properties to be used as ingredients. The chemical composition of bulb, flower, and fruity vegetable processing by-products varies due to plant differences, harvest time, and manufacturing methods.

In recent years, reports have focused on utilizing agro-industrial processing wastes in edible and non-food products. Some studies and reference books on waste utilization are available, and their emphasis is mainly on waste treatment. However, none of those sources comprehensively deal with the bioactive phytochemicals from bulb, flower, and fruity vegetable processing by-products. This book will be structured into chapters covering an overview of bulb, flower, and fruity vegetable processing bio-wastes' phytochemical profile, nutritional value, and biochemical and physicochemical traits. This book will serve as a comprehensive reference book for researchers, educators, students, food and industry processors, and policy developers, providing up-to-date insight.

This book aims to establish a discussion forum on the phytochemistry, functional properties, and health-promoting impacts of bioactive compounds in bulb, flower, and fruity vegetable processing wastes and the applications of those by-products in edible and non-food items. The book also goes through pertinent recovery-related issues and different techniques for developing new products.

The book is a valuable scientific source for researchers researching the recycling of agro-industrial bio-wastes from bulb, flower, and fruity vegetable processing by-products. Moreover, phytochemists, technicians, food scientists, nutritionists, and pharmacists will find this book helpful in creating novel drugs and food items. With a focus on the connection between several diseases and the health-enhancing impact of the plant-active compounds in the bulb, flower, and fruity vegetable bio-wastes, the authors of this work contribute the functional traits of plant-active compounds from bulb, flower, and fruity vegetable bio-wastes that might be responsible for health-promoting potential.

The book contains chapters that cover the following topics:

1. Economic values of by-products from bulb, flower, and fruity vegetable processing
2. Chemical composition and active compounds of extracts from the bulb, flower, and fruity vegetable bio-wastes processing by-products
3. Functional and biological traits of active compounds and extracts from the bulb, flower, and fruity vegetable bio-wastes processing by-products
4. Edible and non-food uses of phyto-extracts and active compounds from bulb, flower, and fruity vegetable bio-wastes processing by-products

The editor thanks all the contributors for their valuable contributions and collaboration during the book's production. The Springer staff's help and support, especially *Sofia Costa* and *Neelofar Yasmeen*, was essential for completing the task and is appreciated.

Makkah, Saudi Arabia — Mohamed Fawzy Ramadan

About the Book

Utilizing main goals, resource optimization, and by-product reduction, the circular economy aims to reuse plant resources for advantageous uses. The sustainability of industrialization and the economy are topics covered by agro-industrial waste management. The circular economy was something the European Commission aimed to promote. Businesses predicted that when the Green Deal was unveiled, which aimed to make Europe carbon neutral by 2050, bio-waste and its harmful environmental effects would decline. The "*Farm to Fork*" plan also aims to integrate circular economy concepts into food processing successfully.

Several studies have demonstrated the recovery, composition, and functionality of bioactive compounds from vegetable processing by-products. Furthermore, phytochemicals extracted from different vegetable species, such as bulb, flower, and fruity vegetables, have been shown to have health-promoting impacts and biological traits in various models. Therefore, transforming vegetable processing by-products into added-value natural additives is promising.

Bulb, flower, and fruity vegetables processing by-products and wastes represent a significant disposal problem for the industry. However, they are promising sources of value-added phytochemicals, emphasizing retrieving bioactive phytochemicals and technologically important secondary metabolites. This makes them highly suitable as raw materials for producing secondary metabolites of industrial significance. Besides, manufacturing conditions affect the by-products generated during vegetable processing. Therefore, these processes might also affect valuable phytochemicals, including antioxidants and vitamins.

This book characterizes phytochemicals in bulb, flower, and fruity vegetables and the health-enhancing traits of those phyto-compounds. As a result of this book's findings, (1) the health-enhancing impacts and applications of phyto-compounds from bulb vegetable, flower vegetable, and fruity vegetable production bio-wastes are discussed, (2) the consumer comprehends the health-enhancing effects of phyto-compounds from bulb vegetables, flower vegetables, and fruity vegetable processing bio-wastes supported by scientific evidence, and (3) the chapters review the recovery methods and special techniques to develop novel applications from specific phyto-compounds, and (4) the economy will be strengthened by incorporating bioactive chemicals from the bio-wastes obtained during the processing of bulb, flower, and fruity vegetables into industrial applications (i.e., edible products, livestock, drugs, and cosmetics).

Bioactive Phytochemicals in By-Products from Bulb, Flower, and Fruit Vegetables aims to provide a discussion on nutritional properties, phytochemistry, and health-enhancing effects of phyto-compounds in bulb, flower, and fruity vegetable processing bio-wastes and the use of these by-products in food and non-food applications. The benefits of using vegetable processing by-products to boost health are outlined.

Key Features

- Explores biochemistry and functional properties of phyto-compounds in bulb, flower, and fruity vegetable manufacturing by-products
- Includes literature reviews and coverage of recent developments
- Authored by international scientists and industry experts
- Addresses the growing application areas, including nutrition, pharmaceuticals, functional foods, and cosmetics

Readership

- Scientists, researchers, and students interested in pharmacology, chemistry, environmental sciences, food science, and technology
- Developers of novel nutraceuticals, pharmaceuticals, cosmetics, and R&D researchers working in phytochemical sectors

Contents

About the Editor and Contributors

About the Editor

Mohamed Fawzy Ramadan is Professor of Food Chemistry and Biochemistry at the Department of Clinical Nutrition, Faculty of Applied Medical Science, Umm Al-Qura University, Makkah, Saudi Arabia.

Prof. Ramadan obtained his Ph.D. (*Dr. rer. nat.*) in Food Chemistry from the Berlin University of Technology (Germany, 2004). He continued his post-doctoral research at ranked universities such as the University of Helsinki (Finland), Max Rubner Institute (Germany), Berlin University of Technology (Germany), and the University of Maryland (USA). In 2012, he was appointed as Visiting Professor (100% teaching) in the School of Biomedicine, Far Eastern Federal University, Vladivostok, Russian Federation.

Prof. Ramadan published over 400 research papers and reviews in international peer-reviewed journals. He also edited and published several books and book chapters (Scopus *h*-index is 52 and more than 9500 citations). He was an invited speaker at several international conferences. Since 2003, Prof. Ramadan is a reviewer and editor in several highly cited international journals, such as *Earth Systems and Environment*, *Journal of Medicinal Food*, *eFood*, *Journal of Umm Al-Qura University for Applied Sciences*, and *Journal of Advanced Research*. He is the editor-in-chief of *Journal of Umm Al-Qura University for Medical Science*.

Prof. Ramadan received several prizes, including Abdul Hamid Shoman Prize for Arab Researcher in Agricultural Sciences (2006), the Egyptian State Prize for Encouragement in Agricultural Sciences (2009), European Young Lipid Scientist Award (2009), AU-TWAS Young Scientist National Awards (Egypt) in Basic Sciences, Technology and Innovation (2012), TWAS-ARO Young Arab Scientist (YAS) Prize in Scientific and Technological Achievement (2013), and Atta-ur-Rahman Prize in Chemistry (2014).

Contributors

Banzeer Ahsan Abbasi Department of Botany, Rawalpindi Women University, Rawalpindi, Pakistan

Mohamed S. Abd El Hafeez Department of Pharmacognosy, Faculty of Pharmacy, Egyptian Russian University, Badr City, Cairo, Egypt

Adnan Ahmed Department of Botany, University of Sargodha, Sargodha, Pakistan

Habib Ali Department of Agricultural Engineering, Khwaja Fareed University of Engineering and Information Technology, Rahim Yar Khan, Pakistan

Qurat Ul An Institute of Food Science and Nutrition, University of Sargodha, Sargodha, Pakistan

Afifa Asif Institute of Food Science and Nutrition, University of Sargodha, Sargodha, Pakistan

Marta Barberis Department of Drug Sciences, University of Pavia, Pavia, Italy

M. Z. Bauleth Department of Human, Biological & Translational Medical Sciences, Faculty of Health Sciences and Veterinary Medicine, University of Namibia, Windhoek, Namibia

Maggy M. Casey Nairenge Department of Biochemistry, Microbiology & Biotechnology, Faculty of Agriculture, Engineering and Natural Sciences, University of Namibia, Windhoek, Namibia

Ahmad Cheikhyoussef Multidisciplinary Research Services, Centre for Research Services, University of Namibia, Windhoek, Namibia

Science and Technology Division, Multidisciplinary Research Services, Centre for Research Services, University of Namibia, Windhoek, Namibia

Natascha Cheikhyoussef Ministry of Higher Education, Technology, and Innovation (MHETI), Windhoek, Namibia

Raffaella Colombo Department of Drug Sciences, University of Pavia, Pavia, Italy

C. N. Cupido Department of Botany, University of Fort Hare, Alice, South Africa

Haya Fatima PMAS Arid Agriculture University, Rawalpindi, Punjab, Pakistan

Faiza Iftikhar Gorsi Institute of Food Science and Nutrition, University of Sargodha, Sargodha, Pakistan

Farhat Gul Department of Plant Sciences, Faculty of Biological Sciences, Quaid-i-Azam University Islamabad, Islamabad, Pakistan

Mostafa M. Hegazy Department of Pharmacognosy and Medicinal Plants, Faculty of Pharmacy, Al-Azhar University, Cairo, Egypt
Department of Pharmacognosy, Faculty of Pharmacy, Sinai University - Arish Branch, Arish, Egypt

Ashiq Hussain Institute of Food Science and Nutrition, University of Sargodha, Sargodha, Pakistan

Ahmed A. Hussein Chemistry Department, Cape Peninsula University of Technology, Cape Town, South Africa

Shumaila Ijaz School of Biomedical Engineering, Shenzhen University, Shenzhen, China

Martha Ikechukwu Sam Nujoma Marine and Coastal Resources Research Centre, Sam Nujoma Campus University of Namibia, Henties Bay, Namibia

Javed Iqbal Department of Botany, Bacha Khan University, Charsadda, Khyber Pakhtunkhwa, Pakistan

Muhammad Jamil Department of Biotechnology and Genetic Engineering, Kohat University of Science and Technology, Kohat, Pakistan

Khurram Kabir PMAS Arid Agriculture University, Rawalpindi, Punjab, Pakistan

Sobia Kanwal Department of Biology and Environmental Sciences, Allama Iqbal Open University, Islamabad, Pakistan

Tariq Mahmood Department of Plant Sciences, Faculty of Biological Sciences, Quaid-i-Azam University Islamabad, Islamabad, Pakistan

Mahmoud A. Mansour Department of Pharmacognosy, Faculty of Pharmacy, Egyptian Russian University, Badr City, Cairo, Egypt

María del Mar Contreras Department of Chemical, Environmental and Materials Engineering, Universidad de Jaén, Jaén, Spain

Reham Hassan Mekky Department of Pharmacognosy, Faculty of Pharmacy, Egyptian Russian University, Badr City, Cairo, Egypt

Chiara Milanese Pavia Hydrogen Lab, Department of Chemistry & C.S.G.I, University of Pavia, Pavia, Italy

M. Makhaba Chemistry Department, Cape Peninsula University of Technology, Cape Town, South Africa

Giulia Moretto Department of Drug Sciences, University of Pavia, Pavia, Italy

Rizwan Nisar Institute of Food Science and Nutrition, University of Sargodha, Sargodha, Pakistan

Adele Papetti Department of Drug Sciences & C.S.G.I, University of Pavia, Pavia, Italy

Vanessa Pellicorio Department of Drug Sciences, University of Pavia, Pavia, Italy

Mohamed Fawzy Ramadan Department of Clinical Nutrition, Faculty of Applied Medical Sciences, Umm Al-Qura University, Makkah, Saudi Arabia

Ghadir Ali Sayed Department of Biochemistry, Faculty of Pharmacy, Egyptian Russian University, Badr City, Cairo, Egypt

R. Sharma Chemistry Department, Cape Peninsula University of Technology, Cape Town, South Africa

J. Sheehama Department of Human, Biological & Translational Medical Sciences, Faculty of Health Sciences and Veterinary Medicine, University of Namibia, Windhoek, Namibia

Hifikepunye Pohamba and Oshakati Campuses, University of Namibia, Oshakati, Namibia

Sherouk Hussein Sweilam Department of Pharmacognosy, Faculty of Pharmacy, Egyptian Russian University, Badr City, Cairo, Egypt

Department of Pharmacognosy, College of Pharmacy, Prince Sattam Bin Abdulaziz University, Al-Kharj, Saudi Arabia

Chin Xuan Tan Department of Allied Health Sciences, Faculty of Science, Universiti Tunku Abdul Rahman, Jalan Universiti Bandar Barat, Kampar, Perak, Malaysia

Hui Ling Tan School of Hospitality and Service Management, Sunway University, Petaling Jaya, Selangor, Malaysia

Seok Shin Tan Jeffrey Cheah School of Medicine and Health Sciences, Monash University Malaysia, Jalan Lagoon Selatan, Bandar Sunway, Selangor, Malaysia

Seok Tyug Tan Jeffrey Cheah School of Medicine and Health Sciences, Monash University Malaysia, Jalan Lagoon Selatan, Bandar Sunway, Selangor, Malaysia

Siraj Uddin Department of Plant Sciences, Faculty of Biological Sciences, Quaid-i-Azam University Islamabad, Islamabad, Pakistan

Zakir Ullah Department of Plant Sciences, Faculty of Biological Sciences, Quaid-i-Azam University Islamabad, Islamabad, Pakistan

Tabassum Yaseen Department of Botany, Bacha Khan University, Charsadda, Khyber Pakhtunkhwa, Pakistan

Shazia Yqaub Institute of Food Science and Nutrition, University of Sargodha, Sargodha, Pakistan

Part I
General Aspects

Introduction to Bioactive Phytochemicals in By-Products from Bulb, Flower, and Fruit Vegetables

Mohamed Fawzy Ramadan

Abstract Consumer interest in the health-promoting traits of bio-wastes has grown in recent years. The authors of this book describe the biological characteristics of active phytochemicals from bulb, flower, and fruity vegetable processing by-products that may be responsible for health-improving effects, given the link between age-related diseases and the health-improving effects of the active phytochemicals. The benefits of phytoconstituents derived from vegetable processing by-products were discussed, along with the state-of-the-art methods for designing novel food and non-food products by characterizing the active compounds in that bio-waste and paying particular attention to antioxidants. The present book's findings have led to the following outcomes: (1) the book discusses the health benefits and applications of phytochemicals derived from bulbs, flowers, and fruity vegetable processing by-products; (2) it makes consumers aware that these benefits are backed by scientific research; (3) it covers recovery issues and various methods for creating new uses for phytochemicals derived from vegetable processing by-products; and (4) it boosts the economy by utilizing active chemicals from processing bio-wastes to industrial applications (new food items, livestock, pharmaceuticals, and cosmetics).

Keywords Bio-waste · SDGs · Environment · Circular economy · Recycling · Functional food · Food processing · Social benefits · Waste valorization

---

M. F. Ramadan (✉)
Department of Clinical Nutrition, Faculty of Applied Medical Sciences, Umm Al-Qura University, Makkah, Saudi Arabia
e-mail: mfhassanien@uqu.edu.sa

M. F. Ramadan (ed.), *Bioactive Phytochemicals in By-products from Bulb, Flower and Fruit Vegetables*, https://doi.org/10.1007/978-3-031-77399-0_1

1 Introduction

Recovery of safe by-products may contribute to the overall profitability of food operations and help contribute to the future vision for sustainable agriculture. Many of the bioactive compounds in vegetables have been studied sufficiently. However, to date, little is known about the occurrence and levels of bioactive phytochemicals present in the by-products of vegetables. Research has focused on identifying the bioactive compounds present in plants with an emphasis on phenolic compounds due to the numerous functions of these compounds in edible plants (Yu et al. 2021; Wink 2022; Drabińska et al. 2021; Cucu et al. 2022; Guan et al. 2021; Sudheer et al. 2022; Swamy 2020; Wang et al. 2023).

Primary production of edible crops presents the opportunity to recover nutrient and bioactive-rich by-products alongside product harvesting. However, little is known about the phytochemical and potential nutritive profiles of many vegetable by-products. This work categorizes some vegetable by-products according to the edible plant part, covering subterranean, aerial, and true fruits used for vegetable purposes. Given the numerous potential bioactivities of phytochemicals, the findings of this work support moving forward with determining the phenolic compound profiles of the described by-products to facilitate making decisions concerning the potential upcycling of any or some of these by-products in the food industry (Espinosa-Alonso et al. 2020; Wang et al. 2023; Cano-Lamadrid and Artes-Hernandez 2021; Tlais et al. 2020; Jalal et al. 2023; Santos et al. 2022; Sorrenti et al. 2023; Reguengo et al. 2022; Teshome et al. 2023).

Caldeira et al. (2019) calculated that the total amount of food bio-waste at the processing stages was approximately 30.6 million tons yearly. Figure 1 shows the estimated contribution of the top five food subsectors at the processing stages. It is crucial to determine which bio-waste should be given priority for valorization

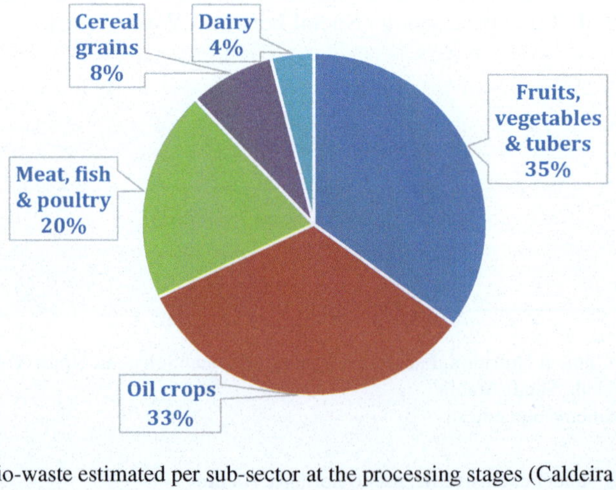

Fig. 1 Food bio-waste estimated per sub-sector at the processing stages (Caldeira et al. 2019)

because different food items are discarded in substantial quantities during the processing stages (Rao et al. 2021; Ramadan Hassanien 2023a, b).

2 SDGs and Environmental-Friendly Food Production

Legal limitations on the disposal of bio-waste have worsened matters because of the harmful effects of by-products on the environment. However, because the recovery of these bioactive substances may be profitable, there is interest in exploiting such bio-wastes to create edible supplements or additions with nutritional value.

The 2030 Agenda for Sustainable Development Goals (SDGs), available at https://sustainabledevelopment.un.org, reflects the attention the world has shown to the problem of food bio-waste. SDG aims to minimize food waste throughout the processing and supply chains and cut global per capita food waste (European Commission 2020; Smeu et al. 2022). Aspects of the value-added chain, such as how plants are grown, moved, processed, sold, and eaten, are the primary connection between people and a sustainable circular economy in the food industry (Ramadan Hassanien 2023a, b). Bio-waste from the inexpensive, effective, and ecologically friendly processing of vegetables could reduce the adverse effects. This strategy aligns with the SDGs for "*Good Health and Well-Being*," which seek to improve human well-being through environmentally friendly food production methods and plants that promote health.

3 By-Products from Bulb Vegetables

Bulb vegetables are those mainly consumed with their swollen underground part, with the reserve derived from the leaves for polysaccharides, which are the primary energy source during their growth. These vegetables belong to the Amaryllidaceae family, which includes 59 genera and over 810 species. The main types of bulb vegetables are (a) *Allium*, bulb onions (*A. cepa* L.) and garlic (*A. sativum* L.), often referred to as white and red onions for their tunics' color.

In the food industry, bulbs are mainly used. However, their production generates considerable by-products, such as leaves, roots, and foliage, discarded in the industry's processing. In a few cases, they are used as animal feeds. Information about the chemical composition of bulbs and leaves is somewhat scattered. Studies have been carried out on bulb by-products or leaves. Bulb by-products show values for dietary fiber higher than 58% dw, ash content between 1 and 6% dw, and protein values between 4 and 14% dw. Leaves have a protein range between 15 and 27% dw, dietary fiber values above 50% dw, and ash from 13 to 34%. When the protein extract from *A. ampeloprasum* has been studied, values of 61.2–95.7% were reported, probably affected by the extraction method but also by the chemical composition of differently prepared leaves (Espinosa-Alonso et al. 2020; Lau et al.

2021; Majumder and Annegowda 2021; Taghian Dinani and van der Goot 2023; Vojvodić Cebin et al. 2020; Ali et al. 2023; Mateos-Aparicio 2021; Salem et al. 2023; Sharma et al. 2021a, b; Ganesh et al. 2022).

3.1 Types of Bulb Vegetables

Bulb vegetables owe their name to the part of the plant that is harvested to be consumed. They are the underground storage organs of certain plants and typically have a delicate flavor and a soft texture. Bulb vegetables are not a homogeneous group; they belong to different botanical families and differ in composition and nutrients. Concerning their edible bulb, these vegetables are usually divided into seven subtypes.

1. True bulb vegetables include plants from the Alliaceae family. They develop from a short stem, the basal plate, which is covered by numerous fleshy scales that grow from the tip of the stem downwards. Garlic is amongst the best-known representatives of bulb vegetables. Other alliaceous crops include onions, shallots, and scallions.
2. Stem and stem-leaf vegetables develop from an underground stem. The best-known representatives are celery and kohlrabi.
3. Tuberous root vegetables are swollen parts of a plant's root system. One of this group's most influential members is the potato plant's tuber. Yam is another example of a tuberous root vegetable; crosne is a lesser-known one eaten in the Western world but is growing in popularity due to its earthy taste and medicinal properties.
4. Bulbous stem vegetables harbor the bulb at the base of a leafy top or stem. Amongst the families that accommodate plants that yield bulbous stem vegetables is Amaryllidaceae (Chourasia et al. 2022; Enyiukwu et al. 2020; Atif et al. 2020; Khosa et al. 2021; Rabinowitch and Goldstein 2020; Sheikh et al. 2022; Chauhan et al. 2021; Wu et al. 2020; Ajayi et al. 2022; Manafi and Nazari 2021).

3.2 Common By-Products from Bulb Vegetables

To know if and from which processed products one can expect by-products, it is essential to know which massive residues, in terms of weight and/or volume, are generated during or after the processing and/or consumption of a specific plant part not suitable or ready for direct consumption and which, therefore, remain little considered or are candidates for disposal on site and/or in landfills. There are less available data produced by the primary sources of the total production of burned vegetables, which is first made up of shallot and, secondly, by the closely related and widely used semi-spicy-variety garlic "*Allium sativum.*"'

Consequently, the small onion family contributes significantly to generating brassica by-products, with a less significant contribution from leek (lew). Fewer, also if still more than 10 million tonnes, the cultivation of suspecting sexual bulbs covers areas, but even smaller is the quantity of vegetables from fruits containing cucumber and pumpkin for the production of unripe bulbs neither immediately consumed nor canned; the same occurs for the fleshy reproductive stems of globe and roman artichoke with, among the first three plant complement, the bulky pack of acid leaf bases, and green and unripe artichoke bracts and the outermost protective leaves (Vojvodić Cebin et al. 2020; Rais and Gautam 2023; Celano et al. 2021; Črnivec et al. 2021; Gnanasundari et al. 2023; Thivya et al. 2023; John et al. 2020; Santiago et al. 2020; Kiran et al. 2024; Stoica et al. 2023).

In addition to colored and sweet or bitter juices from the flesh, the golden carrot-colored skin and the onion skin in shades ranging from pale yellow to brown-red are also used to decorate foods eaten directly. The full or partial bitter net of the bulbs of today's rapidly emerging vegetable crops to improve water economy and fertilization in combination with stress treatments, soil disinfection, and pest management, are left to complete their cycle for rapid reproduction (Wang et al. 2022; Kussmann et al. 2023; Balasubramaniam et al. 2021; Sławińska and Olas 2022; Coelho et al. 2023; Manicardi et al. 2023; Pérez et al. 2021; Mutalipassi et al. 2021; Yang et al. 2023; Fonseca et al. 2021).

4 By-Products from Flower Vegetables

Like other vegetables being studied, flower vegetables are rich in fiber, minerals, and vitamins C and K. The results of reduced-intensity extraction suggest tamarillo peel and seed meal are potentially antihyperglycemic agents. Red cabbages are seasonal vegetables grown worldwide from the Brassicaceae family and are rich in vitamin C, β-carotene, and other bioactive compounds. This biopharmaceutical investigation aims to find bioactive compounds such as vitamins, minerals, and dietary fibers that contain anti-nutrients in the shoots, leaves, and immature old purple sweet corn pods. These fragrant and medicinal benefits were known thousands of years ago and can now be addressed with modern science and technology using advanced biopharmaceutical methods (Orqueda et al. 2020; Amin et al. 2021; Orqueda et al. 2023; Orqueda et al. 2022; Kumar et al. 2024; Medina-López et al. 2022; Añibarro-Ortega et al. 2022; Ritika et al. 2024; Pangestu et al. 2021; Jafari et al. 2023).

4.1 *Types of Flower Vegetables*

Flower vegetables are so named because they form aboveground plant structures on their stems, like the loose-decker in a football game. These plants produce flowers or parts of a flower that, when properly prepared, are usually eaten. The definition of flower vegetables is broad. In Japan, China, and Western countries, some indeterminate and determinate leafy green plants or head Paleo plants are listed together as flowers or particular vegetables, such as artichokes and certain leafy green plants.

The U.S. Department of Agriculture (USDA) suggested categorizing vegetable crops into nine major groups: bulb, fruit, pod, root, stem, leaf, floriculture, tuber, and spice-yielding. In that classification, flowering vegetables are part of bulb, stem, and floriculture crops. Thus, this classification may show the major types of flower vegetables grown and eaten worldwide. Some of the onion and garlic crowd are loosely embraced by the group bulb crops and included as subtending flower vegetables. The onion clan encompasses bulb genera from two plant families, Liliaceae and Alliaceae, and the garlic group includes two *Allium* species (Singh et al. 2020; Faust and Dole 2021; Anumala and Kumar 2021; Rehana and Bala 2022; Garba and Yunusa 2024; Attri et al. 2022; Proietti et al. 2022).

Plants classified as early leafy green bulbing crops offer edible leaf petioles that are valid flower vegetable components. Unfortunately, they spawn hardy growths that are unsuitable for food as potential mature plants and thus are not classified as flower vegetables: four types of land potherbs and three forms of water potherbs cannot capture the same as published representatives of the green bulbing crops (Phillips and Rix 2021; Malik et al. 2021; Dehgan 2023; Sahu et al. 2020; Modarelli; Olotu et al. 2024; Brewster and Rabinowitch 2022; Osafo 2021).

4.2 *Common By-Products from Flower Vegetables*

The residual materials generated from processing bulb vegetables are related to direct waste and by-products from human consumption. As waste material directly results from preparation, human by-product material will not be considered. The common by-products generated from flower vegetables are divided into two main types: waste and a combination of waste and by-products. From a horticultural classification perspective, flowering vegetables (which produce fruiting vegetables) are annual plants in which the edible portion is consumed for the flesh of the vegetative parts. The same plant can set flowers and, unless harvested, will produce seed or fruit; they are exclusively harvested for their unique reproductive organs. The remaining stems, roots, or seeds become waste material (Taghian Dinani and van der Goot 2023; Ueda et al. 2022; Ganesh et al. 2022; Nguyen et al. 2020; Esparza et al. 2020; Tlais et al. 2020; Nardella et al. 2022; Moodley and Trois 2024; Mateos-Aparicio 2021; Fayaz et al. 2023).

5 By-Products from Fruity Vegetables

The last category of vegetables examined in this work is the commercial one of "fruity vegetables," mainly gathered from the fruit botanic family of "Solanaceae." However, results in the present work also comprise vegetables included in the category "bulb and flowery vegetables" previously evaluated and "brassica as well as leafy vegetables" examined in depth beforehand. Tomatoes seemed to be the most largely studied "fruity vegetables" group, pursued by the pepper one, because of their broad global usage and growing consumer interest. Moreover, not much on the complete vegetable has been recorded in comparison to the other two categories, except for an application of a solid-liquid extraction to enhance the extraction yield of trans-resveratrol and the in vitro antioxidant abilities of extracts in the presence of their stainless-steel by-products, screened to improve the characterization of bioactive content. Fungal spoilage on tomato waste during storage showed the occurrence of non-toxic pantogenol or ketolactone (Wu et al. 2022; Amr and Raie 2022; Collins et al. 2022; Vélez-Terreros et al. 2021; Mostafidi et al. 2020; Rai et al. 2022; Bhatkar et al. 2021; Zurbau et al. 2020).

The significant parts of tomato by-products are derived from treating these constituents or the design to assess and enhance their chemical and functional features. The detected bioactive potentials included high α-orac, phenolic content, and antioxidant capacity. The most common part of the pepper trait examined in this fragrance was the pericarp. The treatments they were subjected to (non-treated, plucked, steamed, and sealed) were different than indicated in the tomato trait, except for the pepper trim. They were listed: antioxidant potential, vitamin C and E contents, a sharp scent, color strength, pungency, and phenolic content.

5.1 *Types of Fruity Vegetables*

The botanic characteristic of several vegetables explains the most common terminology, "onion," which is widely accepted for all *Allium cepa* L. species. The nomenclatural distinction between the onion and shallot refers to the treatment of *Ocimum basilicum* L. The bulb is alternatively exhibited in two types: crunchy sweet-type bulbs or red/purple-bulb onions and common onions or yellow-bulb onions. Approximately 33% of shallots mature into multicloves, developing several offset cloves, particularly in the banquet or culinary exhibited Allium species. The Filanziada smoke-dried-onion bulb exhibits a round shape of sweet multi-clove-onions. The macro-table sweet Spanish onions are big globe typical bulb varieties.

Fruity vegetables are often called pepo, squashes, marrows, and courgettes. Watermelons are cucurbits presented in the list of fruits. Tomatoes can range from small cherry tomatoes to larger beefsteak varieties. There are vegetables collectively referred to as "fruity vegetables" including tomatoes, eggplants, zucchinis, and cucumbers that ripen seeds in the fruit so they can grow from the seed. Because of its vital functions, bioactive phytochemicals have become critical for human health. Low bioavailability and absorption, mediated by the enzyme, limit the absorption of some bioactive phytochemicals such as allicin from garlic, the solubility of some essential oil monoterpenes and sesquiterpenes in carrot essential oil (Sansan et al. 2024; Fernandes et al. 2020; Hammad et al. 2023; Arena et al. 2024; Yusupov et al. 2021; Brahimi et al. 2022; Ochar and Kim 2023; Ricciardi et al. 2020; Manjunathagowda and Anjanappa 2020; Saleh et al. 2021).

5.2 Common By-Products from Fruity Vegetables

Sustainability in the agro-food industry exceeds respect for the environment and society, encompassing reducing production costs and obtaining economic returns. With this in mind, various studies have sought to identify the leftovers from industrial food production and their value. In that sense, the by-products or residues obtained after processing activities are essential, as they can reach approximately 50–80% of the starting material, depending on the processing technology. In recent years, the phytochemical and sensory profile of by-products from different vegetable sources has been summarized, and various strategies for using these compounds have also been discussed, such as obtaining high-added-value extracts, obtaining bioactive compounds with technological applications, and padding with health-promoting compounds.

The provided key summary of these vegetable processing by-products (often formed in large amounts) would benefit researchers. Exploring bioactive phytochemicals in this work aims to provide an overview of possible fruitful vegetable processing bioactive residues, which would precede the broader expert review into vegetable processing by-products (Sganzerla et al. 2022; Ribeiro et al. 2022; Eliche-Quesada et al. 2021; Zhang et al. 2023; Murtaza and Ameer 2022; Liu et al. 2022; Nawaz et al. 2020; Eslami et al. 2022; Zhou et al. 2021).

6 Methods for Extraction and Analysis of Phytochemicals

Vegetable by-products have been reported to contain a variety of bioactive phytochemicals beneficial for their re-usage as foods, beverages, weight-control agents, pharmaceuticals, etc., which deserve more in-depth study. Exploring bioactive compounds from by-products is a promising strategy to maximize the overall economic benefits of vegetable crop production and reduce the environmental impact caused by the refuse of waste (Marcillo-Parra et al. 2021; Sorrenti et al. 2023; Coman et al. 2020; Espinosa-Alonso et al. 2020; Majumder and Annegowda 2021; ‘Aqilah et al. 2023; Ramadan Hassanien 2023a, b; Oleszek et al. 2023).

The extraction of phytochemicals from by-products often employs a combination of liquid-solid and liquid-liquid extraction techniques to ensure a high yield efficiency. Ultrasonication, microwave, and high-temperature extraction are common liquid-solid techniques for vegetable by-product-based phytochemicals. Soxhlet extraction is still used in some laboratories because it can provide complete extraction. Supercritical extraction techniques and accelerating solvent extraction-based techniques can provide lipophilic/thin-water bioactive compound-enriched extracts in an efficient and eco-friendly way. Solid phase extraction or microextraction is often used as a pre-treatment or projection tool for eliminating impurities or analyzing targeted compounds. Some peptides (prebiotic activity) and pigments could be extracted using enzymatic hydrolysis and fermentation techniques. A new extraction technique (like a gas-phase transport-cold trap system) has been developed to extract volatile compounds such as essential oils from plant tissues. An alternative, cost-effective technique to produce essential oil and by-products of water extraction can be hydrodistillation. Different extraction techniques may produce biochemical constituents with various biological activities (Dias et al. 2021; Castellano et al. 2022; Huang and Huang 2020; Biris-Dorhoi et al. 2020; Zhang et al. 2020a, b; Bodoira and Maestri 2020; Sabir et al. 2020; Chávez-González et al. 2020; Batiha et al. 2020).

7 Applications and Potential Benefits of Bioactive Phytochemicals from Vegetable By-products

The physical appearance (e.g., shape, color) and chemical and textural characteristics of vegetable by-products determines the potential application into the food, serves as an ingredient in animal feed, or is a source for bioactive compound extraction using various methods. In practice, bioactive phytochemicals from vegetable by-products can be used in the pharmaceutical and cosmetics industries (e.g., as natural antioxidant and antimicrobial agents, anti-inflammatory and anti-cancer agents), in the food industry (as a natural food preservative and flavoring agents, and also as a food supplement for functional foods or nutraceuticals due to various health benefits), and in the agrochemical and wastewater treatment processes. The applications of bioactive phytochemicals could be medium-cost-effective due to the pristine or minimal processing and extraction development, the potential of multi-stakeholder partnership, and increasing consumer health awareness.

The major selling point of using bioactive phytochemicals from vegetable by-products would be its organic and eco-friendly perception despite the health, financial, environmental, and social benefits. Additionally, using bioactive compounds from vegetable by-products would help significantly reduce vegetable waste disposal and mitigate the environmental impact of raw material production. This will attract interest and promote change in the production chain or impact the current supply chain soon when the integration and implementation of utilities and waste conversion technologies from the bioeconomy concept of the supply chain and region are rising (Jalal et al. 2023; Lau et al. 2021; Trigo et al. 2022; Rațu et al. 2023; Taghian Dinani and van der Goot 2023; Malenica et al. 2022; Dilucia et al. 2020; Socas-Rodríguez et al. 2021; Rao et al. 2021; Ueda et al. 2022).

8 Aims and Features of the Book

This book aims to establish a multidisciplinary forum for discussion on the phytochemistry, functional properties, health-promoting effects of bioactive compounds in bulb, flower, and fruity vegetable processing by-products, and the food and non-food applications of these by-products. The book unveils the scientific underpinnings of these by-products' health-promoting benefits and applications, primarily from a phytochemistry perspective. It also reviews the processing techniques used to develop novel applications. It will support the economy by integrating bulb, flower, and fruity vegetable processing by-products into applied activities (food production, livestock production, pharmaceuticals, and cosmetics). The work is a valuable contribution to the researchers studying agro-industrial waste recycling from bulb, flower, and fruity vegetable processing by-products. The tentative book chapter submission has a diversity of developments in phytochemistry and environmental chemistry.

Each chapter covers the following topics:

1. The introduction and economic values of each discussed bio-waste from bulb, flower, and fruity vegetable processing.
2. Chemical composition and bioactive compounds of extracts from bulb, flower, and fruity vegetable processing by-products.
3. Biological and functional properties of extracts and bioactive compounds from bulb, flower, and fruity vegetable processing by-products.
4. Food and non-food applications of extracts and bioactive compounds from bulb, flower, and fruity vegetable processing by-products.
5. Valorization of bulb, flower, and fruity vegetable processing by-products for non-health purposes.

References

Ajayi EO, Akin-Idowu PE, Aderibigbe OR, Ibitoye DO, Afolayan G, Adewale OM, Adesegun EA, Ubi BE (2022) Nigeria root vegetables: production, utilization, breeding, biotechnology and constraints. In: Advances in root vegetables research. IntechOpen

Ali Z, Gupta N, Singh J, Bhat A, Sood M, Bandral JD, Reshi M (2023) Utilization of by-products of fruits and vegetables in food industries. Chem Sci Rev Lett 12(45):49–54

Amin T, Naik HR, Naseer B, Hussain SZ (2021) Therapeutic potential of anthocyanin against diabetes. In: Health benefits of secondary Phytocompounds from plant and marine sources. Apple Academic Press, pp 109–133

Amr A, Raie W (2022) Tomato components and quality parameters. A review. Jordan journal of. Agric Sci 18:199

Añibarro-Ortega M, Pinela J, Alexopoulos A, Petropoulos SA, Ferreira IC, Barros L (2022) The powerful Solanaceae: food and nutraceutical applications in a sustainable world. In: Advances in food and nutrition research, vol 100. Academic Press, pp 131–172

Anumala NV, Kumar R (2021) Floriculture sector in India: current status and export potential. J Hortic Sci Biotechnol 96(5):673–680

Aqilah NMN, Rovina K, Felicia WXL, Vonnie JM (2023) A review on the potential bioactive components in fruits and vegetable wastes as value-added products in the food industry. Molecules 28

Arena D, Ben Ammar H, Major N, Kovačević TK, Goreta Ban S, Al Achkar N, Rizzo GF, Branca F (2024) Diversity of the morphometric and biochemical traits of Allium cepa L. Varieties. Plants 13(13):1727

Atif MJ, Ahanger MA, Amin B, Ghani MI, Ali M, Cheng Z (2020) Mechanism of allium crops bulb enlargement in response to photoperiod: a review. Int J Mol Sci 21(4):1325

Attri M, Kashmir J, Sharma J, Bochalaya RS, Sandhu R, Nesar NA (2022) Weed management in vegetables and flowers crops in India. J Pharm Innov 11(2):652–661

Balasubramaniam V, Gunasegavan RDN, Mustar S, Lee JC, Mohd Noh MF (2021) Isolation of industrial important bioactive compounds from microalgae. Molecules 26(4):943

Batiha GES, Alqahtani A, Ojo OA, Shaheen HM, Wasef L, Elzeiny M, Ismail M, Shalaby M, Murata T, Zaragoza-Bastida A, Rivero-Perez N (2020) Biological properties, bioactive constituents, and pharmacokinetics of some *Capsicum* spp. and capsaicinoids. Int J Mol Sci 21(15):5179

Bhatkar NS, Shirkole SS, Mujumdar AS, Thorat BN (2021) Drying of tomatoes and tomato processing waste: a critical review of the quality aspects. Dry Technol 39(11):1720–1744

Biris-Dorhoi ES, Michiu D, Pop CR, Rotar AM, Tofana M, Pop OL, Socaci SA, Farcas AC (2020) Macroalgae—A sustainable source of chemical compounds with biological activities. Nutrients 12(10):3085

Bodoira R, Maestri D (2020) Phenolic compounds from nuts: extraction, chemical profiles, and bioactivity. J Agric Food Chem 68:927

Brahimi A, Landschoot S, Bekaert B, Hajji L, Hajjaj H, Audenaert K, Haesaert G, Mazouz H (2022) Exploring the genetic and phenotypic diversity within and between onion (Allium cepa L.) ecotypes in Morocco. J Genet Eng Biotechnol 20(1):96

Brewster JL, Rabinowitch HD (2022) Onions and allied crops: 3 volume set

Caldeira C, De Laurentiis V, Corrado S, van Holsteijn F, Sala S (2019) Quantification of food waste per product group along the food supply chain in the European Union: a mass flow analysis. Resour Conserv Recycl 149:479–488

Cano-Lamadrid M, Artes-Hernandez F (2021) By-products revalorization with non-thermal treatments to enhance phytochemical compounds of fruit and vegetables derived products: a review, vol 11. Foods

Castellano JM, Ramos-Romero S, Perona JS (2022) Oleanolic acid: extraction, characterization and biological activity. Nutrients 14

Celano R, Docimo T, Piccinelli AL, Gazzerro P, Tucci M, Di Sanzo R, Carabetta S, Campone L, Russo M, Rastrelli L (2021) Onion peel: turning a food waste into a resource. Antioxidants 10(2):304

Chauhan A, Sharma D, Kumar R, Shiwani K, Sharma N (2021) Methods of propagation in vegetable crops. Recent trends in propagation of forest and horticultural crops. Taran Publication, New Delhi, pp 270–281

Chávez-González ML, Sepúlveda L, Verma DK, Luna-García HA, Rodríguez-Durán LV, Ilina A, Aguilar CN (2020) Conventional and emerging extraction processes of flavonoids. PRO 8(4):434

Chourasia KN, More SJ, Kumar A, Kumar D, Singh B, Bhardwaj V, Kumar A, Das SK, Singh RK, Zinta G, Tiwari RK (2022) Salinity responses and tolerance mechanisms in underground vegetable crops: an integrative review. Planta 255(3):68

Coelho MC, Rodrigues AS, Teixeira JA, Pintado ME (2023) Integral valorisation of tomato by-products towards bioactive compounds recovery: human health benefits, vol 410. Food Chem, p 135319

Collins EJ, Bowyer C, Tsouza A, Chopra M (2022) Tomatoes: an extensive review of the associated health impacts of tomatoes and factors that can affect their cultivation. Biology 11

Coman V, Teleky BE, Mitrea L, Martău GA, Szabo K, Călinoiu LF, Vodnar DC (2020) Bioactive potential of fruit and vegetable wastes. Adv Food Nutr Res 91:157–225

Črnivec IGO, Skrt M, Šeremet D, Sterniša M, Farčnik D, Štrumbelj E, Poljanšek A, Cebin N, Pogačnik L, Možina SS, Humar M (2021) Waste streams in onion production: bioactive compounds, quercetin and use of antimicrobial and antioxidative properties. Waste Manag 126:476–486

Cucu AA, Baci GM, Cucu AB, Dezsi Ş, Lujerdean C, Hegeduş IC, Bobiş O, Moise AR, Dezmirean DS (2022) Calluna vulgaris as a valuable source of bioactive compounds: exploring its phytochemical profile, biological activities and apitherapeutic potential. Plan Theory 11(15):1993

Dehgan B (2023) Angiosperms: flowering plants. In: Garden plants taxonomy: volume 1: ferns, gymnosperms, and angiosperms (monocots). Springer International Publishing, Cham, pp 173–603

Dias MC, Pinto DCGA, Silva AMS (2021) Plant flavonoids: chemical characteristics and biological activity. Molecules 26

Dilucia F, Lacivita V, Conte A, Del Nobile MA (2020) Sustainable use of fruit and vegetable by-products to enhance food packaging performance. Food Secur 9

Drabińska N, Jeż M, Nogueira M (2021) Variation in the accumulation of phytochemicals and their bioactive properties among the aerial parts of cauliflower. Antioxidants 10

Eliche-Quesada D, Felipe-Sesé MA, Fuentes-Sánchez MJ (2021) Biomass bottom ash waste and by-products of the acetylene industry as raw materials for unfired bricks. J Build Eng 38:102191

Enyiukwu DN, Bassey IN, Nwaogu GA, Chukwu LA, Maranzu JO (2020) Postharvest spoilage and management of fruits and vegetables: a perspective on small-holder agricultural systems of the tropics. Greener Trend Plant Pathol Entomol 3:001–017

Eslami E, Carpentieri S, Pataro G, Ferrari G (2022) A comprehensive overview of tomato processing by-product valorization by conventional methods versus emerging technologies. Food Secur 12

Esparza I, Jiménez-Moreno N, Bimbela F, Ancín-Azpilicueta C, Gandía LM (2020) Fruit and vegetable waste management: conventional and emerging approaches. J Environ Manag 265:110510

Espinosa-Alonso LG, Valdez-Morales M, Aparicio-Fernandez X, Medina-Godoy S, Guevara-Lara F (2020) Vegetable by-products. Food wastes and by-products: nutraceutical and health potential, pp. 223–266

European Commission. Ensure Sustainable Consumption and Production Patterns (2020) Available online: https://ec.europa.eu/sustainable-development/goal12_en#target-12-3. Accessed 9 Aug 2024

Faust JE, Dole JM (2021) Major cut flowers. Cut Flowers and Foliages

Fayaz U, Bashir I, Fayaz J, Bashir O (2023) Food processing byproducts: their applications as sources of valuable bioenergy and recoverable products. Integrated Waste

Fernandes S, Gois A, Mendes F, Perestrelo R, Medina S, Câmara JS (2020) Typicality assessment of onions (Allium cepa) from different geographical regions based on the volatile signature and chemometric tools. Food Secur 9(3):375

Fonseca LR, Silva GR, Luís Â, Cardoso HJ, Correia S, Vaz CV, Duarte AP, Socorro S (2021) Sweet cherries as anti-cancer agents: from bioactive compounds to function. Molecules 26(10):2941

Ganesh KS, Sridhar A, Vishali S (2022) Utilization of fruit and vegetable waste to produce value-added products: conventional utilization and emerging opportunities-A review. Chemosphere 287:132221

Garba U, Yunusa AK (2024) Allium flowers. Edible Flowers

Gnanasundari K, Krishna KR, Srivignesh S, Kumar MANISH, Harish A, Kumar AR (2023) Processing and value addition in onion. Planta 6:1097–1103

Guan R, Van Le Q, Yang H, Zhang D, Gu H, Yang Y, Sonne C, Lam SS, Zhong J, Jianguang Z, Liu R (2021) A review of dietary phytochemicals and their relation to oxidative stress and human diseases. Chemosphere 271:129499

Hammad M, Abbas A, Hussain N, Rehman A, Javed M (2023) Economically important traits of onion. J Phys, Biomed Biol Sci 2023(1):10–10

Huang H, Huang G (2020) Extraction, separation, modification, structural characterization, and antioxidant activity of plant polysaccharides. Chem Biol Drug Des 96:1209

Jafari S, Jafari SM, Ebrahimi M, Kijpatanasilp I (2023) A decade overview and prospect of spray drying encapsulation of bioactives from fruit products: characterization, food application and in vitro gastrointestinal …. Food hydrocolloids

Jalal H, Giammarco M, Lanzoni L, Akram MZ, Mammi LM, Vignola G, Chincarini M, Formigoni A, Fusaro I (2023) Potential of fruits and vegetable by-products as an alternative feed source for sustainable ruminant nutrition and production: a review. Agriculture 13(2):286

John V, Murmu R, Maurya AK, Simon S, Pant H (2020) Chapter-7 eco-friendly Management of pre and Post-Harvest Diseases of onion (*Allium cepa* L.). Current research and innovations. Plant Pathol:113

Khosa J, Bellinazzo F, Kamenetsky Goldstein R, Macknight R, Immink RG (2021) Phosphatidylethanolamine-binding proteins: the conductors of dual reproduction in plants with vegetative storage organs. J Exp Bot 72(8):2845–2856

Kiran PR, Aradwad P, Kumar TVA, Nayana NP, CS R, Sahoo M, Urhe SB, Yadav R, Kar A, Mani I (2024) A comprehensive review on recent advances in postharvest treatment, storage, and quality evaluation of onion (Allium cepa): current status, and challenges. Future Postharvest Food 1(1):124–157

Kumar S, Shree B, Sharma S, Sharma A (2024) Tree tomato: underutilized vegetable for sustainable nutritional and economic security, vol 327. Scientia Horticulturae, p 112867

Kussmann M, Abe Cunha DH, Berciano S (2023) Bioactive compounds for human and planetary health. Front Nutr 10

Lau KQ, Sabran MR, Shafie SR (2021) Utilization of vegetable and fruit by-products as functional ingredient and food. Front Nutr 8

Liu Y, Mo Z, Su Y, Chen Y (2022) State-of-the-art controlled low-strength materials using incineration industrial by-products as cementitious materials. Constr Build Mater 345:128391

Majumder P, Annegowda HV (2021) Fruit and vegetable by-products: novel ingredients for a sustainable society. In: Valorization of Agri-food wastes and by-products. Academic Press, pp 133–156

Malenica D, Kass M, Bhat R (2022) Sustainable management and valorization of Agri-food industrial wastes and by-products as animal feed: for ruminants, non-ruminants and as poultry feed. Sustainability.

Malik G, Singh DB, Malik AA, Gulzar A (2021) Underutilized vegetable species of Indian Himalayas. Neglected and underutilized crops-towards nutritional security and sustainability, pp. 113–132

Manafi H, Nazari F (2021) The effect of storage temperature on biochemical changes in autumn daffodil bulbs (*Sternbergia lutea* (L.) Ker Gawl. Ex Spreng.) and its impact on interactions with …. Acta Physiol Plant 43

Manicardi T, Baioni e Silva G, Longati AA, Paiva TD, Souza JP, Pádua TF, Furlan FF, Giordano RL, Giordano RC, Milessi TS (2023) Xylooligosaccharides: a bibliometric analysis and cur-

rent advances of this bioactive food chemical as a potential product in biorefineries' portfolios. Food Secur 12(16):3007
Manjunathagowda DC, Anjanappa M (2020) Identification and development of male sterile and their maintainer lines in short-day onion (Allium cepa L.) genotypes. Genet Resour Crop Evol 67(2):357–365
Marcillo-Parra V, Tupuna-Yerovi DS, González Z, Ruales J (2021) Encapsulation of bioactive compounds from fruit and vegetable by-products for food application–A review. Trends Food Sci Technol 116:11–23
Mateos-Aparicio I (2021) Plant-based by-products. Food Waste Recovery
Medina-López SV, Zuluaga-Domínguez CM, Fernández-Trujillo JP, Hernández-Gómez MS (2022) Nonconventional hydrocolloids' technological and functional potential for food applications. Foods 11(3):401
Moodley P, Trois C (2024) Roots, tubers, and bulb crops wastes: residue utilization for industrial biotechnology. In: Roots, Tubers, and bulb crop wastes: management by biorefinery approaches. Springer Nature Singapore, Singapore, pp 1–17
Mostafidi M, Sanjabi MR, Shirkhan F, Zahedi MT (2020) A review of recent trends in the development of the microbial safety of fruits and vegetables. Trends Food Sci Technol 103:321–332
Murtaza MA, Ameer K (2022) Food processing industrial byproducts as raw material for the production of plant protein foods. Plant Protein Foods
Mutalipassi M, Esposito R, Ruocco N, Viel T, Costantini M, Zupo V (2021) Bioactive compounds of nutraceutical value from fishery and aquaculture discards. Foods 10(7):1495
Nardella S, Conte A, Del Nobile MA (2022) State-of-art on the recycling of by-products from fruits and vegetables of Mediterranean countries to prolong food shelf life. Foods 11
Nawaz A, Li E, Irshad S, Xiong Z, Xiong H, Shahbaz HM, Siddique F (2020) Valorization of fisheries by-products: challenges and technical concerns to food industry. Trends Food Sci Technol 99:34–43
Nguyen TT, Rosello C, Bélanger R, Ratti C (2020) Fate of residual pesticides in fruit and vegetable waste (FVW) processing. Foods 9
Ochar K, Kim SH (2023) Conservation and global distribution of onion (Allium cepa L.) germplasm for agricultural sustainability, vol 12. Plants
Oleszek M, Kowalska I, Bertuzzi T, Oleszek W (2023) Phytochemicals derived from agricultural residues and their valuable properties and applications. Molecules 28
Olotu PN, Olotu IA, Datok T, Famojuro TI (2024) A study on 74 varieties of palm trees used around the world as food and ornaments. J Pharmacogn Phytochem 13(1):102–119
Orqueda ME, Zampini IC, Torres S, Isla MI (2023) Functional characterization and toxicity of pectin from red Chilto fruit waste (peels). Plants 12
Orqueda ME, Torres S, Zampini IC, Cattaneo F, Di Pardo AF, Valle EM, Jiménez-Aspee F, Schmeda-Hirschmann G, Isla MI (2020) Integral use of Argentinean Solanum betaceum red fruits as functional food ingredient to prevent metabolic syndrome: effect of in vitro simulated gastroduodenal digestion. Heliyon 6(2):e03387
Orqueda ME, Zampini IC, Bravo K, Osorio E, Isla MI (2022) Potential use of native fruits waste from Argentina as nonconventional sources of cosmetic ingredients. J Cosmet Dermatol 21(10):5058–5065
Osafo R (2021) Nutrient composition and health benefits of underutilized green leafy vegetables in Ghana: a review. Euro J Nutr Food Saf:43
Pangestu OO, Angelica C, Kurniawan J, Meindrawan B, Widyaningrum D (2021) Design formula and product of beverages made from Tamarillo fruit (*Solanum betaceum*) and Job's tears (Coix lacryma-jobi L.) using Kano method. In: IOP Conference Series: earth and Environmental Science. (Vol. 794, No. 1, p. 012139). IOP Publishing
Pérez JP, Muñoz AA, Figueroa CP, Agurto-Muñoz C (2021) Current analytical techniques for the characterization of lipophilic bioactive compounds from microalgae extracts. Biomass Bioenergy 149:106078
Phillips R, Rix M (2021) Vegetables: the definitive guide for gardeners

Proietti S, Scariot V, De Pascale S, Paradiso R (2022) Flowering mechanisms and environmental stimuli for flower transition: bases for production scheduling in greenhouse floriculture, vol 11. Plants

Rabinowitch HD, Goldstein RK (2020) Allium crops. In: The physiology of vegetable crops. CABI, Wallingford UK, pp 421–456

Rai A, Kumari K, Vashistha P (2022) Umbrella review on chilling injuries: post-harvest issue, cause, and treatment in tomato. Scientia. Horticulturae 293:110710

Rais M, Gautam P (2023) Waste management in tomato, onion, and potato crops towards a cleaner environment: insights and future perspectives. Int J Agric Resour Gov Ecol 19(3):247–255

Ramadan Hassanien MF (2023a) Introduction to bioactive phytochemicals from vegetable oil and oilseed processing by-products. In: Ramadan Hassanien MF (ed) Bioactive phytochemicals from vegetable oil and oilseed processing by-products. Reference series in Phytochemistry. Springer, Cham. https://doi.org/10.1007/978-3-030-91381-6_1

Ramadan Hassanien MF (2023b) Bioactive phytochemicals from vegetable oil and oilseed processing by-products. In: Ramadan Hassanien MF (ed) Springer International Publishing, Cham, ISBN: 978-3-030-91380-9, https://doi.org/10.1007/978-3-030-91381_6

Rao M, Bast A, De Boer A (2021) Valorized food processing by-products in the EU: finding the balance between safety, nutrition, and sustainability. Sustain For 13:4428. https://doi.org/10.3390/su13084428

Rațu RN, Veleşcu ID, Stoica F, Usturoi A, Arsenoaia VN, Crivei IC, Postolache AN, Lipşa FD, Filipov F, Florea AM, Chițea MA (2023) Application of Agri-food by-products in the food industry. Agriculture 13(8):1559

Reguengo LM, Salgaço MK, Sivieri K, Júnior MRM (2022) Agro-industrial by-products: valuable sources of bioactive compounds. Food Res Int 152:110871

Rehana S, Bala M (2022) Under exploited ornamental crops: treasure for floriculture industry. Ann Hortic 15:43

Ribeiro TB, Voss GB, Coelho MC, Pintado ME (2022) Food waste and by-product valorization as an integrated approach with zero waste: future challenges. Future Foods

Ricciardi L, Mazzeo R, Marcotrigiano AR, Rainaldi G, Iovieno P, Zonno V, Pavan S, Lotti C (2020) Assessment of genetic diversity of the “Acquaviva red onion” (Allium cepa L.) apulian landrace. Plan Theory 9(2):260

Ritika B, Ismail BB, Garba U, Mishra S, Jha AK, Naik B, Kumar V, Rather MA, Rizwana, Preet MS (2024) Himalayan fruit and circular economy: nutraceutical potential, traditional uses, challenges and opportunities. Food Prod Process Nutr 6(1):71

Sabir SM, Zeb A, Mahmood M, Abbas SR, Ahmad Z, Iqbal N (2020) Phytochemical analysis and biological activities of ethanolic extract of Curcuma longa rhizome. Braz J Biol 81(3):737–740

Sahu P, Biswal M, Mishra N (2020) Use of underutilized green leafy vegetables as food nutrition and ethnobotanical among rural community of Odisha: a review. Int J Chem Stud 8(5):851–859

Saleh M, Bashir MA, Khan KA, Mahmood R, Sarwar G, Rafiq K, Khan MH, Ghramh HA, Ahmad Z, Al-Solami HMA (2021) Onion flowers anthesis and insect pollinators preferences on onion (Allium cepa L.) crop. Fresenius Environ Bull 30(3):2580–2585

Salem MA, Mansour HEA, Mosalam EM, El-Shiekh RA, Ezzat SM, Zayed A (2023) Valorization of by-products derived from onions and potato: extraction optimization, metabolic profile, outstanding bioactivities, and industrial applications. Waste Biomass Valori 14(6):1823–1858

Sansan OC, Ezin V, Ayenan MAT, Chabi IB, Adoukonou-Sagbadja H, Saïdou A, Ahanchede A (2024) Onion (Allium cepa L.) and drought: current situation and perspectives. Scientifica 2024(1):6853932

Santiago B, Calvo AA, Gullon B, Feijoo G, Moreira MT, Gonzalez-Garcia S (2020) Production of flavonol quercetin and fructooligosaccharides from onion (Allium cepa L.) waste: an environmental life cycle approach. Chem Eng J 392:123772

Santos D, da Silva JAL, Pintado M (2022) Fruit and vegetable by-products' flours as ingredients: a review on production process, health benefits and technological functionalities. Lwt 154:112707

Sganzerla WG, Tena-Villares M, Buller LS, Mussatto SI, Forster-Carneiro T (2022) Dry anaerobic digestion of food industry by-products and bioenergy recovery: a perspective to promote the circular economy transition, vol 13. Waste Biomass Valori, pp 1–15

Sharma M, Usmani Z, Gupta VK, Bhat R (2021a) Valorization of fruits and vegetable wastes and by-products to produce natural pigments. Crit Rev Biotechnol 41(4):535–563

Sharma S, Katoch V, Kumar S, Chatterjee S (2021b) Functional relationship of vegetable colors and bioactive compounds: implications in human health. J Nutr Biochem 92:108615

Sheikh FR, Jose-Santhi J, Kalia D, Singh K, Singh RK (2022) Sugars as the regulators of dormancy and sprouting in geophytes. Ind Crop Prod 189:115817

Singh P, Shree S, Bhardwaj A (2020) Overview of the floriculture sector: performance, problems, and prospects. Sustainable Agriculture

Sławińska N, Olas B (2022) Selected seeds as sources of bioactive compounds with diverse biological activities. Nutrients 15

Smeu I, Dobre AA, Cucu EM, Mustățea G, Belc N, Ungureanu EL (2022) Byproducts from the vegetable oil industry: the challenges of safety and sustainability. Sustainability 14:2039. https://doi.org/10.3390/su14042039

Socas-Rodríguez B, Álvarez-Rivera G, Valdés A, Ibáñez E, Cifuentes A (2021) Food by-products and food wastes: are they safe enough for their valorization? Trends Food Sci Technol 114:133–147

Sorrenti V, Burò I, Consoli V, Vanella L (2023) Recent advances in health benefits of bioactive compounds from food wastes and by-products: biochemical aspects. Int J Mol Sci 24(3):2019

Stoica F, Rațu RN, Veleșcu ID, Stănciuc N, Râpeanu G (2023) A comprehensive review on bioactive compounds, health benefits, and potential food applications of onion (Allium cepa L.) skin waste. Trends Food Sci Technol 141:104173

Sudheer S, Gangwar P, Usmani Z, Sharma M, Sharma VK, Sana SS, Almeida F, Dubey NK, Singh DP, Dilbaghi N, Kashani HRK (2022) Shaping the gut microbiota by bioactive phytochemicals: an emerging approach for the prevention and treatment of human diseases. Biochimie 193:38–63

Swamy MK (2020) Plant-derived bioactives

Taghian Dinani S, van der Goot AJ (2023) Challenges and solutions of extracting value-added ingredients from fruit and vegetable by-products: a review. Crit Rev Food Sci Nutr 63(25):7749–7771

Teshome E, Teka TA, Nandasiri R, Rout JR, Harouna DV, Astatkie T, Urugo MM (2023) Fruit by-products and their industrial applications for nutritional benefits and health promotion: a comprehensive review. Sustainability 15(10):7840

Thivya P, Bhanu Prakash Reddy N, Bhosale Yuvraj K, Sinija VR (2023) Recent advances and perspectives for effective utilization of onion solid waste in food packaging: a critical review. Rev Environ Sci Biotechnol 22(1):29–53

Tlais AZA, Fiorino GM, Polo A, Filannino P, Di Cagno R (2020) High-value compounds in fruit, vegetable and cereal byproducts: an overview of potential sustainable reuse and exploitation. Molecules 25

Trigo JP, Alexandre EM, Saraiva JA, Pintado ME (2022) High value-added compounds from fruit and vegetable by-products–characterization, bioactivities, and application in the development of novel food products. Crit Rev Food Sci Nutr 60(8):1388–1416

Ueda JM, Pedrosa MC, Heleno SA, Carocho M, Ferreira IC, Barros L (2022) Food additives from fruit and vegetable by-products and bio-residues: a comprehensive review focused on sustainability. Sustainability 14(9):5212

Vélez-Terreros PY, Romero-Estévez D, Yánez-Jácome GS, Simbaña-Farinango K, Navarrete H (2021) Comparison of major nutrients and minerals between organic and conventional tomatoes. A review. J Food Compos Anal 100:103922

Vojvodić Cebin A, Šeremet D, Mandura A, Martinić A, Komes D (2020) Onion solid waste as a potential source of functional food ingredients. Eng Power: Bull Croat Acad Eng 15(3):7–13

Wang C, Li M, Duan X, Abu-Izneid T, Rauf A, Khan Z, Mitra S, Emran TB, Aljohani AS, Alhumaydhi FA, Thiruvengadam M (2023) Phytochemical and nutritional profiling of tomatoes; impact of processing on bioavailability-a comprehensive review. Food Rev Intl 39(8):5986–6010

Wang P, Ding M, Zhang T, Wu T, Qiao R, Zhang F, Wang X, Zhong J (2022) Electrospraying technique and its recent application advances for biological macromolecule encapsulation of food bioactive substances. Food Rev Intl 38(4):566–588

Wink M (2022) Current understanding of modes of action of multicomponent bioactive phytochemicals: potential for nutraceuticals and antimicrobials. Annu Rev Food Sci Technol 13:337

Wu L, Li X, Ma L, Blom J, Wu H, Gu Q, Borriss R, Gao X (2020) The "pseudo-pathogenic" effect of plant growth-promoting bacilli on starchy plant storage organs is due to their α-amylase activity which is stimulating endogenous opportunistic pathogens. Appl Microbiol Biotechnol 104:2701–2714

Wu X, Yu L, Pehrsson PR (2022) … tomato products as nutritious as fresh tomatoes? Scoping review on the effects of industrial processing on nutrients and bioactive compounds in tomatoes. Adv Nutr 13:138

Yang N, Zhang Q, Chen J, Wu S, Chen R, Yao L, Li B, Liu X, Zhang R, Zhang Z (2023) Study on bioactive compounds of microalgae as antioxidants in a bibliometric analysis and visualization perspective. Front Plant Sci 14:1144326

Yu M, Gouvinhas I, Rocha J, Barros AI (2021) Phytochemical and antioxidant analysis of medicinal and food plants towards bioactive food and pharmaceutical resources. Sci Rep 11

Yusupov Z, Deng T, Volis S, Khassanov F, Makhmudjanov D, Tojibaev K, Sun H (2021) Phylogenomics of allium section Cepa (Amaryllidaceae) provides new insights on domestication of onion. Plant Divers 43(2):102–110

Zhang J, Wen C, Zhang H, Duan Y, Ma H (2020a) Recent advances in the extraction of bioactive compounds with subcritical water: a review. Trends Food Sci Technol 95:183–195

Zhang R, Zhang X, Tang Y, Mao J (2020b) Composition, isolation, purification and biological activities of Sargassum fusiforme polysaccharides: a review. Carbohydr Polym 228:115381

Zhang Y, Wang X, Zhu W, Zhao Y, Wang N, Gao M, Wang Q (2023) Anaerobic fermentation of organic solid waste: recent updates in substrates, products, and the process with multiple products co-production. Environ Res 233:116444

Zhou J, Gullon B, Wang M, Gullon P, Lorenzo JM, Barba FJ (2021) The application of supercritical fluids technology to recover healthy valuable compounds from marine and agricultural food processing by-products: a review. Processes 9(2):357

Zurbau A, Au-Yeung F, Blanco Mejia S, Khan TA, Vuksan V, Jovanovski E, Leiter LA, Kendall CW, Jenkins DJ, Sievenpiper JL (2020) Relation of different fruit and vegetable sources with incident cardiovascular outcomes: a systematic review and meta-analysis of prospective cohort studies. J Am Heart Assoc 9(19):e017728

Part II
Phytochemicals from Bulb Vegetable By-products

Bioactive Phytochemicals from Garlic (*Allium sativum* L.) By-Products

Farhat Gul, Zakir Ullah, Javed Iqbal, Banzeer Ahsan Abbasi, Shumaila Ijaz, Siraj Uddin, Tabassum Yaseen, Habib Ali, Muhammad Jamil, Sobia Kanwal, and Tariq Mahmood

Abstract *Allium sativum* (garlic) is well known for its medicinal and culinary properties, and it has a long history of practice spanning centuries. Beyond its bulb, garlic yields valuable by-products that are potent sources of bioactive phytochemicals. This chapter examines the wide range of bioactive compounds in garlic by-products and considers their potential in various domains, including pharmaceuticals,

---

Farhat Gul, Zakir Ullah and Javed Iqbal contributed equally with all other contributors.

F. Gul · Z. Ullah · S. Uddin · T. Mahmood
Department of Plant Sciences, Faculty of Biological Sciences, Quaid-i-Azam University Islamabad, Islamabad, Pakistan
e-mail: zakirullah@bs.qau.edu.pk; tmahmood@qau.edu.pk

J. Iqbal (✉) · T. Yaseen
Department of Botany, Bacha Khan University, Charsadda, Khyber Pakhtunkhwa, Pakistan
e-mail: tabassumyaseen@bkuc.edu.pk

B. A. Abbasi
Department of Botany, Rawalpindi Women University, Rawalpindi, Pakistan

S. Ijaz
School of Biomedical Engineering, Shenzhen University, Shenzhen, China

H. Ali
Department of Agricultural Engineering, Khwaja Fareed University of Engineering and Information Technology, Rahim Yar Khan, Pakistan

M. Jamil
Department of Biotechnology and Genetic Engineering, Kohat University of Science and Technology, Kohat, Pakistan
e-mail: dr.jamil@kust.edu.pk

S. Kanwal
Department of Biology and Environmental Sciences, Allama Iqbal Open University, Islamabad, Pakistan
e-mail: sobia.kanwal@aiou.edu.pk

M. F. Ramadan (ed.), *Bioactive Phytochemicals in By-products from Bulb, Flower and Fruit Vegetables*, https://doi.org/10.1007/978-3-031-77399-0_2

agriculture, and nutraceuticals. Key topics discussed include the components of garlic and garlic by-products, such as stems and leaves, peels, and bulbs, as well as their bioactive components, including allicin, flavonoids, saponins, and organosulfur compounds. This chapter also covers the extraction processes, highlighting conventional and modern techniques. In addition, the therapeutic potential of both garlic and garlic by-products is illuminated, including their antioxidant, cardiovascular, antidiabetic, and antimicrobial effects. Furthermore, the study delves into the possible applications of each garlic product's nutraceuticals, non-food, and food applications. The chapter concludes with an emphasis on the necessity for more research into the bioavailability, commercial viability, and safety of garlic by-products as valuable sources of bioactive phytochemicals.

Keywords Garlic by-products · Bioactive phytochemicals · Food and non-food applications

1 Introduction

The inevitable by-products of processing fruits and vegetables are frequently a rich source of bioactive chemicals. The seeds, peel, and skin contain these vital bioactive chemicals; however, they are usually thrown away (Iqbal et al. 2017; Iqbal et al. 2018a). Nevertheless, it has been found that some fruit and vegetable pieces have higher quantities of active ingredients. As a result of this understanding, the food sector, among others, increasingly emphasizes cutting waste and adverse environmental effects (Ali et al. 2022; Trigo et al. 2022). Utilizing discarded plant components to create valuable products like flour and fiber additions is a growing trend in the food business. Using these leftover resources, we can get inexpensive components that promote health (Marcillo-Parra et al. 2021). Functional foods, feeds, and dietary supplements can be made use of these ingredients. This is especially helpful in nations where local production, such as fruits and vegetables, is typically more expensive. Juices and other processed foods can offer the same health advantages at a lower cost (Kowalska et al. 2017; Ali et al. 2023; Iqbal et al. 2018c).

The phytochemicals found in plants have become the subject of intense research as potential protective agents against various diseases (Gul et al. 2022; Iqbal et al. 2018b). Bioactive compounds derived from fruits and vegetables have been utilized by humans for centuries, serving as remedies for a wide range of illnesses and promoting overall health. However, in recent years, scientists have only begun isolating and evaluating these active compounds to understand their effects on the human body (Vasanthi and ShriShriMal 2012; Ayaz et al. 2019). This interest is primarily driven by increasing healthcare expenses and the growing desire of consumers to take responsibility for their health. People are becoming more conscious of the potential health benefits of plant-based bioactive compounds, leading to a surge in interest and demand for these natural remedies (Manach et al. 2017; Abbasi et al. 2018, 2019).

Garlic (*Allium sativum*) is one of the famous plants that have been scientifically investigated for their bioactive components (Londhe et al. 2011). Along with leeks, onions, and shallots, garlic is Allium family member (Asemani et al. 2019). The term "stinking rose" originated from its off-white to purple-red colored bulbs, which have been among the best producers of bioactive phytochemicals that could potentially be therapeutic. For thousands of years, garlic has been greatly valued in many cultures across the globe for both its remarkable flavor and its health-promoting qualities (Ross 2007; Azzini et al. 2014; Batool et al. 2019).

Garlic has been appraised for its capacity to raise immune cell activity. During the Middle Ages, it was also employed for protection against epidemics, demonstrating its potency as a herbal treatment (Garty 1993). Garlic's healing powers have been extensively explored via folklore, and countless studies and applications are still being conducted today. Garlic's adaptability and therapeutic qualities are constantly being discovered by researchers as they continue to explore its possible advantages (Bayan et al. 2014; Batool et al. 2020).Many bioactive substances in garlic contribute to its potential as a therapeutic agent. Among these compounds is allicin, which gives garlic its distinctive aroma and numerous health advantages. Allicin has been found to have antibacterial, antioxidant, and anti-inflammatory qualities (Amagase et al. 2001). Garlic also contains flavonoids, organosulfur compounds, and saponins, which are other bioactive substances linked to many health advantages (Subroto et al. 2021; Tavares et al. 2021; Badshah et al. 2022).

Strengthening cardiovascular health is one of garlic's leading health benefits. According to research, garlic may help lower blood pressure and cholesterol and stop blood clots from forming. Garlic's capacity to reduce artery inflammation and enhance blood vessel function is thought to be the cause of these benefits. Garlic consumption has, therefore, been associated with a lower incidence of heart disease and stroke (Ried 2016; Varshney and Budoff 2016). Garlic is well known for its ability to strengthen the immune system and its cardiovascular effects. It has been discovered that garlic increases the production of immune cells, boosts their activity, and supports immune system health overall (Ried 2016). This may help defend against more dangerous infections and ordinary ailments like the flu and colds (Percival 2016). The antibacterial and antioxidant qualities of garlic are thought to be responsible for strengthening the immune system by preventing harmful microbes, viruses, and free radicals (Harris et al. 2001; Melguizo-Rodríguez et al. 2022; Zahra et al. 2021). Garlic has also been investigated for its anticancer effects. Numerous substances included in garlic, such as diallyl sulfide and allicin, have been demonstrated in investigations to possess anti-tumor properties. These substances may aid in causing cancer cells to die off and stop growing and spreading (Almatroodi 2020).

Garlic is a versatile element that may be utilized in many culinary recipes in addition to its medicinal qualities. Its distinct flavor gives food greater depth and intricacy, which is why cuisines worldwide love to use it. Garlic has the power to enhance the flavor of any food, whether it is roasted and used as a garnish or used in sauces, marinades, or soups (Pandiya and Gojiya 2018; Wong 2020; Ijaz et al. 2024b).

To summarize, garlic is an amazing plant with many possible health advantages. Garlic has been the focus of much research due to its therapeutic potential, which includes its ability to stimulate the immune system, improve cardiovascular health, and possibly even treat cancer. As scientists continue exploring its bioactive ingredients, Garlic is a humble "stinking rose" with much more to offer than flavor and perfume. Garlic may offer a variety of health advantages and enhance general well-being when included in a balanced diet.

2 Overview of Garlic

Since it has been cultivated for thousands of years, garlic is utilized extensively in cooking and medicine (Mansell and Reckless 1991). The history of garlic cultivation and use has varied from ancient times. Garlic was a significant domesticated plant with a variety of functions, including in food, mythology, and medicine, according to archeological evidence from the Late Bronze Age Akrotiri, Thera, Greece (Sarpaki 2021; Zahra et al. 2023). Since ancient times, its therapeutic qualities have been acknowledged; references can be found in Greek temples, Egyptian pyramids, and medical textbooks from Egypt, Greece, Rome, China, and India (Rivlin 2001). Although the plant's use has expanded around the world, it is thought to have originated in Central Asia over 6000 years ago (Ejaz et al. 2003). It is known, nevertheless, that the plant and its extracts have been used for a very long time as a medicine to treat and prevent illness. Many of these traditional applications, such as their potential for treating and preventing a variety of ailments, have been validated by modern research.

Garlic is a functional food that is commonly acknowledged and valued. Numerous studies have been conducted on the plant's pharmacological qualities, which include its biologically active ingredients, which further validate the plant's therapeutic potential (Choudhary et al. 2022; Sarswat et al. 2023; Ijaz et al. 2024a).

Several health benefits of garlic have been demonstrated, including defense against chemical and natural toxins. Various harmful agents and organisms in the body, including bacteria, viruses, protozoa, and fungi, are known to be less active when exposed to them (Rouf et al. 2020). Its organic component, allyl sulfur, has been shown to suppress the growth of tumors; oil-soluble compounds such as diallyl disulfide are especially effective in this regard. Garlic has been associated with a lower risk of cancer through diet, and its compounds have potential applications in cancer treatment. Garlic's bioactive ingredients, like allicin and diallyl sulfide, have been shown to influence various stages of the development of cancer by modifying mitochondrial permeability, preventing angiogenesis, and boosting antioxidative and proapoptotic characteristics (Milner 1996; Zhang et al. 2020; De Greef et al. 2021; Ijaz et al. 2023).

It is believed to shield the body from toxins and pollutants and to have cytotoxic and anticancer properties (Dorrigiv et al. 2020). Potential antioxidants are also abundant in garlic and can be found in many areas of the plant (Belhadj et al. 2015).

Garlic's possible health benefits have been the subject of much research, with studies stressing its antibacterial, cardiovascular, and anticancer characteristics, among other broad-spectrum therapeutic effects. Researchers and medical experts have become interested in the potential health benefits of garlic, in particular, garlic's capacity to alter the biomarkers of several diseases. It notably emphasizes its function as an immune system booster, especially in light of the COVID-19 pandemic. This research points to garlic's potential as a natural treatment for various diseases (Ansary et al. 2020; Chakraborty and Majumder 2020) (Fig. 1).

2.1 Processing of Garlic

Comparing various processed garlic products is the greatest way to illustrate the significance of bioactive phytochemicals in garlic. The phytochemical profile of the garlic product varies depending on the processing technique used. This impacts the product's biological activity and health-promoting properties (Lu et al. 2023). Alliin, found in fresh garlic, is converted into allicin by the enzyme Alliinase when a garlic clove gets damaged by crushing or chopping (Arditti et al. 2005). Allicin is extremely unstable and quickly breaks down into various sulfur compounds. Variables affect that procedure, including pH, solvent type, and temperature. For instance, allicin breaks down more rapidly in solutions containing hydrogen bonds and higher temperatures. Allicin breaks down into various sulfur-containing chemicals, including ajoene, vinyldithiins, and S-allyl-l-cysteine (Mishra et al. 2001; Seki et al. 2018).

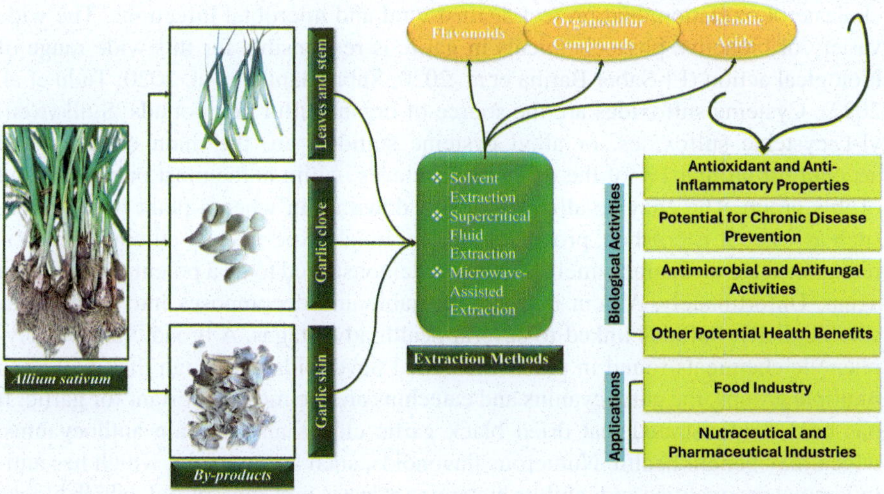

Fig. 1 An overview of garlic by-products, extraction methods, phytochemicals, and their various applications

Garlic is aged by combining chemical and enzymatic reactions that yield these chemicals. These substances show a range of therapeutic actions, such as cardioprotective, antioxidant, and anticancer activities. They are also believed to be stable. However, the processing techniques impact these compounds' chemistry (Abe et al. 2020; Kodera et al. 2020). Many other garlic products do not match the catabolites produced by fresh garlic since they are aged differently (Setiyoningrum et al. 2022).

Similarly, garlic's total phenolics, flavonoids, and antioxidant activity can increase by aging it for extended periods and at higher temperatures (Shin et al. 2008). According to Kang et al. (2011), processing red garlic results in a change in its physicochemical properties. The product has higher crude protein and ash levels and less moisture than fresh garlic. These findings imply that aging processes are responsible for the distinct catabolites of fresh garlic and that these processes can be altered to improve the functional qualities of garlic products.

3 Bioactive Phytochemicals Present in Garlic

Compounds that impact isolated cells, tissues, or living organisms are called bioactive. Non-nutritive substances in plant food called bioactive phytochemicals can prevent and treat disease. In vitro, animal, and, more recently, human research has been conducted on them. Garlic's phytochemicals have been connected to a variety of biological processes. Like the effects of numerous fruits and vegetables, these activities are linked to the prevention of chronic diseases. It has been demonstrated that using old garlic extract, which has an allyl sulfide soluble in water, lowers the chance of developing several types of cancer. Garlic's components also have anti-hypertensive and antioxidant properties, prevent cholesterol synthesis and LDL-cholesterol oxidation, and protect against viral and microbial infections. The wide variety of bioactive phytochemicals in garlic is responsible for this wide range of biological action (El-Saber Batiha et al. 2020; Subramanian et al. 2020; Tudu et al. 2022). Cysteine sulfoxides are the source of organosulfur compounds. S-alkyl(en)yl-L-cysteine sulfoxides, or alkyl cysteine sulfides, are the main category and account for around 1% of the oil in garlic cloves. Alliin is the most prevalent kind of this group. The enzyme alliinase breaks down alliin when a plant is disturbed, such as when it is crushed, producing allicin. In lab experiments, allicin—which is recognized for its distinct smell—has been demonstrated to be a potent antibacterial agent. Unfortunately, Allicin is highly unstable and decomposes into several oil-soluble diallyl sulfides linked to several health advantages. A broad class of polyphenolic chemicals found in plants are called flavonoids. Although they belong to multiple groups, the anthocyanins and catechins are the most important for garlic. It has been demonstrated that dried black garlic cloves also contain anthocyanins, which may benefit health. Numerous flavonoids, such as quercetin, which has anti-inflammatory qualities and inhibits histamine release, and kaempferol, which blocks the activity of enzymes that break down the articular cartilage, are present in aged garlic aqueous extracts (Miękus et al. 2020; Yang et al. 2020; Bhatwalkar et al. 2021).

3.1 Organosulfur Compounds

One important characteristic that sets garlic apart from other members of the Alliaceae family is the presence of organosulfur compounds. Garlic's volatile oil imparts a unique flavor and aroma. It is also one of garlic's special qualities and benefits humans. Crushing a garlic clove causes chemical and biological reactions in the sulfur-containing molecules. Allicin, among all the chemicals containing sulfur, has the strongest antibacterial action, based on investigations. The sulfur compounds diallyl sulfide and diallyl disulfide are formed when allicin, which is highly volatile and unstable, rapidly decomposes (Abe et al. 2020). It has been demonstrated that diallyl disulfide lowers serum cholesterol levels. In various assays, diallyl disulfide dramatically lowered the cholesterol level more than clofibrate did. The rats experienced no adverse side effects from diallyl disulfide, its finest feature. This outcome offers great promise for discovering a novel cholesterol-lowering treatment. Diallyl disulfide (DADS) has been demonstrated to increase the hepatic LDL receptor and LDL degradation activity with a potency comparable to diallyl trisulfide while inhibiting HMG-CoA reductase activity. According to this study, DADS is a potent inhibitor of cholesterol synthesis, and part of its cholesterol-lowering impact can be attributed to elevated LDL receptor activity and improved LDL decomposition. This suggests that DADS is a valuable mediator of garlic's cholesterol-lowering actions. The effects of garlic on cholesterol are consistent with those described for allicin and DATS, as are the changes in liver enzyme regulation (Lammi et al. 2020; Suppressa et al. 2020; Wu et al. 2021; Panyod et al. 2022).

The relevance of these various classes may vary depending on how the compounds' varying bioactivities, biosorptions, and metabolisms reflect the garlic's antibacterial, lipid-lowering, and antiplatelet aggregation properties and its flavor and scent. The fact that odor of chopped or crushed garlic can be revived by soaking in 20 °C water for 60 min, which converts these compounds to the volatile dialk(en)yl disulfides and trisulfides, and then simmering in olive oil for 1–2 min before the garlic is added to a dish, indicates the significance of the flavor precursors alliin, S-allyl cysteine, and S-methyl cysteine. The sulfur compounds undergo this transformation more easily than when they initially transform to allicin in the intact garlic because heat readily degrades allicin (Abe et al. 2020; Ma et al. 2021; Li et al. 2015). The first class consists of two-carbon allyl sulfur compounds, which include allicin and the transformation products that contain allyl and methyl. The ajoenes, dithiine, and 1,3 dithiolane rings containing sulfur compounds with an extra sulfur side chain constitute another sub-class. E/Z- and A/Z-1, 2 S-(methylethyl) 1-propenyl and 3-methylthiopropenyl sulfoxide and sulfone, as well as the associated polychalcogenides, belong to a third class of compounds called vinyldithiins. Cyclic polysulfide compounds, or trithiins and tetrasulfanes, constitute a fourth class. The fourth class consists of the more heat and light stable S-alk(en)yl cysteine sulphoxide compounds, such as S-allyl cysteine and S-methyl cysteine, which are odorless and taste precursors, and alliin, the main sulfur-containing component in garlic supplements (Bhatwalkar et al. 2021; Ramirez et al. 2021; Tudu et al. 2022).

3.2 *Flavonoids*

According to an early study, 40–60% of the flavonoids in garlic juice can be dialyzed, and another 40–60% are probably linked to other sulfur compounds. This comparison was made between the flavonoid content of garlic juice and the dialysate and retentate fraction. This implied that some garlic's sulfur-containing chemicals and flavonoids might combine to form complexes. Analysis of the dialysate fraction revealed that it could suppress platelet aggregation generated by ADP more effectively than the original garlic juice sample, which showed evidence of this. This dialysate fraction might be helpful for future research to separate these putative antiplatelet aggregation components because it might contain specific flavonoid sulfur conjugates. Throughout the plant kingdom, flavonoids—polyphenolic compounds—are extensively dispersed. Different kinds of flavonoids, including anthocyanins, flavonols, and flavones, are present in garlic (*Allium sativum*). Hesperetin, quercetin, kaempferol, and eriodictyol are a few of the primary flavonoids in garlic bulbs. In numerous domains of garlic research, quercetin—the most researched flavonoid in garlic—has been found to exhibit numerous activities (Alide et al. 2020; Ansary et al. 2020; Yang et al. 2020; Pham and Pham 2022). Garlic contains relatively high quantities of myricetin, quercetin, and apigenin flavonoids, according to Beato et al. (2011). Furthermore, flavonoids were found in all the garlic cultivars that Calimpang et al. (2023) examined. Compared to mature garlic bulbs, Thach (2022) found that garlic husk extracts had a much higher total flavonoid concentration.

Garlic produces flavonols, flavones, flavanones, and dihydroflavonols, which are sub-classes of flavonoids. According to earlier studies, garlic can make compounds that vary depending on the species in comparatively large concentrations. Quercetin is a component in garlic that belongs to the flavonoid class, flavonols. Numerous types of garlic have been shown to contain quercetin. Garlic produces isorhamnetin as a flavone. Isorhamnetin has been demonstrated in vitro to prevent the retronuclear synthesis of Hepatitis B viral DNA; these findings have clinical significance; garlic use may have health importance (Silva et al. 2020; Yang et al. 2020).

3.3 *Phenolic Acids*

Phenolic acids are substances with a variety of biological functions. They belong to a class of secondary plant metabolites that have been linked to a variety of effects, from poisonous to therapeutic. The most well-known characteristic is their antioxidative ability, which not only protects the plant from infections and autotoxicity but also improves human health when the plant is eaten or used as a natural medication (Subroto et al. 2021; Sunanta et al. 2023). The effects of antioxidants at the cellular level can alleviate autotoxicity. Phenolics' antioxidative qualities may shield plants from the damaging effects of ultraviolet light. Since they can be triggered in reaction to UV radiation and have varying degrees of absorption and/or reflection of UV, they also have the potential to function as UV screens. Using garlic phenolics' ability to absorb UV light can be advantageous for human health, as it can be used to make sunscreens that can help reduce photocarcinogenesis and erythema.

The oxidative impacts of phenolic acids can be useful and harmful when they reach the point of oxidative stress. They can have detrimental effects on both plant and human health. This is the other end of the beneficial-to-harmful spectrum (Avgeri et al. 2020; Sasi et al. 2021; Chaudhary et al. 2022). Phenolic acids found in garlic have been measured in both the peels and cloves. However, the types and concentrations of these chemicals vary. Garlic peels and cloves contain high concentrations of hydroxycinnamic acid derivatives and low concentrations of benzoic acid derivatives; however, garlic's varieties and preparation techniques can vary. The most common benzoic acid is syringic acid, and ester conjugates of hydroxycinnamic acids are predominant.

Some types of garlic contain more phenolic compounds than others, and it is thought that the more phenolic acids there are in the garlic, the more health benefits it would provide. Garlic phenolics are probably unstable and reactive in the cells because of the instability of the enzymes involved in phenolic production and the phenolics' vulnerability to oxidation and enzymatic polymerization. This could impact garlic preparations with a high phenolic content and extracted phenolics. Structure–activity connections, nevertheless, it would be prudent for the garlic industry to endeavor to conserve the phenolic components of garlic, given their potential health benefits. Avoiding overprocessing garlic with a high phenol content could help achieve this, as garlic with a high phenol content could be deemed a healthy food (Perez-Ortiz et al. 2020; Dinu et al. 2023; Martínez-Inda et al. 2023; Wongsa et al. 2023).

Garlic's phenolic component composition has been thoroughly studied. Most phenolic compounds are free and bound hydroxycinnamic acids and the associated aldehydes and acids. 5-O-caffeoylquinic acid was found to be the predominant hydroxycinnamic acid in garlic, but minor levels of its 3-O- and 4-O-methyl isomers were also present. Garlic includes a diverse range of bound phenolic compounds attached to different sugar moieties through ester linkages. Despite the presence of additional phenolic compounds in garlic, hydroxycinnamic acids appear to constitute most of all phenolic compounds. Such findings may shed light on the possible health benefits of garlic and offer valuable information for the future characterization of these chemicals (Dinu et al. 2023; dos Santos et al. 2022) (Table 1).

3.4 Saponins

Garlic also contains phytochemicals known as saponins. Research has indicated that saponin chemicals could play a role in garlic's ability to withstand pests. Saponins are substances having a diverse array of biological functions. Within plants, they

Table 1 Garlic phytochemicals with their chemical formula, major class, and sub-class

Phytochemical	Chemical formula	Major class	Sub-class
Allicin	$C_6H_{10}OS_2$	Sulfur compound	Organosulfur
Alliin	$C_6H_{11}NO_3S$	Sulfur compound	Organosulfur
Diallyl sulfide	$C_6H_{10}S$	Sulfur compound	Organosulfur
Diallyl disulfide	$C_6H_{10}S_2$	Sulfur compound	Organosulfur
Diallyl trisulfide	$C_6H_{10}S_3$	Sulfur compound	Organosulfur
Ajoene	$C_9H_{14}OS_3$	Sulfur compound	Organosulfur
Diallyl tetrasulfide	$C_6H_{10}S_4$	Sulfur compound	Organosulfur
Allyl methyl trisulfide	$C4H_8S_3$	Sulfur compound	Organosulfur
S-Allylcysteine	$C_6H_{11}NO_2S$	Sulfur compound	Organosulfur
S-allylmercaptocysteine	$C_6H_{11}NO_2S_2$	Sulfur compound	Organosulfur
1,2-Vinyldithiin	$C_6H_6S_2$	Sulfur compound	Organosulfur
Γ-glutamylcysteine	$C_8H_{14}N_2O_5S$	Sulfur compound	Organosulfur
Flavonoids		Polyphenol	Flavones, flavonols, etc.
Phenolic acids		Polyphenol	Hydroxybenzoic acids, hydroxycinnamic acids, etc.
Saponins		Terpenoids	Triterpenoid saponins, steroidal saponins
Organoselenium compounds		Selenium compounds	Selenocompounds, selenoamino acids

serve as a defense mechanism against diseases and pests (Zaynab et al. 2021). Research on the green peach aphid (GPA) of insects suggests that saponins may deter certain types of insects. Insects that have evolved to consume plants that contain saponins typically possess detoxifying mechanisms for saponins and may even exhibit an attraction toward them. Protozoa, fungi, and yeasts are all inhibited by saponins derived from various plant sources. Garlic saponins have been found to have a variety of biological effects in studies involving mammalian cells, including cytotoxicity and effects on lipid metabolism (Abdelrahman and Jogaiah 2020; De Greef et al. 2021; Ramdani et al. 2023).

Two distinct saponin subgroups have been found in garlic recently. For one of these saponins, the aglycone has been discovered as benzyl-(or phenylmethyl)-2-O-β-D-glucopyranosyl-β-D-glucopyranoside; for the other, it is a glucopyranoside with an α-l-rhamnopyranosyl-(1 → 2)-[β-D-xylopyranosyl-(1 → 6)]-glucopyranosyl moiety (β-D). Its synthesis was carried out by hydrolyzing the macrocyclic ether that was produced after thiourea-(1 → 6)-β-D-glucopyranosyl-D-fructofuranoside and 2,4-dihydroxy-6-methylbenzaldehyde condensed. A combination of 2-O-β-D-glucopyranosyl-D-fructofuranoside and its 2-epimer was obtained by hydrolyzing periodic acid and then reducing it with Raney nickel. This was phenylated with benzyl chloride to yield the appropriate glucopyranoside precursor for the synthetase. A combination of the fucose and rhamno isothiocyano derivatives was produced by the periodic cleavage of this precursor and its subsequent reaction with thiourea and base. Using 1,2-ethanedithioliation and anomer separation, a good yield of the former was produced. The required saponin aglycone was obtained by cyclizing this chemical with sodium in liquid ammonia and then demethylating it. Under acidic conditions, these moieties bound to a trisaccharide broke up the macrocyclic ether. Disaccharides in an appropriate configuration for direct attachment to the aglycone by cyclization with silver carbonate in pyridine were generated by controlling the anomer of the thiourea condensation product. The aglycone was produced via a transposition to the 1,2 2,3 dicarbamate using dimethylaminomethylene chloride and a periodic cleavage of the thiourea link. As of currently, no synthesizing or even conjecturing has been done on the aglycone from the other saponin (Qasim et al. 2020; Sari et al. 2022; Tudu et al. 2022; Okoro et al. 2023) (Fig. 2).

4 Properties of Garlic and its Phytochemicals

4.1 Antioxidant Properties

Garlic has several potential health benefits due to its many active components. The enzymes in the crushed clove give antioxidant action. Garlic gets its distinct flavor and odor from a range of allyl sulphides produced when the thiol-containing molecule allicin breaks down, which has been demonstrated to have antioxidant and possible cardiovascular properties (Jones et al. 2007). The antioxidant properties of garlic are thought to be attributed to these organosulphur molecules (Chen 2012). Several research studies have proven this, some of which are listed below.

Furthermore, compounds in the crushed clove, flavonoids and phenolic acid exhibit antioxidative qualities. Studies have indicated that crushed cloves contain antioxidative chemicals such as β-caryophyllene, eugenol, and eugenyl acetate (Gaspar et al. 2018). These substances belong to the flavonoid and polyphenolic classes, which are well known for their antioxidant properties (Huyut et al. 2017; Ullah et al. 2023c). Crushed clove's potential as a natural antioxidant is therefore supported by the presence of these chemicals in it. Garlic contains a flavonoid called quercetin, which has been demonstrated to have strong antioxidant qualities (Khan et al. 2022). It has also been discovered that quercetin may prevent heme polymerization, a process essential to developing malaria parasites (Manu et al. 2013).

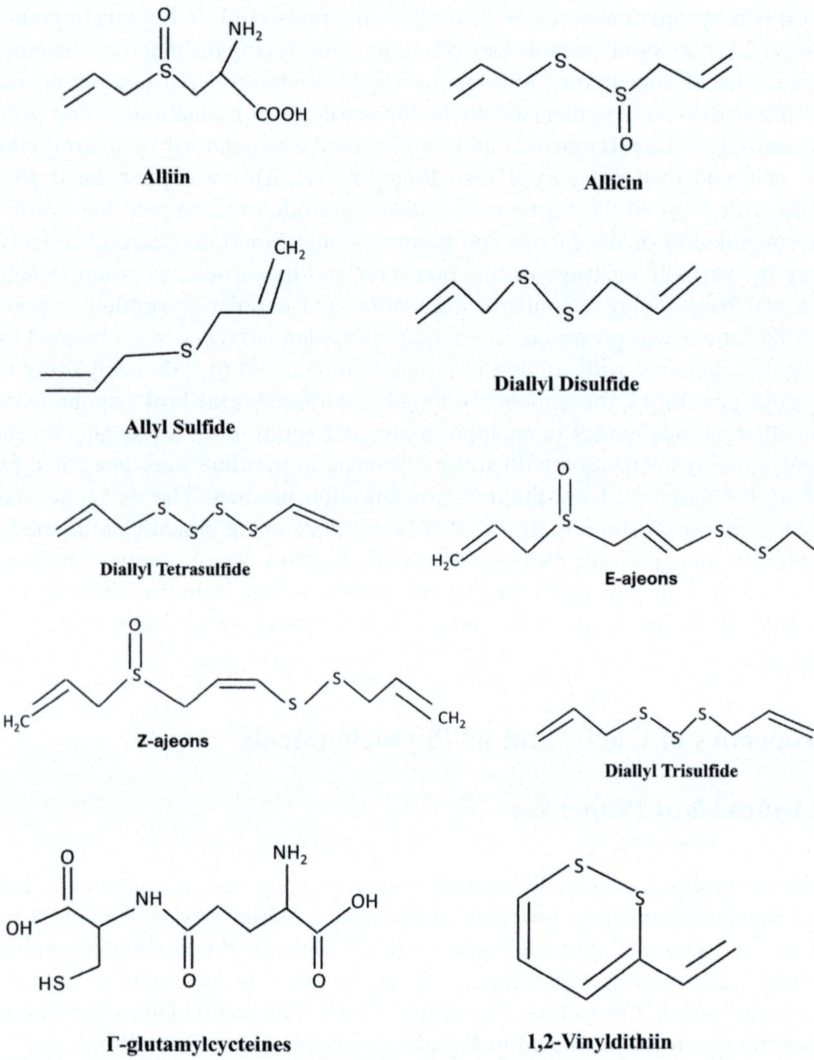

Fig. 2 Chemical structures of some organosulfur compounds

Moreover, it has demonstrated significant antioxidant action across a range of studies. Quercetin has been the focus of numerous studies because of its anti-inflammatory and antioxidant properties (Ahmeda et al. 2009).

Garlic's primary function is thought to be antioxidant activity. Allicin's valuable biological qualities are mainly ascribed to its capacity to "donate" sulfur to cysteine residues in different enzymes, which makes it a powerful reducing agent (Nadeem et al. 2021). Numerous phase II detoxification enzyme systems are subsequently induced as a result. These systems, essential to cells' defense against reactive oxygen species damage, are typically activated as a defensive mechanism against chemically produced toxicities, which can lead to a range of human disorders (Munday et al. 2005). The body's defense against xenobiotic-mediated toxicity depends on these enzymes, which include quinone reductases, glutathione S-transferases, and UDP-glucuronosyltransferases (Chen 2012). Due to its pro-oxidant qualities, which can impair a cell's electrical potential and cause apoptosis, allicin is assumed responsible for inducing these enzymes (Gruhlke et al. 2010). Furthermore, it has been demonstrated that allicin inhibits the catalytic activity of the enzyme streptolysin O, which is generated by Streptococcus pyogenes, by forming disulfide bonds with cysteine residues (Arzanlou and Bohlooli 2010). All these results suggest that allicin is a potent reducing agent because of its ability to transfer sulfur to cysteine residues, which is the source of its biological properties.

4.2 Cardiovascular Health

Cardiovascular disorder (CVD) covers a range of disorders, but it is usually associated with atherosclerosis, the buildup of cholesterol and fats on the inner walls of arteries. Atherosclerosis is a long-term process that leads to the narrowing and hardening of the arteries. In turn, it can cause a variety of CVD problems. High blood pressure, angina, heart attacks, and strokes are all associated with CVD, and they generally revolve around the topic of damage to the arteries (Morel et al. 2022; Tohirova and Shernazarov 2022; Gusev and Sarapultsev 2023).

Garlic and its by-products have been thoroughly investigated, and its benefits for cardiovascular health have been reported in scientific literature. Due to its abundant bioactive components, garlic is well known for having several beneficial impacts on cardiovascular health. It has been demonstrated that allicin, a major bioactive component of garlic, has cardioprotective, anti-inflammatory, and antioxidant qualities (Chan et al. 2013). Research has shown that sulfur-containing components of garlic, such as allicin, may reduce the consequences of inflammation and cardiovascular disease by inhibiting the hydrogen sulfide (H_2S) signaling pathway (Rodrigues and Percival 2019). Garlic and its derivatives are also well known for potentially preventing and treating metabolic and cardiovascular disorders, such as diabetes, hypertension, thrombosis, hyperlipidemia, and atherosclerosis (Banerjee and Maulik 2002). Garlic is said to have several bioactive components, such as allicin, sulfinic acids, and other compounds containing sulfur, which contribute to its cardiovascular effects. The constituents have been associated with vasodilator, cardiotonic, and antioxidant actions, in addition to their capacity to scavenge free radicals and prevent platelet aggregation and angiogenesis.

Furthermore, studies have shown that garlic can modify platelet activities, enhance lipid profiles, and prevent cardiovascular diseases linked to atherosclerosis and platelets. Several clinical trials have shown the beneficial effects of garlic on overall cardiovascular health (Patil and Ravindra 2008; Choi et al. 2011; Chan et al. 2013; Irfan et al. 2019). Allicin, allyl sulfides, and phenylpropanoids are among the bioactive components of garlic that have been linked to anti-inflammatory, cardioprotective, and antioxidant properties. According to reports, these substances increase the antioxidant capacity of cells, have anti-carcinogenic properties, and lower the risk of cardiovascular illnesses (Ou et al. 2010; Jensen et al. 2021). Moreover, in several experimental models, garlic has been shown to guard the cardiovascular system, improve cardiac contractile dysfunction, and prevent platelet aggregation. Garlic's health advantages extend beyond its benefits for the heart. Add to the benefits of garlic's general health-promoting qualities, it has also been shown to have immunomodulatory effects (Xiao et al. 2016; Kunová et al. 2021).

In short, extensive research validates the benefits of garlic and its by-products for cardiovascular health. Garlic is a helpful dietary addition for promoting cardiovascular health since its bioactive chemicals have been linked to cardioprotective, anti-inflammatory, and antioxidant actions.

4.3 Antidiabetic Effects

Deficiencies in insulin secretion and action cause high blood glucose levels, and both define diabetes mellitus as a long-term metabolic illness. Prolonged consequences of the disease might impact multiple organs, such as the kidneys, heart, blood vessels, nerves, and eyes. Type 1 and 2 diabetes are the most prevalent among the various forms of the disease. Insulin shortage results from the autoimmune condition known as type 1 diabetes, caused by the body's immune system attacking and killing the pancreatic beta cells that produce insulin. Conversely, relative insulin insufficiency and insulin resistance are indicators of type 2 diabetes. Furthermore, type 2 diabetes may become more likely to develop later in life if gestational diabetes is present during pregnancy. Diabetes is becoming increasingly common worldwide, which has substantial consequences for public health and healthcare systems worldwide (Ullah et al. 2024; Basevi 2011; Ta 2014).

Numerous studies and publications have examined the potential antidiabetic effects of phytochemicals in garlic by-products. Garlic's antidiabetic properties have been linked to its phytochemical content, which includes S-allylcysteine (SAC), allicin, and other organosulfur compounds (Kalhotra et al. 2020). In human and animal models, these phytochemicals have demonstrated the potential to avert cardiometabolic disorders linked to metabolic syndrome (Amor et al. 2019). Additionally, it has been discovered that garlic extract stimulates the proliferation of skeletal muscle cells and inhibits dipeptidyl-peptidase IV (DPP-4), indicating a possible mode of action for treating diabetes mellitus (Saikat et al. 2021).

Garlic has been shown in frequent studies to have antidiabetic properties in both clinical trials and experimental animal models. For example, oral treatment of raw garlic led to significant improvements in antioxidant enzyme activities in diabetic patients' erythrocytes and a considerable decrease in blood glucose levels and lipid metabolism (Ota and Ulrih 2017).

Garlic oils also have been demonstrated to enhance oral glucose resistance and insulin sensitivity in a dose-dependent manner, suggesting that they may have antidiabetic qualities. Additionally, studies have shown that garlic oil's anti-inflammatory and antioxidant qualities enhance insulin sensitivity and lessen oxidative stress (Saikat et al. 2021).

The antidiabetic effects of aged black garlic, made via pectinase treatment with high hydrostatic pressure processing, have been studied in streptozotocin-treated mice, with promising outcomes (Kim et al. 2017). The potential of garlic in managing metabolic syndrome and its associated complications was highlighted by Madkor et al. (2011), who studied the modulatory effects of garlic on hyperglycemia, dyslipidemia, and oxidative stress in diabetic rats and compared them with those of ginger, turmeric, and their mixture. Additionally, it has been documented that garlic inhibits α-glucosidase, which may be a possible mechanism of garlic's antidiabetic effects (Obih et al. 2019). Garlic and aged black garlic have also been shown to possess antioxidant properties, suggesting that they could reduce oxidative stress linked to diabetes mellitus (Lee et al. 2009).

4.4 Anticancer Potential of Garlic Phytochemicals

Many studies have been conducted on the phytochemicals found in garlic and their by-products to investigate their possible anticancer effects. Many biological activities, such as antioxidant, anti-inflammatory, and anticancer properties, have been reported for garlic and its bioactive components Shang et al. 2019. Garlic's chemical components, including its phytochemicals and sulfinic acids, have been linked to anti-inflammatory effects. Furthermore, it has been demonstrated that aged black garlic can prevent some cardiometabolic changes linked to metabolic syndrome, suggesting it may have anticancer properties (Amor et al. 2019).

Moreover, garlic's phytochemical screening has shown the presence of phenols and flavonoids, which are recognized to have possible anticancer effects (Calimpang et al. 2023). Garlic's bioactive components, including allicin, have been linked to possible anticancer actions due to its antioxidant properties (Lee et al. 2009). Additionally, it has been shown that garlic extracts contain flavonoids, phenolic chemicals, alkaloids, and tannins, all of which may contribute to their potential anticancer effects (El-khamissi et al. 2019). Garlic extracts' antibacterial and perhaps anticancer properties have also been linked to the presence of saponins (Bernaldez and Vicencio 2021).

Garlic's antioxidant power has been associated with its phytochemical composition, especially existence of volatile sulfur compounds such as alliin, allicin, diallyl disulfide, diallyl trisulfide, diallyl sulfide, S-allyl cysteine, ajoene, and allyl mercaptan (Obih et al. 2019), which may help explain some of its possible anticancer benefits (Suleria et al. 2015).

In conclusion, a variety of biological activities, including antioxidant, anti-inflammatory, and possibly anticancer effects, are exhibited by the phytochemicals found in garlic and its by-products, including allicin, sulfinic acids, flavonoids, phenols, and saponins.

4.5 Applications in Functional Foods

Garlic has been used as a functional food for thousands of years. In this context, functional food potentially positively affects health beyond basic nutrition (Das et al. 2010). Garlic has long been considered beneficial to health in traditional systems of medicine; however, until the last decade, little scientific evidence existed to support these claims. This situation is rapidly changing, and there is now an overwhelming amount of scientific and clinical data to validate the health benefits of garlic. This evidence is largely a result of a substantial increase in research on garlic in the 1990s that was driven by a rising interest in natural health products and a rapid rise in the understanding of disease mechanisms. These mechanisms support the possibility that garlic consumption may prevent or retard disease states (Qi 2006).

Two recent trends, a rise in the consumption of functional foods and the rapid growth of the dietary supplement industry, suggest that the future consumption of garlic and garlic products (extracts, oils, etc.) for health-related reasons will be substantial (Suleria et al. 2015; Javed et al. 2021). Globally, functional foods have become an essential part of the health and wellness culture. This trend is driven by an aging population that desires to stay active and healthy and an increased awareness of health and nutrition in young people who strive to prevent age-related diseases. Garlic in its various forms is ideally placed to be an important component of functional foods or nutraceuticals, which can be defined as food or food products that provide health and medical benefits, including the prevention and treatment of disease (Akerele et al. 2016; Javed et al. 2021).

4.6 Garlic Extracts as Natural Food Additives

Garlic is widely used as a functional ingredient and flavoring agent in the food industry. It is particularly popular in Mediterranean, Middle Eastern, and Asian diets. As such, garlic-derived products such as garlic oils and liquid extracts are widely used in these cuisines and often used to prepare processed foods (Valente et al. 2014). The unique flavors and health-promoting effects of garlic are greatly valued in today's health-conscious society (Kaur et al. 2023). However, the sensory attributes of garlic are such that when it is released from its storage—e.g., when minced or sliced—pungent odors are produced as volatile sulfur compounds are released. These odors can often lead to consumer dissatisfaction and breath malodor (Borlinghaus et al. 2021). Efforts have been made to remove or mask such odors from garlic preparations using various chemical treatments and/or microencapsulation. However, these methods are not considered "natural" and often require the addition of other volatile compounds to mask the garlic odor (Tavares et al. 2021).

In seeking natural methods to remove/mask the release of volatile sulfur compounds, it has investigated the addition of allyl sulfides to garlic oil to enhance its flavor while reducing the production of "garlicky" malodors. The addition of diallyl disulfide or diallyl trisulfide to garlic oil increased its sulfur content and altered its volatile composition to that of fresh garlic. Sulfur-containing volatiles arise from alliin and deoxyalliin's thermal degradation (Hwang and Kim 2022).

Further, research has shown that certain water-soluble organosulfur compounds found in aged garlic extracts can neutralize the production of sulfur compounds from fresh garlic (Li et al. 2015; Sharma and Rani 2023). These compounds could effectively prevent the malodor associated with garlic-containing foodstuffs. Using aged garlic extracts or purified garlic organosulfur compounds could present an effective and natural means to improve garlic food preparations' flavor while preventing malodors' production. This could have widespread applications in the food industry, from producing retail products to preparing and pre-packaging culinary ingredients (Adiwijaya 2017).

4.7 *Enhancing Absorption of Bioactive Phytochemicals*

Although the use of dietary supplements is a popular method of increasing the intake of various bioactive compounds, functional foods may be an overall more effective means of increasing the bioavailability of garlic phytochemicals due to their frequent consumption and practical nature (Arshad et al. 2021). Utilizing garlic compounds as functional ingredients in foods is another possible method to increase the absorption of these phytochemicals. A recent in vivo study showed that rats fed aged garlic extract mixed into their food showed increased bioavailability of S-allylcysteine (SAC)—a sulfur-containing compound found in abundance in aged garlic when compared to rats administered SAC alone. Since the consumption of garlic is more common as a food than a dietary supplement, this method may be more practical and effective for increasing the bioactive effects of garlic. Another study found that using garlic oil as a food seasoning increased the bioavailability and metabolism of organosulfur compounds in rats, possibly by enhancing tissue retention (Kodera et al. 2020; Rauf et al. 2022).

Several studies have reported that the bioavailability of garlic compounds may be enhanced by the method of ingestion (e.g., powder, oil, aged extract), which may increase absorption. Other means of improving the absorption of garlic compounds include using various solvents, such as oils or alcohol, which may increase the potency and efficacy of dietary supplements. However, the use of various extraction solvents poses the risk of altering the chemical composition of the garlic product, which may create a product with different biological activity than the raw material (Moreno-Ortega et al. 2020; Bar et al. 2022; Ezeorba et al. 2022).

5 Extraction Methods

Extraction techniques are necessary to extract biochemicals from plants. The process of extracting medicinally active phytoconstituents by applying established techniques and selective solvents can be termed extraction. The type of phytoconstituents used influences the choice of solvent used in the extraction process. There are several approaches used, each with pros and cons of their own. Soxhlet, hydrodistillation, liquid-liquid extraction (LLE), pressurized liquid extraction (PLE), solid-phase extraction (SPE), supercritical fluid extraction (SFE), pulsed electric field (PEF), microwave-assisted extraction (MAE), enzyme-assisted extraction (EAE), ultrasound-assisted extraction (UAE), and high-voltage electrical discharge are examples of extraction techniques (Bhatti et al. 2022). To extract bioactive molecules, especially phenolic compounds, from plant sources, solvent extraction (SE) is a commonly utilized conventional technique. Ultrasonic, microwave, and supercritical fluid extraction are more sophisticated methods that provide faster extraction times without leaving any organic residue in the extracted extracts (Easmin et al. 2015). To better liberate the compounds from the plant matrix, these procedures have been improved with further steps such as enzyme- and instantaneously regulated pressure drop-assisted extractions. The traditional extraction process used polar or non-polar solvents to extract the bioactive components from garlic (Putnik et al. 2019). Among the often-employed extraction techniques are solvent extraction, superficial fluid extraction, microwave-assisted extraction, maceration, infusion, percolation, and decoction (Sindhu et al. 2021).

5.1 Solvent Extraction

The simplest extraction technique is solvent extraction. Mostly, it is an effective method of extracting garlic oil. Because of the related low processing heat, extracting medicinal plants to create functional foods and nutraceuticals is recommended. The process involves pulverizing dried garlic (with less than 5% moisture content), extracting it using petroleum ether, and then soaking it in commercial hexane for 8 to 10 h in a Soxhlet mechanism. This results in oil in a solvent extract with less than 1% humidity that can be refined using various techniques. An altered technique uses an orbital shaker at room temperature, in contact with the garlic and solvent, and macerates the material instead of grinding it. The same amount of solvent is used just once. A better extraction technique would be to create a liquid extraction system that uses a continuous solid discharge, either countercurrent or co-current (Kumar et al. 2023) (Özcan et al. 2021).

Pure garlic oil can be obtained directly from the solid garlic matrix using Soxhlet extraction and organic solvents such as ethanol, methanol, and acetone. Garlic oil is highly soluble in a wide range of organic solvents, which makes this process extremely effective, as demonstrated by yields greater than 90%. Unfortunately, using many solvents and heating them for a long time during Soxhlet extraction can often lead to the molecule degrading thermally. An alternative method involves immersing the mixture, stirring it occasionally, and then centrifuging the crushed garlic. This allows the garlic oil to diffuse into the solvent over a few days slowly. Even though this process takes longer, thermal deterioration is much reduced and can be avoided by keeping the mixture cool. "Cold finger" extraction can imitate this issue (Bar et al. 2022; Ciric et al. 2020; Ezeorba et al. 2022).

Solvent extraction is thought to be among the easiest and most effective techniques for separating and purifying a solid matrix's constituents. The best solvents must be selected to dissolve a given molecule to the greatest extent possible while keeping other compounds intact. To maximize the efficiency of the desired chemical separation from the solid matrix, the perfect solvent for a given process is the one with the lowest affinity for the unwanted component (Liu and Chen 2021; Opare et al. 2021; Chen et al. 2022).

5.2 *Supercritical Fluid Extraction*

Supercritical Fluid Extraction (SFE) has been recommended as a substitute for solvent extraction due to the unfavorable outcomes observed in certain situations. By using CO_2, this non-polar method prevents the possible pyrrolization of allicin and instead extracts more durable and stable sulfur compounds. Numerous intended biological effects, such as antibacterial and cytotoxic properties, have been demonstrated to be exhibited by these. When examined as individual substances, various extracts may have various effects due to the ability to fractionate the sulfur compounds during extraction. For example, diallyl disulfide is a promising chemopreventive drug, and diallyl trisulfide has been proven to stimulate phase II detoxifying enzymes potently. One significant benefit over frequently used solvents is the use of CO_2, which is inexpensive, non-flammable, and non-toxic. Previous research has revealed that solvent extraction residues may remain in the final food product. This may not be ideal, mainly if the extracts will be used as functional food additives. CO_2 can be extracted and recycled during the solvent recovery and purging phases, preventing waste and reducing product contamination (de Jesus and Maciel Filho 2020; Arumugham et al. 2021; Mitra et al. 2022).

5.3 *Microwave-Assisted Extraction*

Extraction of these chemicals using microwave-assisted extraction (MAE) was demonstrated to be an effective technique. An overview of modern sample preparation methods is given by Huie (2002), who highlights the use of MAE, which speeds up the heating process by directly heating the solvent or solvent mixture. This greatly reduces the extraction time. As a quick extraction method that effectively transfers microwave energy to materials through molecular interaction with the electromagnetic field, Proestos and Komaitis (2008) describe MAE as providing a fast transfer of energy to the extraction solvent and raw plant materials. When the antioxidant capacity of extracts produced by microwave assistance and ultrasound-assisted extraction and maceration are compared, Sharma et al. (2008) find that the antioxidant capacity of microwave-assisted extracts is much higher.

For garlic, ethanol was discovered to produce the greatest outcomes when different solvents were used with the irradiation procedure to see which worked best. This was ascertained through alliin and solid content extraction into the liquid phase. The optimum solvent for extracting solid content was 70% aqueous ethanol, whereas pure ethanol was the best for extracting alliin. This is significant since it is known that alliin serves as a precursor to the bioactive molecule allicin. The extraction of alliin into the liquid phase has also been demonstrated to increase with temperature. Although allicin will begin to form at 37 °C, this is an intriguing discovery as it may expedite and improve the process. This suggests that it would be challenging to extract alliin into the liquid phase without losing it before converting it to allicin (Chaudhary et al. 2021; Cheng et al. 2021; Bar et al. 2022). According to Ru et al. (2009), MAE is faster and more effective than traditional methods since it requires less solvent and has a quicker extraction process (Fig. 3).

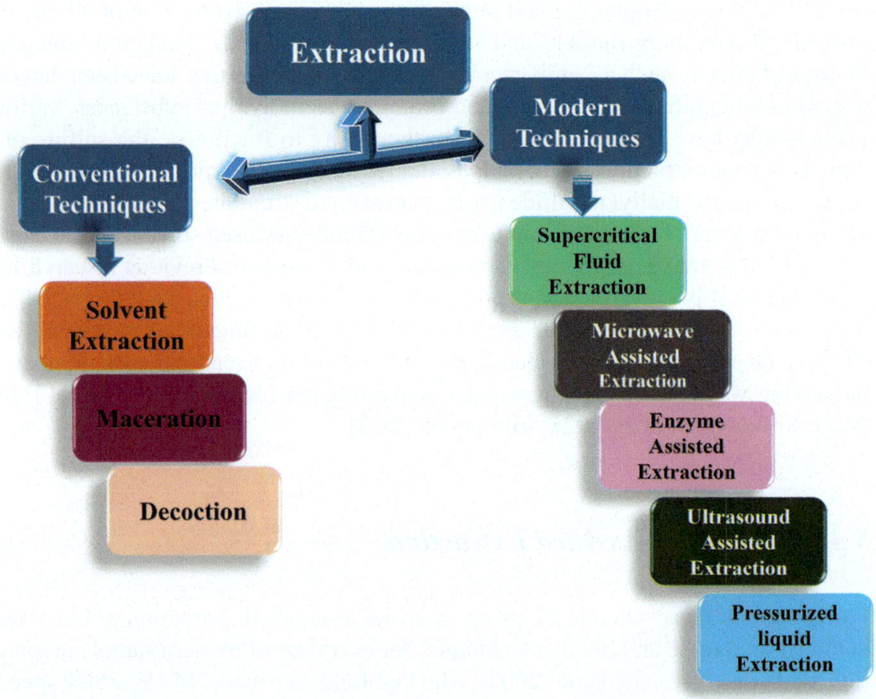

Fig. 3 Some major extraction techniques involved in garlic processing

6 By-Products of Garlic

Garlic has several uses and produces several valuable by-products. Each component of garlic—the peel, leaves, stem, and bulb—has unique qualities and uses. Processing garlic yields enormous amounts of peel, frequently seen as a worthless by-product (Palkovics et al. 2020). According to Blazewicz-Wozniak and Michowska (2011), garlic leaves contain essential oil content, which is rich in aromatic compounds and affected by the ecotype of the garlic plant. Oszmiański et al. identified flavonol derivatives in these leaves, with kaempferol being the most abundant, which may indicate potential health advantages of garlic leaves. Garlic stalks are also essential by-products as stalks can be utilized as soil amendments to boost antioxidant defense systems and promote plant growth after being air-dried and ground into a powder (Ghani et al. 2019; Ghani et al. 2022;). Furthermore, garlic bulbs—frequently used as a spice—are essential to many recipes (Bede and Zaixiang 2020).

Garlic's by-products have a variety of functions, ranging from medical to agricultural. Improved animal feed formulations, new treatment therapies, and sustainable agricultural practices can all result from understanding and appropriate use of garlic by-products.

6.1 *Garlic Stalk and Leaves*

Plants have traditionally utilized their green portions for flavors and fragrances. The garlic plant is one example of this as well. The taste of the leaves and stalks is milder than that of the garlic bulb. Garlic leaves are utilized similarly to bulb garlic throughout Europe and the Middle East, where they are plucked from garlic previously grown for its bulb. However, garlic leaves are a staple in Eastern Asian cuisine, especially Thailand and Laos. Certain herbs and spices, especially the pungent flavors of garlic and onion, are well known for their prominence in Thai and Lao cuisines (Tiong 2020; Sinsawasdi et al. 2022).

For culinary and medical applications, garlic stalks and bulbs have been utilized since the dawn of recorded history. Garlic leaves and stalks are mostly considered by-products by many individuals (Saif et al. 2020). They are, nevertheless, also utilized in other regions of the world. Garlic leaf consumption is determined by availability, sharp flavors, and contemporaneous cooking activities (Selim et al. 2022). Europe, the Middle East, and Asia use the thin leaves to flavor and garnish. Many garlic leaves produced in Southern European countries are plaited and utilized similarly to bulb garlic. Garlic plants are farmed for their leaves in Asia and the Pacific. Harvesting and selling the stalks and bulbs would be a significant economic benefit during the summer when there is less demand for garlic bulbs (Sarpaki 2021; Sturza and Ghendov-Moşanu 2021).

6.1.1 Garlic Leaves and Stalks Used in Culinary Applications

The garlic stalk and leaves are used in a wide range of ethnic dishes, and their utilization is essential (Zocchi et al. 2024). Garlic stalks and leaves are commonly used in Indian, Thai, Laotian, Vietnamese, and Chinese meals; nevertheless, precise recipes have not been provided. As more material becomes accessible due to the demand for garlic bulbs and the mechanization of bulb harvesting, there is a significant deal of opportunity for improved exploitation of the garlic stalk and leaves (Siu and Mantzioris 2020; Tamang et al. 2020; Sinsawasdi et al. 2022). The stalk and green leaves comprise about 80% of the garlic plant. This substance is mainly used in several ethnic or regional food products. In Korea, where this product is widespread, the primary garlic bulb is peeled and utilized with the green leaves for either fresh or fermented pickles. Another variety of this product is garlic, which has been pickled with some or all of its green leaves (Takagi 2020). Rice flour and water are used to make the dough for the dumplings, which are then filled with a mixture of minced pork, shiitake mushrooms, ginger, garlic, and spice. Boiling or steaming is used to produce the finished product, with boiling being the more popular method (Settapramote et al. 2024).

6.1.2 Garlic Leaves and Stalks: Medicinal Uses

Garlic leaves and stalks are used medicinally, and their therapeutic effects are derived from organosulfur compounds. When garlic is consumed orally, the same molecules that give it its unique scent are primarily expelled through the lungs rather than the skin (Tudu et al. 2022; Walag et al. 2020). This means that the main disincentive to consuming garlic and using garlic-based products is the smell and breath of garlic. Garlic leaves and stalks have the potential to ameliorate this issue partially (Borlinghaus et al. 2021). Even though chlorophyll has no sulfur, the smell can be mitigated by chewing parsley or green leaves. As garlic becomes increasingly well-liked as a medicinal tool, more and more consumers and medical experts will seek knowledge on reducing adverse effects and enhancing its positive impact (Qiu et al. 2020; Afzaal et al. 2021). Garlic's diuretic and anti-hypertensive qualities are mainly attributed to its volatile oil and sodium and potassium ions. In many cases of edema and/or hypertension, skin peeling and sodium elimination from the body are not urgently necessary; in these cases, irrigation therapy with garlic stalk tea may be considered. In addition, individuals who wish to circumvent the systemic impacts of garlic preparations will discover that the stalk and leaves of garlic do not induce any bothersome symptoms related to the digestive or respiratory systems and can be applied topically to treat specific conditions, such as rheumatoid arthritis (Özçelik et al. 2020; Godara et al. 2022).

6.1.3 Garlic Leaves and Stalks' Nutritional Values

Vitamin B6, C, and sulfur content are all very high in garlic. Along with trace amounts of many other nutrients, it has a good amount of calcium, copper, potassium, phosphorus, iron, manganese, vitamin B6, vitamin C, selenium, and fiber. Garlic has extremely little cholesterol, saturated fat, or sodium. Additionally, it includes allicin, which can strengthen the immune system and has antifungal and antibacterial qualities (Lenková et al. 2017; Ayoade et al. 2023). Excessive dosages of garlic supplementation have been demonstrated to dramatically lower oxidative stress in people with high blood pressure and raise antioxidant enzymes in humans. According to Moosavian et al. (2020), the antioxidative qualities and combined benefits of lowering blood pressure and cholesterol may help prevent prevalent brain disorders like dementia and Alzheimer's.

6.1.4 Methods for Processing Garlic Stalks and Leaves

Some value garlic leaves and stalks as significant by-products. They may be a good source of naturally occurring antioxidants that help keep lipids in diet and biological systems stable. The methods used to extract the garlic leaves and stems that affect the antioxidant activity are significant. The drying process impacted the chemical and physical characteristics of the garlic stalk and leaves. The microwave process can produce high-quality dehydrated material when using dehydrated garlic stalk and leaf material. Extraction procedures determine the effectiveness of antioxidant extraction from garlic leaves and stalks. Higher temperatures and shorter extraction times generally produced better results (Kurnia et al. 2021; Skoczylas et al. 2023).

Techniques for Drying

Numerous studies on the drying of garlic leaves and stalks have been published. Garlic stalks were sun-dried in China until their moisture level dropped to 8–10%. After that, it was pulverized to make feed, and the nutrients it contained were compared to those of a fresh garlic stem. According to Lee et al. (2014), sun-dried garlic stalks had similar nutrients and were more digestible than fresh garlic stalks. Nonetheless, an Indian study found that sun-drying garlic leaves resulted in a significant reduction of chlorophyll and a browning of the leaves. This was undesired since the dried leaves are mainly used as a spice and food additive. It was also discovered that the dried leaves have less calcium and phosphorus than the fresh ones. Furthermore, because of the unpredictability of American weather patterns and the possibility of animal damage to the crop, sun drying was found unsuitable for the large-scale production of dried garlic leaves.

One of the earliest and most popular techniques for preserving food is drying. The primary goal of drying is to extend the shelf life of seasonal crops for year-round distribution and use. It is among the most straightforward preservation techniques requiring the least resource input. Effective moisture removal is essential for dried goods to remain of high quality. On the other hand, drying at too low a temperature might lead to spoiling during storage, and too much heat can alter the flavor, color, and nutritional content. There are two types of drying techniques: solar drying and artificial drying. Because sun drying is an affordable method, it is frequently employed in developing nations.

Nevertheless, it is frequently carried out in unclean and contaminated environments, which could make the product dangerous to eat. Dryers, ovens, and other devices that offer greater control over temperature and humidity levels during the drying process are typically used for artificial drying (Feng et al. 2021; Makarichian et al. 2021). Using the appropriate drying technique under ideal drying circumstances is crucial to producing high-quality dehydrated garlic stem and leaf products.

Methods of Extraction

Garlic oil can be extracted with a variety of tools. Whole plants, including the bulb, stalks, and leaves, can be subjected to distillation; typically, the oil from the bulb is extracted simultaneously (Addala et al. 2016). Steam distillation is another method for obtaining garlic oil (Dehariya et al. 2021). Conventional techniques for extracting garlic oil involve using solvents like methanol, ethanol, or petroleum. Research indicates that using pure acetone as a solvent yields an odorless garlic oil with a high level of Allinase (Cañizares et al. 2002; Cavalcanti et al. 2021). One new technique that has been developed is to peel the garlic directly into a steam distillation unit using a "fully enclosed type automatic garlic peeler." Garlic that has been peeled directly yields oil with a higher Allinase concentration than garlic that has been peeled manually (Manjunatha et al. 2014; Guardia et al. 2023). The extraction of high amounts of Allinase from garlic has also been demonstrated using the recently established "Microwave and Enzyme Complex" approach. According to Putra et al. (2020), this technique entails freezing minced garlic, thawing it with enzymes, heating it with microwaves, and pressing it through filter fabric.

6.1.5 Possible Prospects of Garlic Stalk and Leaves for the Future

New garlic goods won't be well-received by consumers unless they have a unique, higher quality than current offerings. Garlic's health-promoting qualities are becoming a big selling point. Thus, new garlic products should be credited with these qualities. Research should focus on sulfur-containing compounds found in stalks and leaves to investigate if they have more health-enhancing qualities than those found in bulbs, as these chemicals have been primarily linked to the health-enhancing qualities of bulbs. It is also essential to investigate the potential nutritional or health-promoting benefits of chlorophyll and other color components present in the tissues of green garlic (Li et al. 2015; Hill et al. 2023).

Only by thoroughly understanding the chemical makeup of garlic by-products and the changes that occur throughout the harvesting, drying, and storage of garlic can higher value-added by-products be generated. Finding possible applications for garlic leaves and stalks requires understanding their chemical makeup. Furthermore, a deeper comprehension of the chemical alterations that take place during the senescence of the stalk and leaves may make it possible to regulate the garlic's flavor and odor characteristics (El-Saber Batiha et al. 2020; Skoczylas et al. 2023) (Fig. 4).

6.2 *Garlic Bulb*

Developing novel processes for turning agricultural leftovers into high-value chemicals and materials for environmental sustainability, waste reduction, and improved crop use efficiency is becoming increasingly crucial. Up until now, not much has been spoken about the garlic by-product. However, given how much garlic is produced globally on a reasonably large scale, the amount of waste has the potential to be a valuable resource.

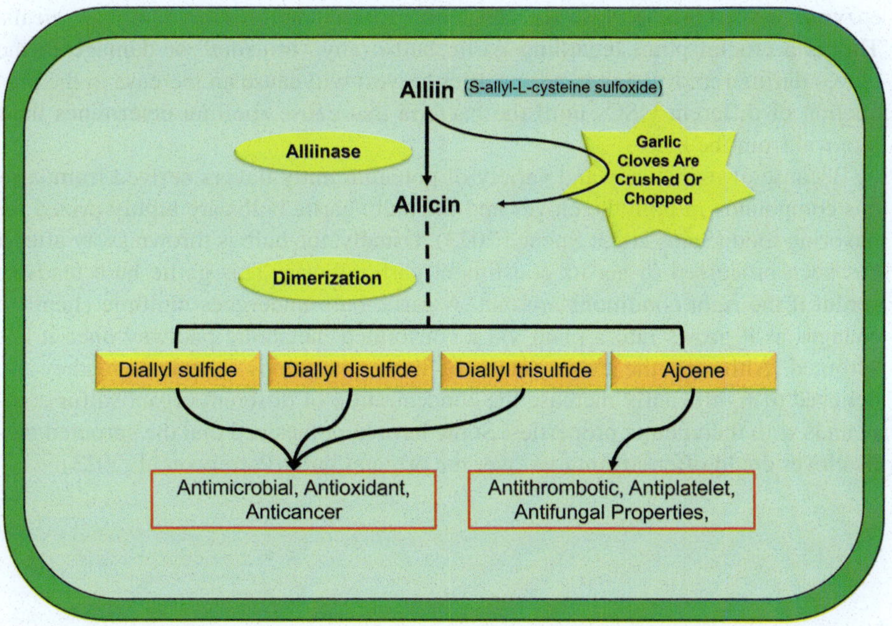

Fig. 4 Demonstration of the alliin conversion into different bioactive chemicals with their therapeutic properties

Allium sativum, or garlic, is a crop widely grown and used for its raw form and products with added value. A significant number of garlic is lost, especially in areas where it is actively grown for its bulbs, even while the majority is utilized as a fresh vegetable or processed into oil and other value-added goods (Loppi et al. 2021). Garlic bulbs comprise several smaller bulbs, or cloves, encircled by the bulb's fleshy scales and leaves and coated in a papery tunic. Garlic bulbs are divided into individual cloves, and the leaves are removed after harvesting once the plant's outer leaves have dried. Some waste materials, such as leaves and bulbs, are unfit for human eating or processing and could be a valuable source of by-products (Liu et al. 2024). Open-type garlic is more readily available and offers a larger window of opportunity for growers if crop performance is predicted to be subpar due to inhibitory diseases or pests. Open-type garlic takes bulbs longer than vegetative propagation varieties. A cytological analysis of the formation of garlic bulbs shows that basal plate cells and, to a lesser extent, meristematic cells divide quickly during the first swelling phase. The older leaves then roll up against the scape, and the leaf bases gradually elongate. This is crucial to realize that to maximize bulb production, a crop's age and maturity efficiency are crucial.

It is necessary to keep in mind that garlic bulbs are collections of storage tissue where the success of each clove's enzyme activity will determine its final flavor and pungency, given the wide variation in stored alliinase activity and OSC composition amongst garlic cultivars and the relative instability of these compounds. Because these precursor molecules are stabilized, cellular disturbance from mechanical harvesting or plant damage frequently results in a significant release of enzymes, which breaks down the OSCs into different sulfur-containing chemicals. This is a crucial point regarding garlic bulbs: any "bruising" or damage to the cloves during removal and handling after harvest will cause an increase in the production of different OSCs until the bacteria that cause spoiling determines their removal from the bulb.

Because of its potency and variety of potential nutty flavors derived from various compounds in both the cloves and the bulb, garlic bulbs are highly prized for flavoring food (Saras 2023; Spence 2023). Usually, the bulb is thrown away after it has been processed to get its constituent garlic cloves. This garlic bulb tends to sprout if the right conditions are met. A garlic bulb undergoes multiple chemical changes as it grows into a plant via a convoluted metabolic pathway once it has sprouted. Although the exact nature of these alterations is unknown, they are believed to significantly increase the concentration of different organosulfur compounds with therapeutic properties. Some have hypothesized that the sprouted garlic cloves could offer advantages over the original bulb (Parreño et al. 2023).

6.2.1 Nutritious and Medicinal Properties of Garlic Bulbs

In terms of nutritional richness, the composition of the garlic bulb is superior to that of other parts of the plant. Larger garlic bulbs with ten or more cloves are available (Loppi et al. 2021). Garlic bulbs have protein, carbs, iron, vitamin C, and fat among their nutrients. However, the most intriguing feature is the highest concentration of fructose-oligosaccharides (FOS). FOS is a low-calorie sugar that is helpful for people with diabetes because it is not soluble in enzymes. Garlic juice extract's FOS has been demonstrated in a study to have the ability to reduce blood glucose levels in vivo. This results from FOS's prebiotic properties, which support the colon's beneficial bacteria's survival and activation (Qiu et al. 2022; Meuninck 2023).

Garlic bulbs are a by-product of harvesting garlic crops, as was previously described. Garlic bulbs are generally only used as a cooking element, with the remainder being thrown away as trash. However, the application of traditional medicine prompts us to investigate garlic bulbs' nutritional and therapeutic qualities, notably their fructose-oligosaccharide (FOS) concentration. Although it has been established that this substance functions as a prebiotic, its application in medical therapy has not been made clear. All this knowledge may inspire new concepts for dietary supplements or products made from garlic bulbs for medical purposes (Espinoza et al. 2020; Naqash et al. 2021).

6.2.2 Nutritional Composition of Garlic Bulbs

The bulb of *Allium sativum* is a tasty culinary component and a significant source of essential nutrients. The macronutrient composition of fresh garlic is relatively modest, mainly consisting of carbohydrates with traces of protein and fat. Garlic does, however, have several essential micronutrients and bioactive substances that support its health advantages. Higher amounts of carbohydrates, mainly fiber and fructose-based polymers, are found in aged garlic. During the drying process of garlic, levels of sulfur amino acids rise while levels of macronutrients drop (Kang et al. 2022; Loghmanifar et al. 2022). There is limited research indicating that people with irritable bowel syndrome (IBS) who limit their intake of foods high in fructans have digestive discomfort, mostly due to the high fructan content of garlic. Patients with IBS should exercise caution when using garlic. Prebiotics, such as fructans, are thought to promote the growth of advantageous Bifidobacterium in the large intestine. This could account for the observation that garlic has been shown, particularly in in vitro experiments, to promote the development of this microbe. Accordingly, garlic may have prebiotic qualities and be a functional food (Lisciani et al. 2017; Aoun et al. 2021; Qiu et al. 2022; Liu et al. 2024).

6.2.3 Potential Benefits of Garlic Bulb

Population research on cardiovascular disorders indicates a positive correlation between garlic and the cardiovascular system. According to a study, regular garlic consumption may reduce the risk of high blood pressure. Another US study discovered that males who consumed much garlic had a decreased chance of acquiring atherosclerosis than their counterparts. These results and others provide consistent evidence that eating garlic can help reduce CVD (Imaizumi et al. 2023). A recent clinical experiment examined a standardized garlic preparation's effects on arterial stiffness, inflammation, and blood fat composition. According to this study's findings and a few others, garlic may help reduce certain CVD risk factors. This is significant because it may have an impact on people who already have CVD and may help prevent someone from having CVD in the future. An abundance of evidence points to the beneficial effects of garlic on several CVD risk variables (Valls et al. 2022; Imaizumi et al. 2023).

Higher garlic consumption has been linked to a lower risk of several malignancies, and population-based studies have demonstrated the beneficial effect of garlic in preventing cardiovascular illnesses. Consequently, research on the health advantages of garlic and its preparations has grown significantly to support the inclusion of garlic consumption in a healthy lifestyle.

One of the more significant recent types of research examined the possible utility of garlic in reducing immune system damage brought on by an infection with *Trypanosoma cruzi*. Chagas disease, a crippling sickness that mainly affects 16–18 million people in the Americas, is caused by this parasite mouse infection with *T. cruzi*. In response to non-parasite antigens, *cruzi* significantly reduces the synthesis of INF-y and IL-2 cytokines and the delayed-type hypersensitive response (DTH). These responses and the preservation of Th1 activity are enhanced when aged garlic extract is administered to sick mice. There is a notable reduction in splenic iNOS and NO generation, which suggests a lower risk of host cell damage and a decrease in macrophage cytotoxicity. Additionally, as a dependable marker of immune system damage, the extract inhibits increased blood sCD8 levels (Vathsala and Krishna Murthy 2020; Khalil and Maher 2022).

The potential benefit of using garlic bulbs as an immunostimulant and antibacterial is reported. Because allicin interacts with thiol-containing proteins, garlic has antibacterial properties. This interaction may inhibit the reduction of cysteine and sulfhydryl compounds in bacteria. In one study, the growth of *Helicobacter pylori* in the stomachs of infected mice is examined in response to raw, crushed garlic. Allicin-containing aqueous extract stops bacterial growth in vitro. In areas where regular treatment is too expensive, this extract or raw garlic may be a cheap and effective substitute for triple therapy.

6.2.4 Methods of Processing Garlic Bulbs

Many studies on the use of garlic in food and cooking have been conducted, leading to the creation of a wide range of processed and prepared goods. Due to their affordability and general availability, raw garlic is still included in many of these recipes, but because of their uniform composition and ease of use, processed garlic extracts may become more common as ingredients. Products made from dried garlic can be used instead of fresh garlic. Typically, this is done by crushing or slicing fresh garlic, which is then dried in a dehydrator or oven. A month of high heat and regulated humidity is needed to age whole raw garlic, creating black garlic, a high-value commodity. This is explained by the Maillard process and enzymatic reactions that alter the chemical makeup of garlic components, producing a sweet, balsamic flavor without any off odors. This procedure has been replicated with fresh garlic slices and garlic paste to get similar flavors. A retort processing study is presently being conducted to generate a shelf-stable product.

Pickled garlic is a traditional product in several countries. It can be made using an acidified brine, much like pickled onions, but there is a chance that the lactic acid bacteria will cause the product to ferment. Another common product that is frequently manufactured at home is garlic oil. However, caution must be used during preparation and storage to prevent the growth of *Clostridium botulinum* spores, which can result in a potentially lethal form of food poisoning. This has prompted research into safe procedures for large-scale commercial product manufacture and small-scale preparation. Another area of interest for the development of garlic products is the inclusion of garlic into frozen and convenience foods.

An effective extraction procedure is necessary to generate high yields of required components without compromising quality, a prerequisite for commercializing any by-product. Plant extracts are complex combinations that frequently contain both volatile and non-volatile components. As such, the conditions under which the extract is extracted can be changed to produce the desired form, such as an oily, powdery preparation, or scent. The following techniques are frequently used in liquid-liquid extraction: maceration, percolation, decoction, and refluxing. These are commonly followed by solvent removal to produce a dry, non-volatile extract and distillation to extract the volatile portion. Steam distillation, microwave extraction, pressurized liquid extraction, supercritical fluid extraction, and ultrasound-assisted extraction are some more methods of extraction that could be used. Since these factors substantially impact extract yield, composition, and quality, it is imperative to ascertain the best extraction method, time, and solvent combination for a given component.

6.2.5 Extraction Methods for Garlic Bulb

At home, two kinds of garlic oil are used. One is a naturally occurring essential oil produced using conventional techniques. This oil is used to season various foods and subtly introduce garlic taste to cooking oil. The oil that is extracted using a solvent is the other kind. These days, the most popular form of this is garlic-infused olive oil, which is also used for seasoning and cooking (Ciğeroğlu et al. 2022; Saras 2023). Food, medicine, and the chemical and flavor industries use garlic oil.

Additionally, it is a component of pharmaceuticals, where it is used to make syrups and capsules for respiratory issues as an antibacterial and a replacement for garlic oil. Garlic clove juice is also utilized (Rouf et al. 2020; Saif et al. 2020; Tesfaye 2021). The main elements that can be used for the many garlic by-products are source material quality, an appropriate extraction method and its surroundings, and packaging that prevents spoiling (An et al. 2022).

Many techniques are used in garlic bulb extraction processes to separate the bulbs' essential elements and beneficial compounds. A well-liked method of aged garlic extract (AGE) extraction involves extracting fresh garlic for up to 20 months to produce an odorless extract with antioxidant-rich, stable, and water-soluble organosulfur components. This tactic can help preserve these compounds, which can scavenge free radicals and prevent oxidative damage (Nencini et al. 2011).

Another extraction method involves the use of multiple solvents and variables. For Golubkina et al. (2020), distilled water was added to dried garlic powder, and the mixture was allowed to remain at room temperature for 2 h to extract garlic extract. In a different study, Golubkina et al. (2020) isolated dry onion bulb homogenates using 70% ethanol at 80 °C. Furthermore, studies have investigated the extraction of essential oils from garlic bulbs using various methods, such as hydrodistillation, sono-hydrodistillation, and ultrasonic-assisted hydrodistillation (Boubechiche et al. 2017). These techniques can extract the volatile compounds and antioxidants that give garlic its distinct smell and health benefits. Furthermore, using ultra-performance liquid chromatography (UPLC) to extract polyphenols and antioxidants from garlic bulbs has allowed researchers to analyze the extracts' qualitative and quantitative composition (Nagella et al. 2014). This method can identify and quantify the polyphenolic compounds that increase the antioxidant activity of garlic bulbs.

Different extraction techniques are utilized for garlic bulbs depending on the desired compounds and characteristics of the extract. Garlic bulbs often extract bioactive compounds, antioxidants, and other critical components for food, medicine, and agriculture use. These procedures include solvent extraction, aged extraction, essential oil extraction, high-performance liquid chromatography (HPLC), and UPLC.

6.2.6 Commercial Uses of Garlic Bulb

In recent years, the pharmaceutical industry has seen a rise in demand for garlic by-products as raw materials due to its exceptional medicinal qualities. Numerous advantages can be obtained from garlic oil produced by steam-distilling crushed garlic (Shaikhiev et al. 2022; Pavlić et al. 2023). According to Espinoza et al. (2020), old garlic juice (digest), aged garlic extract, and aged garlic powder are further by-products of garlic utilized in the pharmaceutical business. The raw materials for many pharmaceutical formulations, including gelatin capsules and liquid and solid orals, are garlic oil and its derivatives. Additionally, they are utilized in creating various medicinal oils and in Ayurvedic formulations (Bar et al. 2022). Although research and development on these products are ongoing, garlic oil may soon be directly prescribed in allopathic medical formulations (Parveen et al. 2020; Azmi and Shukla 2021). While investigations on newer garlic by-products, such as garlic peel oil and garlic black onion dehydrated powder, are still in the initial phase, garlic oil and its derivatives have a strong future in the pharmaceutical business. Because of its high sulfur content, garlic peel oil, derived via solvent extraction of garlic skins, has considerable potential as a natural antioxidant and antibacterial agent.

Similarly, a variety of ready-to-eat food products and spices can benefit from the taste-enhancing properties of garlic black onion powder, which is a rich source of water-soluble sulfur. They thus have much potential in the food processing industry and other related fields. A new avenue for the creation of various ventures in small-scale and agro-based food industries involved in garlic production, garlic products, or other processed food products has been opened by research and development of garlic oil and its other derivatives, thanks to increased garlic production in India and established consumer trust over garlic products (Carreón-Delgado et al. 2023; Hernández-Montesinos et al. 2023).

6.2.7 Future Prospect of Garlic Bulbs

Our understanding of the garlic bulb's physical and chemical features, the plant's main component, is still incomplete. Therefore, the first step toward conducting more research on garlic bulbs is to establish a foundational understanding of their chemical and physical characteristics. Treating the garlic bulb according to specific procedures can maximize the outcomes by understanding its physical and chemical features more deeply. The different processing techniques can determine the quality of the product, and research indicates that there are currently few methods for processing garlic bulbs for several uses. The findings of studies on garlic bulbs that used techniques including silage production, basic fermentation, and extraction to extract chemical components make this clear. But before the product may be used as a substitute, more time will need to pass (Qiu et al. 2022; Parreño et al. 2023).

The benefits of *Allium sativum*, or "garlic," are becoming more well-recognized, and it is now proven that garlic has many potential applications for human health. On the other hand, a large portion of garlic is used to accumulate leftover by-products from the prior form of garlic. There is at least 20% bulb material left over from each 100% total weight of garlic that has not been used and is not being treated, even though some by-products have been studied in the past. This issue presents several concerns regarding the detrimental consequences of accumulating organic material—known to pollute the environment. The writer has been motivated by this phenomenon to conduct additional studies on garlic bulb by-products to minimize their adverse effects and maximize their usefulness (Chakraborty et al. 2021; Garmame Galgaye 2022; Melkamu 2022).

6.2.8 Garlic Bulbs Are Subject to Further Research

The process of producing garlic is based on post-harvest technology, but the literature on this topic is far less abundant than on cultivation. Given that 40–60% of the harvest is frequently lost before it reaches a stable market, this is an area where research could directly benefit farmers (Chekol et al. 2023). Currently, cloves and bulbs undergo a significant physiological change, resulting in a quality loss. The curing process is one example; it can take many forms, such as field drying, drying in a shed, or even drying outdoors for several weeks. Although there is less information on how temperature and time at each stage affect product shelf life, steps in bulb roasting may be carried out to enhance taste and/or "value adding" (Desta et al. 2021; Naqash et al. 2021).

Research on bulb growth's biochemical and physiological components would be very beneficial. Clove quality might be preserved for longer periods, and weight loss could be minimized if it were feasible to stop the sprout elongation process and prevent root formation after harvest. During some seasons, ware bulb storage may even be more profitable than inadequate small-sized cloves. A deeper comprehension of the connection between environmental variables and bulb development might also make it possible to manipulate crop duration, which is crucial in some areas for upland rice rotation (Ahmed et al. 2020; Takagi 2020).

The nature and content of the clove, the main harvested product, need to be more thoroughly characterized in terms of its physical and chemical makeup to demonstrate how the garlic bulb asset could be better utilized. Only when the market trends and customer demands are considered can a judgment be made regarding the primary product's most profitable application be made. In tropical conditions, very little research has been done on the impact of seasonal variations on bulb formation. To help farmers reduce losses, more research is needed on the possibility of clove loss as well as the severity of different forms of bulb damage at various phases of maturation caused by biotic and abiotic variables (Rogers et al. 2021; Tribble et al. 2021).

6.3 *Garlic Peels*

A large part of harvested garlic bulbs is "peel waste," which can come from a manually operated or high-speed peeling operation. The garlic's papery, thin exterior layers, and some of its thicker underlying tissues constitute the "peel waste" produced by the high-speed procedure (Singh 2018). After the garlic head is removed, the peels are typically wasted. They can be an affordable treatment for various illnesses and are an untapped source of therapeutic benefits (Kalse and Swami 2022; Salem et al. 2023). A simulation of this fact would encourage using peels to extract different oils and evaluate each oil or their combination for antibacterial activity. A straightforward, inexpensive device can make garlic peel oil using steam distillation (Teng et al. 2020; Hernández-Montesinos et al. 2023). Another way to produce garlic peel is to transform a regular cooker into a distillation apparatus by changing the lid and adding rubber tubes to the gasket area. The toxicity of oils extracted using different solvents can be evaluated to determine a safe limit for therapeutic usage. These oils can be used to create lotions or ointments, and their larvicidal, antibacterial, and antifungal properties can be assessed (Anis et al. 2021; Manyako et al. 2022; Maleki and Esmaeili 2023).

When making stocks, soups, sauces, and pickles, garlic peels—an uncommon species of vegetable peel with a high phenolic content—can be an aromatic addition. Garlic peels, when dried and powdered, have antibacterial and antioxidant qualities that make them a great addition to meals, flour, baked goods, and meat products. The anticancer and antimetastatic properties of garlic, onion, and their peels are well-established (dos et al. 2022; Hernández-Montesinos et al. 2023). Rats used in the experiment were given B16 melanoma cell implants and a diet high in fat, casein, and carbs like glucose. Garlic decreased the tumor weight by half when added to the diet; it decreased by 35% when combined with casein and 21% when combined with fat. When such tumorigenesis-containing meals were pulverized in freeze-dried onion and garlic n-hexane extract, B16 melanoma was comparatively inhibited (Abdel-Daim et al. 2020; Al-Shehri 2021).

Today, humanity is trying to find and stay on the golden mean. Natural resources should be used as efficiently as possible to ensure success without harming the environment. As a result, efforts are being made worldwide to control waste and use by-products. One of the oldest crops still under cultivation is garlic; the outer peels are thrown away as trash, but the garlic heads are used for food and medicine. Research on its use has benefited a wide range of activities (Chaudhary et al. 2021; Subroto et al. 2021).

6.3.1 Using Garlic Peels in Culinary Items

Many different ethnic cuisines employ extracted garlic in place of garlic cloves. Through small-scale experiments, it was found that garlic and garlic peel could add a vibrant, herbaceous flavor to Japanese-style fried rice. Garlic oil was used, and fried rice was combined with the peel and cloves of garlic. Research revealed that incorporating the peel into the rice and removing it produced a little garlic taste.

Traditional peeling or other mechanical methods are used as the initial stage in extracting garlic flavor from the bulbs. This extracts the "garlic peel," which is the outer layer of the bulb that envelops the cloves and makes up 10–30% of the bulb's total weight. Considering that 250,000 tons of garlic are consumed annually in America alone, this percentage is highly noteworthy (Dinu et al. 2023).

Pickled garlic was a typical food among the English in the seventeenth century. Garlic was common for French women to épicez leur amant or season their partners. Garlic is liked by many since it is used in many cultures as a flavoring for meals intended for a loved one (Spence 2023). Saxons and English used garlic and other herbs to cure the leg of lamb during the Middle Ages, long before garlic was used to heal warts. The leg of lamb would be slow roasted after being steeped in a mixture of garlic, rosemary, and thyme; the garlic and herbs would provide most of the taste.

Flavor is regarded as a dish's most crucial component in all cuisines. It has long been known that garlic enhances various dishes' flavor, aroma, and palatability. The Roman era acknowledged garlic and its peels as a flavoring agent to infuse wine with aroma (Bonasia et al. 2020; Ekşi et al. 2020).

6.3.2 Commercial Uses of Garlic Peel

Garlic's beneficial qualities imply that the peels might work similarly. It has been discovered that garlic peels prevent cell death, superoxide production, and lipid peroxidation. They also significantly improve the antioxidant defense of the cell. This was found when garlic peels were employed in an experiment to lessen cell damage. These results are noteworthy because most of garlic's cytotoxic and genotoxic properties prevent it from being used as a chemoprotective agent. This is important because it might isolate the health benefits of garlic, which has been effective in most situations. The pharmacology and toxicology of garlic and evaluating its potential use as a chemoprotective drug are greatly aided by the findings of chemoprotective cytotoxic and genotoxic effects of garlic or its preparations, especially in extracellular systems. In this sector, finding drugs that prevent specific components of chemotherapy from protecting normal cells is crucial (Khan et al. 2022). The antioxidant properties of garlic peels could be very useful in preventing inflammation and the oxidative stress that goes along with it. This has a lot of pathophysiological consequences. Peels may also be used to slow the progression of atherosclerosis because the condition is triggered by oxidative modification of low-density lipoprotein (LDL) and because garlic extracts and diallyl sulfides have cytostatic solid and cytotoxic effects on smooth muscle proliferation (Subroto et al. 2021).

One common spice is garlic. Garlic peels and cloves need to be separated before using. This separating process produces a significant amount of peel. The availability of peels has expanded along with the growing demand for items made with garlic. These peels contain antioxidants and chemicals that contain sulfur. These substances are what give garlic its health benefits. Considering these facts, research on the industrial uses of garlic peels is highly relevant. Owing to a paucity of knowledge on this subject, examining potential industrial uses based on garlic peels' established characteristics is critical.

6.3.3 Effects of Garlic Peels on the Economy

A product's potential for profitability largely depends on its market demand. Undoubtedly, more research is necessary, given the possibility of such an abundant and affordable chemical supply with a broad range of applications. However, to influence the market demand for them, the results regarding the yield of cysteine from garlic peels still need to be extensively reported. Because of its ease of use and low cost, the sulfur donor approach is now the one that the industry favors, meaning that it has a larger chance of being used in industrial settings. Garlic peel demand would skyrocket if a technique for isolating and concentrating the enzyme alliinase were developed and then used in a "natural" context to create cysteine from alliin. Garlic peels would undoubtedly be valued higher if there was more demand for them, and this is evident when comparing the methods used to extract cysteine. Ultrafiltration and an aqueous combination are two straightforward and affordable methods that can be used with the enzyme process in the home industry without requiring specialized equipment.

6.3.4 Demands and Market Value of Garlic Peels

Income determines the demand for garlic peels without considering the price, assuming that garlic is a lower-quality commodity. The next best alternative is to trade off garlic for other foods at dinnertime comparable in flavor and have a positive cross-elasticity of demand. Because the researchers' data indicates that the price-demand elasticity of garlic in Turkey is −0.48, they argued that garlic is a lower-quality good in Turkey. In India, the rural populace views garlic as a luxury commodity despite having a higher income, making it an inferior good for them. Demand for garlic will rise if food goods in China, India, and Thailand—three of the world's top garlic-consuming nations—become more expensive. In the long run, this will increase demand for garlic peels, especially because these countries' populations are growing. Potential markets for garlic peel oil and meal include high-population nations like China, India, Africa, and Southeast Asia, where garlic is a low-maintenance crop and an essential supply of table garlic. The demand for substitute goods is the positive elasticity of cross-price. The demand for garlic peels will rise with the price of garlic. An illustration from Indonesia demonstrates how research into substitute food crops such as sweet potatoes and cassava was spurred by increases in rice prices and soybeans.

The fundamentals of supply and demand in economics apply to the garlic peel market just as they do to other markets. When a product is in demand in the market, it generally means that consumers buy a specific amount at a given price, all other things being equal. A product's price, the cost of competing goods, consumer income, tastes and preferences, expectations for future prices and product availability, and the total number of purchasers in the market all influence demand for a given good or service. When it comes to garlic peels, the price that maximizes the quantity purchased moves the product down the demand curve. Garlic peels will be in great demand if there is a significant demand for garlic.

6.3.5 Prospects for Garlic By-Products in the Future

Products made from garlic are in demand in the consumer and industrial markets. Garlic peels enhance existing garlic processing operations and provide a sustainable way to cut waste in the garlic sector. It has been demonstrated that garlic peels retain the active components of garlic oil and have antifungal properties. Creating a technique to convert garlic peels back into a comprehensible form for ingestion is one way to use them. In essence, garlic would be compressed into a different shape and then filled with a mixture of garlic peel. This would guarantee that consumers are getting some of the health benefits that garlic peels offer while also giving them an alternate garlic product.

Further investigation into the efficacy of aged garlic peel in producing black garlic would be another line of inquiry. The production of a dehydrated garlic peel with high levels of allicin and other beneficial components would be an ideal option, allowing garlic peel powder to be used as a health supplement or a functional ingredient in various food products. Given the potential health benefits of black garlic, this could provide a valuable by-product of the black garlic production process. The benefits of dehydrated garlic peels include their compactness and odorlessness, eliminating many drawbacks associated with existing nutritional supplements containing garlic. This study also raises the possibility of creating a brand-new functional food category focusing on garlic-based goods. Demand for the creation of functional foods is rising as people become more health aware and look for a way to enjoy food and get health benefits together. Such a food would be a snack food fortified with garlic using a composition of garlic peels. Creating a composition for garlic peels has promised broad application as a processing aid in the food industry. Two common vegetables used in cooking and food manufacturing are garlic and onion.

7 Potential of Garlic's By-Products

By-products are leftover or secondary products from a production process that can be further used (Londero 2001). Many by-products with a wide range of uses are produced during garlic processing. By-products include black garlic, husks, straw, peels, and juice processing debris. It has been determined that these by-products are inexpensive substrates that can be used to produce high-value goods, including feed alimentation, soil amendments, and bioenergy (Kallel and Ellouz Chaabouni 2017). It is usually thrown away in landfills and other waste disposal sites (Mashad et al. 2019). Significant amounts of potassium, phosphorus, sulfur, and other elements can be found in the dry mass of garlic peel (Polyakov et al. 2020).

Garlic peel with surface modifications can efficiently extract crystal violet dye from wastewater, as shown by Raaj et al.'s (2022) research, underscoring its potential to reduce water pollution. Furthermore, Kallel and Ellouz Chaabouni (2017) highlighted the usefulness of garlic by-products, especially the peel, in various industrial uses, including feed alimentation, soil amendment, and bioenergy. According to Singiri et al. (2022), garlic peel has a variety of biological and biochemical functions, including promoting growth and development, thwarting infections, and granting resistance to abiotic stressors. Furthermore, studies on fattening sheep have shown that garlic skin alters the rumen microbiome and metabolome, suggesting that garlic skin may influence animal nutrition and health (Zhu et al. 2020). Additionally, research has been done on extracting pectin from garlic waste in dual-acid conditions, indicating the possibility of obtaining an advantage from garlic by-products (Şen et al. 2022). The bioactive chemicals found in garlic husks have been examined by Thach (2022), suggesting possible uses in the food, pharmaceutical, cosmetic, and nutraceutical industries. Oh et al. (2022) study on the processing waste of garlic juice has been assessed for use as a supplement in fish meals for young fish, demonstrating potential benefits for disease resistance, growth performance, and antioxidant activity. Together, these studies show the wide variety of by-products that may be obtained from the processing of garlic and their potential uses in several fields, such as agriculture, food production, animal nutrition, and health. Together, these studies highlight the significance of garlic peels as valuable resources with various applications and waste products.

7.1 Applications of Garlic By-Products in Food and Non-Food

When processing garlic, the outer tunics turn a color that varies from light purple to white. These are often removed. Numerous studies have reported antioxidant activity in the outer tunics (Sunanta et al. 2023). When the garlic bulbs are harvested, the haulm—the leaves and stem—are extracted mechanically. Its moisture content is 62%. Trimming removes the roots, which are roughly 68% moisture content. Any portion of the garlic plant is recognized to have a strong garlic taste and fragrance, and research has shown that the plant contains a few beneficial chemicals (Takagi 2020). These by-products are thought to have fresh garlic's flavor, taste, and bioactive components. Novel functional foods and medications could be developed because of the extraction of these bioactive substances (Qiu et al. 2020, 2022).

Additionally, garlic has a few water-soluble substances, including fructans, oligosaccharides, pectins, selenium, and arginine. Numerous decades-long investigations have concentrated on these sulfur- and water-soluble compounds' biological activity, identification, and isolation. Most of this research made use of aged garlic preparations or raw garlic. However, it is not cost-effective to dispose of garlic waste in the form of its by-products, such as the bulb's outer tunics, haulm, and roots, in the modern day where environmental consciousness is driving up demand for organic materials.

The most common sulfur component in garlic is alliin, which has no smell. However, when the garlic tissue is broken, the enzyme alliinase, kept in a different compartment, transforms alliin into the fragrant molecule allicin. Garlic's fresh-state characteristics are attributed to allicin. Allicin content can be increased by crushing fresh garlic and letting it stand for a few minutes before cooking. Diallyl sulfide, diallyl disulfide, diallyl trisulfide (DATS), and ajoene are among the various sulfur-containing chemicals quickly formed from the extremely unstable molecule allicin. When garlic is processed, these molecules are changed into different sulfur compounds (El-Saber Batiha et al. 2020; Subramanian et al. 2020).

Allium sativum is a spice or herbal remedy. It has been used to enhance and flavor various foods in practically every form of cuisine globally. Numerous bioactivities of garlic, including preventing bacterial and fungal infections, diabetes, atherosclerosis, lipemia, and tumors, have also been linked to its anti-tumor properties. These actions are thought to be connected to garlic's distinctively strong flavor and smell, which is believed mainly brought on by organosulfur compounds (Al-Shehri 2021; Rauf et al. 2022).

7.2 Function of Bioactive Substances

Recently, much research has been conducted on the many uses of garlic outside of food and medicine. This is because the garlic business's by-products, or waste, are regulated. Regarding the use of garlic oil as a flavoring agent in the culinary, pharmaceutical, and cosmetic industries, a wealth of knowledge is accessible. Numerous allyl sulfide compounds found in garlic oil, such as diallyl disulfide (DADS), diallyl trisulphide (DATS), and diallyl sulfide (DAS), are known to have impacts on different types of organisms. One organosulphur molecule generated from garlic called ajoene has been suggested to offer potential in the creation of pharmaceuticals. It is commonly recognized that the non-volatile polar chemical fraction, obtained by steam distillation and garlic oil removal, contains a variety of substances that have biological effects. This material, often known as garlic residue, includes several effects, such as antioxidants, antibiotics, and other compounds (Bhatwalkar et al. 2021).

7.3 Function in Food Items

From garlic by-products, beneficial compounds have been created to give a food product a particular quality. An illustration of this is the addition of allinase to soy protein isolate to create a garlic-flavored protein with improved functional attributes. Allinase is encapsulated with different coating materials to increase its tolerance to stomach acidity and enable its controlled release in the intestines (Tavares et al. 2021). Since soy protein isolate has no fat or carbohydrates, it is now possible to create food products with reduced fat and carbohydrates that taste like garlic and have improved functional qualities. Garlic's ability to act as a preservative in food and food supplements and its antibacterial action against microorganisms that cause food spoilage has been extensively studied. Organosulfur compounds produced from garlic have a wide range of antibacterial activity against Gram-positive and Gram-negative bacteria. The compound's kind, concentration, and the bacteria it is fighting against all significantly impact this action (Nakamoto et al. 2020; Bhatwalkar et al. 2021). High relative efficiency against bacteria that cause food poisoning and are known to be a major cause of food deterioration has been demonstrated. Various food products have been artificially contaminated or spontaneously ruined with different microorganisms in the presence of garlic, garlic extracts, or chemicals produced from garlic to evaluate the potential of garlic as a preservative (Bagheri et al. 2020).

The most common application for by-products (especially garlic oil) is as a flavoring in food. This is because of its strong flavor and scent, which may give many meals a unique flair. Compared to fresh garlic bulbs, garlic oil may be manufactured inexpensively and has a longer shelf life without significantly losing flavor. The oil from the bulb and the macerated garlic have been extracted using steam distillation and microwave techniques. Then, various food products frequently used in the culinary business, like oils, condiments, canned goods, soups, sauces, ready meals, and snack items, can have garlic oil added to them. In the home, garlic oil is also used to provide the proper flavor to food while cooking. Other oils created for organosulfur compounds obtained from garlic can also be used in various food applications to impart the appropriate functional qualities. An illustration of this is the addition of diallyl disulfide, an oil-soluble substance, to a variety of food items to capitalize on its recognized ability to act as an anti-carcinogen (Kishimoto and Kashiwagi 2020; Verma et al. 2023).

7.4 Using Garlic By-Products in Culinary Applications

It is widely accepted that *Allium sativum* has a wide range of culinary applications across numerous cuisines. Garlic has powerful sensory qualities. Depending on how much and in what form it is given, it can add heat (from the chemical compound allicin), pungency (closely tied to spiciness), and a hint of sweetness. In traditional Chinese medicine, garlic is thought to have a "warming" thermal character due to its pungent and hot properties. Although individuals in Western societies tend to associate garlic with a single distinct flavor, it is essential to understand that garlic has a wide range of flavor intensities. This is mostly explained by the need to produce components from garlic that are used as seasoning and give foods distinct garlic flavors (Ekşi et al. 2020; Saif et al. 2020; Spence 2023).

The difficult issue that garlic processors encounter when attempting to eliminate significant amounts of by-products produced in creating value-added items like black garlic or garlic paste has been examined in culinary applications of garlic by-products. These by-products build up rapidly, degrade swiftly, and are frequently stored for a long time before being disposed of permanently. Because garlic by-products are organic, there is a significant risk of environmental contamination and health risks when disposing of them. Therefore, finding practical applications for these by-products is preferable. Garlic's distinct flavor and perfume make it an ideal application for its by-products, and the culinary arts are no exception. The production of soup stocks, savory biscuits, and snacks (such as cheese straws), composite meat products, sauces, and gravies are among the specific culinary applications for garlic by-products that have been identified (Larousse 2022; Figueroa et al. 2023; Culinary Institute of America (CIA) 2024).

7.5 *Possible Ingredients in Food Items*

Because garlic leaves and stalks are rich in bioactive compounds and may offer health benefits, they are useful functional additives in food products. Flavonoids, phenolic compounds, and antioxidants are thought to be responsible for the functional properties of these garlic plant components (Nencini et al. 2011; Lachowicz et al. 2017). Aged garlic extracts have been shown to have antioxidant activity, indicating their capacity to lessen oxidative damage. They contain persistent organosulfur compounds that scavenge free radicals (Nencini et al. 2011). Because of their high phenolic content and potential as antioxidants, garlic leaves and stalks are also helpful in producing food products and dietary supplements (Lachowicz et al. 2017; Ullah et al. 2023a). The research highlights landrace garlic's chemical makeup and nutritional advantages, with a special focus on the mineral, organic acid, fatty acid, and free sugar content (Petropoulos et al. 2018). These findings demonstrate the diversity of nutrients found in garlic, suggesting that it is a promising addition to increase the nutritional content of food products. Because they have been shown to contain triterpenoids, carotenoids, and chlorophylls, garlic stalks may also be utilized as functional food ingredients (Lachowicz et al. 2017). Studies by Devarajan and Mohanmarugaraja (2017) and Ashfaq et al. (2021a, b) have shown that garlic leaves and stalks can improve the bioactive chemical content and antioxidant properties of a range of food products, such as soups, sauces, and functional meals.

The utilization of garlic by-products, especially stalks, is consistent with the trend of sustainable food production by minimizing waste and optimizing the nutritional value of raw materials (Alvarez-Jubete et al. 2014). Furthermore, garlic may be used to strengthen antioxidant defense systems and stimulate plant growth, which could be beneficial in creating functional food products due to its allelopathic effects on plant physiology (Ding et al. 2016; Ghani et al. 2019). Due to their abundance of bioactive compounds, antioxidants, and minerals, garlic leaves and stalks have the potential to be helpful food additions. They are valued additions to the food business, helping to meet the growing need for tasty and healthy functional meals due to their potential to increase the nutritional profile, antioxidant capacity, and health-promoting aspects of food compositions.

7.6 Antimicrobial Capability for Food Preservation

Studies have demonstrated the ability of garlic extracts to inhibit the growth of certain bacteria, fungi, and viruses, indicating the possibility of using them as a natural food preservative (Ross 2007). When garlic is crushed, minced, or bruised, the enzyme alliinase releases allicin from alliin, giving it antibacterial, antifungal, and antibiotic qualities. Numerous bacterial strains, including streptococci, mycobacterium, bacillus, and staphylococci, as well as multiple pathogenic fungi and yeasts, have been demonstrated to be susceptible to the action of allicin (Abdulrasheed et al. 2015; Bhatwalkar et al. 2021). Garlic aqueous extracts are also antimicrobial active, and studies have demonstrated that garlic oil, with its higher concentration of sulfur compounds, has even more antibacterial properties. Garlic's distinct smell is also caused by these sulfur compounds, which has prompted studies on deodorizing garlic extracts for use in culinary applications (Bhatwalkar et al. 2021; Bar et al. 2022). Aqueous garlic extracts have been shown to have a potent antibacterial action on both Gram-positive and Gram-negative bacteria. Increased garlic extract concentrations have been demonstrated to be beneficial in stopping bacterial development and prolonging the shelf life of various food products. These studies suggest that garlic extracts have the potential as natural preservatives for a variety of food products, even though the preservation of garlic products using garlic extracts may be undesirable due to the intensification of flavor (Lee et al. 2014; Ismail et al. 2020; Abidullah et al. 2021).

The three leading causes of food spoiling are oxidation, enzymes, and microbes. Oxidation is the leading cause of food quality degradation. By slowing the reaction pace, antioxidants stop oxidation and the subsequent rancidity (Bensid et al. 2022; Chib et al. 2020; Ullah et al. 2023c). Heat treatment can inactivate enzymes; however, for some goods, this causes the desired texture and flavor to be lost (Shao et al. 2021; Deak and Mohácsi-Farkas 2023). Perhaps employing microwave therapy or ionizing radiation to immobilize or inactivate the enzymes might be a more successful approach (Shahi et al. 2021; Umair et al. 2022). However, the most popular technique for preserving food is either eliminating the microbes that cause food deterioration or cultivating an environment that prevents them from proliferating and causing the alterations associated with deterioration. This can be accomplished by adjusting pH and water activity levels or applying different chemical preservatives. Public worries about the safety of chemical preservatives have heightened interest in natural preservatives in recent years, with many studies concentrating on compounds derived from plants that have antibacterial qualities (Rahman 2020; Batiha et al. 2021). Food can be preserved using the abundant bioactive compounds found in garlic leaves and stalks, which have potent antibacterial properties. Garlic's antibacterial properties against many illnesses are attributed to its diverse range of compounds, such as diallyl sulfides and allicin. Because it prevents the growth of bacteria that cause food to decay, garlic is an essential component of food preservation techniques and a useful natural way to extend the shelf life of food products.

7.7 Non-Food Applications of Garlic By-Products

Garlic leaves and stem remnants can be used for much more than cooking. We refer to these as "non-food applications." Garlic's bioactive constituents, such as sulfides and allicin, are responsible for its medicinal properties, rendering it a valuable tool for various uses (Capasso 2013). Garlic leaves can also be used for purposes other than food, such as making biofertilizers and organic amendments. Studies have shown that vermicomposting grape marc and garlic leaves together results in excellent organic biofertilizers with a high concentration of bioactive polyphenols. These fertilizers may enhance plant development and soil fertility (Domínguez et al. 2014).

Moreover, it has been found that incorporating raw garlic stems into soil has an impact on the soil microbiome, which promotes the development of beneficial microbial communities and enhances soil health (Ghani et al. 2019). Furthermore, flavonoids and phenolic chemicals found in garlic leaves and stalks could be helpful in the production of herbal remedies and natural health supplements. Antibacterial, anti-inflammatory, and antioxidant products can be formulated using these bioactive compounds.

In addition to being utilized in agriculture and medicinal, garlic leaves and stalks can also be employed in environmental sustainability projects. Because they can alter soil properties, nutrient dynamics, and enzyme activities, garlic stalks have the potential to be used as organic soil amendments. Garlic by-products, including leaves and stalks, can also produce high-grade cellulose and cellulose nanocrystals, which offer greener alternatives for various industrial uses (Rovera et al. 2023). Garlic by-products reduce environmental impact and support the circular economy by being utilized in waste valorization and pulp and paper manufacturing.

Beyond food applications, garlic leaves and stem leftovers can be used for various purposes. Because of its beneficial properties and bioactive compounds, garlic has several uses outside cooking. It can be used in environmental sustainability, medicine, and agriculture.

7.8 Nutraceutical Products

Research in alternative medicine has produced many notable products, including nutraceuticals and functional foods. Anything from isolated nutrients to dietary supplements, herbal products, and particular diets used for illness prevention or treatment can be categorized as a "nutraceutical." It is commonly known that garlic and its preparations can be used as medications to prevent and treat cancer and cardiovascular disease. These views have prompted research into the possible applications of garlic products for other disease prevention and treatment and the creation of chemically defined functional foods for long-term ingestion and novel medicinal foods (Espinoza et al. 2020). Apart from these compounds, a plethora of information indicates that garlic has several beneficial impacts on the immune system. Thus, it is quite likely that in the upcoming years, garlic products that regulate immune function will also be created (Subroto et al. 2021; Melguizo-Rodríguez et al. 2022).

Garlic and preparations made with garlic have a great deal of potential to become part of health and preventative medicine. It is affordable, and the myth that synthetic medications with high levels of toxicity are the sole way to treat infectious diseases is no longer valid (Qiu et al. 2022). Over the past 10 years, there has been a noticeable surge in the usage of alternative medicine in Western nations. Growing consumer dissatisfaction with allopathic treatment and their desire for an alternative medical care model are some factors propelling this trend.

In its bulb and leaves, garlic contains bioactive compounds with a host of health benefits and nutraceutical properties. Due to its associations with anti-inflammatory, neuroprotective, and antioxidant properties, garlic is a common ingredient in pharmaceutical and nutraceutical products.

7.9 *Garlic By-Products: Applications in Medicine*

Pharmaceutical applications employ a specific chemical or plant material that effectively achieves a therapeutic or preventive health outcome (Kanfer and Patnala 2021). While garlic is one of the most well-studied herbal products, it has been a preventative health agent throughout many nations and civilizations. Considering the vast amount of research and usage interest in bulb garlic, this is especially true for garlic by-products, which are not a priority general research issue (Panuccio et al. 2022; Chiavaroli et al. 2024). However, functional foods—foods or ingredients that offer health benefits above and beyond those of essential nutrition—do fall under the definition of specified substances for prevention or therapy. Future studies on garlic by-products may concentrate on this field given the potentially wide range of applications of garlic by-products in pharmaceutical goods, from the nutraceutical application of diallyl sulfide to the prebiotic usage of fructans (Espinoza et al. 2020; Qiu et al. 2022).

Scientifically verified data on garlic's role in this field has emerged as a significant addition to historical and folkloric accounts of the herb's uses, as interest in using natural food ingredients for illness prevention and treatment has grown (Ansary et al. 2020). Numerous components of garlic have been added to a wide range of health and nutraceutical products; this has happened occasionally due to scientific evidence but more frequently because product manufacturers want to profit from garlic's reputation as a health food (Vinayagam et al. 2023; Verma et al. 2023). Aged garlic extract is a commercial preparation of garlic that is frequently used and discussed. Much scientific literature has been written on it, resulting from studies using the same substance. Although the non-bulbous components of garlic would have promise, given their availability as by-products from processed garlic, the potential use of garlic by-products in these goods has not been thoroughly studied (Abe et al. 2020; Kodera et al. 2020).

Another way to boost the effects of synthetic medicines and other medications is by combining them with old garlic extract. This is done because research has demonstrated that aged garlic extract does not inhibit the growth of HepG2 cells and does not have pre-cancerous effects when used with gamma-glutamyl transpeptidase. As a result, patients can use aged garlic extract to improve their cardiovascular and immune systems without experiencing any adverse side effects (Sripanidkulchai 2020).

Garlic is used in pharmaceutical form to extract the active components needed to create a specific product. This substance can then be utilized to investigate how garlic affects specific biological systems or to treat a specific ailment. Research on the dialysable leukotriene synthesis inhibitor, an anti-inflammatory drug that prevents leukotriene synthesis and is made by purifying dialysates from rats' kidneys and spleen after they have been given 1.5% diallyl sulfide, is one example of this. The goal of the study is to ascertain whether DADS and other organosulphur garlic compounds have the potential to treat bronchial asthma and to find out if the anti-inflammatory agent can counteract the effects of leukotrienes in asthmatic conditions (Sun et al. 2020; Rupani et al. 2021; Ezeamuzie et al. 2022).

The cytotoxic effects of garlic extract on cancer cells have been proven to exhibit anticancer properties (Islam et al. 2011). Furthermore, research has indicated that garlic extract possesses antigenotoxic qualities, highlighting its capacity to prevent DNA damage (Morales-González et al. 2019). Similarly, studies have examined garlic extract's role in bone health by examining its effects on calcium transport and skeletal health (Mukherjee et al. 2006). Garlic extract has many pharmacological applications that offer various health benefits. Its bioactive components, which may aid in preventing and treating several illnesses, make it a valuable natural remedy with promising therapeutic effects.

7.10 Garlic By-Products: Applications in Agriculture and Horticulture

Additionally, garlic may be able to help reduce crop diseases. An age-old practice in crop rotation that suppresses disease is using alliums as companion crops. Alliums' volatile chemicals are believed to combat infections transmitted through soil effectively. More recently, several organosulphur compounds, including ajoene, an antiplatelet drug generated from allicin, have exhibited potent efficacy against a range of fungal diseases that affect both plants and humans. Using allium compounds as a treatment for diseased seeds is the main focus of current research (Nishioka et al. 2022).

The process of extracting oil from garlic discards a significant amount of its bioactive components, some of which have insecticidal properties. One particularly interesting compound is hydrophobic diallyl trisulphide, made from allicin. Research has shown that this fumigant is highly effective at killing a wide range of insects, but it is far less hazardous to mammals than many fumigants that are already in use (Ezeorba et al. 2022; Samuel and Ariyo 2022). Due to the higher cost and lower efficacy of biopesticides, these allium extracts may offer a perfect substitute in regions of the world where conventional fumigants based on phosphine and chlorinated hydrocarbons are being phased out due to environmental concerns. Nevertheless, these extracts are necessary for pest control in certain crops (Lengai et al. 2020; Qari et al. 2020).

For millennia, human agriculture has employed plant-derived insecticides, including garlic and allied alliums (shallots, onions, and a few others) (Hill et al. 2023). Their application has increased recently, especially in organic farming, since the risks associated with synthetic pesticides for the environment and human health have come to light. Dialyl disulphide, the main active ingredient in garlic, is a good candidate for pesticide development because of its quick environmental degradation, acute toxicity to a variety of insects, mites, and nematodes, and relatively low toxicity to mammals (Song et al. 2021). Due to research, Dialyl disulphide has been formulated into various pesticides, and all of them have demonstrated extreme activity (Gong et al. 2021).

8 Investigation and Originality

Given that it usually takes 10 to 15 years for a new plant variety to be created and made available to the general population, it is possible to develop garlic types with improved functional and health-promoting qualities. There will probably be a rise in demand for garlic products with established health benefits as consumers become more health-conscious and gravitate toward functional foods and nutraceuticals. Genetic modification and molecular breeding technologies will probably be used to meet this demand. It is impossible to forecast how the public will respond to genetically modified garlic, and considering how consumers currently feel about "natural" and "organic" produce, this point is still unclear (Lu and Young 2020; Suffness and Wall 2021; Swarup et al. 2021).

Research into realizing the full potential of garlic is still in its early stages, despite the crop's recent recognition as a vitamin-rich health food and its historical use by many civilizations more for its medicinal than culinary qualities. Only a small number of the several hundred garlic cultivars known to exist have been thoroughly tested and characterized, and only in the last 30 years have scientific studies been carried out to determine and explain the qualities that give garlic its pungent flavor and potential therapeutic uses. Our knowledge of the genetic and biochemical processes involved in the growth, development, and flavor formation of garlic is minimal when compared to other well known and frequently consumed Allium species like onions, leeks, and shallots (Bonasia et al. 2020; Kıraç et al. 2022; Casals et al. 2023).

9 Conclusion

Based on value addition, garlic is a significant crop with many uses that might be commercially exploited in the food and non-food industries. The entire garlic plant has a long history of usage for treating a wide range of human and animal maladies, but the garlic bulb is the main product of trade due to its usefulness as food and medicine. Researchers have been studying garlic for a greater portion of the twentieth century, and as a result, they have a better understanding of which sections of the plant to process for diverse purposes. On the bulb, a plethora of knowledge on its culinary applications has led to the growth of a garlic business to satisfy consumer demand for unique items like fresh and black garlic.

References

Abbasi BA, Iqbal J, Ahmad R, Bibi S, Mahmood T, Kanwal S, Hameed S (2019) Potential phytochemicals in the prevention and treatment of esophagus cancer: a green therapeutic approach. Pharmacol Rep 71(4):644–652

Abbasi BA, Iqbal J, Mahmood T, Khalil AT, Ali B, Kanwal S, Ahmad R (2018) Role of dietary phytochemicals in modulation of miRNA expression: natural swords combating breast cancer. Asian Pac J Trop Med 11(9):501–509

Abdel-Daim MM, Abdel-Rahman HG, Dessouki AA, Ali H, Khodeer DM, Bin-Jumah M, Aleya L (2020) Impact of garlic (Allium sativum) oil on cisplatin-induced hepatorenal biochemical and histopathological alterations in rats. Sci Total Environ 710:136338

Abdelrahman M, Jogaiah S (2020) Bioactive molecules in plant defense. Bioactive Molecules in Plant Defense

Abdulrasheed M, Ibrahim IH, Mubarak MA, Umar FA (2015) Comparison of antimicrobial activity of seed oil of garlic and Moringa oleifera against some food-borne microorganisms. Bayero J Pure Appl Sci 8(2):196–201

Abe K, Hori Y, Myoda T (2020) Volatile compounds of fresh and processed garlic. Exp Ther Med 19(2):1585–1593

Abidullah M, Jadhav P, Sujan SS, Shrimanikandan AG, Reddy CR, Wasan RK (2021) Potential antibacterial efficacy of garlic extract on *Staphylococcus aureus, Escherichia coli,* and *Klebsiella pneumoniae*: an in vitro study. J Pharm Bioallied Sci 13(Suppl 1):S590–S594

Addala R, Fong L, Vasavada M, Subramanian S (2016) Influence of distillation temperature in the determination of added sulfites in dehydrated garlic powders using the modified optimized Monier-Williams method. J AOAC Int 99(3):721–724

Adiwijaya Z (2017) The effect of natural compound in binding and reducing off-flavors components of garlic

Afzaal M, Saeed F, Rasheed R, Hussain M, Aamir M, Hussain S, Anjum FM (2021) Nutritional, biological, and therapeutic properties of black garlic: a critical review. Int J Food Prop 24(1):1387–1402

Ahmed HA, Yu-Xin T, Qi-Chang Y (2020) Optimal control of environmental conditions affecting lettuce plant growth in a controlled environment with artificial lighting: a review. S Afr J Bot 130:75–89

Ahmeda A, Hossain MA, Zhari I (2009) Antioxidant properties of the isolated flavonoids from the medicinal plant Phyllanthus niruri. Asian J Food Agro-Industry 2(3):373–381

Akerele D, Ajoseh S, Sanusi R, Oyawole F (2016) Healthier foods in perspective: what drives garlic purchase decisions and consumption in urban South-West Nigeria? Open Agric 1(1):79–88

Ali S, Khan MR, Iqbal J, Shah SA, Abbasi BA, Yaseen T, Kazi M (2022) Chemical characterization and evaluation of the nephroprotective potential of Parrotiopsis jacquemontiana (Decne) Rehder and Periploca hydaspidis Falc crude extract in CCl4-induced male Sprague-Dawley rats. Saudi J Biol Sci 29(2):702–712

Ali S, Sayed AS, Saeed RF, Iqbal J, Ijaz S, Munazir M (2023) Ethnomedicinal plant use value in Lower Swat, Pakistan. Ethnobot Res Appl 25:1–22

Alide T, Wangila P, Kiprop A (2020) Effect of cooking temperature and time on total phenolic content, total flavonoid content and total in vitro antioxidant activity of garlic. BMC Res Notes 13:1–7

Almatroodi SA (2020) Bioactive compounds of garlic: role in the Management of Various Types of cancer. Pharmanest Int J Adv Pharm Sci 11:1–13

Al-Shehri SA (2021) Efficacy of black garlic extract on anti-tumor and anti-oxidant activity enhancement in rats. Clin Nutr Open Sci 36:126–139

Alvarez-Jubete L, Valverde J, Kehoe K, Reilly K, Rai DK, Barry-Ryan C (2014) Development of a novel functional soup rich in bioactive sulforaphane using broccoli (*Brassica oleracea* L. ssp. italica) florets and byproducts. Food Bioprocess Technol 7:1310–1321

Amagase H, Petesch BL, Matsuura H, Kasuga S, Itakura Y (2001) Intake of garlic and its bioactive components. J Nutr 131(3):955S–962S

Amor S, González-Hedström D, Martín-Carro B, Inarejos-García AM, Almodóvar P, Prodanov M, Granado M (2019) Beneficial effects of an aged black garlic extract in the metabolic and vascular alterations induced by a high fat/sucrose diet in male rats. Nutrients 11(1):153

An Q, Zhao S, Zhao C, Wei R, Zhang L, Li J, Zheng J (2022) Identification of the key emulsifying components from the byproducts of garlic oil distillation. Food Hydrocoll 122:107043

Anis S, Alhakim R, Khoiron AM, Kusumastuti A (2021) Microwave-assisted pyrolysis and distillation of cooking oils for liquid bio-fuel production. J Anal Appl Pyrolysis 154:105014

Ansary J, Forbes-Hernández TY, Gil E, Cianciosi D, Zhang J, Elexpuru-Zabaleta M, Battino M (2020) Potential health benefit of garlic based on human intervention studies: a brief overview. Antioxidants 9(7):619

Aoun A, Maalouf J, Khalil S, Khoury VE (2021) The role of nutrition in the treatment of irritable bowel syndrome. Family Med Primary Care Rev 23(4):495

Arditti FD, Rabinkov A, Miron T, Reisner Y, Berrebi A, Wilchek M, Mirelman D (2005) Apoptotic killing of B-chronic lymphocytic leukemia tumor cells by allicin generated in situ using a rituximab-alliinase conjugate. Mol Cancer Ther 4(2):325–332

Arshad MS, Khalid W, Ahmad RS, Khan MK, Ahmad MH, Safdar S, Suleria HAR (2021) Functional foods and human health: an overview. Functional foods phytochem health promoting potential, 3

Arumugham T, Rambabu K, Hasan SW, Show PL, Rinklebe J, Banat F (2021) Supercritical carbon dioxide extraction of plant phytochemicals for biological and environmental applications–A review. Chemosphere 271:129525

Arzanlou M, Bohlooli S (2010) Inhibition of streptolysin O by allicin–an active component of garlic. J Med Microbiol 59(9):1044–1049

Asemani Y, Zamani N, Bayat M, Amirghofran Z (2019) Allium vegetables for possible future of cancer treatment. Phytother Res 33(12):3019–3039

Ashfaq F, Ali Q, Haider MA, Hafeez MM, Malik A (2021a) Therapeutic activities of garlic constituent phytochemicals. Biol Clin Sci Res J 2021(1)

Ashfaq F, Bilal A, Butt MS, Tehseen S, Akhtar MN, Suleria HAR (2021b) Comparing the application of conventional and supercritical CO_2 extracts of green tea; storage stability and sensory acceptance of coriander sauce. J Food Process Preserv 45(12):e16058

Avgeri I, Zeliou K, Petropoulos SA, Bebeli PJ, Papasotiropoulos V, Lamari FN (2020) Variability in bulb organosulfur compounds, sugars, phenolics, and pyruvate among greek garlic genotypes: association with antioxidant properties. Antioxidants 9(10):967

Ayaz M, Ullah F, Sadiq A, Ullah F, Ovais M, Ahmed J, Devkota HP (2019) Synergistic interactions of phytochemicals with antimicrobial agents: potential strategy to counteract drug resistance. Chem Biol Interact 308:294–303

Ayoade WG, Gbadamosi L, Ajayi MG, Badmus MO (2023) Assessment of minerals and vitamin constituents of some commonly consumed spices. Int J Sci Res Arch 10(1):053–061

Azmi L, Shukla I (2021) Perspectives of phytotherapeutics: diagnosis and cure. In: Phytomedicine. Academic Press, pp 225–250

Azzini E, Durazzo A, Foddai MS, Temperini O, Venneria E, Valentini S, Maiani G (2014) Phytochemicals content in Italian garlic bulb (Allium sativum L.) varieties. J Food Res 3(4):26

Badshah H, Mumtaz AS, Hussain I, Ali B, Iqbal J, Ali S (2022) Wild morels in Pakistan: environmental and trading statues. In: Biodiversity, conservation and sustainability in Asia: volume 2: prospects and challenges in south and middle Asia. Springer International Publishing, Cham, pp 691–702

Bagheri L, Khodaei N, Salmieri S, Karboune S, Lacroix M (2020) Correlation between chemical composition and antimicrobial properties of essential oils against most common food pathogens and spoilers: in-vitro efficacy and predictive modelling. Microb Pathog 147:104212

Banerjee SK, Maulik SK (2002) Effect of garlic on cardiovascular disorders: a review. Nutr J 1:1–14

Bar M, Binduga UE, Szychowski KA (2022) Methods of isolation of active substances from garlic (*Allium sativum* L.) and its impact on the composition and biological properties of garlic extracts. Antioxidants 11(7):1345

Basevi V (2011) Diagnosis and classification of diabetes mellitus. Diabetes Care 34:S62–S69. https://doi.org/10.2337/dc11-S062

Batiha GES, Hussein DE, Algammal AM, George TT, Jeandet P, Al-Snafi AE, Cruz-Martins N (2021) Application of natural antimicrobials in food preservation: recent views. Food Control 126:108066

Batool R, Aziz E, Iqbal J, Salahuddin H, Tan BKH, Tabassum S, Mahmood T (2020) In vitro antioxidant and anti-cancer activities and phytochemical analysis of Commelina benghalensis L. root extracts. Asian Pac J Trop Biomed 10(9):417–425

Batool R, Aziz E, Salahuddin H, Iqbal J, Tabassum S, Mahmood T (2019) Rumex dentatus could be a potent alternative to treatment of microbial infections and of breast cancer. J Tradit Chin Med 39(6)

Bayan L, Koulivand PH, Gorji A (2014) Garlic: a review of potential therapeutic effects. Avicenna J Phytomed 4(1):1

Beato VM, Orgaz F, Mansilla F, Montaño A (2011) Changes in phenolic compounds in garlic (*Allium sativum* L.) owing to the cultivar and location of growth. Plant Foods Hum Nutr 66:218–223

Bede D, Zaixiang L (2020) Dietary polysaccharides from Allium species: a critical review in dietary polysaccharides from Allium species: extraction, characterization, bioactivity, and potential utilization. Acta Sci Agric 4(2):98–112

Belhadj F, Messaoud C, Hlel TB, Demirtaş I, Marzouki MN (2015) Research Article CODEN: IJPRNK ISSN: 2277-8713 F. Belhadj, IJPRBS, 2015; 4(5): 46-68 IJPRBS

Bensid A, El Abed N, Houicher A, Regenstein JM, Özogul F (2022) Antioxidant and antimicrobial preservatives: properties, mechanism of action and applications in food–a review. Crit Rev Food Sci Nutr 62(11):2985–3001

Bernaldez JL, Vicencio MCG (2021) Antibacterial activity of soap formulated from garlic (Allium sativum L.) Extract. J Adv Microbiol 21(1):63–67

Bhatti SA, Hussain MH, Mohsin MZ, Mohsin A, Zaman WQ, Guo M, Korma SA (2022) Evaluation of the antimicrobial effects of Capsicum, Nigella sativa, Musa paradisiaca L., and Citrus limetta: a review. Front Sustain Food Syst 6:1043823

Bhatwalkar SB, Mondal R, Krishna SBN, Adam JK, Govender P, Anupam R (2021) Antibacterial properties of organosulfur compounds of garlic (*Allium sativum*). Front Microbiol 12:613077

Blazewicz-Wozniak M, Michowska A (2011) The growth, flowering and chemical composition of leaves of three ecotypes of *Allium ursinum* L. Acta Agrobot 64(4):171

Bonasia A, Conversa G, Lazzizera C, Loizzo P, Gambacorta G, Elia A (2020) Evaluation of garlic landraces from Foggia province (Puglia region; Italy). Foods 9(7):850

Borlinghaus J, Foerster J, Kappler U, Antelmann H, Noll U, Gruhlke MC, Slusarenko AJ (2021) Allicin, the odor of freshly crushed garlic: a review of recent progress in understanding allicin's effects on cells. Molecules 26(6):1505

Boubechiche Z, Chihib NE, Jama C, Hellal A (2017) Comparison of volatile compounds profile and antioxydant activity of Allium sativum essential oils extracted using hydrodistillation, ultrasound-assisted and sono-hydrodistillation processes. Indian J Pharm Educ Res 51(3):S281–S285

Calimpang IA, Gentallan Jr RP, Borromeo TH, Cruz PCS, Bartolome MCB, Timog EBS (2023) Morphological, phytochemical and molecular characteristics of NSIC-registered varieties of garlic (*Allium sativum* L.) in The Philippines. Agric Sci Dig

Cañizares P, Gracia I, Gómez LA, de Argila CM, de Rafael L, García A (2002) Optimization of *Allium sativum* solvent extraction for the inhibition of in vitro growth of helicobacter pylori. Biotechnol Prog 18(6):1227–1232

Capasso A (2013) Antioxidant action and therapeutic efficacy of Allium sativum L. Molecules 18(1):690–700

Carreón-Delgado DF, Hernández-Montesinos IY, Rivera-Hernández KN, del Sugeyrol Villa-Ramírez M, Ochoa-Velasco CE, Ramírez-López C (2023) Evaluation of pretreatments and extraction conditions on the antifungal and antioxidant effects of garlic (Allium sativum) peel extracts. Plan Theory 12(1):217

Casals J, Rivera A, Campo S, Aymerich E, Isern H, Fenero D, Roig-Villanova I (2023) Phenotypic diversity and distinctiveness of the Belltall garlic landrace. Front Plant Sci 13:1004069

Cavalcanti VP, Aazza S, Bertolucci SKV, Rocha JPM, Coelho AD, Oliveira AJM, Dória J (2021) Solvent mixture optimization in the extraction of bioactive compounds and antioxidant activities from garlic (*Allium sativum* L.). Molecules 26(19):6026

Chakraborty M, Mahmud NU, Ullah C, Rahman M, Islam T (2021) Biological and biorational management of blast diseases in cereals caused by Magnaporthe oryzae. Crit Rev Biotechnol 41(7):994–1022

Chakraborty D, Majumder A (2020) Garlic (Lahsun)–an immunity booster against SARS-CoV-2. Biotica Res Today 2(8):755–757

Chan JYY, Yuen ACY, Chan RYK, Chan SW (2013) A review of the cardiovascular benefits and antioxidant properties of allicin. Phytother Res 27(5):637–646

Chaudhary BU, Lingayat S, Banarjee AN, Kale RD (2022) Preparation and characterization of antioxidant, antimicrobial, and UV-light protection film based on poly (vinyl alcohol) and garlic peel extract. Waste Biomass Valori 13(12):4717–4734

Chaudhary BU, Lingayat S, Banerjee AN, Kale RD (2021) Development of multifunctional food packaging films based on waste garlic peel extract and chitosan. Int J Biol Macromol 192:479–490

Chekol F, Abetie K, Sirany T (2023) Technical efficiency of garlic production under rain-fed agriculture in Northwest Ethiopia: stochastic frontier approach. Cogent Econ Finance 11(2):2242177

Chen C (2012) Induction and inhibition compounds

Chen Z, Li Z, Chen J, Kallem P, Banat F, Qiu H (2022) Recent advances in selective separation technologies of rare earth elements: a review. J Environ Chem Eng 10(1):107104

Cheng Y, Xue F, Yu S, Du S, Yang Y (2021) Subcritical water extraction of natural products. Molecules 26(13):4004

Chiavaroli A, Masciulli F, Ingallina C, Mannina L, Libero ML, Di Simone SC, Ferrante C (2024) Comprehensive metabolite and biological profile of "Sulmona red garlic" ecotype's aerial bulbils. Food Res Int 175:113654

Chib A, Gupta N, Bhat A, Anjum N, Yadav G (2020) Role of antioxidants in food. Int J Chem Stud 8:2354–2361
Choi YH, Jeong HM, Kyung KH, Ryu BH, Lee KY (2011) Aged garlic extract and its components inhibit platelet aggregation in rat. J Life Sci 21(10):1355–1363
Choudhary S, Noor MU, Hussain MS, Mishra M, Tyagi S (2022) Pharmacological properties and phytoconstituents of garlic (Allium sativum L.): a review. Biol Sci 2(4):338–346
Ciğeroğlu Z, Şahin S, Kurtulbaş E (2022) A comparative study of lipid oxidation in garlic oil (Allium sativum L.): an accelerated oxidation. J Food Process Preserv 46(10):e16874
Ciric A, Krajnc B, Heath D, Ogrinc N (2020) Response surface methodology and artificial neural network approach for the optimization of ultrasound-assisted extraction of polyphenols from garlic. Food Chem Toxicol 135:110976
Das D, Vimala R, Das N (2010) Functional foods of natural origin-An overview
De Greef D, Barton EM, Sandberg EN, Croley CR, Pumarol J, Wong TL, Bishayee A (2021) Anticancer potential of garlic and its bioactive constituents: a systematic and comprehensive review. In: Seminars in cancer biology, vol 73. Academic Press, pp 219–264
de Jesus SS, Maciel Filho R (2020) Recent advances in lipid extraction using green solvents. Renew Sust Energ Rev 133:110289
Deak T, Mohácsi-Farkas C (2023) Thermal treatment. In: Food safety management. Academic Press, pp 405–419
Dehariya N, Guha P, Gupta RK (2021) Extraction and characterization of essential oil of garlic (Allium sativa L.). Int J Chem Stud 9(1):1455–1459
Desta B, Woldetsadik K, Ali M, W. (2021) Effect of harvesting time, curing and storage methods on storability of garlic bulbs. The Open Biotechnol J 15(1):36
Devarajan A, Mohanmarugaraja MK (2017) A comprehensive review on Rasam: a South Indian traditional functional food. Pharmacogn Rev 11(22):73
Ding H, Cheng Z, Liu M, Hayat S, Feng H (2016) Garlic exerts allelopathic effects on pepper physiology in a hydroponic co-culture system. Biology Open 5(5):631–637
Dinu M, Soare R, Băbeanu C, Botu M (2023) Evaluation of productivity components and antioxidant activity of different types of garlic depending on the morphological organs. Horticulturae 9(9):1039
Domínguez J, Martínez-Cordeiro H, Álvarez-Casas M, Lores M (2014) Vermicomposting grape marc yields high quality organic biofertiliser and bioactive polyphenols. Waste Manag Res 32(12):1235–1240
Dorrigiv M, Zareiyan A, Hosseinzadeh H (2020) Garlic (*Allium sativum*) as an antidote or a protective agent against natural or chemical toxicities: a comprehensive update review. Phytother Res 34(8):1770–1797
dos Santos PCM, da Silva LMR, Magalhaes FEA, Cunha FET, Ferreira MJG, de Figueiredo EAT (2022) Garlic (Allium sativum L.) peel extracts: from industrial by-product to food additive. Appl Food Res 2(2):100186
Easmin MS, Sarker MZI, Ferdosh S, Shamsudin SH, Yunus KB, Uddin MS, Khalil HA (2015) Bioactive compounds and advanced processing technology: phaleria macrocarpa (sheff.) Boerl, a review. J Chem Technol Biotechnol 90(6):981–991
Ejaz S, Woong LC, Ejaz A (2003) Extract of garlic (*Allium sativum*) in cancer chemoprevention
Ekşi G, Özkan AMG, Koyuncu M (2020) Garlic and onions: an Eastern tale. J Ethnopharmacol 253:112675
El-khamissi H, Saad Z, Rozan H (2019) Phytochemicals screening, antioxidant and anticancer activities of garlic (*Allium sativum*) extracts. J Agric Chem Biotechnol 10(4):79–82
El-Saber Batiha G, Magdy Beshbishy A, Wasef G, Elewa YH, Al-Sagan A, Abd El-Hack ME, Prasad Devkota H (2020) Chemical constituents and pharmacological activities of garlic (Allium sativum L.): a review. Nutrients 12(3):872
Espinoza T, Valencia E, Albarrán M, Díaz D, Quevedo RA, Díaz O, Bastías J (2020) Garlic (Allium sativum L) and its beneficial properties for health: a review. Agroindu Sci 10(1):103–115

Ezeamuzie CI, Rao MS, El-Hashim AZ, Philip E, Phillips OA (2022) Anti-allergic, anti-asthmatic and anti-inflammatory effects of an oxazolidinone hydroxamic acid derivative (PH-251)–A novel dual inhibitor of 5-lipoxygenase and mast cell degranulation. Int Immunopharmacol 105:108558

Ezeorba TPC, Chukwudozie KI, Ezema CA, Anaduaka EG, Nweze EJ, Okeke ES (2022) Potentials for health and therapeutic benefits of garlic essential oils: recent findings and future prospects. Pharmacol Res-Modern Chinese Med 3:100075

Feng Y, Xu B, Yagoub AEA, Ma H, Sun Y, Xu X et al (2021) Role of drying techniques on physical, rehydration, flavor, bioactive compounds and antioxidant characteristics of garlic. Food Chem 343:128404

Figueroa V, Farfán M, Aguilera JM (2023) Seaweeds as novel foods and source of culinary flavors. Food Rev Intl 39(1):1–26

Garmame Galgaye G (2022) Revealing determinants that affects garlic production in Ethiopia using PRISMA methodology. Cogent Food Agric 8(1):2132845

Garty BZ (1993) Garlic burns. Pediatrics 91(3):658–661

Gaspar EM, Duarte R, Santana JC (2018) Volatile composition and antioxidant properties of clove products. Biomed J Sci Tech Res 9(4):1–7

Ghani MI, Ali A, Atif MJ, Ali M, Amin B, Anees M et al (2019) Changes in the soil microbiome in eggplant monoculture revealed by high-throughput Illumina MiSeq sequencing as influenced by raw garlic stalk amendment. Int J Mol Sci 20(9):2125

Ghani MI, Ali A, Atif MJ, Ali M, Amin B, Cheng Z (2022) Arbuscular mycorrhizal fungi and dry raw garlic stalk amendment alleviate continuous monocropping growth and photosynthetic declines in eggplant by bolstering its antioxidant system and accumulation of osmolytes and secondary metabolites. Front Plant Sci 13:849521

Godara SL, Kumhar DR, Chawla N (2022) Prevalence and management of major garlic diseases is a global level: a review. Agric Rev 43(3):312–319

Golubkina N, Amagova Z, Matsadze V, Zamana S, Tallarita A, Caruso G (2020) Effects of arbuscular mycorrhizal fungi on yield, biochemical characteristics, and elemental composition of garlic and onion under selenium supply. Plan Theory 9(1):84

Gong X, Su X, Liu H (2021) Diallyl trisulfide, the antifungal component of garlic essential oil and the bioactivity of its nanoemulsions formed by spontaneous emulsification. Molecules 26(23):7186

Gruhlke MC, Portz D, Stitz M, Anwar A, Schneider T, Jacob C, Slusarenko AJ (2010) Allicin disrupts the cell's electrochemical potential and induces apoptosis in yeast. Free Radic Biol Med 49(12):1916–1924

Guardia LAL, Caballero JMH, Acosta HDC, Quijano SAC, Casaño RDLC (2023, October) Design of a semi-automatic machine to improve garlic processing for the optimization of procedures (cutting, shelling and peeling) in agricultural areas. In: 2023 IEEE 14th Annual Ubiquitous Computing, Electronics & Mobile Communication Conference (UEMCON). IEEE, pp 0788–0792

Gul F, Khan I, Iqbal J, Abbasi BA, Shahbaz A, Capasso R, Mahmood T (2022) Phytochemistry, biological activities and in silico molecular docking studies of Oxalis pes-caprae L. compounds against SARS-CoV-2. J King Saud Univ-Sci 34(6):102136

Gusev E, Sarapultsev A (2023) Atherosclerosis and inflammation: insights from the theory of general pathological processes. Int J Mol Sci 24(9):7910

Harris JC, Cottrell SL, Plummer S, Lloyd D (2001) Antimicrobial properties of *Allium sativum* (garlic). Appl Microbiol Biotechnol 57:282–286

Hernández-Montesinos IY, Carreón-Delgado DF, Ocaranza-Sánchez E, Ochoa-Velasco CE, Suárez-Jacobo Á, Ramírez-López C (2023) Garlic (*Allium sativum*) peel extracts and their potential as antioxidant and antimicrobial agents for food applications: influence of pretreatment and extraction solvent. Int J Food Sci Technol 58(12):6794–6805

Hill CR, Shafaei A, Balmer L, Lewis JR, Hodgson JM, Millar AH, Blekkenhorst LC (2023) Sulfur compounds: from plants to humans and their role in chronic disease prevention. Crit Rev Food Sci Nutr 63(27):8616–8638

Huie CW (2002) A review of modern sample-preparation techniques for the extraction and analysis of medicinal plants. Anal Bioanal Chem 373:23–30

Huyut Z, Beydemir Ş, Gülçin İ (2017) Antioxidant and antiradical properties of selected flavonoids and phenolic compounds. Biochem Res Int, 2017

Hwang IS, Kim MK (2022) Determination of aroma characteristics of commercial garlic powders distributed in Korea via instrumental and descriptive sensory analyses. Preven Nutr Food Sci 27(4):464

Ijaz S, Iqbal J, Abbasi BA, Kanwal S, Tavafoghi M, Ahmed MZ, Mahmood T (2024a) Investigation of bioactive constituents and evaluation of different in vitro antimicrobial, antioxidant, and cytotoxicity potentials of different *Portulacaria afra* extracts. J King Saud Univ-Sci 36(2):103033

Ijaz S, Iqbal J, Abbasi BA, Tufail A, Yaseen T, Uddin S, Sharifi-Rad J (2024b) Current stage of preclinical and clinical development of guggulsterone in cancers: challenges and promises. Cell Biol Int 48(2):128–142

Ijaz S, Iqbal J, Abbasi BA, Ullah Z, Yaseen T, Kanwal S, Cho WC (2023) Rosmarinic acid and its derivatives: current insights on anticancer potential and other biomedical applications. Biomed Pharmacother 162:114687

Imaizumi VM, Laurindo LF, Manzan B, Guiguer EL, Oshiiwa M, Otoboni AMMB, Barbalho SM (2023) Garlic: a systematic review of the effects on cardiovascular diseases. Crit Rev Food Sci Nutr 63(24):6797–6819

Iqbal J, Abbasi BA, Ahmad R, Mahmood T, Kanwal S, Ali B, Badshah H (2018c) Ursolic acid a promising candidate in the therapeutics of breast cancer: current status and future implications. Biomed Pharmacother 108:752–756

Iqbal J, Abbasi BA, Batool R, Mahmood T, Ali B, Khalil AT, Ahmad R (2018a) Potential phytocompounds for developing breast cancer therapeutics: nature's healing touch. Eur J Pharmacol 827:125–148

Iqbal J, Abbasi BA, Khalil AT, Ali B, Mahmood T, Kanwal S, Ali W (2018b) Dietary isoflavones, the modulator of breast carcinogenesis: current landscape and future perspectives. Asian Pac J Trop Med 11(3):186–193

Iqbal J, Abbasi BA, Mahmood T, Kanwal S, Ali B, Shah SA, Khalil AT (2017) Plant-derived anticancer agents: a green anticancer approach. Asian Pac J Trop Biomed 7(12):1129–1150

Irfan M, Kim M, Kim KS, Kim TH, Kim SD, Hong SB, …, Rhee MH (2019) Fermented garlic ameliorates hypercholesterolemia and inhibits platelet activation. Evidence-Based Complementary and Alternative Medicine, 2019

Islam MS, Kusumoto Y, Al-Mamun MA (2011) Cytotoxicity and cancer (HeLa) cell killing efficacy of aqueous garlic (Allium sativum) extract. J Sci Res 3(2):375–382

Ismail RM, Saleh AHA, Ali KS (2020) GC-MS analysis and antibacterial activity of garlic extract with antibiotic. J Med Plants Stud 8(1):26–30

Javed M, Ahmed W, Mian R, Ahmad AMR (2021) Garlic as a potential nominee in functional food industry. In: Herbs and spices-new processing technologies. IntechOpen

Jensen K, Goto K, Giampaoli J, Holland J (2021) Factors associated with consumer knowledge of garlic health properties, garlic preparation knowledge and garlic intake. Californian J Health Promotion 19(1):44–53

Jones MG, Collin HA, Tregova A, Trueman L, Brown L, Cosstick R et al (2007) The biochemical and physiological genesis of alliin in garlic. Med Aromatic Plant Sci Biotechnol 1(1):21–24

Kalhotra P, Chittepu VC, Osorio-Revilla G, Gallardo-Velazquez T (2020) Phytochemicals in garlic extract inhibit therapeutic enzyme DPP-4 and induce skeletal muscle cell proliferation: a possible mechanism of action to benefit the treatment of diabetes mellitus. Biomol Ther 10(2):305

Kallel F, Ellouz Chaabouni S (2017) Perspective of garlic processing wastes as low-cost substrates for production of high-added value products: a review. Environ Prog Sustain Energy 36(6):1765–1777

Kalse SB, Swami SB (2022) Recent application of jackfruit waste in food and material engineering: a review. Food Biosci 48:101740

Kanfer I, Patnala S (2021) Regulations for the use of herbal remedies. In: Herbal medicine in andrology. Academic Press, pp 189–206

Kang MJ, Kang JR, Woo MS, Kang D, Shin JH (2022) Alterations in the physicochemical properties and antioxidant activity during aging of stored raw garlic. Foods 11(10):1390

Kang MJ, Yoon HS, Jeong SH, Sung NJ, Shin JH (2011) Physicochemical characteristics of red garlic during processing. Korean J Food Preserv 18(6):898–906

Kaur T, Rawat M, Kathayat K (2023) Value-added products of garlic and their potential health benefits–A review. Asian J Microbiol Biotechnol Environ Sci 25:707

Khalil AM, Maher ZM (2022) Garlic extract enhances the peripheral blood cellular immune responses in male ICR mice. Sch Int J Tradit Complement Med 5(8)

Khan MW, Otaibi AA, Alsukaibi AK, Alshammari EM, Al-Zahrani SA, Sherwani S, Ahmed N (2022) Biophysical, biochemical, and molecular docking investigations of anti-glycating, antioxidant, and protein structural stability potential of garlic. Molecules 27(6):1868

Kim DG, Kang MJ, Hong SS, Choi YH, Shin JH (2017) Antiinflammatory effects of functionally active compounds isolated from aged black garlic. Phytother Res 31(1):53–61

Kıraç H, Dalda Şekerci A, Coşkun ÖF, Gülşen O (2022) Morphological and molecular characterization of garlic (Allium sativum L.) genotypes sampled from Turkey. Genet Resour Crop Evol 69(5):1833–1841

Kishimoto N, Kashiwagi A (2020) Evaluation of the process of flavoring olive oil with garlic. Food Sci Technol Res 26(5):605–610

Kodera Y, Kurita M, Nakamoto M, Matsutomo T (2020) Chemistry of aged garlic: diversity of constituents in aged garlic extract and their production mechanisms via the combination of chemical and enzymatic reactions. Exp Ther Med 19(2):1574–1584

Kowalska H, Czajkowska K, Cichowska J, Lenart A (2017) What's new in biopotential of fruit and vegetable by-products applied in the food processing industry. Trends Food Sci Technol 67:150–159

Kumar A, Kumar M, Jose A, Tomer V, Oz E, Oz F (2023) Major phytochemicals: recent advances in health benefits and extraction method. Molecules 28(2):887

Kunová S, Haščík P, Lopašovský Ľ, Kačániová M (2021) The microbiological quality of minced pork treated with garlic in combination with vacuum packaging. Slovak J Food Sci 15:453

Kurnia D, Ajiati D, Heliawati L, Sumiarsa D (2021) Antioxidant properties and structure-antioxidant activity relationship of Allium species leaves. Molecules 26(23):7175

Lachowicz S, Kolniak-Ostek J, Oszmiański J, Wiśniewski R (2017) Comparison of phenolic content and antioxidant capacity of bear garlic (Allium ursinum L.) in different maturity stages. J Food Process Preserv 41(1):e12921

Lammi C, Bellumori M, Cecchi L, Bartolomei M, Bollati C, Clodoveo ML, Mulinacci N (2020) Extra virgin olive oil phenol extracts exert hypocholesterolemic effects through the modulation of the LDLR pathway: in vitro and cellular mechanism of action elucidation. Nutrients 12(6):1723

Larousse L (2022) Larousse Gastronomique: the World's greatest culinary encyclopedia. Completely Revised and Updated, Clarkson Potter

Lee YM, Gweon OC, Seo YJ, Im J, Kang MJ, Kim MJ, Kim JI (2009) Antioxidant effect of garlic and aged black garlic in animal model of type 2 diabetes mellitus. Nutr Res Pract 3(2):156

Lee SY, Kim DS, Kyung KH (2014) Factors influencing the stability of garlic thiosulfinates. Food Sci Biotechnol 23:1593–1600

Lengai GMW, Muthomi JW, Mbega ER (2020) Phytochemical activity and role of botanical pesticides in pest management for sustainable agricultural crop production. Scientific African 7:e00239

Lenková M, Bystrická J, Vollmannová A, Tóth T, Kovarovič J (2017) Evaluation and comparison of the content of total polyphenols and antioxidant activity in garlic (*Allium sativum* L.). Potravinárstvo 11(1):65

Li WQ, Zhou H, Zhou MY, Hu XP, Ou SY, Yan RA et al (2015) Characterization of phenolic constituents inhibiting the formation of sulfur-containing volatiles produced during garlic processing. J Agric Food Chem 63(3):787–794

Lisciani S, Gambelli L, Durazzo A, Marconi S, Camilli E, Rossetti C, Marletta L (2017) Carbohydrates components of some Italian local landraces: garlic (Allium sativum L.). Sustainability 9(10):1922

Liu T, Chen J (2021) Extraction and separation of heavy rare earth elements: a review. Sep Purif Technol 276:119263

Liu P, Wang Y, Xu Y, Weng R, Qiu J, Qian Y (2024) Monitoring changes in biochemical and metabolite profiles in garlic cloves during storage. J Agric Food Chem 72:10117

Loghmanifar S, Roozbeh Nasiraie L, Nouri H, Jafarian S (2022) Comparison of fresh and aged garlic extracts in terms of antioxidative power and allicin content. J Med Plants By-Prod 11(Special):29–35

Londero E (2001) By-products. Econ Syst Res 13(1):35–45

Londhe VP, Gavasane AT, Nipate SS, Bandawane DD, Chaudhari PD (2011) Role of garlic (Allium sativum) in various diseases: an overview. Angiogenesis 12(13):129–134

Loppi S, Fedeli R, Canali G, Guarnieri M, Biagiotti S, Vannini A (2021) Comparison of the mineral and nutraceutical profiles of elephant garlic (Allium ampeloprasum L.) grown in organic and conventional fields of Valdichiana, a traditional cultivation area of Tuscany, Italy. Biology 10(10):1058

Lu J, Li N, Li S, Liu W, Li M, Zhang M, Chen H (2023) Biochemical composition, antioxidant activity and antiproliferative effects of different processed garlic products. Molecules 28(2):804

Lu Y, Young S (2020) A survey of public datasets for computer vision tasks in precision agriculture. Comput Electron Agric 178:105760

Ma Y, Liu P, Guo S, Lu R, Zhao X, Wang D, Zhang Y (2021) Nutritional quality and volatile flavor substances of "laba" garlic products produced by either soaking or fumigating with acetic acid. J Food Process Preserv 45(2):e15116

Madkor HR, Mansour SW, Ramadan G (2011) Modulatory effects of garlic, ginger, turmeric and their mixture on hyperglycaemia, dyslipidaemia and oxidative stress in streptozotocin–nicotinamide diabetic rats. Br J Nutr 105(8):1210–1217

Makarichian A, Chayjan RA, Ahmadi E, Mohtasebi SS (2021) Assessment the influence of different drying methods and pre-storage periods on garlic (allium Sativum L.) aroma using electronic nose. Food Bioprod Process 127:198–211

Maleki B, Esmaeili H (2023) Ultrasound-assisted conversion of waste frying oil into biodiesel using Al-doped ZnO nanocatalyst: box-Behnken design-based optimization. Renew Energy 209:10–26

Manach C, Milenkovic D, Van de Wiele T, Rodriguez-Mateos A, de Roos B, Garcia-Conesa MT, Morand C (2017) Addressing the inter-individual variation in response to consumption of plant food bioactives: towards a better understanding of their role in healthy aging and cardiometabolic risk reduction. Mol Nutr Food Res 61(6):1600557

Manjunatha M, Samuel DVK, Anurag RK, Gaikwad N (2014) Development and performance evaluation of a garlic peeler. J Food Sci Technol 51:3083–3093

Mansell P, Reckless JP (1991) Garlic. BMJ: Br Med J 303(6799):379

Manu S, Rohitas Deshmukh RD, Prasad KMN, Vishal Trivedi VT (2013) Screening and characterization of antimalarial heme polymerase inhibitors from garlic cloves, vol 3, p 474

Manyako KE, Chiyanzu I, Mulopo J, Abdulsalam J (2022) Pilot-scale evaluation of concentrating solar thermal technology for essential oil extraction and comparison with conventional heating sources for use in agro-based industrial applications. ACS Omega 7(24):20477–20485

Marcillo-Parra V, Tupuna-Yerovi DS, González Z, Ruales J (2021) Encapsulation of bioactive compounds from fruit and vegetable by-products for food application–A review. Trends Food Sci Technol 116:11–23

Martínez-Inda B, Esparza I, Moler JA, Jiménez-Moreno N, Ancín-Azpilicueta C (2023) Valorization of Agri-food waste through the extraction of bioactive molecules. Prediction of their sunscreen action. J Environ Manag 325:116460

Mashad HM, Zhang R, Pan Z (2019) Onion and garlic. Integrated processing technologies for food and agricultural by-products

Melguizo-Rodríguez L, García-Recio E, Ruiz C, De Luna-Bertos E, Illescas-Montes R, Costela-Ruiz VJ (2022) Biological properties and therapeutic applications of garlic and its components. Food Funct 13(5):2415–2426

Melkamu Y (2022) Survey and Management of Garlic Rust (Puccinia Allii) disease on garlic (allium Sativuml) using different rates and frequency of tilt fungicide applications in Enebsie Sarmidr and Gonchasisoenesie District. Amhara National Regional State. *Ethi* (Doctoral dissertation)

Meuninck J (2023) Basic illustrated edible wild plants and useful herbs. Rowman & Littlefield

Miękus N, Marszałek K, Podlacha M, Iqbal A, Puchalski C, Świergiel AH (2020) Health benefits of plant-derived sulfur compounds, glucosinolates, and organosulfur compounds. Molecules 25(17):3804

Milner JA (1996) Garlic: its anticarcinogenic and antitumorigenic properties. Nutr Rev 54(11):S82

Mishra R, Upadhyay SK, Maheshwari PN (2001) Stability of allicin in garlic—a kinetic study

Mitra S, Das R, Emran TB, Labib RK, Islam F, Sharma R et al (2022) Diallyl disulfide: a bioactive garlic compound with anticancer potential. Front Pharmacol 13:943967

Moosavian SP, Arab A, Paknahad Z, Moradi S (2020) The effects of garlic supplementation on oxidative stress markers: a systematic review and meta-analysis of randomized controlled trials. Complement Ther Med 50:102385

Morales-González JA, Madrigal-Bujaidar E, Sánchez-Gutiérrez M, Izquierdo-Vega JA, Valadez-Vega MDC, Álvarez-González I et al (2019) Garlic (Allium sativum L.): a brief review of its antigenotoxic effects. Foods 8(8):343

Morel S, Bijlenga P, Kwak BR (2022) Intracranial aneurysm wall (in) stability–current state of knowledge and clinical perspectives. Neurosurg Rev 45(2):1233–1253

Moreno-Ortega A, Pereira-Caro G, Ordóñez JL, Moreno-Rojas R, Ortíz-Somovilla V, Moreno-Rojas JM (2020) Bioaccessibility of bioactive compounds of 'fresh garlic ' and 'black garlic' through in vitro gastrointestinal digestion. Foods 9(11):1582

Mukherjee M, Das AS, Das D, Mukherjee S, Mitra S, Mitra C (2006) Role of oil extract of garlic (Allium sativum Linn.) on intestinal transference of calcium and its possible correlation with preservation of skeletal health in an ovariectomized rat model of osteoporosis. Phytother Res 20(5):408–415

Munday R, Munday CM, Munday JS (2005) Hemolytic anemia and induction of phase II detoxification enzymes by Diprop-1-enyl sulfide in rats: dose-response study. J Agric Food Chem 53(25):9695–9700

Nadeem MS, Kazmi I, Ullah I, Muhammad K, Anwar F (2021) Allicin, an antioxidant and neuroprotective agent, ameliorates cognitive impairment. Antioxidants 11(1):87

Nagella P, Thiruvengadam M, Ahmad A, Yoon JaeYeon YJ, Chung IllMin CI (2014) Composition of polyphenols and antioxidant activity of garlic bulbs collected from different locations of Korea, vol 26, p 897

Nakamoto M, Kunimura K, Suzuki JI, Kodera Y (2020) Antimicrobial properties of hydrophobic compounds in garlic: allicin, vinyldithiin, ajoene and diallyl polysulfides. Exp Ther Med 19(2):1550–1553

Naqash S, Naik HR, Dar BN, Makroo HA (2021) Different methods for curing of bulb crops: principle, mechanism and effects on crop quality and its storage. Sci Hortic 289:110483

Nencini C, Menchiari A, Franchi GG, Micheli L (2011) In vitro antioxidant activity of aged extracts of some Italian allium species. Plant Foods Hum Nutr 66:11–16

Nishioka T, Suga H, Shimizu M (2022) The stimulation of indigenous bacterial antagonists by γ-Glutamyl-S-allyl-l-cysteine increases soil Suppressiveness to fusarium wilt. Appl Environ Microbiol 88(24):e01554–e01522

Obih P, Obih JC, Arome O (2019) Is alpha-glucosidase inhibition a mechanism of the antidiabetic action of garlic (*Allium sativum*)? J Biosci Med 7(10):42–49

Oh HY, Lee TH, Lee DY, Lee CH, Sohn MY, Kwon RW et al (2022) Evaluation of garlic juice processing waste supplementation in juvenile black rockfish (Sebastes schlegelii) diets on growth performance, antioxidant and digestive enzyme activity, growth-and antioxidant-related gene expression, and disease resistance against streptococcus iniae. Animals 12(24):3512

Okoro BC, Dokunmu TM, Okafor E, Sokoya IA, Israel EN, Olusegun DO, Ugbogu EEA (2023) The ethnobotanical, bioactive compounds, pharmacological activities and toxicological evaluation of garlic (Allium sativum): a review. Pharmacol Res - Mod Chin Med 100273:100273

Opare EO, Struhs E, Mirkouei A (2021) A comparative state-of-technology review and future directions for rare earth element separation. Renew Sust Energ Rev 143:110917

Ota A, Ulrih NP (2017) An overview of herbal products and secondary metabolites used for management of type two diabetes. Front Pharmacol 8:224659

Ou HC, Tzang BS, Chang MH, Liu CT, Liu HW, Lii CK, Kuo WW (2010) Cardiac contractile dysfunction and apoptosis in streptozotocin-induced diabetic rats are ameliorated by garlic oil supplementation. J Agric Food Chem 58(19):10347–10355

Özcan MM, Ghafoor K, Al Juhaimi F, Uslu N, Babiker EE, Mohamed Ahmed IA, Almusallam IA (2021) Influence of drying techniques on bioactive properties, phenolic compounds and fatty acid compositions of dried lemon and orange peel powders. J Food Sci Technol 58:147–158

Özçelik H, Taştan Y, Terzi E, Sönmez AY (2020) Use of onion (Allium cepa) and garlic (Allium sativum) wastes for the prevention of fungal disease (Saprolegnia parasitica) on eggs of rainbow trout (Oncorhynchus mykiss). J Fish Dis 43(10):1325–1330

Palkovics A, Szilágyi A, Nagy A, Jójárt F (2020) Economic aspects of the processing of vegetable plants. GRADUS 7(2):62–65

Pandiya P, Gojiya DK (2018) Study on preparation and preservation of garlic paste

Panuccio MR, Marra F, Maffia A, Mallamaci C, Muscolo A (2022) Recycling of agricultural (orange and olive) bio-wastes into ecofriendly fertilizers for improving soil and garlic quality. Resour Conserv Recycl Adv 15:200083

Panyod S, Wu WK, Chen PC, Chong KV, Yang YT, Chuang HL, Sheen LY (2022) Atherosclerosis amelioration by allicin in raw garlic through gut microbiota and trimethylamine-N-oxide modulation. NPJ Biofilms Microbiomes 8(1):4

Parreño R, Rodríguez-Alcocer E, Martínez-Guardiola C, Carrasco L, Castillo P, Arbona V, Candela H (2023) Turning garlic into a modern crop: state of the art and perspectives. Plan Theory 12(6):1212

Parveen B, Parveen A, Parveen R, Ahmad S, Ahmad M, Iqbal M (2020) Challenges and opportunities for traditional herbal medicine today, with special reference to its status in India. Ann Phytomed 9(2):97–112

Patil R, Ravindra R (2008) Formulation development and evaluation from garlic oil macerate. Internet J Nutr Wellness 8(1):1–6

Pavlić B, Aćimović M, Sknepnek A, Miletić D, Mrkonjić Ž, Kljakić AC, Teslić N (2023) Sustainable raw materials for efficient valorization and recovery of bioactive compounds. Ind Crop Prod 193:116167

Percival SS (2016) Aged garlic extract modifies human immunity. J Nutr 146(2):433S–436S

Perez-Ortiz JM, Galan-Moya EM, de la Cruz-Morcillo MA, Rodriguez JF, Gracia I, Garcia MT, Redondo-Calvo FJ (2020) Cost effective use of a thiosulfinate-enriched Allium sativum extract in combination with chemotherapy in colon cancer. Int J Mol Sci 21(8):2766

Petropoulos SA, Fernandes Â, Ntatsi G, Petrotos K, Barros L, Ferreira IC (2018) Nutritional value, chemical characterization and bulb morphology of Greek garlic landraces. Molecules 23(2):319

Pham HD, Pham TN (2022) Polyphenol, flavonoid content and antioxidant activity of the garlic (*Allium sativum* L.) extract obtained by two different extraction processes. Advanced materials and manufacturing engineering II, 493

Polyakov A, Alekseeva T, Muravieva I (2020) The elemental composition of garlic (*Allium sativum* L.) and its variability. In: E3S web of conferences. (Vol. 175, p. 01016). EDP Sciences
Proestos C, Komaitis M (2008) Application of microwave-assisted extraction to the fast extraction of plant phenolic compounds. LWT-Food Sci Technol 41(4):652–659
Putnik P, Gabrić D, Roohinejad S, Barba FJ, Granato D, Mallikarjunan K, Kovačević DB (2019) An overview of organosulfur compounds from *Allium* spp.: from processing and preservation to evaluation of their bioavailability, antimicrobial, and anti-inflammatory properties. Food Chem 276:680–691
Putra AFP, Nabila NN, Salsabila SP (2020). Extraction of garlic oil using microwave ultrasonic assisted extraction method
Qari SH, Khalil AH, Abdelfattah NA, Shehawy AA (2020) Effect of different plant extracts and nanoparticles on *Thrips tabaci* (Lind.)(Thysanoptera: Thripidae) under field conditions and their allelopathic potential on the onion, *Allium cepa* L. using bioassays and RAPD analysis. Egypt J Biol Pest Control 30:1–8
Qasim M, Islam W, Ashraf HJ, Ali I, Wang L (2020) Saponins in insect pest control. Co-evolution of secondary metabolites, 897–924
Qi C (2006) Advanced research on garlic. Food and Drug
Qiu Z, Qiao Y, Zhang B, Sun-Waterhouse D, Zheng Z (2022) Bioactive polysaccharides and oligosaccharides from garlic (*Allium sativum* L.): production, physicochemical and biological properties, and structure–function relationships. Compr Rev Food Sci Food Saf 21(4):3033–3095
Qiu Z, Zheng Z, Zhang B, Sun-Waterhouse D, Qiao X (2020) Formation, nutritional value, and enhancement of characteristic components in black garlic: a review for maximizing the goodness to humans. Compr Rev Food Sci Food Saf 19(2):801–834
Raaj EP, Bhuvaneshwari K, Lakshmipathy R, Devi VV, Rico ILR (2022) Garlic peel surface modification and fixed-bed column investigations towards crystal violet dye. Adsorpt Sci Technol 2022:2022
Rahman MS (2020) Food preservation: an overview. Handbook of food preservation, 7–18
Ramdani D, Yuniarti E, Jayanegara A, Chaudhry AS (2023) Roles of essential oils, polyphenols, and saponins of medicinal plants as natural additives and anthelmintics in ruminant diets: a systematic review. Animals 13(4):767
Ramirez DA, Altamirano JC, Camargo AB (2021) Multi-phytochemical determination of polar and non-polar garlic bioactive compounds in different food and nutraceutical preparations. Food Chem 337:127648
Rauf A, Abu-Izneid T, Thiruvengadam M, Imran M, Olatunde A, Shariati MA, Kazhybayeva G (2022) Garlic (*Allium sativum* L.): its chemistry, nutritional composition, toxicity, and anticancer properties. Curr Top Med Chem 22(11):957–972
Ried K (2016) Garlic lowers blood pressure in hypertensive individuals, regulates serum cholesterol, and stimulates immunity: an updated meta-analysis and review. J Nutr 146(2):389S–396S
Rivlin RS (2001) Historical perspective on the use of garlic. J Nutr 131(3):951S–954S
Rodrigues C, Percival SS (2019) Immunomodulatory effects of glutathione, garlic derivatives, and hydrogen sulfide. Nutrients 11(2):295
Rogers CD, Ting M, Li C, Kornhuber K, Coffel ED, Horton RM, Singh D (2021) Recent increases in exposure to extreme humid-heat events disproportionately affect populated regions. Geophys Res Lett 48(19):e2021GL094183
Ross SM (2007) The stinking rose and its potent healing properties. Holist Nurs Pract 21(2):101–102
Rouf R, Uddin SJ, Sarker DK, Islam MT, Ali ES, Shilpi JA, Sarker SD (2020) Antiviral potential of garlic (Allium sativum) and its organosulfur compounds: a systematic update of pre-clinical and clinical data. Trends Food Sci Technol 104:219–234
Rovera C, Carullo D, Bellesia T, Büyüktaş D, Ghaani M, Caneva E, Farris S (2023) Extraction of high-quality grade cellulose and cellulose nanocrystals from different lignocellulosic Agrifood wastes. Front Sustain Food Syst 6:1087867
Ru QM, Zhang LR, Chen JD, Pei ZM, Zheng HL (2009) Microwave–assisted extraction and identification of polysaccharide from *Lycoris aurea*. Chem Nat Compd 45:474–477

Rupani H, Fong WCG, Kyyaly A, Kurukulaaratchy RJ (2021) Recent insights into the management of inflammation in asthma. J Inflamm Res 14:4371–4397
Saif S, Hanif MA, Rehman R, Riaz M (2020) Garlic. In: Medicinal plants of South Asia. Elsevier, pp 301–315
Saikat ASM, Hossain R, Mina FB, Das S, Khan IN, Mubarak MS, Islam MT (2021) Antidiabetic effect of garlic. Rev Bras:1–11
Salem MA, Mohamed OG, Mosalam EM, Elberri AI, Abdel-Bar HM, Hassan M, Abo Mansour HE (2023) Investigation of the phytochemical composition, antioxidant, antibacterial, anti-osteoarthritis, and wound healing activities of selected vegetable waste. Sci Rep 13(1):13034
Samuel AT, Ariyo AL (2022) Garlic products for sustainable organic crop protection. In: Organic farming for sustainable development. Apple Academic Press, pp 165–182
Saras T (2023) Garlic: nature's flavorful wonder. Tiram Media
Sari NF, Ray P, Rymer C, Kliem KE, Stergiadis S (2022) Garlic and its bioactive compounds: implications for methane emissions and ruminant nutrition. Animals 12(21):2998
Sarpaki A (2021) The archaeology of garlic (Allium sativum): the find at Akrotiri, Thera, Greece. Documenta Praehistorica, 48, 48
Sarswat S, Yadav DS, Prakash A (2023) Effect of garlic on human health. Know Res: Multidiscip J 1(07):36–46
Sasi M, Kumar S, Kumar M, Thapa S, Prajapati U, Tak Y, Mekhemar M (2021) Garlic (*Allium sativum* L.) bioactives and its role in alleviating oral pathologies. Antioxidants 10(11):1847
Seki K, Ishikawa J, Okada Y (2018) Contribution of 2-propenesulfenic acid to the antioxidant activity of allicin. J Food Sci 83(5):1265–1270
Selim S, Albqmi M, Al-Sanea MM, Alnusaire TS, Almuhayawi MS, AbdElgawad H, El-Saadony MT (2022) Valorizing the usage of olive leaves, bioactive compounds, biological activities, and food applications: a comprehensive review. Front Nutr 9:1008349
Şen E, Göktürk E, Uğuzdoğan E (2022) Pectin extraction from garlic waste under dual acid condition. J Food Process Preserv 46(12):e17150
Setiyoningrum F, Priadi G, Afiati F (2022) Chemical properties of solo black garlic fermented by Saccharomyces cerevisiae. In: IOP conference series: earth and environmental science. (Vol. 976, No. 1, p. 012044). IOP Publishing
Settapramote N, Kawee-ai A, Phakawan J, Wannasawad K (2024) Effect of heating method on some characteristics of chive garlic dumplings. Appl Food Res 100410:100410
Shahi S, Khorvash R, Goli M, Ranjbaran SM, Najarian A, Mohammadi Nafchi A (2021) Review of proposed different irradiation methods to inactivate food-processing viruses and microorganisms. Food Sci Nutr 9(10):5883–5896
Shaikhiev IG, Kraysman NV, Sverguzova SV (2022) Use of garlic processing by-products to remove pollutants from aqueous media. Biointerface Res Appl Chem 12(4):4518
Shang A, Cao SY, Xu XY, Gan RY, Tang GY, Corke H et al (2019) Bioactive compounds and biological functions of garlic (Allium sativum L.). Food Secur 8(7):246
Shao L, Zhao Y, Zou B, Li X, Dai R (2021) Ohmic heating in fruit and vegetable processing: quality characteristics, enzyme inactivation, challenges and prospective. Trends Food Sci Technol 118:601–616
Sharma K, Rani V (2023) Therapeutic potential of stable organosulfur compounds of aged garlic. Curr Med Chem: Cardiovasc Hematol Agents 21(2):84–95
Sharma UK, Sharma K, Sharma N, Sharma A, Singh HP, Sinha AK (2008) Microwave-assisted efficient extraction of different parts of *Hippophae rhamnoides* for the comparative evaluation of antioxidant activity and quantification of its phenolic constituents by reverse-phase high-performance liquid chromatography (RP-HPLC). J Agric Food Chem 56(2):374–379
Shin JH, Choi DJ, Chung MJ, Kang MJ, Sung NJ (2008) Changes of physicochemical components and antioxidant activity of aged garlic at different temperatures. J Korean Soc Food Sci Nutr 37(9):1174–1181

Silva AS, Reboredo-Rodríguez P, Süntar I, Sureda A, Belwal T, Loizzo MR, Nabavi SM (2020) Evaluation of the status quo of polyphenols analysis: part I—phytochemistry, bioactivity, interactions, and industrial uses. Compr Rev Food Sci Food Saf 19(6):3191–3218

Sindhu RK, Kaur P, Manshu S, Goyal A, Bala R, Sandhu A (2021) Phytochemicals: extraction, isolation methods, identification and therapeutic uses: a review. Plant Arch 21(1):174–184

Singh G (2018) Cells from garlic peel. The Winnower

Singiri JR, Swetha B, Ben-Natan A, Grafi G (2022) What worth the garlic peel? Int J Mol Sci 23(4):2126

Sinsawasdi VK, Toschka HY, Rattanapanone N (2022) Chemical and functional properties of notable ingredients. In: The science of Thai cuisine. CRC Press, pp 65–108

Siu FPL, Mantzioris E (2020) Cultural perspectives. In: Nutrition for sport, exercise and performance. Routledge, pp 325–342

Skoczylas J, Jędrszczyk E, Dziadek K, Dacewicz E, Kopeć A (2023) Basic chemical composition, antioxidant activity and selected polyphenolic compounds profile in garlic leaves and bulbs collected at various stages of development. Molecules 28(18):6653

Song X, Yue Z, Nie L, Zhao P, Zhu K, Wang Q (2021) Biological functions of diallyl disulfide, a garlic-derived natural organic sulfur compound. Evid Based Complement Alternat Med 2021

Spence C (2023) Garlic: tracing its changing popularity in British cuisine. Int J Gastron Food Sci 100854

Sripanidkulchai B (2020) Benefits of aged garlic extract on Alzheimer's disease: possible mechanisms of action. Exp Ther Med 19(2):1560–1564

Sturza R, Ghendov-Moşanu A (2021) Food, nutrition, and health in Moldova. In: Nutritional and health aspects of food in the Balkans. Academic Press, pp 249–262

Subramanian MS, Nandagopal MS, Amin Nordin S, Thilakavathy K, Joseph N (2020) Prevailing knowledge on the bioavailability and biological activities of Sulphur compounds from Alliums: a potential drug candidate. Molecules 25(18):4111

Subroto E, Cahyana Y, Tensiska M, Lembong F, Filianty E, Kurniati E et al (2021) Bioactive compounds in garlic (*Allium sativum* L.) as a source of antioxidants and its potential to improve the immune system: a review. Food Res 5(6):1–11

Suffness M, Wall ME (2021) Discovery and development of taxol. In: Taxol. CRC Press, pp 3–26

Suleria HAR, Butt MS, Khalid N, Sultan S, Raza A, Aleem M, Abbas M (2015) Garlic (Allium sativum): diet based therapy of 21st century–a review. Asian Pac J Trop Dis 5(4):271–278

Sun X, Hou T, Cheung E, Iu TNT, Tam VWH, Chu IMT, Wong CK (2020) Anti-inflammatory mechanisms of the novel cytokine interleukin-38 in allergic asthma. Cell Mol Immunol 17(6):631–646

Sunanta P, Kontogiorgos V, Pankasemsuk T, Jantanasakulwong K, Rachtanapun P, Seesuriyachan P, Sommano SR (2023) The nutritional value, bioactive availability and functional properties of garlic and its related products during processing. Front Nutr 10

Suppressa P, Carbonara C, Scialpi N, Ciavarella A, Sabbà C (2020) Homozygous familial hypercholesterolemia in a young woman with dual gene mutations of low-density lipoprotein receptor and proprotein convertase subtilisin/kexin type 9. J Clin Lipidol 14(2):192–196

Swarup S, Cargill EJ, Crosby K, Flagel L, Kniskern J, Glenn KC (2021) Genetic diversity is indispensable for plant breeding to improve crops. Crop Sci 61(2):839–852

Ta S (2014) Diagnosis and classification of diabetes mellitus. Diabetes Care 37(1):81–90

Takagi H (2020) Garlic Allium sativum L. In: Onions and allied crops. CRC Press, pp 109–146

Tamang JP, Cotter PD, Endo A, Han NS, Kort R, Liu SQ, Hutkins R (2020) Fermented foods in a global age: east meets west. Compr Rev Food Sci Food Saf 19(1):184–217

Tavares L, Santos L, Noreña CPZ (2021) Bioactive compounds of garlic: a comprehensive review of encapsulation technologies, characterization of the encapsulated garlic compounds and their industrial applicability. Trends Food Sci Technol 114:232–244

Teng Z, Han K, Li J, Gao Y, Li M, Ji T (2020) Ultrasonic-assisted preparation and characterization of hierarchical porous carbon derived from garlic peel for high-performance supercapacitors. Ultrason Sonochem 60:104756

Tesfaye A (2021) Revealing the therapeutic uses of garlic (Allium sativum) and its potential for drug discovery. Sci World J 2021:1

Thach NA (2022) Investigation of the effects of extraction temperature and time on bioactive compounds content from garlic (*Allium sativum* L.) husk. Front Sustain Food Syst 6:1004281

The Culinary Institute of America (CIA) (2024) The professional chef. John Wiley & Sons

Tiong S (2020) Sweet, savory, spicy: exciting street market food from Thailand, Cambodia, Malaysia and More. Page Street Publishing

Tohirova J, Shernazarov F (2022) Atherosclerosis: causes, symptoms, diagnosis, treatment and prevention. Sci Innov 1(D5):7–12

Tribble CM, Martínez-Gómez J, Howard CC, Males J, Sosa V, Sessa EB et al (2021) Get the shovel: morphological and evolutionary complexities of belowground organs in geophytes. Am J Bot 108(3):372–387

Trigo JP, Alexandre EM, Saraiva JA, Pintado ME (2022) High value-added compounds from fruit and vegetable by-products–characterization, bioactivities, and application in the development of novel food products. Crit Rev Food Sci Nutr 60(8):1388–1416

Tudu CK, Dutta T, Ghorai M, Biswas P, Samanta D, Oleksak P et al (2022) Traditional uses, phytochemistry, pharmacology and toxicology of garlic (*Allium sativum*), a storehouse of diverse phytochemicals: a review of research from the last decade focusing on health and nutritional implications. Front Nutr 9:949554

Ullah Z, Gul F, Iqbal J, Abbasi BA, Kanwal S, Chalgham W, Mahmood T (2023c) Biogenic synthesis of multifunctional silver oxide nanoparticles (Ag_2ONPs) using Parieteria alsinaefolia Delile aqueous extract and assessment of their diverse biological applications. Microorganisms 11(4):1069

Ullah Z, Iqbal J, Abbasi BA, Akhtar W, Kanwal S, Ali I, Mahmood T (2023b) Assessment of Gus expression induced by anti-sense Os PPO gene promoter and antioxidant enzymatic assays in response to drought and heavy metal stress in transgenic Arabidopsis thaliana. Sustainability 15(17):12783

Ullah Z, Iqbal J, Gul F, Abbasi BA, Kanwal S, Elsadek MF, Mahmood T (2024) Biogenic synthesis, characterization, and in vitro biological investigation of silver oxide nanoparticles (AgONPs) using Rhynchosia capitata. Sci Rep 14(1):10484

Ullah R, Ullah Z, Iqbal J, Chalgham W, Ahmad A (2023a) Aspartic acid-based Nano-copper induces resilience in *Zea mays* to applied Lead stress *via* conserving photosynthetic pigments and triggering the antioxidant biosystem. Sustainability 15(16):12186

Umair M, Jabeen S, Ke Z, Jabbar S, Javed F, Abid M, Conte-Junior CA (2022) Thermal treatment alternatives for enzymes inactivation in fruit juices: recent breakthroughs and advancements. Ultrason Sonochem 86:105999

Valente C, Aboua G, Du Plessis SS (2014) Garlic and its effects on health with special reference to the reproductive system. ABOUA G. Antioxidant-antidiabetic agents and human health. 3ª edição. Africa do Sul: Intech, 259-577

Valls RM, Companys J, Calderón-Pérez L, Salamanca P, Pla-Pagà L, Sandoval-Ramírez BA et al (2022) Effects of an optimized aged garlic extract on cardiovascular disease risk factors in moderate hypercholesterolemic subjects: a randomized, crossover, double-blind, sustained and controlled study. Nutrients 14(3):405

Varshney R, Budoff MJ (2016) Garlic and heart disease. J Nutr 146(2):416S–421S

Vasanthi HR, ShriShriMal N, Das DK (2012) Phytochemicals from plants to combat cardiovascular disease. Curr Med Chem 19(14):2242–2251

Vathsala PG, Krishna Murthy P (2020) Immunomodulatory and antiparasitic effects of garlic–arteether combination via nitric oxide pathway in Plasmodium berghei-infected mice. J Parasit Dis 44(1):49–61

Verma T, Aggarwal A, Dey P, Chauhan AK, Rashid S, Chen KT, Sharma R (2023) Medicinal and therapeutic properties of garlic, garlic essential oil, and garlic-based snack food: an updated review. Front Nutr 10:1120377

Vinayagam R, Eun Lee K, Ambati RR, Gundamaraju R, Fawzy Ramadan M, Gu Kang S (2023) Recent development in black garlic: nutraceutical applications and health-promoting phytoconstituents. Food Rev Intl 39(6):3534–3554

Walag AMP, Ahmed O, Jeevanandam J, Akram M, Ephraim-Emmanuel BC, Egbuna C, Uba JO (2020) Health benefits of organosulfur compounds. Functional foods and nutraceuticals: bioactive components, formulations and innovations, 445–472

Wong S (2020) Science of cooking: bring out the best in garlic. New Sci 245(3273):51

Wongsa P, Bhuyar P, Tongkoom K, Spreer W, Müller J (2023) Influence of hot-air drying methods on the phenolic compounds/allicin content, antioxidant activity and α-amylase/α-glucosidase inhibition of garlic (Allium sativum L.). Eur Food Res Technol 249(2):523–535

Wu YR, Li L, Sun XC, Wang J, Ma CY, Zhang Y et al (2021) Diallyl disulfide improves lipid metabolism by inhibiting PCSK9 expression and increasing LDL uptake via PI3K/Akt-SREBP2 pathway in HepG2 cells. Nutr Metab Cardiovasc Dis 31(1):322–332

Xiao H, Zhong Z, Low ZX, Huang Y, Sun Y, Yao Z (2016) Separation of sulfoether compounds in garlic oil by integrated membrane technologies. J Food Process Eng 39(6):591–600

Yang D, Dunshea FR, Suleria HA (2020) LC-ESI-QTOF/MS characterization of Australian herb and spices (garlic, ginger, and onion) and potential antioxidant activity. J Food Process Preserv 44(7):e14497

Zahra SA, Iqbal J, Abbasi BA, Shahbaz A, Kanwal S, Shah SL, Mahmood T (2021) Antimicrobial, cytotoxic, antioxidants, enzyme inhibition activities, and scanning electron microscopy of *Lactuca orientalis* (Boiss.) Boiss. seeds. Microscopy Res Tech 84(6):1284–1295

Zahra N, Iqbal J, Arif M, Abbasi BA, Sher H, Nawaz AF, Calina D (2023) A comprehensive review on traditional uses, phytochemistry and pharmacological properties of Paeonia emodi Wall. Ex Royle: current landscape and future perspectives. Chin Med 18(1):23

Zaynab M, Sharif Y, Abbas S, Afzal MZ, Qasim M, Khalofah A, Li S (2021) Saponin toxicity as key player in plant defense against pathogens. Toxicon 193:21–27

Zhang Y, Liu X, Ruan J, Zhuang X, Zhang X, Li Z (2020) Phytochemicals of garlic: promising candidates for cancer therapy. Biomed Pharmacother 123:109730

Zhu W, Su Z, Xu W, Sun H, Gao J, Tu D, Cao H (2020) Garlic skin induced shifts in the rumen microbiome and metabolome of fattening sheep

Zocchi DM, Sulaiman N, Prakofjewa J, Sõukand R, Pieroni A (2024) Local wild food plants and food products in a multi-cultural region: an exploratory study among diverse ethnic groups in Bessarabia, Southern Moldova. Sustainability 16(5):1968

Bioactive Phytochemicals from Onion (*Allium cepa* L.) By-Products

Chin Xuan Tan, Seok Shin Tan, Hui Ling Tan, and Seok Tyug Tan

Abstract Bioactive phytochemicals derived from agro-industrial residues, mainly vegetable and fruit waste, are increasingly popular in the food industry. Onions (*Allium cepa* L.) are the most commonly consumed crops worldwide. The increased market for processed onions has resulted in a significant accumulation of waste. This chapter provides an overview of the bioactive compounds and proximate and mineral composition of onion by-products. The application of onion by-products is also compiled and discussed.

Keywords Onion · By-product · Bioactive compound · Flavonoid · Sulfur

Abbreviations

ACSOs	Alk(en)yl cysteine sulfoxides
GAE	Gallic acid equivalents
IDF	Insoluble dietary fiber
QE	Quercetin equivalents
TA	Total ACSOs
TDF	Total dietary fiber
TFC	Total flavonoid content

C. X. Tan (✉)
Department of Allied Health Sciences, Faculty of Science, Universiti Tunku Abdul Rahman, Jalan Universiti Bandar Barat, Kampar, Perak, Malaysia
e-mail: tancx@utar.edu.my

S. S. Tan · S. T. Tan
Jeffrey Cheah School of Medicine and Health Sciences, Monash University Malaysia, Bandar Sunway, Selangor, Malaysia

H. L. Tan
School of Hospitality and Service Management, Sunway University, Petaling Jaya, Selangor, Malaysia

M. F. Ramadan (ed.), *Bioactive Phytochemicals in By-products from Bulb, Flower and Fruit Vegetables*, https://doi.org/10.1007/978-3-031-77399-0_3

TPC	Total phenolic content
TS	Total sulfurs

1 Introduction

Onions (*Allium cepa* L.) belong to the *Amaryllidaceae* family and are one of the most commonly consumed crops worldwide. They are classified according to skin color and taste (Albishi et al. 2013). The overall production of onions has grown by more than 25% in the last few years (Sagar et al. 2022). In 2022, India, China, Egypt, and the United States were the leading onion-producing countries globally (FAO 2024). Onions are the second most commonly harvested fresh vegetables in the European Union (Črnivec et al. 2021). People consume onions for their nutritional, medicinal, and flavor-enhancing properties. The demand for processed onion products, such as canned, dehydrated, and onion pickles, is increasing globally.

Bioactive phytochemicals from agro-industrial residues, especially vegetable and fruit waste, are gaining popularity in the food industry. The outer layers of vegetables and fruits usually contain greater amounts of bioactive compounds than the inner parts, as they are designed to protect the plant parts from various external damaging factors (Črnivec et al. 2021). A notable example is onions, where the skins contain significantly greater amounts of flavonoids than the edible portion (Škerget et al. 2009). It was reported that about 80% of the total flavonoid content in onions was found in the skin (Albishi et al. 2013). The elevated flavonoid levels in onion skins offer protection against the detrimental effects of UV radiation and intracellular hydrogen peroxide (Lee et al. 2008).

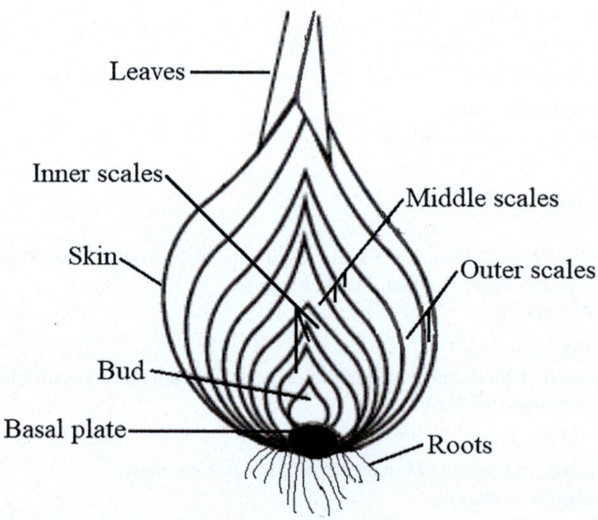

Fig. 1 Anatomy of an onion

The increased market for processed onions has resulted in a significant accumulation of waste. It has been reported that European countries generate approximately 0.6 MT of onion waste annually (Bains et al. 2023). Figure 1 shows the anatomy of an onion. Onion skins, the top and bottom of the bulbs, and the outer fleshy leaves constitute the main by-products resulting from the industrial processing of onions (Črnivec et al. 2021). Due to their strong aroma and high sulfur content, these by-products cannot be valorized as manure and fodder. Among these by-products, researchers widely investigate the bioactive phytochemicals found in onion skins due to their abundance and ease of removal before processing and sale. It was postulated that 70% of the onion waste was contributed by its skin (Sagar et al. 2021). This chapter overviews the bioactive phytochemicals in onion by-products and their applications.

2 Edible and Waste Portions of Onions

Table 1 shows the fresh fraction and water content of edible and waste portions of onions. The waste portion, comprising onion skins, the dry base of the onion scape, and any dry scape and leaves of onions, accounted for about 5% of the total vegetable mass (Črnivec et al. 2021). Meanwhile, the edible bulb tunics contributed to about 95% of the total vegetable mass (Črnivec et al. 2021). The edible and waste portions of white onions contained the highest water content.

3 Proximate Composition

The proximate composition is a crucial parameter for product development and regulatory purposes. As shown in Table 2, the levels of carbohydrate, ash, moisture content, crude protein, fat, and crude fiber of onion skins were in the range of 64.32–76.66%, 6.88–12.24%, 6.27–13.23%, 3.90–5.97%, 0.30–0.47%, and 4.45–8.28%, respectively. Irrespective of the cultivars, onion skins are a good source of carbohydrate, ash, and fiber, but inferior in fat content. Among the 15 onion cultivars investigated, the skins of the NHRDF Red cultivar contained the greatest levels of ash (12.24%), crude fiber (8.28%), and crude protein (5.97%), but the lowest level of carbohydrate (64.32%). Meanwhile, the skins of the Pusa Riddhi cultivar

Table 1 Edible and waste portions of onions

	Red onion		White onion		Yellow onion	
	Edible	Waste	Edible	Waste	Edible	Waste
Fresh fraction (g/kg)	964.1	35.9	964.2	35.8	947.6	52.4
Water content (%)	89.3	46.0	91.7	59.8	87.9	49.0

Source: Črnivec et al. (2021)

Table 2 Proximate composition of onion skins

	Component (%)					
Cultivar	Carbohydrate	Ash	Moisture content	Crude protein	Fat	Crude fiber
NHRDF Red	64.32	12.24	8.76	5.97	0.43	8.28
Agri Found Dark Red	69.08	9.53	8.77	5.73	0.35	6.53
Pusa Red	70.62	8.88	10.33	4.03	0.47	5.67
Pusa Riddhi	76.66	6.88	6.27	4.80	0.36	5.04
Arka Kirtiman	67.44	9.11	10.88	5.27	0.35	6.96
Hisar-2	67.39	10.93	10.76	5.40	0.45	5.07
Sukh Sagar	67.74	10.18	9.98	5.60	0.33	6.17
Hisar-3	69.81	10.07	8.88	4.43	0.41	7.03
Bhima Shakti	74.48	6.90	8.79	4.37	0.41	5.05
Bhima Kiran	70.20	11.60	7.89	4.27	0.34	5.67
Pusa Madhavi	67.02	10.85	13.23	4.10	0.36	4.45
Phursungi Local	70.77	9.84	9.87	4.30	0.43	4.80
Agri Found Light Red	69.37	7.89	11.24	5.83	0.32	5.35
Bhima Shubhra	75.89	7.83	6.46	3.93	0.30	5.49
Udaipur Local	76.43	8.03	6.76	3.90	0.31	4.57

Source: Sagar et al. (2021)

contained the highest level of carbohydrates (76.66%) but the lowest levels of ash (6.88%) and moisture content (6.27%).

Carbohydrates, the primary nutrient in onion skins, can be classified into non-structural and structural carbohydrates. Non-structural carbohydrates such as glucose, sucrose, fructose, and fructans constitute about 51 to 67% of the dry matter of onions (Bains et al. 2023). Glucose, sucrose, and fructose contribute to the calorie and nutritive value of onions, while fructans are considered non-nutritive functional compounds as part of prebiotics. Structural carbohydrates such as pectin, cellulose, and hemicellulose constitute the cellular wall of onions, playing essential roles in their porosity and mechanical strength (Bains et al. 2023).

The soluble, insoluble, and total dietary fiber contents of different portions of onions (skins, inner scales, and outer scales) were reported by Benítez et al. (2011). They observed higher total dietary fiber (TDF) and insoluble dietary fiber (IDF) contents in skins (750 mg/g for TDF and 668–698 mg/g for IDF) compared to the inner scales (169–222 mg/g for TDF and 121–179 for IDF) and outer scales (217–312 mg/g for TDF and 158–270 mg/g for IDF) of onions, suggesting a decrease in TDF and IDF from the outer to the inner layers of the onions. Meanwhile, the soluble dietary fiber content of the skins, outer scales, and inner scales of onions were 52–82 mg/g, 42–59 mg/g, and 43–48 mg/g, respectively.

4 Mineral Composition

Minerals play crucial roles in metabolic processes and are essential for human productivity, proper growth, and health. They are broadly classified into macro and trace minerals. Macro minerals are those required in large quantities by the human body, whereas trace minerals are needed in minimal amounts. The mineral profile of the skins from different onion cultivars is tabulated in Table 3. Potassium, phosphorus, and calcium were the predominant minerals detected in the onion skins, ranging from 877.87–3474.01 mg/kg, 796.22–2873.10 mg/kg, and 762.42–3007.81 mg/kg, respectively. Among all the cultivars, the skins of the NHRDF Red cultivar contained the greatest amounts of macro minerals [potassium (3474.01 mg/kg), calcium (3007.81 mg/kg), phosphorous (2873.10 mg/kg), magnesium (1034.54 mg/kg), and sodium (879.97)] and trace minerals [iron (211.96 mg/kg), zinc (39.27 mg/kg), copper (29.49 mg/kg), and manganese (18.68 mg/kg)]. Both white-skinned onion cultivars, Bhima Shubhra and Udaipur Local, contained the fewest macro and trace minerals. Onion skins contain more minerals than their bulbs because of the lower water content (Sagar et al. 2021). Other than cultivar differences, soil composition and post-harvesting handling methods can affect the mineral composition of onion skins. Ncayiyana et al. (2018) reported that organic manure increased the

Table 3 Mineral composition of onion skins

	Mineral (mg/kg DW)								
Cultivar	K	P	Ca	Mg	Na	Fe	Zn	Cu	Mn
NHRDF Red	3474.01	2873.10	3007.81	1034.54	879.97	211.96	39.27	29.49	18.68
Agri Found Dark Red	3366.17	2418.49	1197.18	762.18	508.34	170.87	34.54	25.52	11.67
Pusa Red	2855.08	2466.60	2699.48	856.01	496.23	181.72	37.46	26.10	13.26
Pusa Riddhi	2662.61	2300.07	2381.53	397.96	404.02	161.35	27.55	21.42	12.27
Arka Kirtiman	2137.99	2056.44	1222.92	207.63	575.46	135.86	21.71	16.52	11.67
Hisar-2	3250.05	2789.47	2892.29	888.78	794.25	197.01	38.39	27.95	17.37
Sukh Sagar	2188.48	2125.58	2107.15	281.13	298.38	143.55	21.39	19.93	11.20
Hisar-3	2667.44	2430.87	2496.18	641.32	599.86	181.07	31.48	24.29	14.09
Bhima Shakti	3140.76	2458.65	1492.81	361.48	488.47	140.82	25.56	23.49	14.03
Bhima Kiran	2614.45	2632.97	1325.89	297.96	674.82	130.74	22.62	18.66	13.44
Pusa Madhavi	2767.56	2362.49	2702.44	743.15	540.69	175.85	35.41	24.41	14.37
Phursungi Local	2233.70	2395.82	2691.22	443.80	431.88	160.53	33.61	21.36	13.33
Agri Found Light Red	3050.16	2551.78	1381.58	335.27	610.89	150.90	29.65	21.78	12.55
Bhima Shubhra	959.05	796.22	770.48	168.57	127.97	110.60	17.95	12.45	10.04
Udaipur Local	877.82	870.27	762.42	188.16	121.07	125.80	19.55	13.42	9.11

Source: Sagar et al. (2021). *DW* Dry weight, *K* Potassium, *P* Phosphorus, *Ca* Calcium, *Mg* Magnesium, *Na* Sodium, *Fe* Iron, *Zn* Zinc, *Cu* Copper, and *Mn* Manganese

concentrations of certain minerals in onions. Meanwhile, improper temperatures used in freezing or drying the harvested onions can reduce mineral content (Petropoulos et al. 2017).

5 Phenolics

Phenolics comprise a large group of approximately 10,000 different compounds, each featuring one or more aromatic rings with attached hydroxyl groups. The sensory properties of plants, including color, taste, astringency, and aroma, are attributed to phenolics, which exist in either free, esterified, or bound form. Free and esterified phenolics are typically extractable using solvolytic solutions such as methanol, ethanol, acetone, and water, while insoluble-bound phenolics are commonly extracted by alkalizing the residue following the isolation of free and esterified phenolics (Yu et al. 2019). Table 4 shows the levels of free, esterified, and bound phenolic and flavonoid contents in onion skins of various colors. Phenolics were mainly present in the free form in onion skins, except in red-skinned onion, where bound phenolics were slightly more abundant than free phenolics. Irrespective of the colors of onion skins, the levels of bound phenolics were higher than those of the esterified forms. White onion contained the lowest amounts of free, esterified, and bound phenolics compared to onions of other skin colors.

The total phenolics and flavonoids of onion skins from different cultivars are summarized in Table 5. It was observed that the white-skinned onion cultivars, Bhima Shubhra and Udaipur Local, exhibited the lowest TPC levels. This finding is consistent with the results reported by Albishi et al. (2013). Other than the onion cultivars, the yield of phenolics in onion skins can be significantly affected by the solvents used for extraction. Absolute acetone proved superior to absolute ethanol, aqueous acetone (35–60%), and aqueous ethanol (35–60%) for isolating phenolics from red onion skins (Škerget et al. 2009). By comparing the retention times of the onion skins with those of standards, Elsebaie and Essa (2018) identified vanillic

Table 4 Free, esterified, and bound phenolics and flavonoids in onion skins

	Skin color			
Component	Red	Yellow	Pearl	White
Phenolics (mg GAE/g dried onion)				
Free	23.67	22.71	62.65	0.54
Esterified	12.50	10.75	22.62	0.01
Bound	25.45	17.96	24.92	0.21
Flavonoids (mg QE/g dried onion)				
Free	20.22	10.69	19.64	0.08
Esterified	2.57	0.56	2.60	0.01
Bound	4.69	1.25	4.62	0.01

Source: Albishi et al. (2013)

Table 5 Skin color, total phenolics, and total flavonoids of different onion cultivars

Cultivar	Skin color[1]	TPC[2] (mg GAE/g DW)	TFC[2] (mg QE/g DW)
NHRDF Red	Dark red	289.04	168.77
Agri Found Dark Red	Dark red	265.10	92.31
Pusa Red	Red	251.71	89.62
Pusa Riddhi	Red	269.74	150.94
Arka Kirtiman	Red	262.69	96.46
Hisar-2	Red	262.78	98.09
Sukh Sagar	Red	254.41	88.11
Hisar-3	Pink	288.74	160.18
Bhima Shakti	Pink	243.68	75.36
Bhima Kiran	Pink	239.91	81.48
Pusa Madhavi	Pink	245.58	66.88
Phursungi Local	Pink	231.73	66.58
Agri Found Light Red	Pink	241.12	83.67
Bhima Shubhra	White	14.55	1.31
Udaipur Local	White	19.74	2.01

Source: [1]Sagar and Pareek (2020a), [2] Sagar et al. (2020). *DW* Dry weight, *TPC* Total phenolic content, *TFC* Total flavonoid content, *GAE* gallic acid equivalents, *QE* quercetin equivalents

(84.78%), ellagic (6.84%), and protocatechuic (2.41%) as the main phenolic compounds in the skins of red onions, while the remaining (5.97%) were contributed by benzoic, salycylic, chlorogenic, catechol, epicatechin, gallic acid, catechin, caffeic, 4-amino benzoic, caffeine, α-coumaric, β-coumaric acid, ferulic, 3,4,5 methoxycinnamic, coumarin, cinnamic, isoferulic, and resveratrol.

6 Flavonoids

Phenolics can be broadly categorized into two groups, namely flavonoids and non-flavonoids. All flavonoids have a phenyl chromene framework (C6-C3-C6) and are divided into various groups based on variations in their side-chain structures. Gallic acid, caffeic acid, chlorogenic acid, tannins, resveratrol, and curcumin are typical examples of non-flavonoids (Akdeniz 2016). Among the 62 edible tropical crops investigated, onion contained the greatest amount of total flavonoids (Miean and Mohamed 2001). Onion skins were reported to contain 20 times more flavonoids than their flesh (Bains et al. 2023).

The flavonoids in onion skins were predominantly found in the free form, with the highest concentration observed in red-skinned onions and the lowest in white-skinned onions (Table 4). This trend is consistent with findings by Zhang et al. (2016), who reported the total flavonoid content of red onion (111.1 mg/100 g fresh weight) as the highest, followed by yellow onion (36.64 mg/100 g fresh weight) and white onion (0 mg/100 g fresh weight). Across different cultivars, the total flavonoid content in onion skins ranged from 1.31 to 168.7 mg QE/g DW, with Bhima Shubhra, a white onion skin cultivar, exhibited the lowest amount, and NHRDF red, a dark red onion skin cultivar, displayed the highest concentration (Table 5).

Selected flavonoids in onion skins of different cultivars were quantified using high-performance liquid chromatography (HPLC), and the results are tabulated in Table 6. The concentrations of quercetin, luteolin, quercetin 3-β-D-glucoside, and kaempferol ranged from 88.40 to 11885.03 mg/kg, 45.48 to 1669.93 mg/kg, 56.18 to 1432.88 mg/kg, and 124.23 to 709.98 mg/kg, respectively. Among the onion cultivars investigated, the highest levels of quercetin, luteolin, quercetin 3-β-D-glucoside, and kaempferol were detected in the skins of NHRDF Red, Pusa Riddhi, Hisar-2, and Bhima Shakti, respectively. Meanwhile, the white-skinned onions, Bhima Shubhra and Udaipur Local contained the lowest amounts of quercetin, luteolin, quercetin 3-β-D-glucoside, and kaempferol. The study of Nuutila et al. (2003) also reported that red onion skins (83, 477 mg/kg) contained a greater level of quercetin than yellow onion skins (34, 430 mg/kg). In contrast to Sagar et al. (2020), Nuutila et al. (2003) did not detect the presence of kaempferol in the skins of red and yellow onions, possibly due to the utilization of different extraction methods and sample pretreatments. Plant samples are commonly subjected to convection, freeze, or microwave vacuum drying prior to flavonoid isolation. Freeze drying has been reported as the most effective pretreatment method for recovering the majority of flavonoids (Jurinjak Tušek et al. 2022). Meanwhile, modern extraction techniques such as ultrasound-assisted extraction, microwave-assisted extraction, supercritical fluids extraction, and pressurized

Table 6 Flavonoids in onion skins

	Concentration (mg/kg DW)			
Cultivar	Quercetin 3-β-D-glucoside	Quercetin	Luteolin	Kaempferol
NHRDF Red	694.73	11885.03	1082.58	427.48
Agri Found Dark Red	471.38	5655.70	317.20	564.85
Pusa Red	378.08	2830.93	956.05	405.78
Pusa Riddhi	631.70	8571.30	1669.93	522.05
Arka Kirtiman	436.28	3738.25	369.15	298.08
Hisar-2	1432.88	11429.30	1032.73	423.15
Sukh Sagar	505.53	6389.13	495.88	470.20
Hisar-3	419.18	4303.00	780.38	520.10
Bhima Shakti	489.55	4558.50	704.68	709.98
Bhima Kiran	439.10	3188.33	466.58	708.23
Pusa Madhavi	68.97	2856.43	897.78	456.68
Phursungi Local	599.70	4642.13	518.40	565.10
Agri Found Light Red	305.63	3496.50	949.58	694.48
Bhima Shubhra	56.18	89.28	46.20	124.23
Udaipur Local	60.65	88.40	45.48	144.85

Source: Sagar et al. (2020). *DW* Dry weight

liquids-assisted extraction could better extract plant flavonoids than conventional Soxhlet techniques (Jurinjak Tušek et al. 2022). Hence, the greater individual flavonoid component detected in the onion skins in the study by Sagar et al. (2020) could be attributed to freeze-drying coupled with ultrasound-assisted extraction during sample preparation.

Quercetin is a citron yellow crystal configuration that is insoluble in water but is soluble in alcohol and lipids. It consists of two aromatic rings and an oxygenated heterocyclic ring, which contains a hydroxyl group at the 3-carbon chain and a carbonyl group at the 4-position. The antioxidant activity of quercetin is known to be the greatest among plant flavonoids (Golubkina and Caruso 2020). Quercetin and its derivatives are the most prevalent flavonoids detected in all parts of the onions. They are responsible for some onion cultivars' yellow and brown pigments (Benítez et al. 2011). Onion skins were reported to contain 48 folds of greater quercetin than the flesh (Kwak et al. 2017). Other than cultivar differences, the levels of quercetin and its derivatives can be affected by postharvest storage and processing conditions. For instance, UV irradiation and high pressure can preserve or even enhance the quercetin levels in onions (Vojvodić Cebin et al. 2020). The quercetin and its derivatives in the skins of red, yellow, and chartreuse onions were reported by Kwak et al. (2017). From the HPLC chromatograms, four peaks corresponding to quercetin-7,4′-diglucoside, quercetin-3-glucoside, quercetin-4′-glucoside, and quercetin were observed in the skins of red and yellow onions, whereas three peaks corresponding to quercetin-7,4′-diglucoside, quercetin-3-glucoside, and quercetin were observed in the skins of chartreuse onions. Potential health benefits associated with the intake of food rich in quercetin and its derivatives include the prevention of cancer and cardiovascular diseases as well as anti-inflammatory (Vojvodić Cebin et al. 2020). Moreover, quercetin has been shown to inhibit the oxidation and cytotoxicity of low-density lipoprotein (Miean and Mohamed 2001).

7 Anthocyanins

The water-soluble anthocyanins are responsible for plants' red, purple, and blue pigments. Several anthocyanins have been isolated from onion skins, especially from onions with red skin color. Most of the anthocyanins detected in red-skinned onions were cyanidin derivatives such as cyanidin-3-glucoside, cyanidin 3-(6″-malonyl glucoside), cyanidin-3-(3″-glucosyl glucoside), and cyanidin 3-(3″-glucosyl-6″-malonyl glucoside), along with minor amounts of peonidin, delphinidin, and petunidin derivatives (Vojvodić Cebin et al. 2020).

The color of anthocyanins in red-skinned onions changes according to pH values (Chadorshabi et al. 2022). At pH levels 3 or lower, anthocyanins appear red or purple, while at pH levels above 5, they appear blue. The total anthocyanins, expressed in mg of anthocyanidin-3-glucoside per 100 g dried onion, in the skins of different onion cultivars followed the order of red color (10.04) > pearl color (6.32) > yellow color (0.21) > white color (0.06) (Albishi et al. 2013).

8 Sulfur

Sulfur is one of the main components present in onions. The alk(en)yl cysteine sulfoxides (ACSOs), sulfur derivatives localized in the cytoplasm, serve as the flavor precursors of onions. The common ACSOs isolated from onions included trans-(+)-S-1-propenyl-L-cysteine sulfoxide (PECSO), (+)-S-propyl-L-cysteine sulfoxide (PCSO), and (+)-S-methyl-L-cysteine sulfoxide (MCSO). Among these, PECSO was the most abundantly present, constituting up to 80% of total ACSOs (Bains et al. 2023). Variations in the flavor and aroma among the onion cultivars are due to differences in ACSOs concentration and composition.

Upon cell damage and the release of alliinase from vacuoles, ACSOs are rapidly hydrolyzed, forming sulfenic acids, which then transform into thiosulfinates (Golubkina and Caruso 2020). The thiosulfinates can further convert into the lachrymatory factor, cepaene, or *cis*, trans-zwiebelanes. These compounds possess antibiotic properties, can inhibit aggregation of thrombocytes, and reduce blood cholesterol (Golubkina and Caruso 2020). The total sulfurs (TS) and total ACSOs (TA) of skins (15.6 μmol/g for TS and 4.6 μmol/g for TA) were drastically lower than the outer scales (100.2 μmol/g for TS and 29.9 μmol/g for TA) and inner scales (153.1 μmol/g for TS and 54.2 μmol/g for TA) of onions (Benítez et al. 2011). These suggest that the sulfur content may increase from the onions' outer to inner layers.

The pungency and/or taste of onions can be assessed by determining the pyruvic acid concentration, a by-product generated from the degradation of ACSOs (Golubkina and Caruso 2020). For instance, onions with pyruvic acid concentrations less than 3 μmol/g are classified as sweet onions, those with 3–7 μmol/g are classified as moderately pungent onions, and those with more than 7 μmol/g are classified as pungent onions. Generally, white-skinned onions are sweeter than red- and yellow-skinned cultivars and onions that ripen early are associated with lower pungency (Golubkina and Caruso 2020).

9 Application of Onion By-Products

Owing to the high bioactive phytochemicals, onion by-products have been valorized into various products. In the food industry, onion skin powder could be incorporated into bread and bakery products, producing products rich in antioxidant properties yet sensory acceptance (Sagar and Pareek 2020b). Moreover, onion skin extracts exhibited inhibitory solid effects against foodborne pathogens such as *E. coli*, *P. fluorescens*, and *B. cereus* (Škerget et al. 2009). This suggests using onion skin as a natural preservative and antimicrobial agent in food products. Anthocyanins isolated from onion skins can serve as coloring agents for food products and in the formulation of modified atmosphere packaging to preserve oxygen-sensitive products (Elsebaie and Essa 2018; Güner et al. 2021).

On the other hand, culture media prepared from the combination of onion, garlic, brinjal, and pumpkin peels can increase the biomass of microalgae, which can later be utilized in biodiesel production (Sagar et al. 2022). Onion skins obtained from biomass production can undergo biorefinery processes involving lactic acid, anaerobic digestion, ethanol fermentation, and composting, thereby becoming sources for bioplastics, biofuels, biofertilizers, and secondary chemicals (Bains et al. 2023).

References

Akdeniz B (2016) Evaluation of different coating materials for encapsulating phenolic compounds extracted from onion (*Allium cepa*) skin. Middle East Technical University

Albishi T, John JA, Al-Khalifa AS, Shahidi F (2013) Antioxidative phenolic constituents of skins of onion varieties and their activities. J Funct Foods 5:1191–1203

Bains A, Sridhar K, Singh BN et al (2023) Valorization of onion peel waste: from trash to treasure. Chemosphere 343:140178

Benítez V, Mollá E, Martín-Cabrejas MA et al (2011) Characterization of industrial onion wastes (*Allium cepa* L.): dietary fibre and bioactive compounds. Plant Foods Hum Nutr Hum Nutr 66:48–57

Chadorshabi S, Hallaj-Nezhadi S, Ghasempour Z (2022) Red onion skin active ingredients, extraction and biological properties for functional food applications. Food Chem 386:132737

Črnivec IGO, Skrt M, Šeremet D et al (2021) Waste streams in onion production: bioactive compounds, quercetin and use of antimicrobial and antioxidative properties. Waste Manag 126:476–486

Elsebaie EM, Essa RY (2018) Microencapsulation of red onion peel polyphenols fractions by freeze drying technicality and its application in cake. J Food Process Preserv 42:e13654

FAO (2024) Countries by commodity. http://www.fao.org/faostat/en/#rankings/countries_by_commodity. Accessed 30 Mar 2024

Golubkina N, Caruso G (2020) Onion. In: Jaiswal AK (ed) Nutritional composition and antioxidant properties of fruits and vegetables. Academic Press, United States, pp 73–87

Güner S, Boz Z, Yağız Y et al (2021) Investigation of phenolic compounds and antioxidant activity in red and yellow onions and a synergistic utilization of skin extract in modified atmosphere packaging of salmon (Salmo salar). Packag Technol Sci 34:371–382

Jurinjak Tušek A, Šamec D, Šalić A (2022) Modern techniques for flavonoid extraction-to optimize or not to optimize? Appl Sci 12:1–34

Kwak JH, Seo JM, Kim NH et al (2017) Variation of quercetin glycoside derivatives in three onion (*Allium cepa* L.) varieties. Saudi J Biol Sci 24:1387–1391

Lee SU, Lee JH, Choi SH et al (2008) Flavonoid content in fresh, home-processed, and light-exposed onions and in dehydrated commercial onion products. J Agric Food Chem 56:8541–8548

Miean KH, Mohamed S (2001) Flavonoid (myricetin, quercetin, kaempferol, luteolin, and apigenin) content of edible tropical plants. J Agric Food Chem 49:3106–3112

Ncayiyana M, Maboko MM, Bertling I (2018) Alterations in yield, physicochemical components and mineral composition of onion following organic manure and inorganic nitrogen application. Acta Agric Scand Sect B-Soil Plant Sci 68:213–219

Nuutila AM, Puupponen-Pimiä R, Aarni M, Oksman-Caldentey KM (2003) Comparison of antioxidant activities of onion and garlic extracts by inhibition of lipid peroxidation and radical scavenging activity. Food Chem 81:485–493

Petropoulos SA, Ntatsi G, Ferreira ICFR (2017) Long-term storage of onion and the factors that affect its quality: a critical review. Food Rev Int 33:62–83

Sagar NA, Khar A, Vikas, et al. (2021) Physicochemical and thermal characteristics of onion skin from fifteen Indian cultivars for possible food applications. J Food Qual 2021:1–11

Sagar NA, Kumar Y, Singh R et al (2022) Onion waste based-biorefinery for sustainable generation of value-added products. Bioresour Technol 362:127870

Sagar NA, Pareek S (2020a) Antimicrobial assessment of polyphenolic extracts from onion (*Allium cepa* L.) skin of fifteen cultivars by sonication-assisted extraction method. Heliyon 6:e05478

Sagar NA, Pareek S (2020b) Dough rheology, antioxidants, textural, physicochemical characteristics, and sensory quality of pizza base enriched with onion (*Allium cepa* L.) skin powder. Sci Rep 10:18669

Sagar NA, Pareek S, Gonzalez-Aguilar GA (2020) Quantification of flavonoids, total phenols and antioxidant properties of onion skin: a comparative study of fifteen Indian cultivars. J Food Sci Technol 57:2423–2432

Škerget M, Majhenič L, Bezjak M, Knez Ž (2009) Antioxidant, radical scavenging and antimicrobial activities of red onion (*Allium cepa* L) skin and edible part extracts. Chem Biochem Eng Q 23:435–444

Vojvodić Cebin A, Šeremet D, Mandura A et al (2020) Onion solid waste as a potential source of functional food ingredients. Eng Power Bull Croat Acad Eng 15:7–13

Yu M, Yang L, Xue Q et al (2019) Comparison of free, esterified, and insoluble-bound phenolics and their bioactivities in three organs of *Lonicera japonica* and *L. macranthoides*. Molecules 24:1–20

Zhang SL, Peng D, Xu YC et al (2016) Quantification and analysis of anthocyanin and flavonoids compositions, and antioxidant activities in onions with three different colors. J Integr Agric 15:2175–2181

Phytochemicals from Fennel (*Foeniculum vulgare*) By-Products

Mohamed Fawzy Ramadan

Abstract Widespread in the Mediterranean region, fennel (*Foeniculum vulgare*) is a biennial, aromatic, and medicinal plant distinguished by its feathery leaves, umbels of yellow flowers, hollow stem, and scented seeds. Phytochemical profiles of the waste components of fennel (*F. vulgare*) were described for the leaves, bulb, and stems. Several compounds from flavonoid glycosides, flavonoid aglycones, hydroxycinnamic acid derivatives, iridoid derivatives, phenolic acids, and lignans were detected. Additionally, there was relief from the presence of several oxylipins. Numerous metabolites found in *F. vulgare* have biological potential, such as immunomodulatory, antioxidant, and anti-inflammatory traits. This chapter highlighted phytochemicals from fennel (*Foeniculum vulgare*) by-products.

Keywords Leaves · Bulb · Stems · Oxylipins · Flavonoid · Hydroxycinnamic acid · Iridoid · Lignans

1 Introduction

The interest in research on phytochemicals has dramatically increased in recent years, driven by the practical applications of these compounds in various fields. Determining their content in various plants and plant products and their biological activity is crucial, as these compounds are included in various fields of nutrition, health, and pharmaceuticals. This growing interest is evident in the numerous studies on the topic (Crescenzi et al. 2021; Marrelli et al. 2024; Baiano et al. 2023; Politeo et al. 2023a, b; Michailidis et al. 2021; Crescenzi et al. 2022; Barakat et al. 2022; El-Beltagi et al. 2023; Castaldo et al. 2021; Veršić Bratinčević et al. 2023).

M. F. Ramadan (✉)
Department of Clinical Nutrition, Faculty of Applied Medical Sciences, Umm Al-Qura University, Makkah, Saudi Arabia
e-mail: mfhassanien@uqu.edu.sa

M. F. Ramadan (ed.), *Bioactive Phytochemicals in By-products from Bulb, Flower and Fruit Vegetables*, https://doi.org/10.1007/978-3-031-77399-0_4

One of the commercially essential plants is fennel (*Foeniculum vulgare*), which belongs to the family Umbelliferae. Fennel is cultivated as an annual or perennial plant in the Mediterranean, Russia, China, and many other regions. For the human diet, two types of fennel plants are cultivated. Fennel bulbs are a mixture of swollen stems and leaves produced in cold climates (USA, North/South Europe, and Japan) and bulbs produced in mild-winter areas (Italy and some Middle Europe countries). The wild part of the plant is generally known as fennel seeds, and the complete plant (leaves, stems, and fruits) (Jadid et al. 2023; Hosseini et al. 2021; Kirici et al. 2023; Zolotilova et al. 2021; Sanli and Ok 2023; Mohammadi et al. 2023; Bahmani et al. 2024; Gharibvandi et al. 2022; Dheebisha and Vishwanath 2020; Šunić et al. 2023).

Fennel fruits are widely known for their high essential oil content and unique composition, making them very important in manufacturing many edible and pharmaceutical products. There is potential for the use of by-products after the hydrodistillation of fennel fruits (feculae and feathery floral oils) to obtain bioactive compounds, particularly dietary fiber and related phenolics. According to the fennel and anise association research, the essential oil of fennel fruits ranged from 2.6% to 4.4%, and the residual fruit tissues ranged from 0.4% to 10.8% (p/p).

Common fennel by-products include stems and leaves. Studies showed that fennel by-products contain levels of lipids, protein, essential oil, fiber, and ash (Bahmani et al. 2024; Akbari et al. 2024; Elbaz et al. 2024; Machado et al. 2023; Borotová et al. 2021; Elsayed et al. 2022; Šunić et al. 2024; Akbari et al. 2023; Sabzi-Nojadeh et al. 2023; El-Serafy and El-Sheshtawy 2020).

2 Botanical Description of Fennel (*Foeniculum vulgare*)

Considering sexual propagation, fennel plants can be described as a group of annuals or perennials with umbrella-type inflorescences, compound and triangular stems, pinnately dissected glabrous leaves covered by a bluish wax, and bi- or even tripinnatisect. Each inflorescence is formed by 12 to 50 umbrella-type of secondary rays, which, in turn, are formed by 25–75 simple rays. Fennel umbels are about 2–18 cm in diameter, with light yellow florets to intensely yellow ones.

Fennel is considered a sub-giant herbaceous plant ranging from 1.1 to 2 m high. It has paniculate inflorescences that are shaped like an umbrella and produce seeds that can help understand the genetic variability of the plant. This variability appears to be high due to the number of regions across the Mediterranean basin and the USA where the plant is successfully growing, even though it is native to Europe. Fennel can also be called "green fennel" (*Foeniculum vulgare* subsp. vulgare) or "bitter fennel" (*Foeniculum vulgare* subsp. vulgare var. vulgare). Domestically, bitter fennel is widely used during culinary practices. This subspecies mainly grows wild or cultivated in the North of Italy, Hungary, Germany, and France. The seeds present an elongated, flattened seed shape and flattened commissural surface. The seeds have 4–5 primary ridges, and the secondary ridges are variably expressed, giving the seed a typical heteromorphism. The size of fennel seeds depends on the habitat and human intervention, but the shape, color, and aromatic properties usually remain the same (Javed et al. 2020; D'Antuono 2020; Verma and Saxena 2024; Dheebisha and Vishwanath 2020; Zafar et al. 2023; Skaldina 2020; Elbaz et al. 2022; Mohammadi et al. 2023; Hosseini et al. 2021; Bhattarai et al. 2021).

3 Phytochemical Composition of *Foeniculum vulgare* By-Products

Fennel (*Foeniculum vulgare*) is an incredibly versatile and highly regarded plant renowned for its exceptional culinary, ornamental, and medicinal qualities. Within the seeds of fennel reside a diverse array of incredibly valuable compounds, encompassing extraordinary polyphenols, exquisite terpenes, flavonoids, and remarkable volatile oils. Captivatingly, these magnificent seeds serve as the foundation for an extensive range of food additives, beverages, seasonings, and even cosmeceuticals crafted on a grand scale. Consequentially, the demand for exquisite fennel oil and extracts perpetually escalates, making them an indulgence that comes at an exorbitant cost. Consequently, it becomes paramount to diligently extract and explore the life-enhancing elements meticulously lurking within fennel by-products, such as its enchanting leaves, resplendent flowers, mesmerizing inflorescences, and dehydrated fruits. By meticulously scrutinizing and harnessing the economically viable constituents residing within these by-products, a novel realm of prosperity awaits (Yadav et al. 2022; Malin et al. 2022; Repajić et al. 2020; Crescenzi et al. 2021; Noreen et al. 2023a, b; Javed et al. 2020; Urango et al. 2021; Rafieian et al. 2023; Noreen et al. 2023a, b; Zafar et al. 2023).

Currently, the phytochemical composition of fennel by-products is poorly studied. Full characterization of the phytochemical composition of fennel biomass is essential to reveal further potentialities. For example, this data can be exploited in health science studies to understand better the numerous benefits associated with consuming fennel as a source of various bioactive compounds, including possible nutritional interventions to mitigate and prevent the onset of a wide range of diseases. Moreover, having comprehensive knowledge about the polyphenols present in fennel by-products is of utmost importance for extraction purposes, including the development of advanced methods and the utilization of suitable solvents, as it can significantly contribute to the future recycling and sustainable utilization of this abundant waste resource. Although a considerable amount of knowledge and research findings are available for fennel seeds, it is crucial to note that, to the best of our knowledge, no such comprehensive data and insights are available exploring inflorescence extracts and the overall composition of fennel pulp, highlighting the urgency and significance of further investigation and research in these domains (Crescenzi et al. 2021; Baiano et al. 2023; Politeo et al. 2023a, b; Badr 2020;

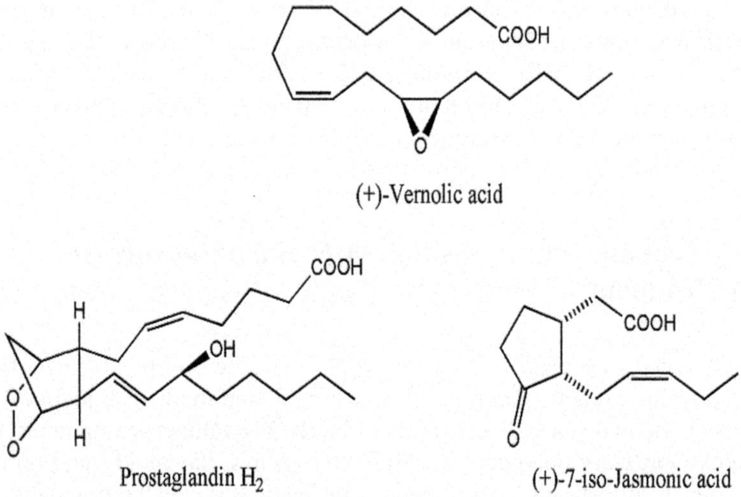

Fig. 1 The structural formulae of some oxylipins

Castaldo et al. 2021; Marrelli et al. 2024; Yadav et al. 2022; Crescenzi et al. 2022; Kraouia et al. 2023; Pencheva et al. 2022).

Crescenzi et al. (2021) characterized of metabolites present in different parts of *F. vulgare* waste. A higher detection of metabolites is obtained through ethanol extraction compared to decoction. The following metabolites have only been reported in hydroalcoholic extracts: glycoside flavonoids, phospholipids, iridoids, phenylpropanoids, and glycolipids. Furthermore, only the decoctions included lignans, oxylipins (Fig. 1), and the flavonoid resveratrol glucoside.

4 Extraction of Phytochemicals from Fennel By-Products

An in-depth investigation into the various extraction techniques, including but not limited to traditional methods, is considered a crucial prerequisite for industries seeking to add value to waste by-products. The extraction of phytochemicals from fennel by-products is of significant interest to researchers. The aim was to attain maximum yields of high-quality phytochemicals that can be utilized in diverse industries such as food, pharmaceuticals, nutraceuticals, cosmetics, and more. Numerous studies have focused on extracting individual phytochemicals from fennel by-products, while others have explored different parts of the plant, such as fruits and aerial parts. These investigations have examined a range of extraction approaches, including physical or mechanical methods, thermal techniques, and solvent-free extraction methods. By delving into the extraction process of the major phytochemicals found in fennel by-products, this discussion aims to provide comprehensive insights and findings for the following variables (Baiano et al. 2023;

Crescenzi et al. 2021; Politeo et al. 2023a, b; Badr 2020; Popović et al. 2024; Kraouia et al. 2023; Castaldo et al. 2021; Yadav et al. 2022).

The extraction technique used for phytochemicals from any plant, part of the plant, by-products, or any substance ensures the maximum recovery of the desired attributes. This process is based on solvent extraction, employing either polar or non-polar solvents for varying durations. The abovementioned techniques encompass a wide range of methods employed in phytochemical extraction, albeit using the same solvent. An alternative approach to extracting gaseous organic compounds, such as essential oils, is through steam distillation. This method allows for the direct extraction of phytochemicals from plants across various fields, showcasing an array of diverse bioactivities (Baiano et al. 2023; Crescenzi et al. 2021; Politeo et al. 2023a, b; Badr 2020; Popović et al. 2024; Kraouia et al. 2023; Castaldo et al. 2021; Yadav et al. 2022).

Additionally, a notably intriguing method involves using a pressurized solvent, specifically CO_2. This technique has garnered substantial interest in recent years due to its mild critical parameters, inertness, abundant availability, ease of removal, and, most notably, its ability to extract solvents without leaving residues. Furthermore, CO_2 can also be employed as an extraction solvent alongside water, alcohols, and/or inert inorganic or organic salts. Overall, these techniques and approaches contribute to the efficient extraction of phytochemicals while maintaining the integrity of the desired compounds (Bitwell et al. 2023; Dheyab et al. 2021; Galali et al. 2020; Tymczewska et al. 2023; Panzella et al. 2020; Aduloju et al. 2023; Sportiello 2024; Calderón-Oliver and Ponce-Alquicira 2021; Frosi et al. 2021; Braga et al. 2022).

5 Health Benefits and Applications of Phytochemicals from Fennel By-Products

Fennel by-products are an excellent source of bioactive compounds, such as dietary fiber, essential oil, fatty acids, macronutrients, carotenoids, amino acids, and phenolic compounds. Growth conditions and environmental factors affect the contents of unidentified carboxylic acids, monosaccharides, monoterpenes, macronutrients, and antioxidant activity. The biological activities of fennel extracts are related to the high amounts of phenolic compounds found in all plant parts. Traditionally, the essential oil of fennel is used as a carminative, a mouth freshener, an antispasmodic, and a digestive aid due to the presence of camphene, fenchone, *p*-cymene, α-terpinene, myrcene, and α-pinene. Administration of fennel powder decreases peroxidation levels and shows an anti-inflammatory effect in heat stress conditions (Baiano et al. 2023; Crescenzi et al. 2021; Castaldo et al. 2021; Politeo et al. 2023a, b; Veršić Bratinčević et al. 2023; Badr 2020; Urango et al. 2021; Yadav et al. 2022; Marrelli et al. 2024).

The main health-beneficial effects of products rich in phenolic compounds are confirmed in vivo, while limited data were obtained in clinical studies. Drinking 20 g dried fennel powder in the form of fennel seed extract in 200 mL hot water during three meals for 8 weeks increased the antioxidant activity of diabetic patients. Incorporating encapsulated fennel into bread makes the bread acceptable to consumers and prolongs its shelf life. By-products and essential oil constituents or their mixtures with an adequate concentration in diet may be used as feed or feed ingredients with a strong antioxidant effect to reduce lipid peroxidation during animal fattening. The dietary inclusion of fennel may improve the oxidative status in meat and milk products, reduce the use of synthetic antioxidants, and provide consumers with healthier and higher-quality foods. In pharmaceutical and cosmetic industries, fennel can be used as a natural source of tyrosinase inhibitors for sunscreens or in treating skin-modeled vitiligo (Yadav et al. 2022; Barakat et al. 2022; Maqbool et al. 2024; Barakat et al. 2023; Akbar and Akbar 2020; Debnath et al. 2023; Mackonochie et al. 2023; Oluwole and Ademuyiwa 2021).

References

Aduloju EI, Yahaya N, Zain NM, Kamaruddin MA, Abd Hamid MA (2023) An overview on the use of DEEP eutectic solvents for green extraction of some selected bioactive compounds from natural matrices. Adv J Chem, Sect A 6(3):253–300

Akbar S, Akbar S (2020) *Trigonella foenum*-graecum L.(Fabaceae/Leguminosae) (Syns.: T. graeca St.-Lag.; T. tibetana (Alef.) Vassilcz.). In: Handbook of 200 medicinal plants: A comprehensive review of their traditional medical uses and scientific justifications, pp 1851–1871

Akbari A, Bahmani K, Kafkas NE, Bilgin OF, Hamijo T, Darbandi AI, Farhadpour M (2024) Evaluation of seed yield, essential oil compositions, and fatty acid profiles in advanced fennel (*Foeniculum vulgare* mill) breeding populations. Biocatal Agric Biotechnol 57:103118

Akbari A, Izadi-Darbandi A, Bahmani K, Farhadpour M, Ebrahimi M, Ramshini H, Esmaeili Z (2023) Assessment of phenolic profile, and antioxidant activity in developed breeding populations of fennel (Foeniculum vulgare Mill). Biocatal Agric Biotechnol 48:102639

Badr AMM (2020) Productive performance of growing rabbits fed diets containing different levels of fennel (*Foenicuium vulagare* L.) Seeds By. Egyptian J Nutr Feed 23:71

Bahmani K, Akbari A, Izadi-Darbandi A, Ghamari T (2024) Phenological traits, seed yield, and essential oil yield of fifty populations of bitter fennel (*Foeniculum vulgare*). Int J Hortic Sci Technol 11(3):391–410

Baiano A, Romaniello R, Giametta F, Fiore A (2023) Optimization of process variables for the sustainable extraction of phenolic compounds from chicory and fennel by-products. Appl Sci 13

Barakat H, Alkabeer IA, Aljutaily T, Almujaydil MS, Algheshairy RM, Alhomaid RM, Almutairi AS, Mohamed A (2022) Phenolics and volatile compounds of fennel (*Foeniculum vulgare*) seeds and their sprouts prevent oxidative DNA damage and ameliorates CCl_4-induced hepatotoxicity and oxidative stress in rats. Antioxidants 11(12):2318

Barakat H, Alkabeer IA, Althwab SA, Alfheeaid HA, Alhomaid RM, Almujaydil MS, Almuziree RS, Bushnaq T, Mohamed A (2023) Nephroprotective effect of fennel (*Foeniculum vulgare*) seeds and their sprouts on CCl_4-induced nephrotoxicity and oxidative stress in rats. Antioxidants 12(2):325

Bhattarai S, Trotter T, Thomson M (2021) Spice technical report-fennel

Bitwell C, Indra SS, Luke C, Kakoma MK (2023) A review of modern and conventional extraction techniques and their applications for extracting phytochemicals from plants, vol 19. Scientific African, p e01585

Borotová P, Galovičová L, Valková V, Ďúranová H, Vuković N, Vukić M, Babošová M, Kačániová M (2021) Biological activity of essential oil from. Acta Horticulturae et Regiotecturae 24(2):148–152

Braga, M. E. M., Seabra, I. J., Dias, A. M. A., & De Sousa, H. C., 2022. Recent trends and perspectives for the extraction of natural products

Calderón-Oliver M, Ponce-Alquicira E (2021) Environmentally friendly techniques and their comparison in the extraction of natural antioxidants from green tea, rosemary, clove, and oregano. Molecules 26

Castaldo L, Izzo L, De Pascale S, Narváez A, Rodriguez-Carrasco Y, Ritieni A (2021) Chemical composition, in vitro bioaccessibility and antioxidant activity of polyphenolic compounds from nutraceutical fennel waste extract. Molecules 26(7):1968

Crescenzi MA, D'Urso G, Piacente S, Montoro P (2021) LC-ESI/LTQOrbitrap/MS Metabolomic analysis of fennel waste (*Foeniculum vulgare* Mill.) as a byproduct rich in bioactive compounds. Food Secur

Crescenzi MA, D'Urso G, Piacente S, Montoro P (2022) UPLC-ESI-QTRAP-MS/MS analysis to quantify bioactive compounds in fennel (*Foeniculum vulgare* mill.) waste with potential anti-inflammatory activity. Meta

D'Antuono LF (2020) Fennel. Carrots and related Apiaceae crops

Debnath S, Kumar H, Sharma A (2023) Foeniculum vulgare from spice to pharma: recent advances in its medicinal value, bioactivities and perspectives. Trad Integrative Med

Dheebisha C, Vishwanath YC (2020) Advances in cultivation of fennel. J Pharmacogn Phytochem 9(2):1295–1300

Dheyab AS, Abu Bakar MF, AlOmar M, Sabran SF, Muhamad Hanafi AF, Mohamad A (2021) Deep eutectic solvents (DESs) as green extraction media of beneficial bioactive phytochemicals. Separations 8(10):176

Elbaz M, Abdesslem SB, Moussa OB, Boulares M, Timoumi M, Hassouna M (2022) Value adding search among a selection of Tunisian fennel (*Foenicuium vulagare* Mill.) cultivars: diversity assessment and selection among a local fennel germplasm. Journal of oasis agriculture and. Sustain Dev 4(2):93–103

Elbaz M, Abdesslem SB, St-Gelais A, Boulares M, Moussa OB, Timoumi M, Hassouna M, Aider M (2024) Essential oils profile, antioxidant and antibacterial potency of Tunisian fennel (*Foenicuium vulagare* Mill.) leaves grown under conventional and organic conditions. Food Chem Adv 4:100734

El-Beltagi HS, Nada RS, Mady E, Ashmawi AE, Gashash EA, Elateeq AA, Suliman AA, Al-Harbi NA, Al-Qahtani SM, Zarad MM, Randhir TO (2023) Effect of organic and bio-fertilization on fruit yield, bioactive constituents, and estragole content in fennel fruits. Agronomy 13(5):1189

Elsayed AA, El-Gohary AE, Khalid KA, Ahmed A (2022) Changes in bitter fennel essential oils exposed to foliar spray with l-phenylalanine. Egypt J Bot 62(1):241–253

El-Serafy RS, El-Sheshtawy AA (2020) Effect of nitrogen fixing bacteria and moringa leaf extract on fruit yield, estragole content and total phenols of organic fennel. Scientia Horticulturae 265:109209

Frosi I, Montagna I, Colombo R, Milanese C, Papetti A (2021) Recovery of chlorogenic acids from Agri-food wastes: updates on green extraction techniques. Molecules 26

Galali Y, Omar ZA, Sajadi SM (2020) Biologically active components in by-products of food processing, vol 8. Food Sci Nutr, p 3004

Gharibvandi A, Karimmojeni H, Ehsanzadeh P, Rahimmalek M, Mastinu A (2022) Weed management by allelopathic activity of *Foeniculum vulgare* essential oil. Plant Biosyst 156(6):1298–1306

Hosseini E, Majidi MM, Ehtemam MH, Ghanadian M, Huyghe C (2021) Variation in a world-wide collection of fennel (*Foeniculum vulgare* var. vulgare). Crop and pasture. Science 72(12):1008–1021

Jadid N, Widodo AF, Ermavitalini D, Sa'adah NN, Gunawan S, Nisa C (2023) The medicinal Umbelliferae plant fennel (*Foeniculum vulgare* Mill.): cultivation, traditional uses, phytopharmacological properties, and application in animal husbandry. Arab J Chem 16(3):104541

Javed R, Hanif MA, Ayub MA, Rehman R (2020) Fennel. Medicinal plants of South Asia

Kirici S, Bayram E, Inan M, Kaya DA, Özel A (2023) Traditional farming of Apiaceae species in Türkiye: *Pimpinella anisum* L., *Foeniculum vulgare* Mill., *Cuminum cyminum* L. In: Medicinal and aromatic plants of Turkey. Springer International Publishing, Cham, pp 241–281

Kraouia M, Nartea A, Maoloni A, Osimani A, Garofalo C, Fanesi B, Ismaiel L, Aquilanti L, Pacetti D (2023) Sea fennel (*Crithmum maritimum* L.) as an emerging crop for the manufacturing of innovative foods and nutraceuticals. Molecules 28(12):4741

Machado AM, Lopes V, Barata AM, Póvoa O, Farinha N, Figueiredo AC (2023) Chemical variability of the essential oils from two Portuguese Apiaceae: *Coriandrum sativum* L. and Foeniculum vulgare Mill. Plants 12(14):2749

Mackonochie M, Rodriguez-Mateos A, Mills S, Rolfe V (2023) A scoping review of the clinical evidence for the health benefits of culinary doses of herbs and spices for the prevention and treatment of metabolic syndrome. Nutrients 15

Malin V, Elez Garofulić I, Repajić M, Zorić Z, Pedisić S, Sterniša M, Smole Možina S, Dragović-Uzelac V (2022) Phenolic characterization and bioactivity of fennel seed (*Foeniculum vulgare* mill.) extracts isolated by microwave-assisted and conventional extraction. PRO 10(3):510

Maqbool M, Chaudhary K, Chalotra R, Chauhan S, Dahiya RS (2024) Phyto-pharmacology of Most common Indian culinary spices and their potential in developing new pharmaceutical therapies. Curr Tradit Med 10(2):117–140

Marrelli M, Lupia C, Argentieri MP, Bava R, Castagna F, Cozza N, Mollace V, Palma E, Statti G (2024) *Foeniculum vulgare* Mill. Aerial parts (Italian 'Finocchio di Isola capo Rizzuto'PGI): valorization of Agri-food waste as a potential source of lipase inhibitors and antioxidants. Resources 13(7):89

Michailidis D, Angelis A, Nikolaou PE, Mitakou S, Skaltsounis AL (2021) Exploitation of *Vitis vinifera, Foeniculum vulgare, Cannabis sativa* and *Punica granatum* by-product seeds as dermo-cosmetic agents. Molecules 26(3):731

Mohammadi M, Pouryousef M, Farhang N (2023) Study on germination and seedling growth of various ecotypes of fennel (*Foeniculum vulgare* Mill.) under salinity stress. J Appl Res Med Aromat Plants 34:100481

Noreen S, Tufail T, Badar Ul Ain H, Awuchi CG (2023a) Pharmacological, nutraceutical, functional and therapeutic properties of fennel (*Foeniculum vulgare*). Int J Food Prop 26(1):915–927

Noreen S, Tufail T, Bader Ul Ain H, Ali A, Aadil RM, Nemat A, Manzoor MF (2023b) Antioxidant activity and phytochemical analysis of fennel seeds and flaxseed. Food Sci Nutr 11(3):1309–1317

Oluwole OB, Ademuyiwa O (2021) Antioxidants in spices: a review of the antioxidant components and properties of some common African spices and their role in human nutrition and plant. In: Antioxidants in plant-microbe interaction

Panzella L, Moccia F, Nasti R, Marzorati S, Verotta L, Napolitano A (2020) Bioactive phenolic compounds from Agri-food wastes: an update on green and sustainable extraction methodologies. Front Nutr 7:60

Pencheva M, Petkova Z, Dincheva I, Kostova I, Damyanova S, Stoyanova A Gaceu L (2022) Phytochemical and biological profiles of fennel fruits (*Foeniculum vulgare* Mill. var. dulce Mill.)

Politeo O, Popović M, Veršić Bratinčević M, Koceić P, Ninčević Runjić T, Mekinić IG (2023b) Conventional vs. microwave-assisted Hydrodistillation: influence on the chemistry of sea fennel essential oil and its by-products. Plan Theory 12(7):1466

Politeo O, Popović M, Veršić Bratinčević M, Kovačević K, Urlić B, Generalić Mekinić I (2023a) Chemical profiling of sea fennel (*Crithmum maritimum* L., Apiaceae) essential oils and their isolation residual waste-waters. Plan Theory 12(1):214

Popović M, Radman S, Generalić Mekinić I, Ninčević Runjić T, Urlić B, Veršić Bratinčević M (2024) A year in the life of sea fennel: annual phytochemical variations of major bioactive secondary metabolites. Appl Sci 14(8):3440

Rafieian F, Amani R, Rezaei A, Karaça AC, Jafari SM (2023) Exploring fennel (Foeniculum vulgare): composition, functional properties, potential health benefits, and safety. Crit Rev Food Sci Nutr 64:1–18

Repajić M, Ekić S, Kruk V, Dragović-Uzelac V (2020) Effect of accelerated solvent extraction conditions on the isolation of bioactive compounds from fennel (*Foeniculum vulgare* Mill.) seeds. Hrvatski časopis za prehrambenu tehnologiju, biotehnologiju i nutricionizam 15(3–4):102–106

Sabzi-Nojadeh M, Aharizad S, Mohammadi SA, Amani M (2023) Screening of several important compounds production in fennel (*Foeniculum vulgare* Mill.) populations. J Med Plants 22(85):98–112

Sanli A, Ok FZ (2023) Determination of optimal harvesting time for essential oil and estragole yield in bitter fennel (*Foeniculum vulgare* Mill.) growing in culture conditions. S Afr J Bot 155:98

Skaldina O (2020) Insects associated with sweet fennel: beneficial visitors attracted by a generalist plant. Arthropod Plant Interact 14:399

Sportiello, L., 2024. Optimization of the of the extraction techniques using natural hydrophobic deep eutectic solvents for the recovery of biomolecules from food and food industry by-products

Šunić L, Ilić ZS, Stanojević L, Milenković L, Lalević D, Stanojević J, Milenković A, Cvetković D (2024) Essential oil content, composition and free radical scavenging activity from different plant parts of Wild Sea fennel (*Crithmum maritimum* L.) in Montenegro. Plants 13(14):2003

Šunić L, Ilić ZS, Stanojević L, Milenković L, Stanojević J, Kovač R, Milenković A, Cvetković D (2023) Comparison of the essential oil content, constituents and antioxidant activity from different plant parts during development stages of wild fennel (*Foeniculum vulgare* Mill.). Horticulturae 9(3):364

Tymczewska A, Klebba J, Szydłowska-Czerniak A (2023) Antioxidant capacity and total phenolic content of spice extracts obtained by ultrasound-assisted extraction using deep eutectic and conventional solvents. Appl Sci 13

Urango ACM, Strieder MM, Silva EK, Meireles MAA (2021) Thermosonication process design for recovering bioactive compounds from fennel: A comparative study with conventional extraction techniques. Appl Sci 11

Verma AK, Saxena SN (2024) Fennel. In: Handbook of spices in India: 75 years of research and development. Springer Nature Singapore, Singapore, pp 2349–2383

Veršić Bratinčević M, Kovačić R, Popović M, Radman S, Generalić Mekinić I (2023) Comparison of conventional and green extraction techniques for the isolation of phenolic antioxidants from sea fennel. PRO 11(7):2172

Yadav U, Kaushik I, Khurana S, Prakash K (2022) Bioactive compounds in fennel seed. In: Spice bioactive compounds. CRC Press, pp 315–338

Zafar S, Khan MK, Perveen S, Iqbal M, Al-Huqail AA (2023) Fennel. In: Essentials of medicinal and aromatic crops. Springer International Publishing, Cham, pp 483–514

Zolotilova OM, Nevkrytaya NV, Zolotilov WA, Ametova ED, Scipor OB, Kravchenko GD (2021) Analysis of the *Foeniculum vulgare* Mill. Collection by the complex of features in the conditions of the Crimea foothills

Part III
Phytochemicals from Flower Vegetable By-products

Phytochemicals from Artichoke (*Cynara carduncolus* var. *scolymus* L.) By-Products

Raffaella Colombo, Giulia Moretto, Chiara Milanese, and Adele Papetti

Abstract Globe artichoke (*Cynara cardunculus* var. *scolymus* L.) belongs to the *Asteraceae* family and represents one of the most diffused vegetables worldwide. Its high production leads to a high amount (almost 60%) of by-products, mainly external bracts, leaves, and stems. Over twenty years ago, artichoke by-products were used primarily as animal feedstuff or compost. Then, the research has focused on extracting nutrients and bioactive compounds, such as carbohydrates (fructooligosaccharides and inulin), vitamins, polyphenols, and terpenes. Over the years, these bioactive compounds received considerable attention for their important role in the reduction of the risk of developing different diseases thanks to their many potential bioactivities: antioxidant, dyspeptic, antimicrobial, hypolipidemic, anticarcinogenic, and anti-glycative. Therefore, artichoke by-product extracts represent promising ingredients for healthy purposes in foodstuffs, food supplements, functional foods, and pharmaceutical and cosmetic preparations because of their biological and/or functional properties. In addition, globe artichoke wastes can be helpful for non-health applications to produce biofuels, polymers, and paper pulps. This chapter discusses all the economic and environmental aspects of the recovery of artichoke by-products, considering their chemical composition and valorization in food and non-food applications.

Keywords Globe artichoke · Artichoke by-products · Circular economy · Environmental sustainability · Phytochemical analysis · Bioactive compounds · Health-promoting effect · Polymers · Bioenergy · Biofuels · Bioethanol

---

R. Colombo · G. Moretto
Department of Drug Sciences, University of Pavia, Pavia, Italy

C. Milanese
Pavia Hydrogen Lab, Department of Chemistry & C.S.G.I, University of Pavia, Pavia, Italy

A. Papetti (✉)
Department of Drug Sciences & C.S.G.I, University of Pavia, Pavia, Italy
e-mail: adele.papetti@unipv.it

M. F. Ramadan (ed.), *Bioactive Phytochemicals in By-products from Bulb, Flower and Fruit Vegetables*, https://doi.org/10.1007/978-3-031-77399-0_5

Abbreviations

a_w	water activity
ABTS	2,2′-azino-bis(3-ethylbenzothiazoline-6-sulfonic acid
AGEs	advanced glycation end products
BHT	butylated hydroxytoluene
CCL2	C-C Motif Chemokine Ligand 2
COX-2	cyclo-oxygenase-2
CQAs	mono-caffeoylquinic acids
CRP	C-reactive protein
DAH	dilute acid hydrolysis
diCQAs	di-caffeoylquinic acids
DM	degree of methoxylation
dm	dry matter
DP	degree of polymerization
DPPH	2,2-Diphenyl-1-picrylhydrazyl
FOS	fructooligosaccharides
FRAP	ferric reducing antioxidant power
GHG	greenhouse gasses
GalA	galacturonic acid
HHV	higher heating value
IL-6	interleukin-6
iNOS	inducible nitric oxide synthase
LDL	low-density lipoprotein
LPS	lipopolysaccharides
M-SHIME®	Mucosal Simulator of the Human Intestinal Microbial Ecosystem
MAE	microwave-assisted extraction
MCP	monocyte chemoattractant protein
MGO	methylglyoxal
MUFAs	monounsaturated fatty acids
NADES	natural deep eutectic solvents
NO	nitric oxide
OHC	oil holding capacity
PEF	pulsed electric field
PLE	pressurized-liquid extraction
PHWE	pressurized-hot water extraction
PN	polymorphonuclear
PUFAs	polyunsaturated fatty acids
SCFAs	short-chain fatty acids
SHIME®	Simulator of the Human Intestinal Microbial Ecosystem
SSF	simultaneous saccharification and fermentation
TGs	triglycerides
TPC	total phenolic content
UAE	ultrasound-assisted extraction
WHC	water holding capacity
5-CQA	5-O-Caffeoylquinic acid

1 Introduction

Asteraceae (formerly Compositae) is considered the largest family of flowering plants with more than 32,000 species, classified in 1,911 genera and 13 subfamilies. *Cynara cardunculus* L. is a genus native to the Mediterranean basin, commonly known as cardoon or wild artichoke, including three botanical varieties, of which the cultivated globe artichoke (var. *Cynara scolymus* L.) is one of the oldest *Asteraceae* cultivated plants (Wiklund 1992; Sonnante et al. 2007).

The globe artichoke production amounts to about 1.50 Mt. according to the Food and Agriculture Organization, and it is widely consumed in the Mediterranean diet, with Italy, Spain, and France being the leading producers. Therefore, for these countries, the issue of artichoke waste disposal is more challenging (FAOSTAT 2022).

In general, the environmental and economic impact of waste disposal is a big issue that has encouraged recovery policies, and circular economy plans, focusing on the objectives of the United Nations 2030 Agenda on food loss and waste reduction (United Nations 2015). FAO estimated that 1.3 BMT of edible food is lost or wasted yearly (2020 data), implying a worldwide cost of $750 billion annually (FAOSTAT 2015). In addition, the agri-food sector generates around 190 MT of by-products yearly, including raw materials, leaves, seeds, brans, pomaces, meals, and cakes (Stahel 2016). Therefore, only a reduction of food loss and waste, which is growing concerning the world's population, that probably will reach 9.8 B by 2050, could have a substantial impact on guaranteeing food security and environmental protection in terms of air and water pollution, and soil erosion (Elimelech et al. 2018). The EU strategy requires operations to obtain an environmentally friendly, resource-efficient, and competitive economy to optimize sustainability and waste disposal costs (The World Bank 2018; European Commission 2020). The most recent trends in the circular economy involve the transformation of biomass waste into bioenergy or biocomposite materials. Nowadays, Italy is the world's leading producer of artichoke (390 Kt/year, 26% of the world production, with 38.500 ha of cultivated land) and, therefore, the leading country in the waste reduction and/or valorization practices (FAO 2022; Rana et al. 2023).

The artichoke edible part consists of a head (flower) and partial stem, while the by-products represent about 80–85% (w/w fresh weight) and are mainly external bracts and stem, but also residual leaves, stalks, and roots, which are generally considered as biomass waste (Pandino et al. 2013; Gominho et al. 2018; Francavilla et al. 2021) (Fig. 1). Both edible part and by-products are rich in bioactive compounds with antioxidant properties, such as polyphenols (caffeic acid, mono- and di-caffeoylquinic acids, caffeoylglucoside derivatives, and flavonoids), and in fiber; in addition, the edible artichoke contains micronutrients (vitamins and minerals), while by-products have been mainly investigated for their content in terpenes and sterols (Pandino et al. 2011; Salem et al. 2015; Mejri et al. 2020; Borsini et al. 2021).

Only 15–30% of artichoke wastes, consisting of bracts and stems, can be valorized for nutraceutical applications, while the remaining parts (70–85%) are residual solid wastes from cultivation (leaves, stalks, and roots) (Gominho et al. 2018)

Fig. 1 Globe artichoke by-products

containing carbon, ash, proteins, nitrogen, carbohydrates (cellulose, hemicellulose, inulin), and lignin (Francavilla et al. 2021), that find few applications in literature in this field, for example, the extraction of caffeoylquinic acids (Pandino et al. 2013) and inulin (Castellino et al. 2020). However, they can be valorized for the production of bioenergy.

The relatively low expensive artichoke by-products (bracts and stems) processing has a positive economic impact because of the easy isolation of phenolic compounds (mainly hydroxycinnamic acids and flavonoids) (Carpentieri et al. 2022), in addition to their use in the last 30–40 years in animal feedstuff and fiber production (Boubaker et al. 2016; Gerschenson et al. 2020; Salami et al. 2019; Domingo et al. 2021; Zaker-Esteghamati et al. 2021). Concerning these last applications, over the years, the direct use of these vegetable food wastes raised an important issue about their safety application for animal feed due to their fast degradation (San Martin et al. 2018).

However, in the last decades, the recovery of artichoke by-products became even more mandatory for the massive amount of produced waste. They are rich in substances with high nutritional values and health-promoting properties (antioxidant, anti-inflammatory, anti-glycative, and prebiotic), thus representing potential functional ingredients suitable to produce enriched foods or food supplements. In addition, they also contain sources of biofuels and polymers that can be used in food and non-food applications (Pandino et al. 2013; Zayed and Farag 2020; Feiden et al. 2023).

Considering the economic aspects of recycling such by-products, the choice of suitable biotechnological processes (i.e., fermentations), analytical procedures and techniques become mandatory to reduce time and save reagents consumption for obtaining high yields. It is fundamental to set-up cheap recovery methodologies and suitable reuse policies to treat high waste disposal (Papargyropoulou et al. 2014). In this framework, the need to overcome the traditional technology limitations, such as environmental and economic impact, and to develop a global circular economy model led to the development of green methodologies as ideal procedures to optimize the extraction of bioactive compounds (Kollia et al. 2017; Stumpf et al. 2020; Frosi et al. 2021; Pagano et al. 2021).

Among the green technologies, the use of green solvents (water, methanol, ethanol, or natural deep eutectic solvents—NADES) (Francavilla et al. 2021; Ozkan 2024) and the application of advanced extraction techniques, such as microwave-assisted extraction (MAE) (Mena-García et al. 2020; Francavilla et al. 2021), pressurized-liquid extraction (PLE), or pressurized-hot water extraction (PHWE) (Pagano et al. 2018; López-Salas et al. 2021; Pagliari et al. 2022;), and ultrasound-assisted extraction (UAE) (Punzi et al. 2014; Rabelo et al. 2016; Saleh et al. 2016; Stumpf et al. 2020; Reche et al. 2021; Ozkan 2024) have become widespread. Recently, more advanced methodologies have also been proposed investigating electrical treatments such as pulsed electric field (PEF)-assisted solid-liquid extraction, which is particularly useful for extracting intracellular compounds from artichoke stems (Carpentieri et al. 2022).

The technology implementation is the first step to increase and speed up the extraction processes and to make them potentially scalable at the industrial level with a great economic impact in terms of efficiency and high yields (Frosi et al. 2021; Holgado et al. 2021; Chilakamarry et al. 2022). Optimizing the extraction process variables promotes extraction efficiency in terms of recovery, solvent consumption, and selectivity, decreasing the total extraction cost with an excellent advantage for sustainability and circular economy (Zayed and Farag 2020). In addition, some methodologies also positively impact bioactivity improvement, thus obtaining better results at biological levels (Punzi et al. 2014).

All these procedures aim to maximize the reuse of globe artichoke by-products as a source of health-beneficial or non-health compounds.

This chapter will discuss the chemical composition of main (bracts and stems) and minor (residual leaves, stalks, and roots) artichokes by-products with their many biological activities correlated to each specific group of compounds. Each bioactivity will be commented explaining the potential mechanisms involved in the health effects. The biological and functional properties of artichokes by-products will be treated by considering food, non-food (pharmaceutical and cosmetic uses), and other applications (mainly biofuels and polymer productions).

2 Artichoke By-Products Chemical Composition

The chemical composition of *C. cardunculus* L. slightly varies according to the different varieties (Fig. 2). The main constituents of artichoke edible parts (leaves and heads) are phenolic acids (caffeoylquinic, di-caffeoylquinic, and hydroxycinnamic acids), flavones (apigenin, luteolin and their glycosides), anthocyanins (cyanidin 3,5-*O*-diglucoside, cyanidin 3-*O*-glucoside, cyanidin 3,5-*O*-malonyldiglucoside, cyanidin 3-*O*-(3″-malonyl) glucoside, and cyanidin 3-*O*-(6″-malonyl-glucoside)),

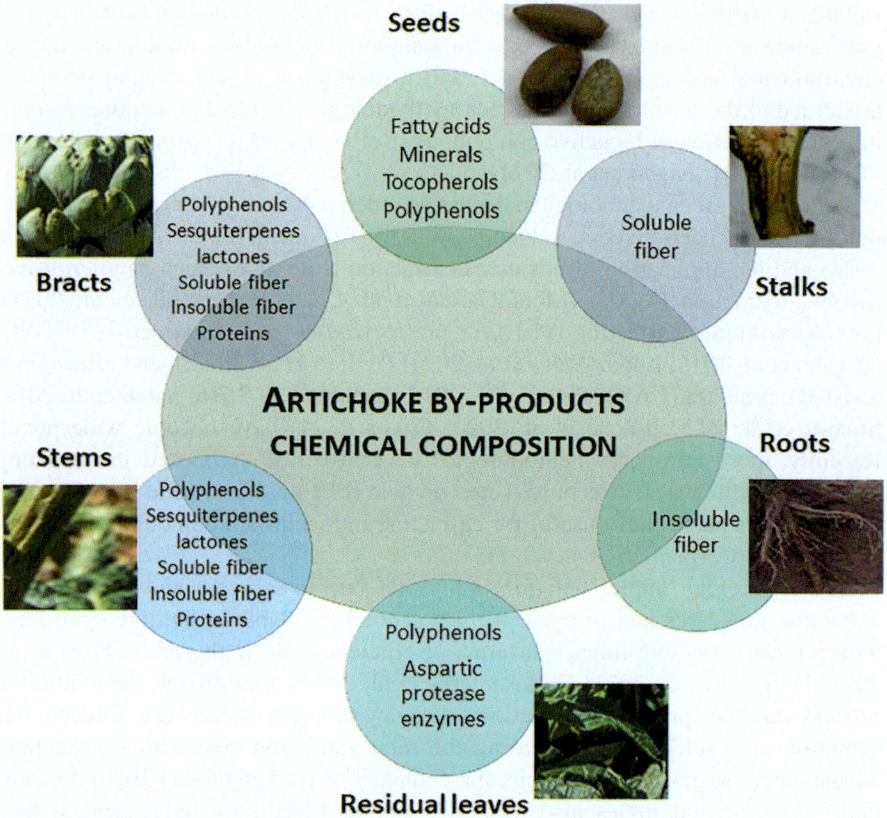

Fig. 2 Chemical composition of each artichoke by-product

terpenoids (mono-, sesqui-, and triterpenes), saturated and unsaturated fatty acids, aspartic endopeptidases (cardosins and cynarases), peroxidases, vitamins (E, C, B3, B6, B8, and B9), fiber (inulin and pectin), and minerals (mainly K, P, Ca, Mg, Na, Zn, Fe, Cr, and Mn). Over the years, this composition has been related to the antioxidant properties (mainly for polyphenols and vitamins E and C content, and to a lesser extent to carotenoids) which have many implications on other biological effects, such as the modulation of lipid metabolism, liver protection, risk reduction of cancer development, and age-related disorders (Lattanzio et al. 2009; Biel et al. 2020; Noriega-Rodríguez et al. 2020; Shallan et al. 2020; Feiden et al. 2023).

Among foods with the best antioxidant properties, artichoke is ranked forth, and regarding artichoke wastes, bracts and stems are the main source of polyphenols and dietary fiber (Jiménez-Moreno et al. 2019; Carpentieri et al. 2022). In particular, bracts located closer to the artichoke heart (inner bracts) are the richest in bioactive compounds, mainly polyphenols, proteins, and inulin (Ruiz-Cano et al. 2014). Bracts and stems are particularly rich in caffeoylquinic acids (CQAs) and flavones (luteolin and apigenin), which are the main compounds responsible for artichoke by-products healthy properties (de Falco et al. 2015; Colantuono et al. 2017). Their qualitative and quantitative profile strictly depends on the extraction process, affecting the corresponding bioactivity, mainly the antioxidant one (Llorach et al. 2002; Rejeb et al. 2020).

The total phenolic content (TPC) of bracts and stems ranges from 8500 mg/kg to 12,700 mg/kg dry matter (dm) depending on different clones and landraces (Pandino et al. 2013), thus highlighting the need to specifically select the most suitable by-products to maximize the recovery of bioactives, in particular at the industrial process level (Pandino et al. 2013; Rejeb et al. 2020). Mono- and di-caffeoylquinic acids (CQAs and di-CQAs) are the most abundant artichoke phenolic compounds (1–8%, w/w dm) and their content is related to the physiological stage of tissues, with the content in young tissues higher than in the senescent ones. In particular, the most abundant compounds both in bracts and stems are 5-*O*-CQA (or chlorogenic acid), 1,5-*O*-di-CQA (Pandino et al. 2013; Pagano et al. 2016), luteolin-7-*O*-rutinoside, and luteolin-7-*O*-glucoside. Flavones constitute less than 10% of TPC (Lattanzio et al. 2009; de Falco et al. 2015). Seeds mainly contain 5-*O*-CQA (22.5–35.8 mg/100 g dm) and 3,5-*O*-di-CQA (197–418 mg/100 g dm), while tocopherols are present in very low concentrations (1.83–4.46 mg/100 g dm) (Petropoulos et al. 2019).

Bracts and stems are also rich in proteins (about 10–15% dm) (Ruiz-Cano et al. 2014). In particular, leaves contain aspartic protease enzymes (cardosins and cyprosins) valuable as rennet for their coagulant properties (Silva and Malcata 2005; Esposito et al. 2016; Bravo Bolívar et al. 2023). Differently, stems contain condensed tannins (about 310–540 µg of catechin equivalent/g dm extract), which are extracted in higher amounts using butanol (Mejri et al. 2020) and peroxidases (recovery yield 7.21%) which are enzymes involved in many metabolic processes with different applications, such as food additives, packaging material component, or diagnostic and medical devices component (Mejiri et al. 2019).

In artichoke wastes (mainly bracts and stems), the insoluble (cellulose, hemicellulose, and lignin) and soluble fiber (inulin, pectin, gums, and β-glucans) contents are approximately around 35 g/100 g and 20 g/100 g dm, respectively (Grigelmo-Miguel and Martín-Belloso 1999; Vaz et al. 2022). In details, bracts and stalks are mainly composed of cellulose (38.5–49.3%, w/w dm), while hemicellulose represents about 20%, w/w dm in bracts, stalks, branches, and puppi, and stalks contain mainly lignin (7.5–13.2%, w/w dm) (Fernández et al. 2006). Among soluble fiber, inulin represents 19–32% w/w dm of artichoke edible part, and notwithstanding its content in bracts is generally lower than 30% but richer than external leaves, these wastes represent a promising source of soluble fiber with interesting sensorial, rheological, and stabilizing properties (Lattanzio et al. 2009; Pagano et al. 2016; Ruiz-Cano et al. 2014). In addition, their inulin content and polymerization degree are strictly related to seasonal and storage conditions (mainly temperature); therefore, careful work is required to preserve the optimal conditions to maximize the inulin amount and its prebiotic properties. Low storage temperatures (3–7 °C) preserve inulin content for a longer period (up to 3 days) than room temperature (only 1 day

(Cavini et al. 2020). Artichoke leaves contain 125 g/kg of total dietary fiber, of which almost 96% is represented by insoluble fibers (Colantuono et al. 2017).

Inulin is also present in roots with a concentration-dependent on cultivars and locations ranging from 6.5 to 20.9 g/100 g dm, with the highest extraction yield obtained using UAE. In addition, this method is time- and cost-saving concerning traditional hot water extraction because of cavitation and mass transfer processes (Raccuia and Melilli 2010; Castellino et al. 2020; Difonzo et al. 2022; El-Kholy et al. 2023).

It is essential to highlight that the prebiotic effect of the soluble fiber on *Lactobacillus plantarum* and *Bifidobacterium bifidum* growth is strictly correlated to the high inulin content and low pectin methoxylation degree, as particularly evident in stems. A low methyl esterification of pectin causes faster fermentation by the intestinal microflora. Pectin content highly changes (0.8–44.3%, expressed as the percentage of neutral sugars) depending on the by-product considered and generally, a rapid heating pre-treatment (70 °C, 5 min) before the extraction doubled the yield (about 10% for bracts and 20% for stems), also increasing the extracted inulin amount and favoring the low degree of methyl esterification (Fissore et al. 2014, 2015).

Seeds represent a small portion of artichoke by-products; however, they have a significant nutritional value due to their crude protein content (25.7–30.4 g/100 g dm) and carbohydrates (44.0–52.2 g/100 g dm). In addition, they are a good source of fats (17.3–23.7 g/100 g dm), of which monounsaturated fatty acids (MUFAs) and polyunsaturated fatty acids (PUFAs) represent 13.08–19.71% and 60.43–70.64%, respectively (oleic and palmitic acids are the main representative of MUFAs, while linoleic acid of PUFAs). Considering minerals composition, K (493–880 mg/100 g dm), Ca (734–1583 mg/100 g dm), Mg (241–809 mg/100 g dm), and Fe (9.2–16 mg/100 g dm) are present in higher concentration than Na (12–24 mg/100 g dm), Zn (1.4–5.2 mg/100 g dm), and Mn (4.21–6.2 mg/100 g dm) (Petropoulos et al. 2019).

Sesquiterpene lactones, such as cynaropicrin, grosheimin, 11β,13-dihydrocynaropicrin, 3β-hydroxy-8α-[(S)-3-hydroxy-2-methylpropionyloxy] guaia-4(15),10(14),11(13)-trien1α,5α,6βH-12,6-olide, 3β-hydroxy-8α-[2-methoxymethyl-2-propenoyloxy]guaia-4(15),10(14),11(13)-trien-1α,5α,6βH-12,6-olide, and deacylcynaropicrin are responsible for the particular artichoke taste and represent another abundant class of compounds present mainly in the external leaves and only in low percentage in stems. Their presence adds to artichoke by-products other important bioactivities primarily related to gastrointestinal digestion, motility, and different intracellular signaling pathways (Colantuono et al. 2018; Matsumoto et al. 2021). In addition, recently, a new (−)-α-bisabolol synthase, able to produce enantiopure (−)-α-bisabolol (a monocyclic sesquiterpene alcohol) has been identified in globe artichoke. This enzyme could be promising in engineered bacteria for producing (−)-α-bisabolol, as already reported in the literature (Lim et al. 2021).

Considering the nutrient and non-nutrient composition of the different by-products, optimizing the different extraction parameters according to the chemical class of the compound is mandatory to obtain high yields. Heat and modifier percentages are very important parameters in the extraction procedures, as they affect the qualitative and quantitative molecular profiles, also causing potential degradation or formation of new compounds (Borsini et al. 2021; Frosi et al. 2021). Polyphenols, proteins, and inulin concentration are strictly related to thermal treatments, differently from fat and fiber content. For example, heating up to 60–70 ° C can cause the loss of soluble proteins but increase the total phenolic acids, flavonoids, and inulin content, probably due to inactivation of oxidases and peroxidases, thus preserving them from oxidation and degradation (Ruiz-Cano et al. 2014) and, especially for CQAs this extraction temperature increases the water diffusivity and therefore the phenolic compounds solubility (Maietta et al. 2017). Similarly, an optimized ethanol percentage (50–75%) allows high yields of phenolic acids, flavonoids, and tannins, mainly from artichoke bracts and leaves (Salem et al. 2017; Mena-García et al. 2020).

Conversely, a higher ethanol percentage (96%) can decrease TPC, as recently demonstrated for bracts and stems. The drying temperature moderately affects the drying rate, while the effective diffusivity and mass transfer coefficient are strongly influenced. This is a key point to be considered to speed up the extraction procedure at the industrial level (Borsini et al. 2021).

Conversely, a high ethanol percentage (85–100%) can be positively applied coupled with medium-high temperature (> 60 °C) by using PLE. This is an accelerated solvent extraction performed at high pressures (around 100 atm), and its parameters can maximize the polyphenol yields also in the presence of high volumes of ethanol (López-Salas et al. 2021). Similarly, the application of PHWE using low ethanol percentage (≤10%) in aqueous solutions allows an optimized CQAs extraction from bracts and external leaves, working on the set-up of temperature and numbers of extraction cycles (Pagano et al. 2018). The use of ultrasound-assisted extraction needs the optimization of time and ultrasonic power to recover phenolic compounds better, as demonstrated for bracts for which 60 min and 240 W resulted in better parameters to obtain promising extracts to be used as food additives (Punzi et al. 2014; Rabelo et al. 2016; Pasqualone et al. 2017).

3 Biological and Functional Properties of Artichoke By-Products Bioactive Compounds

As mentioned above, artichoke by-products are characterized by a high content of secondary metabolites, mainly CQA derivatives and soluble dietary fiber, known for their beneficial properties. Therefore, several authors studied the biological properties of both artichoke by-product extracts and isolated bioactive compounds, which mainly present antioxidant, anti-inflammatory, prebiotic, antimicrobial, and antiproliferative activity, and also found potential applications in treating metabolic disorders. Bioactivities are strictly related to compounds' bioaccessibility and bioavailability. Therefore, in vitro, simulated gastrointestinal digestion processes are increasingly used today to investigate the metabolic fate of potentially active components. Different procedures have been set up by using static or dynamic models and mimicking the digestion conditions with appropriate electrolytes and enzyme mixtures (Kong and Singh 2010; Mazzei et al. 2011; Wickham et al. 2012; Colombo et al. 2021; Colombo et al. 2023). In particular, static approaches are widely used for their simplicity and low cost, and today, a standardized protocol following the work of COST Action Infogest network and further updates is widely used (Minekus et al. 2014; Egger et al. 2016; Brodkorb et al. 2019; Menard et al. 2023).

Conversely, the application of the most recent miniaturized dynamic systems remains less explored because they are very difficult to harmonize and standardize, even if they could represent promising platforms more predictive of molecule fate in vivo (Dupont et al. 2019). As concerns 5-*O*-CQA in artichoke bracts and stems, it has an attractive anti-glycative capacity in in vitro tests (see paragraph 3.1.4), which is strongly affected by the digestion simulation both in static and dynamic systems as evident by its activity loss (Maietta et al. 2017; Colombo et al. 2023). In addition to the biological activities, the rheological properties, color parameters, and oxidative effects of artichoke by-products have been studied to investigate the potential use of such by-products as additives in food or non-food (pharmaceutical and cosmetic industries) applications.

3.1 Biological Properties

3.1.1 Antioxidant Activity

Free radicals are highly reactive species capable of damaging biological systems because of oxidative reactions, which are neutralized by endogenous antioxidant defense systems. The imbalance between antioxidants and free radicals generates oxidative stress, a risk factor for developing chronic diseases like diabetes, cardiovascular diseases, and neurodegenerative disorders. Therefore, the role of exogenous antioxidant compounds, such as polyphenols, is to restore this balance by reacting directly with Reactive Oxygen Species (ROS)/Reactive Nitrogen Species (RNS) to eliminate their unpaired condition by donating electrons and acting as chain-breakers (Malik et al. 2020). Artichoke by-products are a source of antioxidant compounds that can be potentially used to prevent or treat various diseases resulting from oxidative damage (Biel et al. 2020). For example, the antioxidant properties of residual leaves and bracts extract were attributed to 5-*O*-CQA, 1,3-di-*O*-CQA, and luteolin, and the putative protective effect toward liver damage deriving from oxidative processes was evaluated on human hepatocarcinoma HepG2 cells. Data showed that extracts and their isolated compounds have high antioxidant activity and are closely correlated to the polyphenol content. In particular, significant correlations with the level of diCQA derivatives and the high activity of isolated 1,3-di-*O*-CQA suggested that the antioxidant activity was mainly due to the presence of di-CQA isomers (Pagano et al. 2016). This correlation was also confirmed by Jiménez-Moreno et al. (2019) for a 60% v/v methanolic extract obtained by external bracts, leaves, and stems, rich in 5-*O*-CQA, in 2,2-Diphenyl-1-picrylhydrazyl (DPPH), 2,2′-azino-bis(3-ethylbenzothiazoline-6-sulfonic acid (ABTS), and ferric reducing antioxidant power (FRAP) assays.

Furthermore, the same author investigated the capacity of the hydroxycinnamic derivatives fraction (5-*O*-CQA and cynarin) isolated from this extract to counteract the damage induced by hydrogen peroxide (H_2O_2, an oxidative agent used to simulate the pro-oxidative environment that characterizes degenerative diseases) in a Caco-2 cells system, which is valid in vitro model of the intestinal barrier. The data indicated a significant reduction in ROS production induced by H_2O_2 in the presence of 1000 μg/mL of extract after 2 h of incubation, highlighting a potential use of these compounds in the prevention of gastrointestinal tract pathologies caused by oxidative stress (Jiménez-Moreno et al. 2019). Similarly, artichoke by-product extracts can reduce low-density lipoprotein (LDL) oxidation in healthy patients in presence of copper ions, reducing the risk of diseases related to a hypercholesterolemia state. For this purpose, the ethanol and ethylacetate extracts obtained from the leaves were more active than the outer bract extracts in reducing malondialdehyde (MDA, a LDL oxidation product) (Coinu et al. 2007). Finally, aqueous outer bract, leaf, and stem extracts, which were extremely rich in 5-*O*-CQA and 1,5-di-*O*-CQA, had antioxidant properties in in vitro tests (DPPH and ABTS radical assays) and in vivo assays on tobacco BY-2 cells. In particular, the extracts increased cellular polyphenolic content, antioxidant properties, and cell viability by reducing heat stress-induced cell death (Punzi et al. 2014).

3.1.2 Anti-Inflammatory Activity

Inflammation is a response to a pathogen or tissue damage, which involves activating immune cells and releasing various pro-inflammatory mediators, such as cytokines, nitric oxide (NO), histamine, prostaglandins, and leukotrienes. However, like oxidative processes, prolonged inflammatory reactions lead to chronic inflammation, a significant risk factor for several diseases, such as arthritis, inflammatory bowel disease, and cardiovascular and neurodegenerative disorders (Dvorakova and Landa 2017). Several studies showed the anti-inflammatory properties of artichoke by-product extracts, which can counteract the formation of important pro-inflammatory mediators through which chronic diseases can be prevented or delayed. For example, the anti-inflammatory properties of an aqueous extract obtained from PEF-treated artichoke stems were investigated using an in vitro model of cellular inflammation human THP-1 macrophages stimulated with lipopolysaccharides (LPS), which induce the expression of Interleukin-6 (IL)-6 and monocyte chemoattractant protein (MCP)-1/C-C motif chemokine ligand 2 (CCL2) in M1 macrophages. The data indicated a reduction in the expression and secretion of pro-inflammatory mediators (IL-6, CCL2) by human THP-1 macrophages in the presence of the extract, which is characterized by a high content of polyphenols, of which 5-*O*-CQA is the most abundant (53% TPC) (Carpentieri et al. 2022). Instead, the Carr-induced paw edema model was used to evaluate the anti-inflammatory properties of an ethanolic artichoke bract extract. In particular, the injection of a Carrageenan solution (1%) into the subplantar surface of the rats' paw induced an acute inflammatory response characterized by an initial phase, in which there is an enhancement in the expression of Cyclo-oxygenase (COX)-2 and synthesis of type 2 prostaglandins, and a second phase characterized by an increase in vascular permeability, prostaglandins release, and infiltration of polymorphonuclear (PN) cells. The extract reduced the release of inflammatory mediators (prostaglandins) and the migration of PN cells, as confirmed by histopathological examination of the edematous tissue, showing overall anti-inflammatory activity higher than that registered for rats treated with indomethacin (73 *vs* 53%). In addition, the promising anti-inflammatory properties were also confirmed by a reduction in acute inflammation biomarkers, such as C-reactive protein (CRP) and fibrinogen (Salem et al. 2017).

In a recent study, Matsumoto et al. (2021) investigated the influence of six sesquiterpene lactones (including cynaropicrin) isolated from artichoke leaves on the production of NO and/or the induction of inducible nitric oxide synthase (iNOS) in macrophage-like RAW264.7 cell stimulated by LPS. It emerged that sesquiterpene lactones inhibited NO production and/or induction of iNOS by suppressing the JAK-STAT and NF-κB signaling pathways. The acyl portion in sesquiterpenes with an α, β-unsaturated carbonyl group at position 8 and an α-methylene-γ-butyrolactone chain have been identified as functional groups involved in the inhibition of NO production (Matsumoto et al. 2021).

Finally, pectins isolated from artichoke by-products (bracts, stems, and leaves) have been studied for their anti-inflammatory activity. Pectins, characterized by a high content of arabinose and galactose, were tested in a mice colitis model, treated with dextran sulfate sodium as pro-inflammatory agent. It was evident that pectins' functional and biological properties are influenced by structural features, such as degree of esterification, molecular weight (MW), and monomeric content. Some studies highlighted that the presence of arabinose and galactose may affect the biological activity of pectin. Artichoke leaves pectins reduced pro-inflammatory parameters, such as cytokines (TNF-α, IL-1β, and IL-6), enzymes (iNOS), and transmembrane receptors (ICAM-I and TLR-4) in a dose-dependent manner at the intestinal level. Whereas the removal of galactose led to a loss of pectin bioactivity, demonstrating that galactose content is relevant to preserving the anti-inflammatory properties, differently from arabinose, whose removal did not result in a drastic loss of bioactivity (Sabater et al. 2019).

3.1.3 Prebiotic Activity

Prebiotics are a group of non-digestible nutrients able to influence the composition, function, and availability of beneficial microorganisms at the colon level, which obtain energy from the degradation of prebiotics. It is well known that probiotics promote the balance of intestinal flora, preventing or treating several diseases, such as hypercholesterolemia and irritable bowel syndrome, protecting against pathogenic bacteria, and restoring intestinal flora after antibiotic therapy. The inulin-type fructans were widely studied and used as prebiotics because they stimulate the growth of Lactobacilli and Bifidobacteria in the colon, correlated with good health, as from their fermentations short-chain fatty acids are formed and the absorption of calcium, magnesium, and other nutrients is increased (Davani-Davari et al. 2019). Artichoke by-products are an excellent source of inulin, as shown by Zeaiter et al. (2019), who obtained from stems and bracts extracts rich in 70% inulin-type fructans with a high degree of polymerization (DP = 32–42). In vitro studies have shown that artichoke inulin can stimulate the growth of various microorganisms belonging to the *Bifidobacterium* and *Lactobacillus* genera more than commercial fructooligosaccharides (FOS), used as a positive control. Differently from artichoke inulin, for which a comparable level of growth was observed for all strains, commercial FOS were more active on five *Lactobacillus* species (*L. plantarum*, *L. fermentum*, and *L. acidophilus*). At the same time, *L. reuteri* and *L. rhamnosus* grew to a lesser extent (Zeaiter et al. 2019). Similarly, the prebiotic properties of pectin and enzymatically modified pectin extracted from artichoke by-products (stems, bracts, and leaves) were investigated for their effect on the growth of *Bifidobacteria*, *Lactobacilli*, and *Bacteroides/Prevotella* and production of short-chain fatty acids (Ferreira-Lazarte et al. 2018). Artichoke by-product pectin generally stimulated the growth of all tested bacterial strains after 24 h incubation, thus showing bifidogenic properties. Enzymatically modified pectin, characterized by a reduced MW, higher arabinose and galactose content, and higher Gal: Rha ratio, promoted significantly higher growth of Bifidobacteria and Bacteroides/Prevotella populations, which can degrade pectin by different pectinolytic enzymes, while no difference was observed in Lactobacillus growth. In contrast, structural properties did not affect the production of short-chain fatty acids, for which production was increased for all the samples tested. Acetate was the most abundant short-chain fatty acid, followed by propionic and butyric acids in all substrates, agreeing with the results. Therefore, artichoke pectin could be considered an efficient prebiotic for human consumption, similar to inulin, as they stimulate the in vitro growth of beneficial intestinal bacteria and the production of short-chain fatty acids, which are metabolized by the epithelium to provide energy, promote colon well-being, and regulate cell growth. (Ferreira-Lazarte et al. 2018). These results have also been confirmed by Fissore et al. (2015), who investigated the prebiotic properties of various fractions rich in soluble fiber obtained from artichoke by-products (stems and bracts) by acid or enzymatic hydrolysis. In particular, a fraction obtained from artichoke stems with hemicellulase had greater potential in promoting the growth of *L. plantarum* 8114 and *B. bifidum* 11863 than glucose, probably due to the higher content of pectins and inulin.

Overall, artichoke by-products have great prebiotic properties due to the high dietary fiber content, which needs purification, as the presence of other secondary metabolites with antimicrobial properties may counteract their bifidogenic properties (Holgado et al. 2021).

3.1.4 Hypolipidemic, Hypoglycemic, and Antiglycation Activities

Hyperlipidemia, hyperglycemia, and high levels of advanced glycation end products (AGEs) are risk factors for various chronic diseases, such as cardiovascular diseases, diabetes, obesity, and atherosclerosis. Several studies investigated the hypolipidemic, hypoglycemic, and anti-glycative activity of artichoke by-products (Maietta et al. 2017; Mahboubi 2018; Santos et al. 2018;). In particular, the lipid-lowering properties of an extract rich in dietary fiber obtained from stems and bracts were tested on Syrian hamsters fed a high-fat diet. Results indicated that a diet supplemented with 20% artichoke fiber reduced plasma levels of triglycerides (TGs), LDL, and total cholesterol. In addition, a reduction in the accumulation of liver lipids was observed, showing great potential in preventing steatosis. These results are probably related to the structural properties of artichoke by-products inulin and polysaccharide fibers, such as arabinans, arabinogalactans, and xyloglucans, which are responsible for the swelling, oil holding, and cation-exchange capacities, thus reducing lipids and cholesterol absorption at the intestinal level, promoting total fats, TGs, and bile acids elimination through feces, and also reducing the endogenous cholesterol synthesis (Villanueva-Suárez et al. 2019).

Furthermore, the hypolipidemic activity of artichoke leaves is also associated with sesquiterpene derivatives, as evident in a study by Shimoda et al. (2003). Cynaropicrin, aguerin B, grosheimin, and sesquiterpenes isolated from a methanolic leaf extract reduced TGs serum levels in an appropriate animal model (olive oil-loaded mice). This activity is related to slowing down gastric emptying (GE) and reducing absorption since sesquiterpenes also significantly inhibit GE at the same doses (Shimoda et al. 2003). In addition, artichoke leaf extracts inhibit digestive enzymes, such as α-amylase, α-glucosidase, and pancreatic lipase, showing beneficial effects for the treatment of metabolic disorders, regulating blood glucose levels, and satiety signals in patients with obesity (Mahboubi 2018). 5-CQA is one of the main responsible compound for the hypoglycemic activity, as it inhibits the translocase of glucose-6-phosphate, an enzyme involved in the formation of endogenous glucose during the gluconeogenesis process, and reduces glucose intestinal absorption (Rondanelli et al. 2013). Finally, aqueous and ethanolic artichoke extracts obtained from outer bracts and stems can act as potential inhibitors of the glycation process, as demonstrated by Maietta et al. (2017). The anti-glycative properties were investigated at different steps of the glycation process by several in vitro assays. In particular, the extracts inhibited the AGEs formation, interfering with the glycation process at the initial, intermediate, and final stages in the presence of different glycating agents, such as glucose, fructose, ribose, and methylglyoxal (MGO). The ethanolic extract was more active at the initial stage of the reaction, decreasing the formation of the Amadori products in the presence of ribose and MGO. In addition, this extract had high MGO trapping capacity, unlike the aqueous extract, which was more active in the presence of glucose and fructose. This difference is probably due to the different content of secondary metabolites since aqueous extract is richer in 1-*O*-CQA, 3-*O*-CQA, and 1,3-di-*O*-CQA, and ethanolic extract in 5-*O*-CQA, caffeic acid, 1,5-di-*O*-CQA. Therefore, artichoke by-product extracts represent a promising source of anti-glycative agents potentially useful to prevent or reduce AGEs-related diseases, such as atherosclerosis, diabetes, and neurodegenerative diseases.

3.1.5 Antimicrobial and Anti-Proliferative Activity

Food microbial contamination is the leading risk factor for developing foodborne illness, which is of concern at different levels, from the food industry and food safety authorities to consumers. Therefore, identifying new natural antimicrobial agents is important to improve food quality and safety (Gyawali and Ibrahim 2014). At the same time, the spread of antibiotic resistance is an increasingly important public health problem (González-Lamothe et al. 2009). In addition, because of the high mortality rate associated with cancer and the severe side effects of chemotherapy and radiation therapy, recent research is moving toward the identification of natural cytotoxic plant-derived products designed for further drug improvement and development (Unnati et al. 2013). Therefore, several authors investigated the antimicrobial and antiproliferative properties of artichoke by-products. For example, the antimicrobial activity of different extracts obtained with ethyl acetate, methanol, butanol, and ethanol from artichoke floral stems was compared when tested against three Gram-negative bacterial strains (*Enterococcus faecium*, *Salmonella typhimurium*, and *Escherichia coli*), two strains of Gram-positive bacteria (*S. aureus* and *Streptococcus agalactiae*), and yeast *Candida albicans*. All the extracts, although in different percentages, had high antimicrobial properties toward all bacteria, except for *E. coli*, *S. typhimurium*, and *C. albicans*, which were resistant to all extracts. The methanol extract, rich in CQAs and flavones, was the most effective, showing higher inhibitory properties toward Gram-positive bacteria. *S. aureus* and *E. faecium* strains were the most sensitive, showing the widest zone of inhibition (30–31 mm) and the lowest minimum inhibitory concentration (1 mg/mL) and minimum bactericidal concentration (1.5–2 mg/mL) (Mejri et al. 2020). Similarly, the bract ethanolic extracts, characterized by a high polyphenols content, inhibited the microbial growth of foodborne pathogenic bacteria strains, such as *Bacillus cereus*, *S. aureus*, *S. sciuri*, *Salmonella typhi*, *S. enterica*, *E. coli,* and *Pseudomonas aeruginosa*, showing greater activity against *S. enterica* with MIC value of 0.08 mg/mL followed by *E. coli* with MIC value of 0.15 mg/mL. Furthermore, the ethanolic extract had antifungal properties toward all the fungal species used (*Aspergillus flavus*, *A. ochraceus*, *A. niger*, *Fusarium proliferatum*, *A. carbonarius*, *A. parasiticus*, and *Penicillium verrucosum*) except for *A. westerdijikiae*. The highest antifungal activity was registered against *A. ochraceus,* followed by *A. niger* and *F. proliferatum*, while the lowest activity was observed against *A. carbonarius* (Shallan et al. 2020). Finally, Noriega-Rodríguez et al. (2020) studied the antiproliferative properties of stems and bracts hydroalcoholic extracts on cancer cell lines and fibroblasts. Specifically, a reduction in MCF-7 breast cancer cells and Caco-2 cancer colon cell proliferation was observed at high concentrations (750 mg/mL), while fibroblasts showed increased resistance, for which at least 1500 mg/mL was required to decrease cell proliferation, as the pro-oxidant properties of the extract increase. In contrast, increased viability was observed at low concentrations, probably due to other compounds stimulating cell growth. These results were also confirmed by Moselhy et al. (2023), who studied the cytotoxic effects of the ethanol bracts extract on human colon carcinoma, human intestinal carcinoma, and mammalian cells from African green monkey kidney cell lines, which were sensitive to the antiproliferative properties of the artichoke. Therefore, artichoke by-products could also be a potential source of antimicrobial and antiproliferative agents.

In Table 1, all the biological properties of artichoke by-products are summarized.

3.2 Functional Properties

3.2.1 Rheological Properties

As mentioned above, artichoke by-products are characterized by a high content of soluble fiber, such as pectin and inulin, well known for their rheological properties, which can be used in the food or pharmaceutical industry as rheological modifiers. In general, inulin is a fructan constituted by β-D-fructosyl units linked together by (2 → 1) glycosidic bonds, which usually ends with a (1↔2) α-D-glucose unit. It has different DPs that affect rheological properties and potential applications, such as a gelling agent or emulsion stabilizer, since rheological performances increase with the increase in MW (Mensink et al. 2015). Conversely, pectins are mainly composed of three polysaccharide regions: homogalacturonic (HG), rhamnogalacturonic-I (RG-I), and rhamnogalacturonic-II (RG-II) domains, which are linear or branched chains of galacturonic acid (GalA) residues with α-(1 → 4) linkage, which may be present as acid or as methyl ester. The content of GalA methyl esters defines

Table 1 Biological properties of artichoke by-products

Bioactivity	Artichoke by-product	Reference
Antioxidant activity	Bracts, outer bracts, leaves, and stems	Pagano et al. (2016), Jiménez-Moreno et al. (2019)
Anti-inflammatory activity	Stems and leaves	Carpentieri et al. (2022), Salem et al. (2017),
Prebiotic activity	Stems, bracts, and leaves	Zeaiter et al. (2019), Ferreira-Lazarte et al. (2018)
Hypolipidemic activity	Stems, bracts, and leaves	Villanueva-Suárez et al. (2019), Shimoda et al. (2003)
Hypoglycemic activity	Leaves	Mahboubi (2018)
Anti-glycative activity	Outer bracts and stems	Maietta et al. (2017)
Antimicrobial activity	Stems and bracts	Mejri et al. (2020), Shallan et al. (2020)
Antiproliferative activity	Stems and bracts	Noriega-Rodríguez et al. (2020)

pectin degree of methoxylation (DM), which affects gelling properties since gelling occurs in the presence of calcium ions or co-solutes (sucrose, acidifiers) depending on DM. Unesterified GalA units in pectins also affect gel properties and calcium ions' sensitivity (Fissore et al. 2014).

The extraction conditions and the source are important factors that impact pectin composition, structure, and rheological properties and, therefore, its use as a gelling, thickening, or stabilizing agent. For example, preheating stems before the extraction or using enzymatic hydrolysis increases fiber recovery, obtaining about 70% inulin and a lower percentage of low-methoxyl pectins (DM =33–15). The stem aqueous extract containing soluble fiber had Newtonian viscosity when a low shear rate was applied, while a reduction in viscosity was observed at a higher shear rate, thus showing pseudoplastic behavior. In addition, although characterized by a lower pectin content than inulin, artichoke fiber forms a rigid gel in the presence of 40 mg of Ca^{2+}/g pectin, as free carboxylic groups bind calcium ions, cross-linking low DM pectins. However, the presence of inulin and a few methyl-esterified groups probably affects the formation of the gel network, providing greater rigidity.

Conversely, fiber isolated from the preheated artichoke heart forms a creamy, homogeneous gel in the presence of calcium ions, attributable to a different distribution of demethylated blocks in pectin (Fissore et al. 2014). Similarly, Domingo et al. (2019) compared fiber obtained from artichoke stems and bracts by acid hydrolysis and by acid hydrolysis combined with protease and cellulase. High content of low-methoxyl pectins and inulin with different DP characterizes all fibers independently from the method applied. However, the use of hydrolytic enzymes probably induces a different breakdown of the cell wall, releasing polysaccharides with lower thickening capacity, as evident by the lower viscosity values obtained and differently from fibers isolated by acid hydrolysis, which show higher viscosity. In addition, all fibers were sensitive to calcium ions, forming gel-like networks at different calcium cross-linking degrees with different strengths. In particular, the fractions rich in inulin and pectin characterized by high galacturonic acid content and low DM showed the best rheological properties, forming strong and well-structured gels (Domingo et al. 2019). Inulin isolated from artichoke roots, even if characterized by a higher DP (=32) than commercial chicory inulin, has similar gel properties affected by the extraction method. Inulin obtained by UAE forms gels with lower hardness than that obtained by hot water extraction, probably because UAE also increases the release of impurities that interfere with the formation of inter-chain interactions (Castellino et al. 2020).

The chlorophyll content also influences the functional properties of artichoke powder obtained from stem bracts; when removed, the water activity (aw), the water holding capacity (WHC), and the oil holding capacity (OHC) improved. Artichoke by-products have an aw value lower than 0.06, which is considered optimal to increase shelf-life, as it reduces microbial growth; in addition, they have high WHC and OHC values, highlighting their potential applications as food additives to improve water retention during baking processes and to reduce loss of flavor, improving palatability (Ruiz-Cano et al. 2015). Artichoke bracts powder also has excellent emulsifying properties, as shown by the study of Umaña et al. (2021), in which emulsions prepared with 2% artichoke bracts increased their stability with a consequent reduction in coalescence and flocculation. The steric hindrance of inulin and pectin probably improves the emulsion stability, while the presence of amphiphilic proteins and insoluble fiber stabilizes the oil–water (O/W) interface, reducing coalescence. Therefore, artichoke bracts can potentially be used for the preparation of polyunsaturated fatty acid microcapsules, which require dispersion of lipids in O/W emulsions before spray drying since the addition of artichoke increases the stability of the emulsions, does not affect the microencapsulation process, and reduces lipid oxidative processes.

3.2.2 Color Parameters

The color parameters are essential for the potential application as an additive or active ingredient in food and non-food formulation because they can affect the appearance of the final products. Colorimetric measurements of bracts and stem extracts rich in pectin and inulin and obtained by acid or enzymatic hydrolysis revealed different color parameters, lightness (L*), tones (a*, b*), and color (ΔE*). In particular, stem extracts have L* value 50% higher than that registered for bracts, so they are darker. Both extracts have a yellowish tone, indicated by high b* values and a reddish tone, indicated by low a* values, although stem extracts were more yellow and red than bracts. Enzymatic hydrolysis only increased L* and ΔE*, but not in a tone change (Domingo et al. 2021). In addition, the removal of chlorophyll increases L* and a* value as a consequence of green color reduction due to chlorophyll (Ruiz-Cano et al. 2015). Overall, bracts and stems rich in fiber were bright and slightly colored (Domingo et al. 2021).

In Table 2, the functional properties are summarized.

Table 2 Functional properties of artichoke by-products

Rheological property	Artichoke by-product	Reference
Emulsifier properties	Fibers (inulin and pectin) from artichoke bracts	Umaña et al. (2021)
Gelling properties	Fibers (inulin and pectin) from artichoke stems and bracts, inulin from artichoke roots	Fissore et al. (2014), Domingo et al. (2019), Castellino et al. (2020)
WHC, OHC	Fibers from artichoke stems and bracts	Ruiz-Cano et al. (2015)
Color parameters		
a* b*	Stems and bracts	Domingo et al. (2021)
L*	Stems	Domingo et al. (2021)

4 Food and Non-food Applications of By-Products Extracts

Artichoke by-products can be used as food additives or functional food ingredients or as additives or bioactive ingredients in pharmaceutical and cosmetic industries (non-food application) due to their high content of metabolites with nutritional and/or healthy benefits and technological properties.

4.1 Food Applications

4.1.1 Food Additives in Bakery Products

Due to the high dietary fiber and polyphenols content, several authors have studied the potential applications of artichoke stems and bracts in preparing bakery products to improve final products' rheological properties and shelf-life (Frutos et al. 2008; Boubaker et al. 2016). An artichoke by-product extract extremely rich in fiber, especially in insoluble fiber, characterized by a low moisture content, a protein content like other commercially available fiber, and a low-fat content, presenting a high-WHC (8.17 g/g dry weight) and OHC (16.17 g oil/g sample), was suitable for the preparation of bread, not only for its chemical composition but also for the physical properties. Following the incorporation of 2% of this concentrated extract into wheat flour, an improvement in the rheological properties of dough has been observed. In particular, this dough added with fiber had a greater capacity to absorb water, higher stability and tenacity (due both to the high number of hydroxyl groups in the fiber structure that can bind water and to the increased interactions between fiber, water, and gluten), and a more reduced extensibility and softening (due to the presence of cellulose, which promotes a strong interaction between fiber and flour protein) than the dough without fiber addition. Finally, the resulting bread has a slightly lower specific volume, greater hardness and chewiness caused by the high content of dietary fiber, which breaks down the gluten network, thus reducing the available water needed for gluten development, and a darker color than the control, as a consequence of Maillard and caramelization reactions (Boubaker et al. 2016; Canale et al. 2022). Sensory analysis indicated that adding less than 9% artichoke by-products changes the texture properties of the bread, which is acceptable in terms of hardness, chewiness, cohesiveness, and resilience (Frutos et al. 2008). In addition, the bread enriched with flour obtained from artichoke by-products has a better shelf-life due to the high WHC, which reduces water loss during storage and the possible fiber–starch interactions, delaying starch degradation (Boubaker et al. 2016). The bread combined with artichoke flour also has greater structural stability than the bread prepared with re-milled semolina, for which a clear weight, volume, and height reduction is observed over the 4-day storage period. Therefore, artichoke by-product flours improve the structural properties of bread, increase the yield of the final product thanks to the greater ability to hydrate, and improve the shelf-life, guaranteeing a fresh product for several days, which can potentially be used as a fresh product and as a short-life industrial product (Canale et al. 2023). Similarly, the addition of 4% artichoke by-product extract rich in fiber to prepare whole-wheat biscuits with a high-fiber content has been investigated. The final product was acceptable in terms of flavor and appearance and comparable to biscuits formulated with commercially available pea fiber. Indeed, the product was darker and harder due to following the typical artichoke color and high-fiber content compared to biscuits without added fiber or pea fiber-based biscuits. In addition, sensory qualities remained unchanged during normal and accelerated storage conditions, similar to biscuits containing pea fiber but differently from biscuits without fiber, for which a reduction in acceptability, changes in color and flavor were observed. Adding fiber probably increases the product's shelf-life or masks the change in color or flavor (San José et al. 2018). In addition, biscuits enriched with artichoke by-product fiber also have a higher content of polyphenolic compounds than biscuits formulated with pea fiber or without fiber, both at the beginning of preparation and following normal or accelerated storage conditions. These biscuits also exhibit greater antioxidant activity due to the higher polyphenolic content, which is unchanged even after the simulated in vitro digestive process. Therefore, the high polyphenol content increases the products' shelf-life, delaying oxidative degradation such as fat rancidity (San José et al. 2023). Furthermore, these biscuits have a higher moisture content after cooking than biscuits without fiber, similar to what has been observed in bread, thanks to the higher fiber content capable of retaining water. This aspect can be advantageous at the industrial level, as it allows greater control of weight loss and excessive browning, facilitating cooking processes and therefore proving to be a valid alternative to pea fiber both for its rheological properties and for its sensorial acceptability (San José et al. 2023).

Artichoke root represents an alternative and sustainable source of inulin with a high degree of polymerization, which can be used to replace durum wheat semolina to produce fresh pasta due to its technological properties. Adding inulin affects the cooking properties, color, and firmness of fresh pasta. In particular, pasta enriched with inulin requires less cooking time, as the presence of fibers breaks the gluten network by improving water penetration, and has a lower swelling index, as water tends to bind mainly fiber hydroxyl groups, limiting their penetration into the starch granules and the consequent swelling, thus increasing the firmness of the final product. In addition, the final product is acceptable to the consumer overall because the fresh pasta enriched with inulin shows no significant differences in taste, bulkiness, adhesiveness, or hardness. Indeed, adding inulin decreases brightness and yellowness, often linked to high-quality pasta, although consumers nowadays associate darker colors with high-fiber products (Difonzo et al. 2022). Also, the addition of artichoke by-product extract (deriving from a mixture of stems and bracts), rich in polyphenols with antioxidant properties, increases the shelf-life of fresh egg pasta, as evident from the result obtained by Amoriello et al. (2022). The formation of mold colonies was observed after 7–8 days at 5 °C, although normally 3 mm colonies are formed as early as 6 days.

Finally, bracts artichoke has been studied as a potential wheat replacer for cake production. The optimization of the process led to the identification of 16.4% as the ideal percentage for optimal physical and sensory properties because this addition provided high volume, soft cake texture, high total phenolic content with high antioxidant activity, and better cake properties, which consumers have preferred. The optimal addition of artichoke bracts improves the ability to hold air bubbles, thereby increasing volume and reducing hardness, which is inversely related to the specific volume. An increase in volume results in an improvement in softness. Conversely, adding artichoke bract in high percentage disorganizes the starch-gluten matrix and decreases gas holding capacity, increasing hardness (Dadalı 2023).

4.1.2 Food Additives in Dairy Products

Several studies evaluate the application of artichoke by-products in the production of dairy products, such as yogurt, milk, and cheese, to improve rheological properties, probiotic bacteria content, and as a milk-clotting agent in cheese production (Ehsani et al. 2015). In particular, the addition of plant extracts in the production of yogurt follows the need of food companies to identify new food additives that can ensure the survival and activity of probiotic microorganisms, reducing their loss during fermentation and storage (Mortazavian et al. 2007). For this purpose, a leaf artichoke extract was investigated. Its addition at 0.5% increases the availability of *Lactobacillus acidophilus* LA-5 and *Bifidobacterium lactis* BB-12 during fermentation and storage due to a more nutritious and stimulating medium for lactic acid and probiotic bacteria growth containing high amount of artichoke metabolites (i.e., proteins, carbohydrates, fiber, exopolysaccharide, adenine, hypoxanthine, free amino acids, and essential vitamins and minerals). In addition, this artichoke by-product extract improves yogurt physical characteristics, decreasing the syneresis probably due to the presence of inulin, which acts as a thickener, reducing the serum separation through the formation of protein aggregates in the yogurt (H-bridge formation). Although adding this artichoke extract is technologically advantageous, the final product has low acceptability among the panelists due to the yellow color and the bitter flavor, probably due to sesquiterpene lactones (Ehsani et al. 2015).

Similarly to artichoke leaf extract, inulin isolated from artichoke roots improves the viability of probiotic bacteria in yogurt at the end of fermentation, but the final product has an unpleasant flavor and poor acceptability (Ehsani et al. 2018). However, adding inulin to stirred bio-yogurt positively affected the viscosity, slightly increasing it after 21 days of storage. This increase can be attributed to the fact that the viscosity is affected by the number, strength, structure, and spatial distribution of casein micelle bonds since inulin tends to form micelles, as opposed to hydrocolloids, which result in increased viscosity through the formation of bonds between chains. In addition, inulin fortification reduced syneresis, thus confirming what was previously reported by Ehsani et al. (2015), and increases creaminess; conversely, in control without inulin addition, the reduced fat content results in a loss of texture, thus highlighting the potential application of inulin as a fat replacement (El-Kholy et al. 2023).

On the other hand, preliminary studies investigate the potential applications of artichoke by-products in cheese production as a milk-clotting agent. In fact, given the high production and consumption of cheese and the high cost of calf rennet, the search for vegetable coagulants is mandatory to meet market and consumer demands. Several studies have already demonstrated this property for artichoke by-products, attributing it to the presence of cardosin A and cardosin B, characterized by high clotting and intense proteolytic activity able to cleave bovine k-casein, producing cheeses with a creamy soft texture (Bravo Bolívar et al. 2023). Therefore, some authors have studied the enzyme activity of the leaves and roots of artichokes. For example, Llorente et al. (2004) compared the proteolytic and milk-clotting activity of flower, leaf, root, and rhizome extracts using casein and standardized skim milk powder from cows as substrates. The results indicated that only flower and leaf extracts are active. Esposito et al. (2016) deepened this preliminary study in more detail by isolating the enzyme responsible for the leaf's proteolytic activity and confirming its belonging to the aspartic protease family, to which most vegetable coagulant enzymes belong. This enzyme is active at acidic pH, inhibited by pepstatin A and endows with two aspartic acid residues responsible for their catalytic activity. Indeed, further studies are needed to investigate the application in the cheese-making process as a milk-clotting agent. Adding artichoke by-products as a rheological agent in Primosale cheese has also been studied. In particular, the incorporation of flour obtained from the artichoke by-products positively affects the physicochemical cheese characteristics, increasing moisture content and reducing weight loss, due to the high fibers WHC as has also been observed in bread products, and thus resulting in increased yield of the final product. Moreover, the addition of artichoke by-products changes sensory attributes, increasing odor and aroma intensity, tactile sensations (astringent, fibrous, crispness, stickiness, and moisture), and cheese hardness, probably because artichoke by-products give the cheese a homogeneous and compact body (Costa et al. 2018).

4.1.3 Other Applications as Food Additives

Nowadays, the use of natural antioxidants as preservatives is widespread to satisfy the demand of consumers who require safer products without any addition of synthetic additives (Santos-Sánchez et al. 2017). As mentioned earlier, artichoke by-product extracts are characterized by a high content of phenolic acids, such as cynarin (1,5-di-*O*-CQA) and chlorogenic acid (5-*O*-CQA), known for their antioxidant properties, which, therefore, can be used as natural antioxidants in food matrix to prolong their shelf-life. Food degeneration is often caused by oxidative reactions in lipids or proteins (Claus et al. 2015). Several authors have investigated the potential applications of plant extracts rich in antioxidant compounds to replace standard synthetic antioxidants, such as butylated hydroxyltoluene (BHT).

The OXITEST method, which was used to evaluate the preservation of canola oil from oxidation by using bracts, spikes, and receptacles, indicated that these by-products were able to retard the oxidation of the oil by 15%, 24%, and 72%, respectively, showing an antioxidant activity higher than that of BHT, a synthetic preservative. These results highlighted the remarkable preservative action of artichoke by-products (Claus et al. 2015). Likewise, adding an extract of artichoke by-products to tomato juice increases the antioxidant properties, as shown by DPPH· and ABTS assays results, improving the product's shelf-life (Larrosa et al. 2002).

BHT is also used, for example, in the meat industry, and it was replaced with an artichoke by-product extract that was investigated in raw beef patties, monitoring the antioxidant effect during meat refrigerated. Adding the extract increased the antioxidant activity of the raw beef patties tested with the DPPH·assay, which remained constant during the 7 days of storage. It also better-reduced lipid peroxidation and protein oxidation processes than BHT, resulting from the lower content of primary and secondary lipid oxidation products (i.e., hydroperoxides and malondialdehyde) and carboxylic compounds produced by protein oxidation. These results were also supported by the detection of lower iron non-heme content due to the protective effect of artichoke outer bracts metabolites on heme structure by inhibiting the oxidative processes involving lipids and proteins, thus reducing iron release from myoglobin. As a result of oxidation in meat systems, heme iron easily converts to non-heme iron, causing a rancid aroma and a red-to-brownish color change (Ergezer and Serdaroğlu 2018).

Similarly, outer bract extracts can inhibit the growth of psychrophilic and coliform bacteria in a concentration-dependent manner in beef patties. Adding 1000 ppm of extract decreased the development of these bacteria during meatball storage (a reduction of 2.5 log CFU/g, higher than that registered for the control sample). Moreover, the addition of the same extract to beef patties also inhibited the growth of pathogenic bacteria, such as *E. coli* and *Listeria monocytogenes* inoculated during patties storage, with better efficacy on *E. coli* than *L. monocytogenes* (Ergezer et al. 2018). These results highlighted the potential use of artichoke by-product extract as a natural additive that improves the preservation and safety of meat products, improving the stability and quality of meat through its antioxidant and antimicrobial properties.

Another interesting application involves the use of artichoke by-products (stalks, bracts, and leaves) as promising biosorbents for mycotoxins such as aflatoxin B1 (AFB1), zearalenone (ZEA), and ochratoxin A (OTA), which are possible contaminants of foods and feeds. These mycotoxins are responsible for various diseases and involve different animal species, including humans. Therefore, using additives with absorptive properties aims to reduce their absorption from the gastrointestinal tract and distribution to various tissues and organs. For this purpose, the artichoke by-products absorptive properties of different mycotoxins under different pH conditions were investigated under simulated physiological conditions. The extract has higher absorptive capacities at pH 7 toward AFB1 and ZEA, likely due to the presence of acidic groups in toxins structure that promote interaction under neutral conditions; in addition, it can sequester simultaneously different mycotoxins, thus being a promising multi-toxins binder. This property often depends on the chemical composition, as the presence of non-degradable fibers and polyphenols (such as condensed tannins, which have poor absorption) enhances the absorptive properties of by-products along the gastrointestinal tract (Greco et al. 2019).

Artichoke stems have been studied as a potential source of peroxidases (PODs), enzymes that can be used in the food and beverage industry as color modifiers, wine stabilizers, and sequestrants of beer off-flavor. Peroxidase was isolated from stems and purified, obtaining a yield of 7%. The best enzyme activity was registered at 50 °C, pH 6–8, using *o*-dianisidine, catechol, and guaiacol as substrates. Therefore, artichoke stems represent an alternative source of PODs, potentially applied in the food industry and chemical and biotechnological industries (Mejiri et al. 2019).

4.1.4 Functional Food Ingredients

Artichoke by-products benefit humans due to their high content of bioactive compounds, such as dietary fiber and polyphenols, with anti-inflammatory, antioxidant, hypolipidemic, anti-obesity, and hypoglycemic properties. Therefore, several authors investigate the potential application of artichoke by-products as functional food ingredients to enhance the beneficial properties of foods (Feiden et al. 2023).

For example, after a simulated digestion process, the effect of adding 3, 6, and 9% w/w artichoke stems on bread's antioxidant and hypoglycemic properties was studied. Interestingly, the antioxidant activity of the bioaccessible fraction is preserved during digestion and increases as the artichoke stem percentage in the bread increases. According to the data, 82% polyphenols were completely bioavailable at the duodenal level, of which 74% was represented by cynaropicrin, while 88% of caffeic acid was bioaccessible at the colon level. Similarly, this artichoke stem's bioaccessible fraction inhibited α-glucosidase in a dose-dependent, again highlighting a correlation between stem polyphenols content and the bioactivity of enriched bread. Therefore, artichoke stem could be used as a functional ingredient to create a reducing environment in the gut and modulate glucose absorption to prevent diabetes or improve metabolic syndrome (Colantuono et al. 2018). Another application for diabetes prevention is the use of bract extract (containing luteolin, apigenin, and their glycosylated derivatives, mono-CQAs, and di-CQAs) in association with omega-3-rich fish oil to prepare wheat flour-based chips with high nutritional and functional properties for diabetic patients. The incorporation of these ingredients improves the nutritional quality of the product, as it is richer in polyunsaturated essential fatty acids such as linoleic, linolenic, eicosapentaenoic, and docosahexaenoic acids, which are important in the prevention of cardiovascular diseases. Further investigation using diabetic mice also underlined the improvement of serum, liver, and kidney parameters, such as alanine aminotransferase, aspartate aminotransferase, alkaline phosphatase, urea and creatinine, the reduction of blood glucose, triglycerides, total cholesterol, LDL levels and improvement of lipid profile with the increase of HDL level, thus showing potential antihyperglycemic and antihyperlipidemic properties (Benkhoud et al. 2021).

Another recent application reported in the literature is developing a functional food consisting of gluten-free bread fortified with artichoke leaf extract for patients with celiac disease. This study comes from the need to identify plant sources that can increase the intake of antioxidant phenolic compounds to reduce oxidative stress in patients with celiac disease, even following a gluten-free diet. Adding artichoke leaf extract increased the antioxidant activity of bread and protected cells from oxidative stress in the presence of oxidant agents, such as H_2O_2. It reduced the expression of pro-inflammatory cytokines, TNF-α and IL1-β, a, in Caco-2-cells exposed to LPS from *E. coli*, improving the health properties of the gluten-free bread (Vacca et al. 2023).

Similarly, artichoke by-products have been used as functional ingredients to increase the antioxidant properties of chicken soups, fresh pasta, and cheese spreadable (Llorach et al. 2005; Pasqualone et al. 2017; Lucera et al. 2018). Inulin from artichoke roots represents a promising ingredient to produce functional fresh pasta. Beneficial bacteria such as *Lactobacilli* and *Bifidobacteria* ferment inulin at the colon level, showing prebiotic properties, as it increases colonocyte function; in addition, its regular consumption modulates lipid and glucose metabolism and prevents colorectal disorders. An in vitro study shows that adding inulin to a fresh pasta preparation increases probiotics cell density by more than 50% compared to the control, thus reducing *E. coli* cell density. Therefore, the results of this study underlined the prebiotic effect of inulin-enriched pasta and its protective effect against pathogenic microorganisms (Difonzo et al. 2022).

4.1.5 Feed for Animal Consumption

Using artichoke by-products to produce animal feed can be a valid option to reduce waste and disposal costs and meet the needs of intensive and semi-intensive livestock farms, which require large quantities of feed that are not always locally available. In addition, feed production has a significant economic impact on farms, as it requires land, water, etc. Therefore, producing the feed as silage, instead of using it directly as animal feed, would make the product available all year round, regardless of the artichoke season, and would ensure a longer time for feed storage, eliminating the problem related to water content. In addition, bioactive components in the by-products can improve animal health and performance, increasing animal products' nutritional value (Meneses et al. 2020).

Silage is a common crop preservation procedure based on using natural lactic acid fermentation in anaerobic conditions and is very promising for artichoke by-products' long-term preservation (Meneses et al. 2007; Fan et al. 2020). During ensilage, the control of parameters, such as temperature and pH, and the type of fermentative container (plastic or polyethylene materials) can affect by-product silage and its degradation (Meneses et al. 2020). In particular, temperature and water content impact the silage fermentation process and, consequently, organic acids, crude protein, and lignin content. Silage increases the lactic acid bacteria (*Lactobacillus, Lactococcus, Serratia, and Weissella*) in artichoke by-products, inhibiting the harmful bacteria, promoting the formation and accumulation of polyphenols and terpenes, and stabilizing sterol content (stigmasterol and β-sitosterol) (Fan et al. 2020). In addition, the fermentation with lactic acid bacteria (*L. acidophilus* and *B. bifidum*) can increase the artichoke by-product's bioactive concentration and its prebiotic activity, which is related to inulin and fructooligosaccharides content. In addition, this process produces a higher release of short-chain fatty acids (SCFAs), mainly acetic, propionic, and butyric acids, with antimicrobial, antioxidant, anti-inflammatory, and antiproliferative properties (Zeaiter et al. 2019; Van den Abbeele et al. 2020; Holgado et al. 2021), and polyunsaturated fatty acids (PUFAs) (Monllor et al. 2020a, c, 2023; Muelas et al. 2022).

Thus, several authors have evaluated the suitability of artichoke by-products to produce ensilages through the evaluation of some parameters, such as fermentative, chemical, and phytosanitary parameters and the effect of artichoke silage-based feeding for dairy animals on the final food product (Muelas et al. 2017; Monllor et al. 2020c, 2023).

For example, Monllor et al. (2020b) studied the process of ensiling artichoke by-products in commercial-sized silos (300 kg round bales) for 200 days. The obtained results indicated that the product was suitable because of its microbiological and nutritional properties and remained stable during the storage period, thus confirming the production in commercial round bale silos as a suitable practice that can ensure long shelf-life. After 7 days, the product already has a low pH (around 4) and a high lactic acid content, an essential quality indicator, ensuring the growth reduction of potentially harmful microorganisms, anaerobic conditions, and sugar absence. In addition, the poor buffering capacity of plant material allowed the pH to drop more quickly, decreasing the risk of protein and polyphenol loss. In particular, polyphenols increased as β-glucosidase produced by lactic acid bacteria that catalyzes the release of phenols during ensiled fermentation increases. Finally, the low levels of ethanol and butyric acid in ensilages confirm the low growth of enterobacteria, clostridia, and yeasts populations and the concentration of phytosanitary products below the values set by the guidelines makes the product safe for animal and human food. Therefore, the silage has a low pH value, high lactic acid content, which inhibits microbial growth and improves palatability and smell, low levels of ammonia-N as result of low protein degradation, low volatile fatty acid content (propionic, butyric, and isobutyric acids) which indicate low loss of energy and starting material, and very low, if any, phytosanitary levels which are the requirements for the production of quality silage, as confirmed by other authors (Meneses et al. 2005, 2007).

Subsequently, several studies were performed to evaluate the effect of the addition of ensilage in dairy animals' diets (e.g., goats and sheep) on the characteristics of the final product, such as chemical composition, technological properties, and sensory parameters. The incorporation of ensilage obtained from different % (0, 12,5, 25) of artichoke outer bracts in the diet of dairy goats does not affect the milk content in fiber, crude proteins, ash, and casein, although a slight decrease is only observed in the presence of ensilage at 25% (Muelas et al. 2017). In addition, using artichoke by-products ensilage, does not affect the milk's technological properties since the amount of lactose and lactic acid bacteria, whey drainage ability, and coagulating properties are unaffected (Muelas et al. 2017). In addition, the ensiled by-products increase the health properties of milk because of the improvement of the lipid profile due to a higher oleic acid, polyunsaturated fatty acids, and conjugated linoleic acid content, which have anti-atherogenic properties and reduce obesity (Monllor et al. 2020c). Finally, this addition slightly changes the milk color (for which the reason is not yet well understood), while sensory analysis did not detect the presence of off-flavors in milk (Muelas et al. 2017).

Similarly, a diet based on 40% artichoke bracts ensilage during the lactation period of goats does not affect milk production since no changes in milk composition and yield are observed. These data confirm the positive influence of artichoke in higher percentage on the milk lipid profile and, in this case, also on the mineral profile, as higher concentrations of calcium, zinc, manganese, and copper were observed (Monllor et al. 2023). In addition, milk preserves the characteristics necessary for cheese or yogurt production even after undergoing the fermentation process since gel stability, lactic acid bacteria count, and fermentation performance (acidity and organic acids and sugars profile) are not affected by diets (Muelas et al. 2022). The silage of artichoke by-products causes greater degradation in the rumen with lower in vitro digestibility values than in artichoke by-products hay. However, both by-products have intermediate degradability/digestibility values, considering them as alternative feedstuffs in ruminant diets (García-Rodríguez et al. 2019). All studies showed that the ensilage of artichoke by-products can be included in the diet of dairy goats since the resulting product maintains the nutritional and technological properties and characteristics necessary to produce cheese and yogurt without damaging the animal's health. Over the years, long-term preservation procedures have been proposed to have stable and high-quality by-products, preserving and enhancing the content, mainly in polyphenols and terpenes and their antioxidant and anti-inflammatory activities. In particular, silage, which is a common crop preservation procedure, is based on the use of natural lactic acid fermentation in anaerobic conditions, and it seems to be very promising for artichoke by-products long-term preservation (Meneses et al. 2007; Fan et al. 2020).

During ensilage, the control of parameters, such as temperature and pH, and the type of fermentative container (plastic or polyethylene materials) can influence by-product silage and degradation (Meneses et al. 2020). Temperature and water content can particularly affect the silage fermentation process and, consequently, the content of organic acids, crude protein, and lignin. Silage increases the lactic acid bacteria (*Lactobacillus, Lactococcus, Serratia, and Weissella*) in artichoke by-products, inhibiting the harmful bacteria, promoting the formation and accumulation of polyphenols and terpenes, and stabilizing sterol content (stigmasterol and β-sitosterol) (Fan et al. 2020). In addition, the fermentation with lactic acid bacteria (*L. acidophilus* and *B. bifidum*) can increase the amount of bioactive from artichoke by-products (outer bracts, stalks, leaves, and the upper part of inflorescence), and in particular its prebiotic activity, that is related to inulin and fructooligosaccharides content. In addition, this process produces a greater release of short-chain fatty acids (SCFAs), mainly acetic, propionic, and butyric acids, with antimicrobial, anti-oxidant, anti-inflammatory, and antiproliferative properties (Zeaiter et al. 2019; Van den Abbeele et al. 2020; Holgado et al. 2021) and polyunsaturated fatty acids (PUFAs) (Monllor et al. 2020c, 2023; Muelas et al. 2022).

4.1.6 Food Supplements Ingredients

Finally, artichoke by-product extracts are a promising source of active ingredients for preparing food supplements to prevent or treat metabolic disorders. As mentioned above, in vitro studies have shown the hypolipidemic properties of artichoke by-products, supported by clinical studies. For example, a randomized, double-blind, placebo-controlled clinical study highlighted an improvement in the lipid profile of patients with chronic kidney disease after 6 weeks of supplementation with artichoke leaf extract. In particular, a significant reduction in total cholesterol, LDL, and appetite was observed, while no differences were observed between the two groups in HDL and triglyceride levels (Gatmiri et al. 2019). Similarly, an improvement in the lipid profile and serum levels of liver enzymes, such as alanine aminotransferase (ALT) and aspartate aminotransferase (AST), of patients with nonalcoholic steatohepatitis was observed compared to the placebo group (Rangboo et al. 2016).

4.2 Non-food Applications

Artichoke by-products can be used as active ingredients in pharmaceutical and cosmetic products due to the beneficial effects of their bioactive constituents on humans. For example, a study by El-Nashar et al. (2022) investigated the potential applications of artichoke bract extract as a botanical drug for treating Alzheimer's disease (AD). Nanoformulations loaded with extract were tested on mice previously treated with streptozotocin to induce AD; the drug had a potential neuroprotective action due to an improvement in cognitive properties, spatial memory, and a reduction in the biomarkers of inflammation and levels of tau and β-amyloid protein. In a different study, the potential application in cosmetics was evaluated on 20 women between 35 and 55 years old; in fact, an improvement in skin roughness and elasticity was observed following the application of a cream containing 0.002% extract compared with control group treated with a placebo cream without the extract, showing that the extract could be used as an active ingredient with a skin anti-age effect. In addition, the extract improves microcirculation, and endothelial cell vasodilation exerts anti-inflammatory action by inhibiting nitric oxide (NO) production, protecting cells from oxidative damage and improving cell cohesion (D'Antuono et al. 2018).

Several studies demonstrated the potential applications of artichoke by-product extracts as excipients for preparing hydrogels and silver nanoparticles. For example, Órbenes et al. (2021) use artichoke bract extract to formulate starch-based gels in combination with Paulownia extracts, which could be used as a base for special foods for people with special needs (elderly and/or children) or for innovative cosmetics. The extraction techniques and the ratio of artichoke and Paulownia extracts affect the strength of the obtained gels. However, overall, the resulting gel showed suitable mechanical properties and was stable after one and a half months of cold storage, as no water syneresis phenomenon was observed. The 75:25 ratio was optimal for obtaining a soft gel, as the increased content of Paulownia extract reduced the viscoelastic behavior. Finally, artichoke by-product extracts were used for the eco-friendly synthesis of silver nanoparticles (AgNPs) for cancer treatment in combination with photodynamic therapy. Specifically, silver nanoparticles alone exhibit antitumor properties through induction of mitochondrial apoptosis of MCF7 breast cancer cells and increased production of reactive oxygen species (ROS). In contrast, combining photodynamic therapy and silver nanoparticles results in a synergistic effect, thus increasing apoptosis and reducing migration and growth of MCF7 breast cancer cells. Therefore, the green synthesized AgNPs from artichoke leaf extract with photodynamic therapy showed promising application as an anticancer (Erdogan et al. 2019).

In Table 3, food and non-food applications are summarized.

Table 3 Food application and non-food application of artichoke by-products

Food application			
Food additive	**Type of product**	**Artichoke by-product**	**Reference**
Rheological additive	Fresh pasta, cake, biscuits, and bread	Stems, bracts, and roots	Boubaker et al. (2016), Difonzo et al. (2022), San José et al. (2023), Dadalı (2023)
	Yogurt and cheese	Leaves, inulin from artichoke roots, all by-products	Ehsani et al. (2015), Ehsani et al. (2018), Costa et al. (2018)
Milk-clotting agent	Cheese	Leaves	Esposito et al. (2016)
Stabilizer probiotic bacteria content	Yogurt	Leaves	Ehsani et al. (2015)
Preservatives	Fresh egg pasta, bread, biscuits, and raw beef patties	Bracts, stems, and outer bracts	Amoriello et al. (2022), Boubaker et al. (2016), San José et al. (2023), Ergezer et al. (2018)
Biosorbents	Foods and feeds	Stems, bracts, and leaves	Greco et al. (2019)

(continued)

Table 3 (continued)

Food application			
Modifier of the appearance, stabilizer or sequestrant of off-flavor	Foods and beverages	Peroxidase (PODs) from stems	Mejiri et al. (2019)
Functional food ingredients			
Antioxidant and hypoglycemic ingredient	Bread	Stems	Colantuono et al. (2018)
Anti-diabetic ingredient	Wheat-flour based chips	Bracts	Benkhoud et al. (2021)
Antioxidant and anti-inflammatory ingredients	Gluten-free bread	Leaves	Vacca et al. (2023)
Prebiotic ingredient	Fresh pasta	Inulin from artichoke roots	Difonzo et al. (2022)
Antioxidants ingredient	Chicken soups, cheese spreadable, and fresh pasta	Outer bracts, stems, external leaves, all by-products	Pasqualone et al. (2017), Llorach et al. (2005), Lucera et al. (2018)
Animal consumption			
Feed	Ensilage	Bracts and stems	Monllor et al. (2020a, b, c)
Active ingredients			
Hypolipidemic activity	Food supplements	Leaves	Gatmiri et al. (2019)
Non-food application			
Active ingredient	**Type of products**	**Artichoke by-products**	**References**
Neuroprotective activity	Botanical drug	Bracts	El-Nashar et al. (2022)
Skin anti-age activity	Cosmetic formulation	All by-products	D'Antuono et al. (2018)
Excipient			
Rheological additive	Hydrogels	Bracts	Órbenes et al. (2021)
Reactive for synthesis	Silver nanoparticles	Leaves	Erdogan et al. (2019)

5 Other Industrial Applications

Globe artichoke is a versatile plant, and its by-products have some non-health applications, mainly in bioenergy production (solid biofuel, biogas, and bioethanol) and paper pulp (Fernández 1993; Fernández et al. 2006; Hughes et al. 2017). There are still few but very promising other applications, such as polymer composites, herbicides, sorbents, and bio-based materials (microbial and fermentation products) for the recycle and valorization of artichoke by-products in alternative industrial fields (Fiore et al. 2011; Botta et al. 2015; Yegin 2017; Fernández-López et al. 2019; Scavo et al. 2020; Gümüş et al. 2022) (Fig. 3).

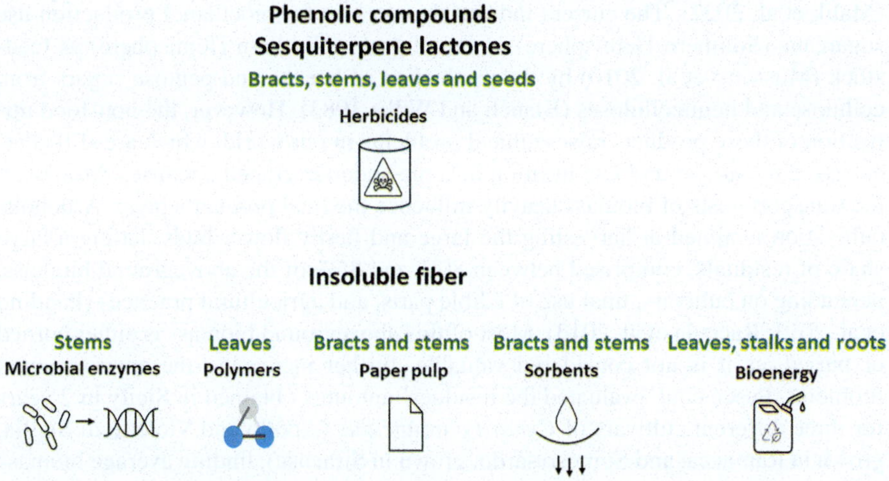

Fig. 3 Main artichoke by-products and non-health industrial applications

5.1 *Bioenergy*

On the environmental side, *Cynara cardunculus* cultivations have a carbon footprint equal to 0.86 kg CO_2eq kg^{-1} of fresh product and a water footprint ranging from 0.81 m^3 kg^{-1} to 0.96 m^3 kg^{-1} of fresh product, so contributing in an important way to the greenhouse gasses emission and to water consumption (Mekonnen and Hoekstra 2011; Martin-Gorriz et al. 2020; Rana et al. 2023).

It has been estimated that globe artichoke cultivation requires more than 85 GJ of primary energy and generates approx. 2400 kg CO_2 eq ha^{-1} of harvested land (Lo Giudice et al. 2014). Compared to other crops, artichoke is considered a great GHGs' emitter: tomatoes release roughly 1645 kg CO_2 eq, cabbage less than 1290 kg CO_2 eq, cucumber 1270 kg CO_2 eq, and spinach 1060 kg CO_2 eq (Yan et al. 2012). Moreover, Beltran et al. (2022) estimated that artichoke and other over-irrigated horticultural crops in Spain are responsible for over 70% of N_2O peri-urban agricultural emissions.

This suggests that reducing the GHG emissions due to this cultivation is essential as well as to "recover" energy from its field residues and food wastes. Ledda et al. (2013) explored the characteristics of artichoke cultivations (variety Spinoso sardo) in 3 years (2007–2010) in Sardinia, finding that the average amount of harvested biomass was about 4 Mg ha^{-1}, with an annual energy yield of 64 GJ ha^{-1}. Also, in the energy field, artichoke can be considered a "multi-purpose crop" (Gominho et al. 2018) since its wastes can be used to produce bioenergy in different sources (bioethanol, solid biofuel, and biomethane) by other techniques (gasification, combustion, and pyrolysis).

In 2018, the total bioethanol production from under-development pilot-scale facilities was 110 billion liters and is expected to reach 140 billion liters in 2025

(Malik et al. 2022). The current industrial processes for bioethanol production use sugarcane (Southern Hemisphere) or cereal grain (Northern Hemisphere) as feedstock (Mussatto et al. 2010) by the production of hexose and pentose sugars from cellulose and hemicelluloses (Blanch and Wilke 1983). However, the non-food utilization of these products raises ethical problems in our world, which are still characterized by hunger and malnutrition in some underdeveloped countries. Moreover, the transport costs of biomass heavily influence the final product's price. Artichoke cultivation is aimed at harvesting the large and fleshy flower buds, letting a large share of residuals, comprised between 80% and 85% of the aboveground biomass, depending on cultivars, final use of edible parts, and agricultural practices (Pandino et al. 2013; Raccuia et al. 2013). Most of this aboveground biomass is either burned or buried, as it is not considered valuable, further increasing the environmental problems. Pesce et al. evaluated the residues' amounts obtained in Sicily in 2 years for three different cultivars of *Cynara cardunculus* (Apollo and Violetto di Sicilia, grown in Ramacca, and Spinoso sardo, grown in Siracusa), finding average biomass productivity of crop residues of around 11 t dm per ha per year: this amount is considerable, especially considering that in Sicily more than 14,000 ha are cultivated with globe artichoke (Pesce et al. 2020). The high amount of lignocellulose in the artichoke residues (stalks, leaves, and unharvested heads) makes them attractive for obtaining polysaccharides and subsequent fermentation to bioethanol. It is to underline that lignocellulosic biomass is the most abundant renewable resource on Earth, constituting about 50% of world biomass. It is outside the human food chain, but its energetic content could often exceed the world's basic energy requirements (Fernandes et al. 2012). Moreover, the globe artichoke cultivations are generally concentrated in productive districts (6 productive districts just Sicily, but the same applies to Spain, in the Region of Murcia and the Valencian Community, and for France, in Brittany), making all the residues available within few kilometers and hence allowing to reduce the transportation costs strongly.

Already in 2006, del Campo et al. (2006) and coworkers showed that artichoke residues could be used as potential biomass for bioethanol production after a dilute acid hydrolysis (DHA) treatment, even if the needed acid concentration and the working temperature were higher (0.5% and 130 °C) than for other residues like tomatoes and red pepper due to the high starch and inulin content.

In 2008, Ballesteros et al. explored the possibility of obtaining bioethanol from artichoke stems and leaves by dilute sulfuric acid in a batch reactor and deeply investigated, by a response surface method employing a Box–Behnken experimental design, the effect of temperature (160–200 °C), acid concentration (0–0.2%, w/w), and solid/liquid ratio (5–10%, w/v) on the overall sugar yields. The working parameters obtained by the DOE to maximize the overall yield in glucose and xylose (70% and 93%, respectively) were 184.5 °C and 0.15% (w/v) acid concentration. Overall, xylose yield reaches a maximum of 90% of the content in raw material at 180 °C and 0.1% (w/v) acid addition and 7.5% (w/v) solid concentration. Regarding overall glucose yield, the best results were achieved at 200 °C, 0.2% (w/w) acid loading, and 7.5% (w/v) solid concentration. These same pre-treatment conditions allowed the highest simultaneous saccharification and fermentation

(SSF) yields (with cellulose-hydrolyzing enzymes), reaching values close to 65% of the theoretical ones, corresponding to about 23 g/L ethanol in the medium. These results are in the order of those reported for steam-exploded sunflower stalks submitted to SSF (ethanol yields close to 68% of theoretical) (Ruiz et al. 2011) and higher than for herbaceous residues, steam exploded and subjected to SSF by using *K. marxianus* (maximum ethanol contents from 16 to 19 g/L in fermentation media) (Ballesteros et al. 2004), thanks to the higher cellulosic content in the artichoke pre-treated material.

For *Spinoso sardo*, Pesce et al. (2020) applied different pre-treatments by neutral and DAH to break the lignocellulosic structure and evaluate the possibility of producing sugars and bioethanol under mild and environmentally friendly conditions. The responses of the biomass to the hydrolysis were evaluated by the saccharification yield of the pre-treated solids: with the harshest applied conditions (H_2SO_4 2% and 121 °C for 1 h), up to 440 kg of fermentable sugars (glucose and xylose) per ton of dry mass were obtained, allowing to calculate a theoretical amount of obtainable bioethanol of around 3,900 kg per cultivated ha. The dry biomass contains free sugars (fructose and glucose), going toward degradation with the formation of furans, a strong inhibitor of the fermentation processes. The decomposition process increased with the acid amount in the solution. To remove the free sugars, a mild pre-treatment (H_2O + 100 °C) was performed before the acid hydrolysis with 2% H_2SO_4: by considering the fermentation of the final biomass and the sugars extracted in the liquid phase and then converted to bioethanol, a theoretical total yield equal to 3361 kg ha^{-1} was obtained. Three years later, a study on the hybrid variety of globe artichoke "Opera F1" (OF1) was performed "from the field to the fermenter" (Pesce et al. 2023) using the wastes collected in 2 years. In this case, the waste's average annual production was 14 t/ha, and the complex treatment made of the two steps of water extraction of the simple sugars plus DAH produced a solid fraction containing 52% glucan and 14% xylan and a slurry containing soluble sugars (17.7%), acetic acid (4.2%), and furanic compounds (0.5%). The slurry was subjected to enzymatic saccharification, resulted in a high yield of glucans (69%), and subsequently to a 24 h ethanolic enzymatic fermentation by the *E. coli* strain MS04, leading to a total efficiency of 76% and volumetric productivity of 0.52 g/L/h, that is around 55% of the value obtained for sugar cane bagasse treated by DHA and alkaline delignification and using recombinant *Saccharomyces cerevisiae* containing the β-glucosidase gene from *Humicola grisea* as enzyme (Ferreira et al. 2010). Considering all the DAH-produced sugars, the total ethanol yield was 283 kg/t dm, equivalent to 2,399 kg per cultivated ha. Benadjemia et al. (2020) demonstrated bioethanol production with a 14.2% yield by putting the sugar-rich juice from the artichoke stalks in contact with yeast (*Saccharomyces cerevisiae*) at pH = 4.5 and 32 °C for 5 days after steam explosion, a treatment commonly used to improve cellulose disruption. To strengthen the concept of circular economy and reuse of wastes, Zuorro et al. (2016) evaluated the elemental composition (in terms of C, H, and N) and the higher heating value (HHV; the amount of heat released by the unit mass or volume of fuel, initially at 25 °C, once it is combusted and the products have returned to a temperature of 25 °C) of artichoke wastes (outer bracts and stems) and of the solid residues obtained after 90% polyphenols extraction by an environmentally friendly solvent procedure (50% v/v aqueous ethanol as solvent, T = 53.3 °C, extraction time = 110.4 min, and liquid to solid ratio = 3.3 mL/g). The carbon and hydrogen atomic content increased after the treatment (from 41% to 44% and from 6% to 6.3%, respectively), leading to an increase in HHV from 19.1 ± 0.3 MJ/kg to 19.6 ± 0.4 MJ/kg, while the N content decreased from 1.8% to 1.4%, making lower the production of nitric oxides during combustion and hence its danger. The obtained values underlined the scarce influence of the extraction pre-treatment on the energy properties of the artichoke waste and resulted in line with those reported for other raw biomasses, lying generally in the range of 15–20 MJ kg^{-1} (Chen 2015; Lavecchia et al. 2019) suggesting the possibility to use this biomass as solid fuel for bioenergy production (as a reference, for coal the HHV range from 25 to 35 MJ kg^{-1}). To facilitate the handling of the solid fuel, the authors tried to prepare pellets by mixing the artichoke residues with different amounts of sawdust (20–25% lignin, 40–45% cellulose, and 20–30% hemicellulose) (Trinh et al. 2013), the substance is chosen because of its low cost and wide availability. The HHV of the mixtures decreased with increasing the additive content, but the decrease was limited to 3–4% for sawdust addition up to 20%, pointing out the possibility of its utilization.

Francavilla et al. determined the HHV of artichoke residues (heads, leaves, stalks, and root collected in the fields) after a sequential extraction process of phenols and inulin by green solvents (water and ethanol)-based MAE, finding a meaningful increase in the C content and hence in the HHV for heads, leaves, and stalks, that allows reaching an average HHV value of 18.92 MJ/kg, with the best performance for the leaves and the heads (19.2 MJ/kg) (Francavilla et al. 2021).

Concerning biomethane, another interesting biofuel, De Menna et al. (2016) evaluated its potential production (BMP) from the aboveground non-food parts (excluding the head) for five artichoke varieties (the most abundant Spinoso sardo, C3, Madrigal, Tema, and Violetto) cultivated in Sardinia in the years spans 2013–2014 and 2014–2015 both in wet basis and with catalytic enzymes application within a 31-days test time, under mesophilic condition (39 ° C). The used enzyme preparation, needed to improve the biodegradability of lignocellulosic feedstocks and the fluidity of the biomass during the mixing process, contained cellulase, xylanase, beta-glucanase, and other carbohydrase activities originated from a selected strain of *Trichoderma reesei*. The best-performing cultivar in terms of by-product yield was Madrigal, a late hybrid characterized by a long growing cycle and a short heads harvesting time (from April to May), while Spinoso sardo showed a statistically significant low yield, being characterized by an early and long harvesting period (from November to March) and, as a consequence, by an early senescence phase. The average by-product yield was 40.1 t ha^{-1}, with a dry mass content of 20.74%, around 90% organic components. The BMP value for the fresh wastes was, on average, equal to 292.2 Nm^3 t_{DOM}^{-1}, which is 34% higher than tomato skins and seeds and similar to wasted tomatoes but ca. 21% lower than maize silage and fiber corn (Bruni et al. 2010; Godin et al. 2014; Calabrò et al. 2015). The enzyme action was limited to a + 1.1% increase in terms of biomethane yield, with positive effects for Spinoso sardo, Violetto, and Tema (+4.0%, +1.9%, and + 1.0%, respectively) and negative variations for Madrigal and C3, (0.3% and _1.3%), probably due to their highest lignin contents. Anyway, the authors observed an improvement in methane production in the beginning days for all the varieties: this result confirms the potential application of this pre-treatment in the co-digestion of slowly biodegradable biomasses, such as livestock waste. Considering the promising results, the authors estimated the possibility of covering 60% of the installed Sardinian biogas production using globe artichoke wastes.

In Table 4, the energy applications are summarized.

5.2 *Paper Pulp*

The use of fiber from non-wood plants represents sustainable pulp production, considering the lack of raw materials and finding new sources. Over the years, the search for non-wood plants has been focused mainly on species with high biomass production (Bajpai 2021).

As mentioned above, artichoke wastes are rich in hemicellulose, cellulose, and lignin, and these components can be used for paper pulp production, representing a potential alternative to eucalyptus pulp, which is a well-known high-quality wood in the pulp industry. Artichoke fiber has a cross-sectional dimension similar to eucalyptus fiber but is longer (about 30%) (Fernández 1993; European Commission 2020; Gominho et al. 2001).

Stems are richer in cellulose (46.4–49.3% dm) and lignin (7.5–13.2% dm) in comparison to bracts (cellulose: 38.5% dm; lignin: 6.6% dm). The content in hemicellulose is similar in both by-products (around 22–24%). Fiber obtained from artichoke stems is longer than 0.5 mm with higher tensile strength and stretch in refined pulp than unrefined ones. In addition, low energy and costs are required for refining artichoke pulp, representing a significant advantage in improving pulp quality without raising costs (Fernández et al. 2006).

Table 4 Energy application of artichoke by-products

Energy	Type of product	Artichoke by-product	Reference
Gaseous biofuel	Biomethane	Aboveground non-food parts	De Menna et al. (2016)
Liquid biofuel	Bioethanol	Harvested biomass, stems, and leaves	del Campo et al. (2006), Ballesteros et al. (2008), Pesce et al. (2020, 2023)
Solid biofuel	Biomass	Harvested biomass, heads, leaves, stalks, and roots collected in the fields	Ledda et al. (2013), Francavilla et al. (2021)
		Residues after extraction of phenols and inulin	Zuorro et al. (2016), Francavilla et al. (2021)

5.3 Polymers

Recently, the lignocellulosic portion of stems has been studied to reinforce poly(lactic acid) (PLA)-based biocomposites (with a fiber loading of 10%), forming polymers with higher tensile strength when unidirectional long artichoke fiber is used. PLA composites are biodegradable polymers with applications ranging from general polymer science to biomedical engineering. Adding artichoke fiber improves PLA matrix properties, reducing their plastic deformation capability (Fiore et al. 2011; Botta et al. 2015). It is well known that thermal behavior is essential in using natural fiber to reinforce biocomposites, as they have low thermal stability. The main thermal degradation of stem fiber ranges between 230 and 295 °C when hemicellulose and pectin depolymerize and at about 352 °C when cellulose decomposes, and therefore, their addition to PLA increases biocomposites stability and their mechanical integrity (Fiore et al. 2011).

Artichoke leaves have also been used as polymers to reinforce high-density polyethylene (HDPE) composites, enhancing their viscoelastic and thermal properties as a function of the amount of artichoke leaf extract added (Gümüş et al. 2022).

5.4 Herbicides

Nowadays, herbicide resistance has increased, so the need for new chemical classes of herbicides has become an important issue. In addition, the search for friendly-environmental compounds overcame the use of synthetic molecules (Peterson et al. 2018).

Phenolic compounds and sesquiterpene lactones are well known for their phytotoxic and antimicrobial activities (see par. 3.1.5), mainly due to the presence of 5-*O*-CQA, 1,5-di-*O*-CQA and monosuccinyl-di-CQA, luteolin and apigenin derivatives, and cynaropicrin in leaves. The main activity is against Gram-positive species (*Bacillus megaterium*, *B. subtilis*, *B. cereus*, *Rhodococcus fascians*, *Listeria innocua*, and *Staphylococcus aureus*). Regarding the action against Gram-negative bacteria (*Xanthomonas perforans*, *Pseudomonas syringae* pv. Tomato, *Escherichia coli*, *Salmonella enterica*, *Pseudomonas fluorescens*) the type of extraction solvent is fundamental; in fact, only ethanolic extracts had antimicrobial properties (Scavo et al. 2019). It was also essential to analyze the allelopathic properties concerning the seed germination of common weed species, such as *Amaranthus retroflexus* L., *Portulaca oleracea* L., and *Stellaria media* (L.) Vill., and *Anagallis arvensis* L. concerning the extraction solvent and different processes applied to the extract. Fresh, dried, or freeze-dried artichoke leaves can have various activities, as the conservation procedures influence the polyphenolic content. The best results on seed germination were obtained for freeze-dried leaves (Scavo et al. 2020).

5.5 Sorbents

Inorganic and organic pollutants are common environmental toxic compounds, and nowadays, heavy metals, such as mainly Cu(II), Cd(II), and Pb(II), represent the largest wastewater contaminants. So, a very important issue is searching for economic strategies to treat water and remove these contaminants. A promising possibility could be using adsorbents from agro-food wastes and by-products (Nguyen et al. 2013). Lignocellulosic biomass is a good sorbent of metal ions thanks to different chemical groups (amino, aldehyde, ketone, hydroxyl, and carboxyl groups) on lignin, cellulose, and hemicellulose skeleton (Salman et al. 2014).

Artichoke bracts and stems have been tested as potential water sorbents of heavy metals. Optimization of their sorption efficiency, which depends on the metal type, sorbent and metal concentration, pH, and temperature, can allow obtaining efficient, selective, rapid, and low-cost sorbents characterized by very high adsorption (93–96%) within 2.5 h (Fernández-López et al. 2019).

5.6 Microbial Enzyme Production

The microbial production of enzymes (xylanases, cellulases, and pectinases) is beneficial in different applications for producing food, animal feed, biofuel, pulp, paper, and textiles. For example, xylanases are hemicellulases produced by different fermentation processes that can hydrolyze the backbone of xylan, the main hemicellulose component. Xylanases are present in nature (mainly in insects) and are produced by different microorganisms (bacteria, fungi, yeast, and algae). The use of agro-industrial wastes as a xylan source, necessary to produce xylanases, can solve the problem of the high cost of commercial xylan (Gautério 2023). As regards the artichoke by-products, stems have been investigated for the production of xylanases by *Aureobasidium* pullulans NRRL Y-2311-1 with potential application in saccharification of lignocellulosic biomass, a sustainable energy resource for biofuel fabrication. The enzyme production was very efficient, and its activity was strictly dependent and directly proportional to stem concentration. Temperature and pH affected xylanase activity, exhibiting its highest value at 30–50 °C and pH 4.0. These processes represent promising sustainable alternatives for producing industrial enzymes from agro-food waste (Yegin 2017).

6 Conclusions and Future Perspectives

The several applications of globe artichoke by-products represent an important example of sustainable reuse, a key step of the circular economy and a big issue of the European recovery policies. The extracts obtained from bracts and stems are the main potential wastes with biological and/or functional properties exploitable in different healthy fields, mainly in food industries, to produce functional foods, foodstuffs, and food supplements. It is also to be mentioned that their use in non-food and non-health applications, such as producing biofuels and bio-based materials, represents modern alternative and renewable energy sources and improves environmental impacts. The progress in the use of efficient green extractions will contribute to the scalability of these processes at industrial level with high yields and rapidity, with the possibility to have industries able to engage in agro-food production and reuse directly.

References

Amoriello T, Mellara F, Ruggeri S, Ciorba R, Ceccarelli D, Ciccoritti R (2022) Artichoke by-products valorization for phenols-enriched fresh egg pasta: a sustainable food design project. Sustainability 14:14778

Bajpai P (2021) Nonwood plant fibers for pulp and paper. Elsevier, Amsterdam

Ballesteros M, Oliva J, Negro MJ, Manzanares P, Ballesteros I (2004) Ethanol from lignocellulosic materials by a simultaneous saccharification and fermentation process (SFS) with *Kluyveromyces marxianus* CECT 10875. Process Biochem 39:1843–1848

Ballesteros I, Manzanares P, Negro MJ, Oliva J, Sáez F (2008) Dilute sulfuric acid pretreatment of cardoon for ethanol production. Biochem Eng J 42:84–91

Beltran AM, Jepsen K, Rufí-Salís M, Ventura S, Madrid Lopez C, Villalba G (2022) Mapping direct N2O emissions from peri-urban agriculture: the case of the Metropolitan Area of Barcelona. Sci Total Environ 20:153514–153524

Benadjemia M, Benderdouche N, Bestani B (2020) Production of bioethanol and activated carbon from artichoke stalks-application of non linear models to dye adsorption. Indian J Environ Prot 40:259–268

Benkhoud H, Baâti T, Njim L, Selmi S, Hosni K (2021) Antioxidant, antidiabetic, and antihyperlipidemic activities of wheat flour-based chips incorporated with omega-3-rich fish oil and artichoke powder. J Food Biochem 45:13297

Biel W, Witkowicz R, Piątkowska E, Podsiadło C (2020) Proximate composition, minerals and antioxidant activity of artichoke leaf extracts. Biol Trace Elem Res 194:589–595

Blanch HW, Wilke CR (1983) Sugars and chemicals from cellulose. Rev Chem Eng 1:71–119

Borsini AA, Llavata B, Umaña M, Cárcel JA (2021) Artichoke by products as a source of antioxidant and fiber: how it can be affected by drying temperature. Food Secur 10:459

Botta L, Fiore V, Scalici T, Valenza A, Scaffaro R (2015) New polylactic acid composites reinforced with artichoke fibers. Materials 8:7770–7779

Boubaker M, Omri AEL, Blecker C, Bouzouita N (2016) Fibre concentrate from artichoke (*Cynara scolymus* L.) stem by-products: characterization and application as a bakery product ingredient. Food Sci Technol Int 22:759–768

Bravo Bolívar MS, Pasini F, Marzocchi S, Ravagli C, Tedeschi P (2023) Future perspective and technological innovation in cheese making using artichoke (*Cynara scolymus*) as vegetable rennet: a review. Food Secur 12:3032

Brodkorb A, Egger L, Alminger M, Alvito P, Assunção R, Ballance S, Bohn T, Bourlieu-Lacanal C, Boutrou R, Carrière F, Clemente A, Corredig M, Dupont D, Dufour C, Edwards C, Golding M, Karakaya S, Kirkhus B, Le Feunteun S, Lesmes U, Macierzanka A, Mackie AR, Martins C, Marze S, McClements DJ, Ménard O, Minekus M, Portmann R, Santos NC, Souchon I, Singh RP, Vergarud GE, Wickham MSJ, Weitschies W, Recio I (2019) Infogest static in vitro simulation of gastrointestinal food digestion. Nat Protoc 14:991–1040

Bruni E, Jensen AP, Pedersen ES, Angelidaki I (2010) Anaerobic digestion of maize focusing on variety, harvest time and pretreatment. Appl Energy 87:2212–2217

Calabrò PS, Greco R, Evangelou A, Komilis D (2015) Anaerobic digestion of tomato processing waste: effect of alkaline pretreatment. J Environ Manag 163:49–52

Canale M, Sanfilippo R, Strano MC, Amenta M, Allegra M, Proetto I, Papa M, Palmeri R, Todaro A, Spina A (2023) Artichoke industrial waste in durum wheat bread: effects of two different preparation and drying methods of flours and evaluation of quality parameters during short storage. Foods 12:3419

Canale M, Spina A, Summo C, Strano MC, Bizzini M, Allegra M, Sanfilippo R, Amenta M, Pasqualone A (2022) Waste from artichoke processing industry: reuse in bread-making and evaluation of the physico-chemical characteristics of the final product. Plan Theory 11:3409

Carpentieri S, Augimeri G, Ceramella J, Vivacqua A, Sinicropi MS, Pataro G, Bonofiglio D, Ferrari G (2022) Antioxidant and anti-inflammatory effects of extracts from pulsed electric field-treated artichoke by-products in lipopolysaccharide-stimulated human THP-1 macrophages. Foods 11:2250

Castellino M, Renna M, Leoni B, Calasso M, Difonzo G, Santamaria P, Gambacorta G, Caponio F, De Angelis M, Paradiso VM (2020) Conventional and unconventional recovery of inulin rich extracts for food use from the roots of globe artichoke. Food Hydrocoll 107:105975

Cavini S, Guzzetti L, Givoia F, Regonesi ME, Di Gennaro P, Magoni C, Campone L, Labra M, Bruni I (2020) Artichoke *(Cynara cardunculus* var. *scolymus* L.) by-products as a source of inulin: how to valorise an agricultural supply chain extracting an added-value compound. Nat Prod Res 36:1–5

Chen W-H (2015) Acidic pretreatment. In: Pandey A, Negi S, Binod P, Larroche C (eds) Pretreatment of biomass processes and technologies. Elsevier, Amsterdam

Chilakamarry CR, Mimi Sakinah AM, Zularisam AW, Sirohi R, Khilji IA, Ahmad N, Pandey A (2022) Advances in solid-state fermentation for bioconversion of agricultural wastes to value-added products: opportunities and challenges. Bioresour Technol 343:126065

Claus T, Maruyama SA, Palombini SV, Montanher PF, Bonafé EG, de Oliveira Santos Junior, O., Matsushita, M., & Visentainer, J. V. (2015) Chemical characterization and use of artichoke parts for protection from oxidative stress in canola oil. Food Sci Technol 61:346–351

Coinu R, Carta S, Urgeghe PP, Mulinacci N, Pinelli P, Franconi F, Romani A (2007) Dose-effect study on the antioxidant properties of leaves and outer bracts of extracts obtained from Violetto di Toscana artichoke. Food Chem 101:524–531

Colantuono A, Ferracane R, Vitaglione P (2018) Potential bioaccessibility and functionality of polyphenols and cynaropicrin from breads enriched with artichoke stem. Food Chem 245:838–844

Colantuono A, Vitaglione P, Ferracane R, Campanella OH, Hamaker BR (2017) Development and functional characterization of new antioxidant dietary fibers from pomegranate, olive and artichoke by-products. Food Res Int 101:155–164

Colombo R, Ferron L, Frosi I, Papetti A (2021) Advances in static in vitro digestion models after the COST action Infogest consensus protocol. Food Funct 12:7619–7636

Colombo R, Paolillo M, Frosi I, Ferron L, Papetti A (2023) Effect of the simulated digestion process on the chlorogenic acid trapping activity against methylglyoxal. Food Funct 14:541–549

Costa C, Lucera A, Marinelli V, Del Nobile MA, Conte A (2018) Influence of different by-products addition on sensory and physicochemical aspects of Primosale cheese. J Food Sci Technol 55:4174–4183

D'Antuono I, Carola A, Sena LM, Linsalata V, Cardinali A, Logrieco AF, Colucci MG, Apone F (2018) Artichoke polyphenols produce skin anti-age effects by improving endothelial cell integrity and functionality. Molecules 23:2729

Dadalı C (2023) Artichoke bracts as fat and wheat flour replacer in cake: optimization of reduced fat and reduced wheat flour cake formulation. J Food Meas Charact 17:98–107

Davani-Davari D, Negahdaripour M, Karimzadeh I, Seifan M, Mohkam M, Masoumi SJ, Berenjian A, Ghasemi Y (2019) Prebiotics: definition, types, sources, mechanisms, and clinical applications. Foods 8:92

de Falco B, Incerti G, Amato M, Lanzotti V (2015) Artichoke: botanical, agronomical, phytochemical, and pharmacological overview. Phytochem Rev 14:993–1018

del Campo I, Alegría I, Zazpe M, Echeverría M, Echeverría I (2006) Diluted acid hydrolysis pretreatment of Agri-food wastes for bioethanol production. Ind Crop Prod 24:214–221

De Menna F, Malagnino RA, Vittuari M, Molari G, Seddaiu G, Deligios PA, Solinas S, Ledda L (2016) Potential biogas production from artichoke byproducts in Sardinia, Italy. Energies 9:92–102

Difonzo G, de Gennaro G, Caponio GR, Vacca M, dal Poggetto G, Allegretta I, Immirzi B, Pasqualone A (2022) Inulin from globe artichoke roots: a promising ingredient for the production of functional fresh pasta. Foods 11:3032

Domingo CS, Gonzàlez CO, Navarro D, Stortz C, Rojas AM, Gerschenson LN, Fissore EN (2021) Enzyme assisted extraction of pectin and inulin enriched fractions isolated from microwave treated *Cynara cardunculus* tissues. Int J Food Sci Technol 56:242–249

Domingo CS, Rojas AM, Fissore EN, Gerschenson LN (2019) Rheological behavior of soluble dietary fiber fractions isolated from artichoke residues. Eur Food Res Technol 245:1239–1249

Dupont D, Arlic M, Blanquet-Diot S, Bornhost G, Cueva C, Deglaire A, Denis S, Ferrua M, Havenaar R, Lelieveld J, Mackie AR, Marzorati M, Menard O, Minekus M, Miralles B, Recio I, Abbeele PVD (2019) Can dynamic in vitro digestion systems mimic the physiological reality? Crit Rev Food Sci Nutr 59:1546–1562

Dvorakova M, Landa P (2017) Anti-inflammatory activity of natural stilbenoids: a review. Pharmacol Res 12:2935

Egger L, Ménard O, Delgado-Andrade C, Alvito P, Assunção R, Balance S, Bradkorb A, Cattenoz T, Clemente A, Comi I, Dupont D, Garcia-Llatas G, Legarda MJ, Le Feunteun S, JanssenDuijghuijsen L, Karakaya S, Lesmes U, Mackie AR, Martins C, Meynier A, Miralles N, Murray BS, Pihlanto A, Picariello G, Santos CN, Simsek S, Recio I, Rigby N, Rioux L, Stoffers H, Tavares A, Tavares L, Turgeon S, Ulleberg EK, Vegarud GE, Vergères G, Portmann R (2016) The harmonized INFOGEST in vitro digestion method: from knowledge to action. Food Res Int 88:217–225

Ehsani J, Mohammad Mortazavian A, Khomeiri M, Ghasem Nejad A (2015) Effects of artichoke (Cynara Scolymus L.) extract addition on microbiological and physico-chemical properties of probiotic yogurt. J Microbiol Biotechnol Food Sci 4:536–541

Ehsani J, Mohesenzadeh M, Khomeiri M, Ghasemnezhad A (2018) Chemical characteristics, and effect of inulin extracted from artichoke (*Cynara scolymus* L.) root on biochemical properties of synbiotic yogurt at the end of fermentation. Iran J Chem Chem Eng 37:291–230

Elimelech E, Ofira A, Eyal E (2018) What gets measured gets managed: a new method of measuring household food waste. Waste Manag 76:68–81

El-Kholy WM, Bisar GH, Aamer RA (2023) Impact of inulin extracted, purified from (chicory and globe artichoke) roots and the combination with maltodextrin as prebiotic dietary fiber on the functional properties of stirred bio-yogurt. Food Nutr Sci 14:70–89

El-Nashar HAS, Abbas H, Zewail M, Noureldin MH, Ali MM, Shamaa MM, Khattab MA, Ibrahim N (2022) Neuroprotective effect of artichoke-based nanoformulation in sporadic Alzheimer's disease mouse model: focus on antioxidant, anti-inflammatory, and amyloidogenic pathways. Pharmaceuticals 15:1202

Erdogan O, Abbak M, Demirbolat GM, Birtekocak F, Aksel M, Pasa S, Cevik O (2019) Green synthesis of silver nanoparticles via *Cynara scolymus* leaf extracts: the characterization, anticancer potential with photodynamic therapy in MCF7 cells. PLoS One 14:e0216496

Ergezer H, Kaya Hİ, ŞiMşek Ö (2018) Antioxidant and antimicrobial potential of artichoke (*Cynara scolymus* L.) extract in beef patties. Czech J Food Sci 36:154–162

Ergezer H, Serdaroğlu M (2018) Antioxidant potential of artichoke (*Cynara scolymus* L.) byproducts extracts in raw beef patties during refrigerated storage. J Food Meas Charact 12:982–991

Esposito M, Di Pierro P, Dejonghe W, Mariniello L, Porta R (2016) Enzymatic milk clotting activity in artichoke (*Cynara scolymus*) leaves and alpine thistle (*Carduus defloratus*) flowers. Immobilization of alpine thistle aspartic protease. Food Chem 204:115–121

European Commission (2020) A European strategy for smart, sustainable and inclusive growth. Available at https://commission.europa.eu/index_en

Fan Z, Chen K, Ban L, Mao Y, Hou C, Li J (2020) Silage fermentation: a potential biological approach for the long-term preservation and recycling of polyphenols and terpenes in globe artichoke (*Cynara scolymus* L.) by-products. Molecules 25:3302

FAOSTAT (2022) Food and agriculture organization: crops and livestock products. Available at http://fenix.fao.org/faostat/internal/en/#data/QCL

FAOSTAT (2015) Food and agriculture organization: technical platform on the measurement and reduction of food loss and waste. Available at https://www.fao.org/platform-food-loss-waste/flw-data/en/

Feiden T, Valduga E, Zeni J, Steffens J (2023) Bioactive compounds from artichoke and application potential. Food Technol Biotechnol 61:312–327

Fernandes DLA, Pereira SR, Serafim LS, Evtuguin D, Xavier AMRB (2012) Second generation bioethanol from lignocellulosics: processing of hardwood sulphite spent liquor. In: Lima MAP, Natalense APP (eds) Bioethanol. Rijeka, Croatia, pp 123–152

Fernández J (1993) Production and utilization of *Cynara cardunculus* L. biomass for energy, paper-pulp and food industry. Final report JOUB 0030-ECCE. Commission of the European Communities, Brussels, p 92

Fernández J, Curt MD, Aguado PL (2006) Industrial applications of *Cynara cardunculus* L. for energy and other uses. Ind Crop Prod 24:222–229

Fernández-López J, Angosto JM, Roca MJ, Doval Miñarro M (2019) Taguchi design-based enhancement of heavy metals bioremoval by agroindustrial waste biomass from artichoke. Sci Total Environ 653:55–63

Ferreira V, de Oliveira Faber M, da Silva Mesquita S, Pereira Jr N (2010) Simultaneous saccharification and fermentation process of different cellulosic substrates using a recombinant Saccharomyces cerevisiae harbouring the β-glucosidase gene. Electron J Biotechnol 13:5–6

Ferreira-Lazarte A, Kachrimanidou V, Villamiel M, Rastall RA, Moreno FJ (2018) In vitro fermentation properties of pectins and enzymatic-modified pectins obtained from different renewable bioresources. Carbohydr Polym 199:482–491

Fiore V, Valenza A, Di Bella G (2011) Artichoke (*Cynara cardunculus* L.) fibres as potential reinforcement of composite structures. Compos Sci Technol 71:1138–1144

Fissore EN, Santo Domingo C, Gerschenson LN, Giannuzzi L (2015) A study of the effect of dietary fiber fractions obtained from artichoke (*Cynara cardunculus* L. var. *scolymus*) on the growth of intestinal bacteria associated with health. Food Funct 6:1667–1674

Fissore EN, Santo Domingo C, Pujol CA, Damonte EB, Rojas AM, Gerschenson LN (2014) Upgrading of residues of bracts, stems and hearts of *Cynara cardunculus* L. var. *scolymus* to functional fractions enriched in soluble fiber. Food Funct 5:463–470

Francavilla M, Marone M, Marasco P, Contillo F, Monteleone M (2021) Artichoke iborefinery: from food to advanced technological applications. Foods 10:112

Frosi I, Montagna I, Colombo R, Milanese C, Papetti A (2021) Recovery of chlorogenic acids from Agri-food wastes: updates on green extraction techniques. Molecules 26:4515

Frutos MJ, Guilabert-Antón L, Tomás-Bellido A, Hernández-Herrero JA (2008) Effect of artichoke (*Cynara scolymus* L.) fiber on textural and sensory qualities of wheat bread. Food Sci Technol Int 14:49–55

García-Rodríguez J, Ranilla MJ, France J, Alaiz-Moretón H, Carro MD, López S (2019) Chemical composition, in vitro digestibility and rumen fermentation kinetics of agro-industrial by-products. Animals 9:861

Gatmiri SM, Khadem E, Fakhrian T, Kamalinejad M, Hosseini H, Ghorat F, Alamdari A, Naderi N (2019) The effect of artichoke leaf extract supplementation on lipid profile of chronic kidney disease patients; a double-blind, randomized clinical trial. J Renal Inj Prev 8:225–229

Gautério GV (2023) Industrial xylanase production using Agri-food wastes through microbial applications microbial bioprocessing of Agri-food wastes. Industrial enzymes. In: Molina G, Usmani Z, Sharma M, Benhida R, Kuhad RC, Gupta VK (eds) Microbial bioprocessing of Agri-food wastes. CRC Press, pp 83–116

Gerschenson LN, Fissore EN, Rojas AM, Bernhardt DC, Domingo CS (2020) Artichoke. In: Jaiswal AK (ed) Nutritional composition and antioxidant properties of fruits and vegetables. Academic Press, pp 55–69

Godin B, Mayer F, Agneessens R, Gerin P, Dardenne P, Delfosse P, Delcarte J (2014) Biochemical methane potential prediction of plant biomasses: comparing chemical composition versus near infrared methods and linear versus non-linear models. Bioresour Technol 175:382–390

Gominho J, Curt MD, Lourenço A, Fernández J, Pereira H (2018) *Cynara cardunculus* L. as a biomass and multi-purpose crop: a review of 30 years of research. Biomass Bioenergy 109:257–275

Gominho J, Fernandez J, Pereira H (2001) *Cynara cardunculus* L.–a new fibre crop for pulp and paper production. Ind Crop Prod 13:1–10

González-Lamothe R, Mitchell G, Gattuso M, Diarra MS, Malouin F, Bouarab K (2009) Plant antimicrobial agents and their effects on plant and human pathogens. Int J Mol Sci 10:3400–3419

Greco D, D'Ascanio V, Santovito E, Logrieco AF, Avantaggiato G (2019) Comparative efficacy of agricultural by-products in sequestering mycotoxins. J Sci Food Agric 99:1623–1634

Grigelmo-Miguel N, Martín-Belloso O (1999) Comparison of dietary fibre from by-products of processing fruits and greens and from cereals. LWT-Food Sci Technol 32:503–508

Gümüş BE, Yağci O, Taşdemir M (2022) High-density polyethylene/artichoke leaf powder polymer composites: dynamic mechanical, morphological and thermal properties. Iran Polym J 31:787–797

Gyawali R, Ibrahim SA (2014) Natural products as antimicrobial agents. Food Control 46:412–429

Holgado F, Campos-Monfort G, de las Heras C, Rupérez P (2021) In vitro fermentability of globe artichoke by-product by *lactobacillus acidophilus* and *Bifidobacterium bifidum*. Bioact Carbohydr Diet Fibre 26:120086

Hughes SR, Qureshi N, López-Núñez JC, Jones MA, Jarodsky JM, Galindo-Leva LA, Lindquist MR (2017) Utilization of inulin-containing waste in industrial fermentations to produce biofuels and bio-based chemicals. World J Microbiol Biotechnol 33:78

Jiménez-Moreno N, Cimminelli MJ, Volpe F, Ansó R, Esparza I, Mármol I, Rodríguez-Yoldi MJ, Ancín-Azpilicueta C (2019) Phenolic composition of artichoke waste and its antioxidant capacity on differentiated Caco-2 cells. Nutrients 11:1723

Kollia E, Markaki P, Zoumpoulakis P, Proestos C (2017) Antioxidant activity of *Cynara scolymus* L. and *Cynara cardunculus* L. extracts obtained by different extraction techniques. Nat Prod Res 31:1163–1167

Kong F, Singh RPA (2010) A human gastric simulator (HGS) to study food digestion in human stomach. J Food Sci 75:625–635

Larrosa M, Llorach R, Espìn JC, Tomás-Barberán FA (2002) Increase of antioxidant activity of tomato juice upon functionalisation with vegetable byproduct extracts. Lebensm-Wiss Technol 35:532–542

Lattanzio V, Kroon PA, Linsalata V, Cardinali A (2009) Globe artichoke: a functional food and source of nutraceutical ingredients. J Funct Foods 1:131–144

Lavecchia R, Maffei G, Paccassoni F, Piga L (2019) Artichoke waste as a source of phenolic antioxidants and bioenergy. Waste Biomass Valoriz 10:2975–2984

Ledda L, Deligios PA, Farci R, Sulas L (2013) Biomass supply for energetic purposes from some *Cardueae* species grown in Mediterranean farming systems. Ind Crop Prod 47:218–226

Lim HS, Kim SK, Woo SG, Kim TH, Yeom SJ, Yong W, Ko YJ, Kim SJ, Lee SG, Lee DH (2021) (−)-α-Bisabolol production in engineered Escherichia coli expressing a novel (−)-α-bisabolol synthase from the globe artichoke *Cynara cardunculus* var. *scolymus*. J Agric Food Chem 69:8492–8503

Llorach R, Espiän JC, Tomás-Barberán FA, Ferreres F (2002) Artichoke (*Cynara scolymus* L.) byproducts as a potential source of health-promoting antioxidant phenolics. J Agric Food Chem 50:3458–3464

Llorach R, Tomás-Barberán FA, Ferreres F (2005) Functionalisation of commercial chicken soup with enriched polyphenol extract from vegetable by-products. Eur Food Res Technol 220:31–36

Llorente BE, Brutti CB, Caffini NO (2004) Purification and characterization of a milk-clotting aspartic proteinase from globe artichoke (*Cynara scolymus* L.). J Agric Food Chem 52:8182–8189

Lo Giudice A, Mbohwa C, Clasadonte MT, Ingrao C (2014) Life cycle assessment interpretation and improvement of the Sicilian artichokes production. Int J Environ Resour Res 8:305–316

López-Salas L, Borrás-Linares I, Quintin D, García Gomez P, Giménez-Martínez R, Segura-Carretero A, Lozano-Sánchez J (2021) Artichoke by-products as natural source of phenolic food ingredient. Appl Sci 11:3788

Lucera A, Costa C, Marinelli V, Saccotelli MA, Del Nobile MA, Conte A (2018) Fruit and vegetable by-products to fortify spreadable cheese. Antioxidants 7:61

Mahboubi M (2018) Cynara scolymus (artichoke) and its efficacy in management of obesity. Bull Fac Pharm Cairo Univ 56:115–120

Maietta M, Colombo R, Lavecchia R, Sorrenti M, Zuorro A, Papetti A (2017) Artichoke (*Cynara cardunculus* L. var. *scolymus*) waste as a natural source of carbonyl trapping and antiglycative agents. Food Res Int 100:780–790

Malik K, Sharma P, Yang Y, Zhang P, Zhang L, Xing X, Yue J, Song Z, Nan L, Yujun S, El-Dalatony MM, Salama E-S, Li X (2022) Lignocellulosic biomass for bioethanol: insight into the advanced pretreatment and fermentation approaches. Ind Crop Prod 188:115569–115582

Malik SM, Azam M, Basra MAR (2020) Impact of natural antioxidants on biological systems. J Life Sci 4:140–162

Martin-Gorriz B, Gallego-Elvira B, Martinez-Alvarez V, Maestre-Valero JF (2020) Life cycle assessment of fruit and vegetable production in the region of Murcia (south-East Spain) and evaluation of impact mitigation practices. J Clean Prod 265:121656–121669

Matsumoto T, Nakashima S, Nakamura S, Hattori Y, Ando T, Matsuda H (2021) Inhibitory effects of cynaropicrin and related sesquiterpene lactones from leaves of artichoke (*Cynara scolymus* L.) on induction of iNOS in RAW264.7 cells and its high-affinity proteins. J Nat Med 75:381–392

Mazzei D, Giusti S, Sbrana T, Ahluwalia A (2011) Multicompartmental modular bioreactor as innovative system for dynamic cell cultures and co-cultures. In: Antolli PG, Liu Z (eds) Bioreactors: design, properties and applications. Nova Science Inc, USA

Mejiri F, Karmali A, Jaoued N, Casabianca H, Hosni K (2019) Purification and partial characterization of peroxidases from three food waste by-products: broad bean pods, pea pods, and artichoke stems. Appl Biochem Biotechnol 189:576–588

Mejri F, Baati T, Martins A, Selmi S, Serralheiro ML, Falé PL, Rauter A, Casabianca H, Hosni K (2020) Phytochemical analysis and in vitro and in vivo evaluation of biological activities of artichoke (*Cynara scolymus* L.) floral stems: towards the valorization of food by-products. Food Chem 333:127506

Mekonnen MM, Hoekstra AY (2011) The green, blue and grey water footprint of crops and derived crop products. Hydrol Earth Syst Sci 15:1577–1600

Mena-García A, Rodríguez-Sánchez S, Ruiz-Matute AI, Sanz ML (2020) Exploitation of artichoke byproducts to obtain bioactive extracts enriched in inositols and caffeoylquinic acids by microwave assisted extraction. J Chromatogr A 1613:460703

Menard O, Lesmes U, Shani-Levi CS, Araiza Calahorra A, Lavoisier A, Morzel M, Rieder A, Feron G, Nebbia S, Mashiah L, Andres A, Bornhorst G, Carrière F, Egger L, Gwala S, Heredia A, Kirkhus B, Macierzanka A, Portman R, Recio I, Santé-Lhoutellier V, Tournier C, Sarkar A, Brodkorb A, Mackie A, Dupont D (2023) Static in vitro digestion model adapted to the general older adult population: an INFOGEST international consensus. Food Funct 14:4569–4582

Meneses M, Madrid M, Martínez-Teruel A, Hernández F (2005) Ensiling capacity, chemical composition and multiresidue evaluation of fresh artichoke (*Cynara scolymus* L.) by-product to be used in ruminant feeding. *Options Méditerranéennes : Série a.* Séminaires Méditerranéens 67:351–354

Meneses M, Martínez-Marín AL, Madrid J, Martínez-Teruel A, Hernández F, Megías MD (2020) Ensilability, in vitro and in vivo values of the agro-industrial by-products of artichoke and broccoli. Environ Sci Pollut Res 27:2919–2925

Meneses M, Megías MD, Madrid J, Martínez-Teruel A, Hernández F, Oliva J (2007) Evaluation of the phytosanitary, fermentative and nutritive characteristics of the silage made from crude artichoke (*Cynara scolymus* L.) by-product feeding for ruminants. Small Rumin Res 70:292–296

Mensink MA, Frijlink HW, Van Der Voort Maarschalk K, Hinrichs WLJ (2015) Inulin, a flexible oligosaccharide I: review of its physicochemical characteristics. Carbohydr Polym 130:405–419

Minekus M, Alminger M, Alvito P, Ballance S, Bohn T, Bourlieu C, Carrière F, Boutrou R, Corredig M, Dupont D, Dufour C, Egger L, Golding M, Karakaya S, Kirkhus B, Le Feunteun S, Lesmes U, Macierzanka A, Mackie A, Marze S, McClements DJ, Ménard O, Recio I, Santos CN, Singh RP, Vegarud GE, Wickham MSJ, Weitschies W, Brodkorb A (2014) A standardised static in vitro digestion method suitable for food an international consensus. Food Funct 5:1113–1124

Monllor P, Muelas R, Roca A, Atzori AS, Díaz JR, Sendra E, Romero G (2020a) Long-term feeding of dairy goats with broccoli by- product and artichoke silages: Milk yield, quality and composition. Animals 10:1670

Monllor P, Romero G, Muelas R, Sandoval-Castro CA, Sendra E, Ramón Díaz J (2020b) Ensiling process in commercial bales of horticultural by-products from artichoke and broccoli. Animals 10:831

Monllor P, Romero G, Sendra E, Atzori AS, Díaz JR (2020c) Short-term effect of the inclusion of silage artichoke by-products in diets of dairy goats on milk quality. Animals 10:339

Monllor P, Zemzmi J, Muelas R, Roca A, Sendra E, Romero G, Díaz JR (2023) Long-term feeding of dairy goats with 40% artichoke by-product silage preserves milk yield, nutritional composition and animal health status. Animals 13:3585

Mortazavian AM, Ehsani MR, Mousavi SM, Rezaei K, Sohrabvandi S, Reinheimer JA (2007) Effect of refrigerated storage temperature on the viability of probiotic micro-organisms in yogurt. Int J Dairy Technol 60:123–127

Moselhy M, Abd-Elhafez K, El-Kholany E, Gohar M, Nasr N (2023) Antimicrobial, antioxidant and anticancer properties of globe artichoke and grape by-products as a source of the bio-active phenolic compounds. Egypt J Chem 66:609–624

Muelas R, Monllor P, Romero G, Sayas-Barberá E, Navarro C, Díaz JR, Sendra E (2017) Milk technological properties as affected by including artichoke by-products silages in the diet of dairy goats. Foods 6:112

Muelas R, Romero G, Díaz JR, Monllor P, Fernández-López J, Viuda-Martos M, Cano-Lamadrid M, Sendra E (2022) Quality and functional parameters of fermented milk obtained from goat milk fed with broccoli and artichoke plant by-products. Foods 11:2601

Mussatto SI, Dragone G, Guimarães PMR, Silva JPA, Carneiro LM, Roberto IC, Vicente A, Domingues L, Teixeira JA (2010) Technological trends, global market, and challenges of bioethanol production. Biotechnol Adv 28:817–830

Nguyen TA, Ngo HH, Guo WS, Zhang J, Liang S, Yue QY, Li Q, Nguyen TV (2013) Applicability of agricultural waste and by-products for adsorptive removal of heavymetals from wastewater. Bioresour Technol 148:574–585

Noriega-Rodríguez D, Soto-Maldonado C, Torres-Alarcón C, Pastrana-Castro L, Weinstein-Oppenheimer C, Zúñiga-Hansen ME (2020) Valorization of globe artichoke (*Cynara scolymus*) agro-industrial discards, obtaining an extract with a selective effect on viability of cancer cell lines. Processes 8:715

Órbenes G, Rodríguez-Seoane P, Torres MD, Chamy R, Zúñiga ME, Domínguez H (2021) Valorization of artichoke industrial by-products using green extraction technologies: formulation of hydrogels in combination with paulownia extracts. Molecules 26:4386

Ozkan G (2024) Valorization of artichoke outer petals by using ultrasound-assisted extraction and naturaldeep eutectic solvents (NADES) for the recovery of phenolic compounds. J Sci Food Agric 104:2744–2749

Pagano I, Campone L, Celano R, Piccinelli AL, Rastrelli L (2021) Green non-conventional techniques for the extraction of polyphenols from agricultural food by-products: a review. J Chromatogr A 1651:462295

Pagano I, Piccinelli AL, Celano R, Campone L, Gazzerro P, de Falco E, Rastrelli L (2016) Chemical profile and cellular antioxidant activity of artichoke by-products. Food Funct 7:4841–4850

Pagano I, Piccinelli AL, Celano R, Campone L, Gazzerro P, Russo M, Rastrelli L (2018) Pressurized hot water extraction of bioactive compounds from artichoke by-products. Electrophoresis 39:1899–1907

Pagliari S, Cannavacciuolo C, Celano R, Carabetta S, Russo M, Labra M, Campone L (2022) Valorisation, green extraction development, and metabolomic analysis of wild artichoke by-product using pressurised liquid extraction UPLC–HRMS and multivariate data analysis. Molecules 27:7157

Pandino G, Lombardo S, Mauromicale G (2013) Globe artichoke leaves and floral stems as a source of bioactive compounds. Ind Crop Prod 44:44–49

Pandino G, Lombardo S, Mauromicale G, Williamson G (2011) Profile of polyphenols and phenolic acids in bracts and receptacles of globe artichoke (*Cynara cardunculus* var. *scolymus*) germplasm. J Food Compos Anal 24:148–153

Papargyropoulou E, Lozano R, Steinberger JK, Wright N, Ujang ZB (2014) The food waste hierarchy as a framework for the management of food surplus and food waste. J Clean Prod 76:106–115

Pasqualone A, Punzi R, Trani A, Summo C, Paradiso VM, Caponio F, Gambacorta G (2017) Enrichment of fresh pasta with antioxidant extracts obtained from artichoke canning by-products by ultrasound-assisted technology and quality characterisation of the end product. Int J Food Sci Technol 52:2078–2087

Pesce GR, Alves-Ferreira J, Hsiao A, Torrado I, Martinez A, Mauromicale G, Fernandes MC (2023) Bioethanol production from globe artichoke residues: from the field to the fermenter. Bioenergy Res 16:1567–1578

Pesce GR, Fernandes MC, Mauromicale G (2020) Globe artichoke crop residues and their potential for bioethanol production by dilute acid hydrolysis. Biomass Bioenergy 134:105471–105479

Peterson MA, Collavo A, Ovejero R, Shivrain V, Walsh MJ (2018) The challenge of herbicide resistance around the world: a current summary. Pest Manag Sci 74:2246–2259

Petropoulos S, Fernandes Â, Pereira C, Tzortzakisc N, Vaz J, Soković M, Barros L, Ferreira ICFR (2019) Bioactivities, chemical composition and nutritional value of *Cynara cardunculus* L. seeds. Food Chem 289:404–412

Punzi R, Paradiso A, Fasciano C, Trani A, Faccia M, de Pinto MC, Gambacorta G (2014) Phenols and antioxidant activity in vitro and in vivo of aqueous extracts obtained by ultrasound-assisted extraction from artichoke by-products. Nat Prod Commun 9:1315–1318

Rabelo RS, Machado MTC, Martínez J, Hubinger MD (2016) Ultrasound assisted extraction and nanofiltration of phenolic compounds from artichoke solid wastes. J Food Eng 178:170–180

Raccuia SA, Melilli MG (2010) Seasonal dynamics of biomass, inulin, and water-soluble sugars in roots of *Cynara cardunculus* L. Field Crop Res 116:47–153

Raccuia SA, Melilli MG, Calderaro P, Scandurra S (2013) Globe artichoke genetic variability for residual biomass production as renewable resources of energy in South Italy. Acta Hortic 983:129–132

Rana RL, Bux C, Lombardi M (2023) Trends in scientific literature on the environmental sustainability of the artichoke (*Cynara cardunculus* L. spp.) supply chain. Br Food J 125:2315–2332

Rangboo V, Noroozi M, Zavoshy R, Rezadoost SA, Mohammadpoorasl A (2016) The effect of artichoke leaf extract on alanine aminotransferase and aspartate aminotransferase in the patients with nonalcoholic steatohepatitis. Int J Hepatol 3:1–6

Reche C, Rosselló C, Umaña MM, Eim V, Simal S (2021) Mathematical modelling of ultrasound-assisted extraction kinetics of bioactive compounds from artichoke by-products. Foods 10:931

Rejeb IB, Dhen N, Gargouri M, Boulila A (2020) Chemical composition, antioxidant potential and enzymes inhibitory properties of globe artichoke by-products. Chem Biodivers 17:e2000073

Rondanelli M, Monteferrario F, Perna S, Faliva MA, Opizzi A (2013) Health-promoting properties of artichoke in preventing cardiovascular disease by its lipidic and glycemic-reducing action. Monaldi Arch Chest Dis 80:17–26

Ruiz E, Miralles J, Cara C, Romero I, Moya M, Castro E (2011) Strategies for bioethanol production from sunflower stalks. *Afinidad.* J Chem Engg Theo Appl Chem 68:417–423

Ruiz-Cano D, Frutos MJ, Hernández-Herrero JA, Pérez-Llamas F, Zamora S (2015) Effect of chlorophyll removal and particle size upon the nutritional and technological properties of powdered by-products from artichoke (*Cynara scolymus*, L.) industrial canning. Int J Food Sci Technol 50:2383–2390

Ruiz-Cano D, Pérez-Llamas F, Frutos MJ, Arnao MB, Espinosa C, López-Jiménez JA, Castillo J, Zamora S (2014) Chemical and functional properties of the different by-products of artichoke (*Cynara scolymus* L.) from industrial canning processing. Food Chem 160:134–140

Sabater C, Molina-Tijeras JA, Vezza T, Corzo N, Montilla A, Utrilla P (2019) Intestinal anti-inflammatory effects of artichoke pectin and modified pectin fractions in the dextran sulfate sodium model of mice colitis. Artificial neural network modelling of inflammatory markers. Food Funct 10:7793–7805

Salami SA, Luciano G, O'Grady MN, Biondi L, Newbold CJ, Kerry JP, Priolo A (2019) Sustainability of feeding plant by-products: a review of the implications for ruminant meat production. Anim Feed Sci Technol 251:37–55

Saleh IA, Vinatoru M, Mason TJ, Abdel-Azim NS, Aboutabl EA, Hammouda FM (2016) A possible general mechanism for ultrasound-assisted extraction (UAE) suggested from the results of UAE of chlorogenic acid from *Cynara scolymus* L. (artichoke) leaves. Ultrason Sonochem 31:330–336

Salem MB, Affes H, Athmouni K, Ksouda K, Dhouibi R, Sahnoun Z, Hammami S, Zeghal KM (2017) Chemicals compositions, antioxidant and anti-inflammatory activity of *Cynara scolymus* leaves extracts, and analysis of major bioactive polyphenols by HPLC. Evid Based Complementary Altern Med, p 4951937

Salem MB, Affes H, Ksouda K, Dhouibi R, Sahnoun Z, Hammami S, Zeghal KM (2015) Pharmacological studies of artichoke leaf extract and their health benefits. Plant Foods Hum Nutr 70:441–453

Salman FM, El-Nomeary YAA, Abedo AA, Abd El-Rahman HH, Mohamed MI, Ahmed SM (2014) Utilization of artichoke (*Cynara scolymus*) by-products in sheep feeding. Am Eurasian J Agric Environ Sci 14:624–630

San José FJS, Collado-Fernández M, Álvarez-Castellanos PP (2023) Variation, during shelf life, of functional properties of biscuits enriched with fibers extracted from artichoke (*Cynara scolymus* L.). Nutrients 15:3329

San José FJS, Collado-Fernández M, López R (2018) Sensory evaluation of biscuits enriched with artichoke fiber-rich powders (*Cynara scolymus* L.). Food Sci Nutr 6:160–167

San Martin D, Bald C, Cebrian M, Iñarra B, Orive M, Ramos S, Zufía J (2018) Principles for developing a safe and sustainable valorisation of food waste for animal feed: second generation feedstuff. In: Preedy VR, Patel VB (eds) Handbook of famine, starvation, and nutrient deprivation. From biology to policy. Springer International Publishing AG, United Kingdom

Santos HO, Bueno AA, Mota JF (2018) The effect of artichoke on lipid profile: a review of possible mechanisms of action. Pharmacol Res 137:170–178

Santos-Sánchez NF, Salas-Coronado R, Valadez-Blanco R, Hernández-Carlos B, Guadarrama-Mendoza PC (2017) Natural antioxidant extracts as food preservatives. Acta Sci Pol Technol Aliment 16:361–370

Scavo A, Pandino G, Restuccia C, Mauromicale G (2020) Leaf extracts of cultivated cardoon as potential bioherbicide. Sci Hortic 261:109024

Scavo A, Pandino G, Restuccia C, Parafati L, Cirvilleri G, Mauromicale G (2019) Antimicrobial activity of cultivated cardoon (*Cynara cardunculus* L. var. *altilis* DC.) leaf extracts against bacterial species of agricultural and food interest. Indus Crops Prod 129:206–211

Shallan MA, Ali MA, Meshrf WA, Marrez DA (2020) In vitro antimicrobial, antioxidant and anticancer activities of globe artichoke (*Cynara cardunculus* var. *scolymus* L.) bracts and receptacles ethanolic extract. *Biocatalysis and agricultural*. Biotechnology 29:101774

Shimoda H, Ninomiya K, Nishida N, Yoshino T, Morikawa T, Matsuda H, Yoshikawa M (2003) Anti-hyperlipidemic sesquiterpenes and new sesquiterpene glycosides from the leaves of artichoke (*Cynara scolymus* L.): structure requirement and mode of action. Bioorg Med Chem Lett 13:223–228

Silva SV, Malcata FX (2005) Studies pertaining to coagulant and proteolytic activities of plant proteases from *Cynara cardunculus*. Food Chem 89:19–26

Sonnante G, Pignone D, Hammer K (2007) The domestication of artichoke and cardoon: from roman times to the genomic age. Ann Bot 100:1095–1100

Stahel WR (2016) The circular economy. Nature 531:435–438

Stumpf B, Künne M, Ma L, Xua M, Yana F, Piepho HP, Honermeier B (2020) Optimization of the extraction procedure for the determination of phenolic acids and flavonoids in the leaves of globe artichoke (*Cynara cardunculus* var. *scolymus* L.). J Pharm Biomed Anal 177:112879

The world bank, World development report (2018) Available at https://www.worldbank.org/en/publication/wdr2018/brief/world-development-report-2018-data

Trinh NT, Jensen AP, Dam-Johansen K, Knudsen NO, Sørensen HR, Hvilsted SA (2013) Comparison of lignin, macroalgae, wood, and straw fast pyrolysis. Energy Fuel 27:1399–1409

Umaña M, Wawrzyniak P, Rosselló C, Llavata B, Simal S (2021) Evaluation of the addition of artichoke by-products to O/W emulsions for oil microencapsulation by spray drying. LWT Food Sci Technol 151:444–454

United Nations (2015) Transforming our world: the 2030 agenda for sustainable development A/RES/70/1. Available at https://sdgs.un.org/2030agenda

Unnati S, Ripal S, Sanjeev A, Niyati A (2013) Novel anticancer agents from plant sources. Chin J Nat Med 11:16–23

Vacca M, Pinto D, Annunziato A, Ressa A, Calasso M, Pontonio E, Celano G, De Angelis M (2023) Gluten-free bread enriched with artichoke leaf extract in vitro exerted antioxidant and anti-inflammatory properties. Antioxidants 12:845

Van den Abbeele P, Ghyselinck J, Marzorati M, Villar A, Zangara A, Smidt CR, Risco E (2020) *In vitro* evaluation of prebiotic properties of a commercial artichoke inflorescence extract revealed bifidogenic effects. Nutrients 12:1552

Vaz AA, Odriozola-Serrano I, Oms-Oliu G, Martín-Belloso O (2022) Physicochemical properties and bioaccessibility of phenolic compounds of dietary fibre concentrates from vegetable by-products. Foods 11:2578

Villanueva-Suárez MJ, Mateos-Aparicio I, Pérez-Cózar ML, Yokoyama W, Redondo-Cuenca A (2019) Hypolipidemic effects of dietary fiber from an artichoke by-product in Syrian hamsters. J Funct Foods 56:156–162

Wickham MJS, Faulks RM, Mann J, Mandalari G (2012) The design, operation, and application of a dynamic gastric model. Dissolut Technol 19:15–22

Wiklund A (1992) The genus *Cynara* L. (*Asteraceae-Cardueae*). Bot J Linn Soc 109:75–123

Yan M, Pan GX, Chen L (2012) An analysis of carbon footprint of vegetable production in Jiangsu, China. Acta Hortic 958:203–210

Yegin S (2017) Xylanase production by *Aureobasidium pullulans* on globe artichoke stem: bioprocess optimization, enzyme characterization, and application in saccharification of lignocellulosic biomass. Prep Biochem Biotechnol 47:441–449

Zaker-Esteghamati H, Seidavi A, Bouyeh M (2021) Effect of *Cynara scolymus* and its derivatives on broilers: an updated review. Anim Biotechnol 32:656–662

Zayed A, Farag MA (2020) Valorization, extraction optimization and technology advancements of artichoke biowastes: food and non-food applications. LWT Food Sci Technol 132:109883

Zeaiter Z, Regonesi ME, Cavini S, Labra M, Sello G, Di Gennaro P (2019) Extraction and characterization of inulin-type fructans from artichoke wastes and their effect on the growth of intestinal bacteria associated with health. Biomed Res Int 2019:1083952

Zuorro A, Maffei G, Lavecchia R (2016) Reuse potential of artichoke (*Cynara scolymus* L.) waste for the recovery of phenolic compounds and bioenergy. J Clean Prod 111(Part A):279–284

Phytochemicals from Cauliflower (*Brassica oleracea*) By-Products

Mohamed Fawzy Ramadan

Abstract Edible plant parts are a rich source of bioactive molecules. By-products from vegetal sources offer the chance to reclaim them by expropriating the bioactive molecules. Phytochemicals in vegetables protect plants, have health benefits, and reduce disease risk. This chapter reported on cauliflower (*Brassica oleracea*) by-products used in the human diet and intends to provide further insight into a nutritional evaluation. Brassica vegetables are a valuable source of phytochemicals, including glucosinolates, phenolic compounds, and carotenoids. Organosulfur compounds are a subgroup of phytochemicals found in cauliflower. In addition, glucoraphanin is a glucosinolate reported in *B. oleracea* var. botrytis and is considered one of the main cauliflower phytochemicals. Bioactivity studies have been conducted using plain phytochemicals of cauliflower and showed potential biological activities. The crude extracts from the cauliflower by-product were used for a radical scavenging assay and exhibited DPPH· radical scavenging potential, lipid peroxidation inhibition, and hydroxyl radical scavenging activity. Future outlooks for researchers on nutrition, processing, packaging, food quality, and industries dealing with agricultural products include expanding research on cauliflower by-products as a hidden source of phytochemicals.

Keywords Brassicaceae · Organosulfur · Glucosinolates · Isothiocyanates · Glucoraphanin

M. F. Ramadan (✉)
Department of Clinical Nutrition, Faculty of Applied Medical Sciences, Umm Al-Qura University, Makkah, Saudi Arabia
e-mail: mfhassanien@uqu.edu.sa

M. F. Ramadan (ed.), *Bioactive Phytochemicals in By-products from Bulb, Flower and Fruit Vegetables*, https://doi.org/10.1007/978-3-031-77399-0_6

1 Introduction

Phytochemicals are non-nutritive components usually found in food of plant origin, including fruits and vegetables. They constitute a wide variety of secondary metabolites with several pharmacological properties. A fund of research on a variety of vegetables illustrates the inherent status of the by-products and their native phytochemicals to elicit specific beneficial medicinal outcomes. Although the "Phytoquarantine" of botanicals is a prerequisite in developed countries, scientifically authenticated "green" phytochemicals have recently emerged as an alternative strategy to curtail agricultural commodities' internal and external infections. The content of different phytochemicals varies among species, varieties, and cultivars, but it also varies over the various edible parts of plants. Several edible plant parts are a rich source of bioactive molecules. The application of agri-pharmaceuticals using co-products from agricultural processing is also presented in the western world for liver disorders, diabetic-related enzymes, and many other indications (Shinali et al. 2024; Espinosa-Alonso et al. 2020; Favela-González et al. 2020; Collado-González et al. 2022; Majumder and Annegowda 2021; Galali et al. 2020; Cano-Lamadrid and Artes-Hernandez 2021).

By-products from vegetal sources offer the chance to reclaim them by expropriating the bioactive molecules. Moreover, nutraceuticals and bioactive principles have been used to supplement various functional foods, as suggested by the European Regulation No. 1924/2006/CE, improving the nutritional aspect of many consumers and consequently assuring better healthiness (Espinosa-Alonso et al. 2020; Majumder and Annegowda 2021; Shinali et al. 2024; Lau et al. 2021; Ben-Othman et al. 2020; Tlais et al. 2020; Rațu et al. 2023).

This chapter focuses on cauliflower peel and stems by-products used in the human diet and intends to provide further insight into a nutritional evaluation. Therefore, the subsequent sections of this article describe cauliflower, its by-product production, and the transplantation of the biomolecules available in the waste materials.

2 Botanical Description of Cauliflower (*Brassica oleracea*)

Head cauliflower (*Brassica oleracea* var. botrytis) is a member of the Brassicaceae family of the order Brassicales. The Latin name *Brassica oleracea* describes a wild cole of the cabbage group. The head cauliflower, also known as "cauliflower," "cabbage flower," "Italian cauliflower," or "broccoli heads," is named botrytis, derived from the Greek word botrus, which means a bunch of grapes. The worldwide area and production of cauliflower were estimated to be 1,586,000 and 25,568,122 tonnes, respectively, according to the Food and Agriculture Organization of the United Nations (FAO) in 2019.

Brassica oleracea is an important amphidiploid species (2n = 18) used to create cultivars like cauliflower and broccoli. Its wild relatives have lost genetic diversity due to backcrosses with kale. The Brassicaceae family consists mainly of herbaceous species. *B. oleracea* and *B. rapa* are closely related, while *B. napus* comes from their hybridization. Morphologically, *B. oleracea* and *B. napus* are similar, with simple leaves and yellow/white flowers. *B. oleracea's* inflorescence varies, being either loosely or flat-shaped. Wild cabbages have thin floral sticks, while sea cabbages have a spiral inflorescence. Pamakules are less-developed cauliflower and broccoflower varieties with top florals like blue cauliflower (Żyła et al. 2021; Kamiński et al. 2020; Ji et al. 2023; Rakshita et al. 2021; Zhu et al. 2022).

3 Phytochemicals in Cauliflower

Phytochemicals in vegetables and fruits protect plants, have health benefits, and reduce disease risk. Phytochemicals in edible plants contribute to their color and health properties. Studies show that a diet rich in fruits and vegetables lowers cancer rates and fights cancer through different pathways.

Cauliflower, like the other familiar edible representatives of the genus (*Brassica oleracea*), has some specific properties due to its phytochemical contents. Phytochemicals are categorized into primary and secondary metabolites. The primary metabolites produce proteins, fats, carbohydrates, and nucleic acids, constituent parts of living organisms. On the other hand, secondary metabolites, known as phytonutrients, are often the bioactive parts of phytochemicals and protect against certain diseases due to their anti-inflammatory, anti-allergic, and pain-relieving properties. These compounds, especially the ones found in broccoli and cauliflower, have attracted pharmacologists' attention, particularly in cancer prevention. It is well known that Brassica vegetables are a valuable source of several phytochemicals, including glucosinolates, phenolic compounds, and/or carotenoids (Soengas et al. 2021; Favela-González et al. 2020; Šalić and Šamec 2022; Ramirez et al. 2020; Wu et al. 2021; Ağagündüz et al. 2022; Kapusta-Duch and Kusznierewicz 2021; Shree et al. 2022).

The organosulfur phytochemicals are found in all tissues of the cauliflower plant, especially the curd. Organosulfur compounds are a main subgroup of phytochemicals found in cauliflower and are known to present several potential human health benefits, including acting as chemo-preventive agents. However, because of the physiologically active compounds, cooking methods could mediate or influence the intestine level of these active molecules. For this reason, precious compounds are claimed to be absorbed in the small intestine without metabolizing (Drabińska et al. 2021; Miękus et al. 2020; Goncharov et al. 2021; Batista and da Silva-Maia 2021; Idrees et al. 2020; Ruhee et al. 2020; Hill et al. 2023; Abdalla et al. 2020; Walag et al. 2020).

4 Utilization of Cauliflower By-Products

Novel technologies devoted to fine raw material utilization are constantly developing. It is expected that further studies oriented on vegetable by-products could expand our knowledge of the existing substances and, consequently, improve the general estimation of their potential benefits for broadly defined human nutrition and pharmacy. Data concerning the possibilities of using cauliflower by-products are still scarce and more affected by different botanical or geographical provenance sources. Cooked cauliflowers are drunk, and a clear loss is realized when the cob is discarded. An interesting possibility would be using the boiling wastewater rich in extracted phytochemicals for some cooking processes or soups.

Cauliflower waste is an example of a by-product of some value. Cauliflower waste consists of leaves and an inflorescence stalk, which form during the early stages of the salable inflorescence head development and constitute a considerable part of cauliflower plants (Fig. 1).

Its weight-based proportion concerning the marketable production can reach 50–70%. The cauliflower inflorescence overgrowing over other plant segments is an adaptation that allows the vegetable to harvest the sunlight sufficiently and to join nutrients reasonably. For this reason, the overgrowing head never has leaves. A logistic convenience occurs in this case because the bipolar heads can be cut from

Fig. 1 Cauliflower (*Brassica oleracea*) leaves and an inflorescence stalk

the stems very near the round interface during the primary sorting process (Lalramhlimi et al. 2022; Rațu et al. 2023; Gaur et al. 2020; Stoica et al. 2023; Parthasarathy et al. 2024).

5 Bioactivity of Cauliflower By-Product Phytochemicals

Many phytochemicals are found in cauliflower, such as sinigrin, glucosinolates, isothiocyanates, polyphenols, and flavonoids. Several bioactivity studies have been conducted using plain phytochemicals of cauliflower and showed potential biological activities. The crude extracts from the cauliflower by-product were used for a radical scavenging assay and exhibited DPPH· radical scavenging activity (IC_{50} = 51 µg/mL), lipid peroxidation inhibition (IC_{50} = 296 µg/mL), and hydroxyl radical scavenging activity (IC_{50} = 382 µg/mL). Moreover, the antioxidant activities of extracts of cauliflower by-product in Rancimat tests showed those extracts slowed down the development of oxidation more effectively than the control (Espinosa-Alonso et al. 2020; Zeng et al. 2023; Le et al. 2020; Tlais et al. 2020; Jovanović et al. 2022; Leporini et al. 2020; Lizárraga-Velázquez et al. 2020).

The in vitro anti-inflammatory effects of the extracts from the cauliflower by-product were tested in RAW 264.7 murine macrophage cells. The extract from the cauliflower by-product decreased the levels of pro-inflammatory mediators by inhibiting their synthesis (iNOS at 200 µM and COX-2 at 50 µg/mL). There are no available reports on the potential of cauliflower by-product phytochemicals to act as an angiogenesis inhibitor. Bioactivity evaluations of cauliflower by-product phytochemicals, including sinigrin, have been performed to date, and most of studies suggested potential biological effects. In addition to finding the quantities of the phytochemicals above, it is essential to investigate their bioactivities, as these properties are directly related to food's ability to deliver these compounds with their associated health claims. The bioactivity studies of the natural vegetables by-products have the potential to confirm their nutritional and functional properties further when used as food ingredients (Olszewska et al. 2020; Coman et al. 2020; Gomez-Molina et al. 2024; Li et al. 2022; Yan et al. 2023; Kamal et al. 2022; Punzo et al. 2021).

6 Applications of Cauliflower By-Product Phytochemicals in Food and Pharmaceutical Industries

The European Food Safety Authority has published several health claims about consuming the main identified bioactive compounds in cauliflower by-products, either as pure compounds or from a placebo-controlled food enrichment. These studies revealed that among all the beneficial properties traditionally related to polyphenol bioactivity, such as anti-inflammatory, gastrointestinal, or activity on the microbiota, cauliflower phytochemicals can also be used to develop nutri-functional properties that would add value to the finished food product. Thus, cauliflower by-product phytochemicals can be incorporated into fruits, dairy and bakery products, meat, and fish. These phytochemicals play a vital role as antioxidants, anti-browning agents for drinks, and antifungal agents for cheese, meat, fish, and bakery products. They can also improve nutritional quality via effects on the glycemic index and the glycemic response (Drabińska et al. 2021; Biondi et al. 2021; Ramirez et al. 2020; Shi et al. 2022; Waseem et al. 2022; Cuellar-Nuñez et al. 2022; Muntean et al. 2021; Ciudad-Mulero et al. 2023).

Glucoraphanin (Fig. 2) is a glucosinolate reported in cauliflower (*B. oleracea* var. botrytis) and is considered one of the main cauliflower phytochemicals. Glucoraphanin has been shown to have an application in the use of ready-to-eat vegetables such as pre-packed bagged salads, the fresh-cut industry, food and

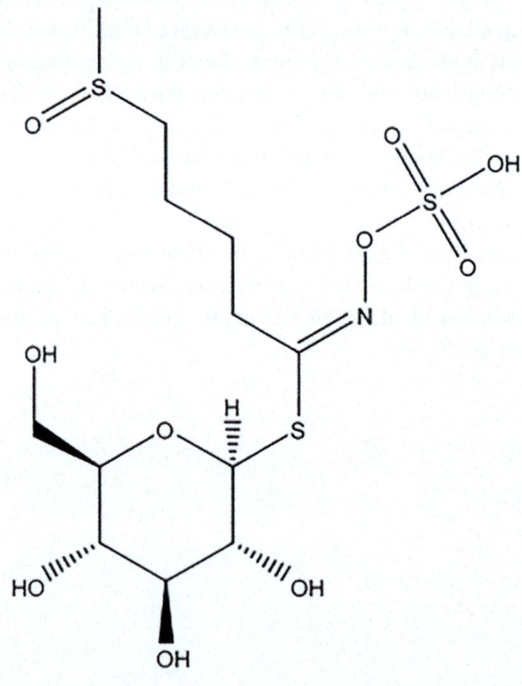

Fig. 2 Strucutue of glucoraphanin

beverage, delicatessen or restaurants, where the intake is a single portion/single meal consumption (up to 200 g) as part of a calorie-controlled diet. Additionally, whole foods traditionally enriched with glucosinolates are the Brassica vegetables, and in this category, it is expected that over 100 g of the finished product will be consumed as part of a balanced diet. In the above applications, the consumption of the food product is spread equally across all meals and, as such, can constitute regular daily intake (Picchi et al. 2020; Kumar et al. 2022; Wieczorek et al. 2022; Biondi et al. 2021; Wang et al. 2022; Sun et al. 2021; Shinali et al. 2024; Radünz et al. 2024).

There is a growing exploration of the applications of these compounds in pharmaceutical formulations. One of the emerging trends in the pharmaceutical industry is the application of dietary plants and/or their components, referred to as phytochemicals, as an adjuvant therapy to enhance the health-promoting potential of pharmaceuticals and manage non-communicable diseases. Significantly aided by the supply of antioxidant phytochemicals, malnutrition and undernutrition are tackled, which affect billions of children and adults in Africa, Asia, and Latin America.

The growing interest in natural products, due to significantly lower health risks (side effects) than synthetic drugs, has led to a commercial shift in these products, allowing companies to meet market demands for healthcare products. Most dietary plants contain allergens, e.g., oilseed rape and mustard seed, that can affect humans and animals, causing different symptoms such as dermatitis, asthma, and allergic inflammation, as their glucosinolates degradation in plants results in pungent-tasting isothiocyanates. In conclusion, one must, however, consider two critical ecological aspects in the case of cauliflowers (Shinali et al. 2024; Espinosa-Alonso et al. 2020; Majumder and Annegowda 2021; Valisakkagari et al. 2024; Cano-Lamadrid and Artes-Hernandez 2021; Teshome et al. 2024; Sharma et al. 2021; Coman et al. 2020).

7 Challenges and Future Perspectives

Due to the high antioxidant capacity of cauliflower by-product phytochemicals, they have drawn attention as potential food additives to increase shelf life. More research is needed on the economic and environmental impacts of using these phytochemicals in industry. Developing more efficient extraction methods is essential to reduce costs. Limited information exists on these phytochemicals' absorbability, distribution, metabolism, and excretion, particularly in vivo experiments on rats. Overcoming consumer preferences for digestates used in animal feed is another challenge, including isolating beneficial compounds and testing for taste. Fermentability research is necessary to enhance the nutritional value of digestates. Chemical structure changes during reduction, ammonia-soluble residues, and retention times also require further investigation. (Drabińska et al. 2021; Espinosa-Alonso et al. 2020; Rafiuddin et al. 2021; Nartea et al. 2023; Gudiño et al. 2022).

8 Conclusion and Future Directions

Cauliflower, *Brassica oleracea*, is a nutrient-rich source, but 66% of its weight is wasted due to by-products from peeling and discarding damaged curds. Research into utilizing these by-products is in the early stages, focusing on third-tier applications like feed and bio-energy. Consumers and researchers aim to minimize waste and replace synthetic ingredients with natural sources, exploring the hidden value of agro-industrial co-products through extraction and conversion technologies. Brassica species, including cauliflower, have high antioxidant potential and can contribute to functional food formulations, adding value to the product and the by-product.

Future outlooks for researchers on nutrition, processing, packaging, food quality, and industries dealing with agricultural products include expanding research on cauliflower by-products as a hidden source of phytochemicals. By enriching and presenting the comprehensive phytochemical profile of by-products from different regions, it will be determined which parts of the cauliflower are most beneficial and in what ways. This understanding will aid in developing accurate plans for extraction processes. To transition from disposing of low-value raw materials, further studies are needed to explore the potential uses of cauliflower co-products in nutraceuticals, functional foods, and sports nutrition. These studies should prioritize high value, scientific strength, and consumer recognition (Drabińska et al. 2021; Espinosa-Alonso et al. 2020; Shinali et al. 2024; Collado-González et al. 2022; Nartea et al. 2023; Gudiño et al. 2022; Hashem et al. 2021; Favela-González et al. 2020).

References

Abdalla MA, Sulieman S, Mühling KH (2020) Regulation of selenium/sulfur interactions to enhance chemopreventive effects: lessons to learn from Brassicaceae. Molecules

Ağagündüz D, Şahin TÖ, Yılmaz B, Ekenci KD, Duyar Özer Ş, Capasso R (2022) Cruciferous vegetables and their bioactive metabolites: from prevention to novel therapies of colorectal cancer. Evid Based Complement Alternat Med 2022(1):1534083

Batista G, da Silva-Maia JK (2021) Generation and alterations of bioactive organosulfur and phenolic compounds. In: Chemical changes during processing and storage of foods. Academic Press

Ben-Othman S, Jõudu I, Bhat R (2020) Bioactives from Agri-food wastes: present insights and future challenges. Molecules 25

Biondi F, Balducci F, Capocasa F, Visciglio M, Mei E, Vagnoni M, Mazzoni L (2021) Environmental conditions and agronomical factors influencing the levels of phytochemicals in Brassica vegetables responsible for nutritional and sensorial properties. Appl Sci 11(4):1927

Cano-Lamadrid M, Artes-Hernandez F (2021) By-products revalorization with non-thermal treatments to enhance phytochemical compounds of fruit and vegetables derived products: a review, vol 11. Foods

Ciudad-Mulero M, Domínguez L, Morales P, Fernández-Ruiz V, Cámara M (2023) A review of foods of plant origin as sources of vitamins with proven activity in oxidative stress prevention according to EFSA scientific evidence. Molecules 28(21):7269

Collado-González J, Piñero MC, Otalora G, López-Marín J, Del Amor FM (2022) Enhancement of bioactive constituents in fresh cauliflower by-products in challenging climate conditions. Antioxidants 11(5):958

Coman V, Teleky BE, Mitrea L, Martău GA, Szabo K, Călinoiu LF, Vodnar DC (2020) Bioactive potential of fruit and vegetable wastes. Adv Food Nutr Res 91:157–225

Cuellar-Nuñez ML, Luzardo-Ocampo I, Lee-Martínez S, Larrauri-Rodríguez M, Zaldívar-Lelo de Larrea G, Pérez-Serrano RM, Camacho-Calderón N (2022) Isothiocyanate-rich extracts from cauliflower (*Brassica oleracea* var. Botrytis) and radish (*Raphanus sativus*) inhibited metabolic activity and induced ROS in selected human HCT116 and HT-29 colorectal cancer cells. Int J Environ Res Public Health 19(22):14919

Drabińska N, Jeż M, Nogueira M (2021) Variation in the accumulation of phytochemicals and their bioactive properties among the aerial parts of cauliflower. Antioxidants 10

Espinosa-Alonso LG, Valdez-Morales M, Aparicio-Fernandez X, Medina-Godoy S, Guevara-Lara F (2020) Vegetable by-products. Food wastes and by-products: nutraceutical and health potential, pp 223–266

Favela-González KM, Hernández-Almanza AY, De la Fuente-Salcido NM (2020) The value of bioactive compounds of cruciferous vegetables (Brassica) as antimicrobials and antioxidants: a review. J Food Biochem 44(10):e13414

Galali Y, Omar ZA, Sajadi SM (2020) Biologically active components in by-products of food processing. Food Sci Nutr 8:3004

Gaur VK, Sharma P, Sirohi R, Awasthi MK, Dussap CG, Pandey A (2020) Assessing the impact of industrial waste on environment and mitigation strategies: a comprehensive review. J Hazard Mater 398:123019

Gomez-Molina M, Albaladejo-Marico L, Yepes-Molina L, Nicolas-Espinosa J, Navarro-León E, Garcia-Ibañez P, Carvajal M (2024) Exploring phenolic compounds in crop by-products for cosmetic efficacy. Int J Mol Sci 25(11):5884

Goncharov NV, Belinskaia DA, Ukolov AI, Jenkins RO, Avdonin PV (2021) Organosulfur compounds as nutraceuticals. In: Nutraceuticals. Academic Press, pp 911–924

Gudiño I, Martín A, Casquete R, Prieto MH, Ayuso MC, Córdoba MDG (2022) Evaluation of broccoli (*Brassica oleracea* var. italica) crop by-products as sources of bioactive compounds. Sci Hortic 304:111284

Hashem A, Mousa E, Adawy E (2021) The use of cauliflower waste extracts as a natural antioxidant to improve quality of beef sausage. Egyptian. J Food Sci 0

Hill CR, Shafaei A, Balmer L, Lewis JR, Hodgson JM, Millar AH, Blekkenhorst LC (2023) Sulfur compounds: from plants to humans and their role in chronic disease prevention. Crit Rev Food Sci Nutr 63(27):8616–8638

Idrees N, Saquib M, Azmi S, Ahmad I, Hussain MK (2020) Anticancer and chemopreventive phytochemicals from cruciferous plants. In: Plant-derived bioactives: chemistry and mode of action, pp 375–385

Ji X, Aflitos SA, Peters S, Schranz ME, Philipse-Berendsen J, Vogelaar A, de Jong H (2023) FISH painting for chromosome identification of aneuploid cauliflower (*Brassica oleracea* L. var. botrytis). Nucleus 66(3):273–288

Jovanović AA, Vajić UJV, Mijin DZ, Zdunić GM, Šavikin KP, Branković S, Bugarski BM (2022) Polyphenol extraction in microwave reactor using by-product of *Thymus serpyllum* L. and biological potential of the extract. J Appl Res Med Aromat Plants 31:100417

Kamal RM, Abdull Razis AF, Mohd Sukri NS, Perimal EK, Ahmad H, Patrick R, Rigaud S (2022) Beneficial health effects of glucosinolates-derived isothiocyanates on cardiovascular and neurodegenerative diseases. Molecules 27(3):624

Kamiński P, Marasek-Ciolakowska A, Podwyszyńska M, Starzycki M, Starzycka-Korbas E, Nowak K (2020) Development and characteristics of interspecific hybrids between *Brassica oleracea* L. and B. napus L. Agronomy 10(9):1339

Kapusta-Duch J, Kusznierewicz B (2021) Young shoots of white and red headed cabbages like novel sources of glucosinolates as well as antioxidative substances. Antioxidants 10

Kumar S, Sharma S, Kumar V, Sharma R, Minhas A, Boddu R (2022) Cruciferous vegetables: a mine of phytonutrients for functional and nutraceutical enrichment. In: Current advances for development of functional foods modulating inflammation and oxidative stress. Academic Press, pp 401–426

Lalramhlimi B, Mukherjee D, Chakraborty I, Ghosh N, Chattopadhyay A, Dey RC (2022) Fruit and vegetable wastes as livestock feeds. In: Fruits and vegetable wastes: valorization to bio-products and platform chemicals. Springer Nature Singapore, Singapore, pp 139–168

Lau KQ, Sabran MR, Shafie SR (2021) Utilization of vegetable and fruit by-products as functional ingredient and food. Food Sci Nutr

Le TN, Sakulsataporn N, Chiu CH, Hsieh PC (2020) Polyphenolic profile and varied bioactivities of processed Taiwanese grown broccoli: a comparative study of edible and non-edible parts. Pharmaceuticals

Leporini M, Loizzo MR, Sicari V, Pellicanò TM, Reitano A, Dugay A, Tundis R (2020) Citrus× Clementina Hort. Juice enriched with its by-products (peels and leaves): chemical composition, in vitro bioactivity, and impact of processing. Antioxidants 9(4):298

Li H, Xia Y, Liu HY, Guo H, He XQ, Liu Y, Gan RY (2022) Nutritional values, beneficial effects, and food applications of broccoli (*Brassica oleracea* var. italica Plenck). Trends Food Sci Technol 119:288–308

Lizárraga-Velázquez CE, Leyva-López N, Hernández C, Gutiérrez-Grijalva EP, Salazar-Leyva JA, Osuna-Ruíz I, Ávalos-Soriano A (2020) Antioxidant molecules from plant waste: extraction techniques and biological properties. Processes 8(12):1566

Majumder P, Annegowda HV (2021) Fruit and vegetable by-products: novel ingredients for a sustainable society. In: Valorization of Agri-food wastes and by-products. Academic Press, pp 133–156

Miękus N, Marszałek K, Podlacha M, Iqbal A, Puchalski C, Świergiel AH (2020) Health benefits of plant-derived sulfur compounds, glucosinolates, and organosulfur compounds. Molecules 25(17):3804

Muntean D, Ştefănuţ MN, Căta A, Buda V, Danciu C, Bănică R, Ienaşcu IM (2021) Symmetrical antioxidant and antibacterial properties of four *Romanian cruciferous* extracts. Symmetry 13(5):893

Nartea A, Fanesi B, Pacetti D, Lenti L, Fiorini D, Lucci P, Falcone PM (2023) Cauliflower by-products as functional ingredient in bakery foods: fortification of pizza with glucosinolates, carotenoids and phytosterols. Curr Res Food Sci 6:100437

Olszewska MA, Granica S, Kolodziejczyk-Czepas J, Magiera A, Czerwińska ME, Nowak P, Owczarek A (2020) Variability of sinapic acid derivatives during germination and their contribution to antioxidant and anti-inflammatory effects of broccoli sprouts on human plasma and human peripheral blood mononuclear cells. Food Funct 11(8):7231–7244

Parthasarathy S, Lakshmidevi P, Yashodha P, Gopalakrishnan C (2024) Pests and diseases in vegetable crops. CRC Press

Picchi V, Fibiani M, Scalzo RL (2020) Cauliflower. In: Nutritional composition and antioxidant properties of fruits and vegetables. Academic Press, pp 19–32

Punzo A, Porru E, Silla A, Simoni P, Galletti P, Roda A, Caliceti C (2021) Grape pomace for topical application: green NaDES sustainable extraction, skin permeation studies, antioxidant and anti-inflammatory activities characterization in 3D human keratinocytes. Biomol Ther 11(8):1181

Radünz M, Camargo TM, Raphaelli CO, Radünz AL, Gandra EÁ, Zavareze R (2024) Chemical composition, antimicrobial and antioxidant activities of broccoli, kale, and cauliflower extracts, vol 79, p 700

Rafiuddin M, Suneetha WJ, Kumari BA (2021) Standardization and evaluation of cauliflower stalks powder incorporated products. Cutting-Edge Res Agric Sci 7:9–16

Rakshita KN, Singh S, Verma VK, Sharma BB, Saini N, Iquebal MA, Behera TK (2021) Agro-morphological and molecular diversity in different maturity groups of Indian cauliflower (*Brassica oleracea* var. botrytis L.). PLoS One 16(12):e0260246

Ramirez D, Abellán-Victorio A, Beretta V, Camargo A, Moreno DA (2020) Functional ingredients from Brassicaceae species: overview and perspectives. Int J Mol Sci 21(6):1998

Rațu RN, Veleșcu ID, Stoica F, Usturoi A, Arsenoaia VN, Crivei IC et al (2023) Application of Agri-food by-products in the food industry. Agriculture 13(8):1559

Ruhee RT, Roberts LA, Ma S, Suzuki K (2020) Organosulfur compounds: a review of their anti-inflammatory effects in human health. Front Nutr 7

Šalić A, Šamec D (2022) Changes in the content of glucosinolates, polyphenols and carotenoids during lactic-acid fermentation of cruciferous vegetables: a mini review. Food Chem: X 16

Sharma M, Usmani Z, Gupta VK, Bhat R (2021) Valorization of fruits and vegetable wastes and by-products to produce natural pigments. Crit Rev Biotechnol 41(4):535–563

Shi M, Gu J, Wu H, Rauf A, Emran TB, Khan Z, Suleria HA (2022) Phytochemicals, nutrition, metabolism, bioavailability, and health benefits in lettuce-a comprehensive review. Antioxidants 11(6):1158

Shinali TS, Zhang Y, Altaf M, Nsabiyeze A, Han Z, Shi S, Shang N (2024) The valorization of wastes and byproducts from cruciferous vegetables: a review on the potential utilization of cabbage, cauliflower, and broccoli byproducts. Food Secur 13(8):1163

Shree B, Kumar S, Sharma S, Katoch V (2022) Functional significance of underutilized high value cruciferous vegetables-an exotic gleam in the gloomy guise of their functional importance. S Afr J Bot 145:420

Soengas P, Velasco P, Fernández JC, Cartea ME (2021) New vegetable brassica foods: a promising source of bioactive compounds. Foods 10

Stoica F, Usturoi A, Arsenoaia VN, Crivei IC, Postolache AN, Lips FD, Filipov F, Florea AM et al (2023) Application of agri-food by-products in the food industry. Agriculture 13:1559

Sun J, Wang Y, Pang X, Tian S, Hu Q, Li X, Lu Y (2021) The effect of processing and cooking on glucoraphanin and sulforaphane in brassica vegetables. Food Chem 360:130007

Teshome E, Teka TA, Urugo MM, Nandasiri R, Gemede HF, Rani I, Astatkie T (2024) Extraction methods, industrial uses, and nutritional benefits of vegetable byproducts. Int J Veg Sci 30(3):334–363

Tlais AZA, Fiorino GM, Polo A, Filannino P, Di Cagno R (2020) High-value compounds in fruit, vegetable and cereal byproducts: an overview of potential sustainable reuse and exploitation. Molecules 25

Valisakkagari H, Chaturvedi C, Rupasinghe HPV (2024) Green extraction of phytochemicals from fresh vegetable waste and their potential application as cosmeceuticals for skin health. Processes 12

Walag AMP, Ahmed O, Jeevanandam J, Akram M, Ephraim-Emmanuel BC, Egbuna C et al (2020) Health benefits of organosulfur compounds. In: Functional foods and nutraceuticals: bioactive components, formulations and innovations, pp 445–472

Wang J, Liu Z, Dou J, Lv J, Jin N, Jin L et al (2022) A comparative study on the nutrients, mineral elements, and antioxidant compounds in different types of cruciferous vegetables. Agronomy 12(12):3121

Waseem M, Akhtar S, Qamar M, Saeed W, Ismail T, Esatbeyoglu T (2022) Effect of thermal and non-thermal processing on nutritional, functional, safety characteristics and sensory quality of white cabbage powder. Food Secur 11(23):3802

Wieczorek MN, Dunkel A, Szwengiel A, Czaczyk K, Drożdżyńska A, Zawirska-Wojtasiak R, Jeleń HH (2022) The relation between phytochemical composition and sensory traits of selected Brassica vegetables. LWT 156:113028

Wu X, Huang H, Childs H, Wu Y, Yu L, Pehrsson PR (2021) Glucosinolates in Brassica vegetables: characterization and factors that influence distribution, content, and intake. Annu Rev Food Sci Technol 12(1):485–511

Yan L, Zhou G, Shahzad K, Zhang H, Yu X, Wang Y, Zhang X (2023) Research progress on the utilization technology of broccoli stalk, leaf resources, and the mechanism of action of its bioactive substances. Front Plant Sci 14:1138700

Zeng Y, Zhou W, Yu J, Zhao L, Wang K, Hu Z, Liu X (2023) By-products of fruit and vegetables: antioxidant properties of extractable and non-extractable phenolic compounds. Antioxidants

Zhu M, Yang L, Zhang Y, Zhuang M, Ji J, Hou X, Wang Y (2022) Introgression of clubroot resistant gene into *Brassica oleracea* L. from *Brassica rapa* based on homoeologous exchange. Horticult Res 9:uhac195

Żyła N, Fidler J, Babula-Skowrońska D (2021) Economic and academic importance of *Brassica oleracea*. The *Brassica oleracea*. Genome

Part IV
Phytochemicals from Fruit Vegetable By-products

Bioactive Phytochemicals from Bell Pepper (*Capsicum annuum*) By-Products

Seok Tyug Tan, Seok Shin Tan, and Chin Xuan Tan

Abstract Bell pepper is one of the most widely cultivated crops worldwide; however, the processing of bell peppers from farm to table has resulted in considerable agricultural waste, including peels, seeds, placentas, stems, and leaves. These agricultural wastes or by-products can pose a significant environmental threat if not repurposed. Therefore, this chapter discusses bell pepper by-products' chemical composition and potential food and non-food applications. Moreover, this chapter also dives into the bioactive phytochemicals that can be extracted from bell pepper by-products and their possible health-promoting effects. Interestingly, despite the abundance of research on bell peppers, there is a scarcity of studies focusing on the by-products derived from them. Nevertheless, emerging evidence indicates that bell pepper by-products are exceptional sources of nutrients and bioactive phytochemicals. Additional research is required to gain deeper insights into the properties and potential food and non-food applications of bell pepper by-products discussed in this chapter.

Keywords Waste · Peels · Seeds · Placentas · Stems · Leaves · Applications

1 Introduction

Bell pepper, also known as sweet pepper or Capsicum is an annual herbaceous plant belonging to the species *Capsicum annuum*. This plant is believed to have originated in Mexico, but it was introduced to Asia in the sixteenth century (CABI 2019).

S. T. Tan (✉) · S. S. Tan
Jeffrey Cheah School of Medicine and Health Sciences, Monash University Malaysia, Jalan Lagoon Selatan, Bandar Sunway, Selangor, Malaysia
e-mail: Tan.SeokTyug@monash.edu

C. X. Tan
Department of Allied Health Sciences, Faculty of Science, Universiti Tunku Abdul Rahman, Jalan Universiti Bandar Barat, Kampar, Perak, Malaysia

M. F. Ramadan (ed.), *Bioactive Phytochemicals in By-products from Bulb, Flower and Fruit Vegetables*, https://doi.org/10.1007/978-3-031-77399-0_7

Bell peppers are currently cultivated in regions with warm climates (18–30 °C), such as Asia, America (southern North and northern South regions), Europe (southern and central areas), and Africa (subtropical and tropical regions) (CABI 2019; Baenas et al. 2019). The short and bushy bell pepper plant has dark green and lance-shaped leaves alternating along the light green woody stems. At the apex of the stem, the plant bears approximately 1-inch (2.54 cm) diameter white or purple bell-shaped flowers that grow into edible fruits (North Carolina Extension Gardener Plant Toolbox 2024). Even though the color of bell peppers can be influenced by their genotypes, maturity stages, and growing conditions, green, orange, red, and yellow bell peppers are the predominant varieties in the market (Sood et al. 2023).

It has been previously reported that bell pepper is one of the most widely cultivated crops worldwide. According to statistics from the Food and Agriculture Organization of the United Nations (FAO), the global production of fresh chilies and peppers stood at 36 million metric tonnes in 2021. With 16 million metric tonnes of production, China was the leading producer of fresh chilies and peppers in the same year (Food and Agriculture Organization of the United Nations 2024). Although the United States has been the largest importer of fresh bell pepper since 2016, it is also observed that Poland, the Netherlands, and Switzerland recorded more than a 20% surge in import values in 2023 compared to the preceding year (2022) (Tridge 2024). The growing demand is unsurprising since literature has consistently reported that fresh bell peppers contain many vitamins (especially vitamin C and vitamin B) and bioactive phytochemicals (Wong and Tan 2020; The United States Department of Agriculture 2021).

Table 1 shows the nutrient composition of fresh bell peppers in various colors. Interestingly, it is observed that orange bell pepper has the highest levels of vitamin C (158 mg/100 g), niacin (1.08 mg/100 g), vitamin B6 (0.332 mg/100 g), and biotin (0.58 μg/100 g) compared to other varieties. Even though the orange bell pepper has the same thiamin content as the red bell pepper (0.055 mg/100 g), its thiamin content is still the highest compared to the yellow and green bell peppers. Concerning the bioactive compounds in bell peppers, evidence-to-date suggests that bell peppers are good sources of phenolic acids (hydroxybenzoic acids and hydrocinnamic acids), flavonoids (quercetin, luteolin, and apigenin), and carotenoids (capsanthin, β-carotene, and zeaxanthin) (Anaya-Esparza et al. 2021).

Although bell peppers can be consumed raw, cooked in various ways (such as grilling, roasting, or sautéing), or incorporated into a wide range of dishes (such as salads, sandwiches, wraps, or soups), the processing of bell peppers from farm to table has resulted in a considerable amount of agricultural waste (Anaya-Esparza et al. 2021; Yasin et al. 2023). If not repurposed, agricultural waste from bell peppers, including peels, seeds, placentas, stems, and leaves, can pose a significant environmental threat. Bell pepper seeds, for instance, have comprised more than 30% of the dry weight; however, they are typically discarded as waste during processing (Li et al. 2018). Since it has been previously reported that agricultural waste is a potent source of bioactive phytochemicals (Tan et al. 2020; Nirvesha Nair and Tan 2021), repurposing agricultural waste into nutraceuticals may help mitigate environmental pollution caused by poor waste disposal. Therefore, this chapter aims

Table 1 Nutrient composition of fresh bell peppers in various colors

Nutrient/100 g bell pepper (unit)[a]	Green bell pepper	Orange bell pepper	Red bell pepper	Yellow bell pepper
Water (g)	94.1	91.9	91.9	92.1
Energy (kcal)	23	32	31	31
Carbohydrates (g)	4.78	6.7	6.65	6.6
Protein (g)	0.72	0.88	0.9	0.82
Fat (g)	0.11	0.16	0.13	0.12
Fiber (g)	0.9	1	1.2	1.1
Calcium (mg)	7	5	6	7
Iron (mg)	0.19	0.37	0.35	0.36
Magnesium (mg)	9	10.4	11	10.3
Phosphorus (mg)	17	27	27	25
Potassium (mg)	163	201	213	197
Sodium (mg)	<2.49	<2.49	<2.5	<2.5
Zinc (mg)	0.13	0.24	0.2	0.19
Copper (mg)	0.058	0.035	0.04	0.045
Manganese (mg)	0.1	0.149	0.133	0.142
Selenium (μg)	<2.5	0.3	<2.5	<2.5
Vitamin C (mg)	99.5	158	142	139
Thiamin (mg)	0.039	0.055	0.055	0.049
Riboflavin (mg)	0.089	0.102	0.142	0.098
Niacin (mg)	0.636	1.08	1.02	0.91
Vitamin B6 (mg)	0.247	0.332	0.303	0.292
Biotin (μg)	<3.7	0.58	0.427	0.46
Folate (μg)	22	NA	47	42

NA Not Available

[a] Data is retrieved from The United States Department of Agriculture (2021)

to discuss the bioactive phytochemicals in bell pepper by-products (peels, seeds, stems, and leaves) and explore their potential food and non-food applications.

2 Chemical Composition of Bell Pepper By-Products

It is worth noting that peels, seeds, placentas, stems, and leaves are the primary by-products derived from the processing of fresh bell peppers. Even though bell peppers can be consumed whole, some individuals prefer to remove the peels before consuming them. This practice leaves bell pepper peels as one of the food wastes (or by-products) generated in the kitchen, alongside seeds and stems. To the best of the authors' knowledge, bell pepper seeds are the most extensively studied among the previously mentioned by-products. In general, the chemical composition and bioactive phytochemicals of bell pepper by-products are associated with the maturity stage, variety, color, growing conditions, and post-harvest storage conditions of the bell peppers (Cvetković et al. 2022; Colombari et al. 2022; Sood et al. 2023).

Bell pepper seeds are usually processed into powdered seeds, seed flour, or seed oil in the food industry. The macronutrient composition of bell pepper seeds in various forms (defatted seed flour and powdered seeds) is presented in Table 2. Although defatted seed flour has higher carbohydrate (56.28% dry weight), crude fat (19.57% dry weight), and moisture content (70.95% dry weight) than powdered seeds (carbohydrate = 44.2% dry weight, crude fat = 15.1% dry weight, and moisture = 7.2% dry weight), its crude protein (19.28% dry weight) is comparatively lower than that reported in powdered seeds (29.1% dry weight). The discrepancies in macronutrient content may be attributed to variations in analytical methods and the diversity of bell pepper cultivars. For instance, the vast difference in the moisture content could be because the powdered seeds were subjected to a longer drying duration (30 °C for three days) compared to the defatted seed flour (50 °C for 12 h) (Embaby and Mokhtar 2011; Eskander et al. 2023). In addition to those previously mentioned, a recent study by Colombari et al. (2022) also reported that bell pepper seeds are good sources of micronutrients, including phosphorus (4.4 to 4.7 g/kg dry matter), potassium (12.3 to 12.4 g/kg dry matter), calcium (0.87 to 1.02 g/kg dry matter), magnesium (2.4 g/kg dry matter), nitrogen (28 to 31 g/kg dry matter), and sulfur (1.8 to 2.1 g/kg dry matter).

It is reported that bell pepper seed oils contain more than 80% unsaturated fatty acids (Cocan et al. 2021; Embaby et al. 2022). Owing to their substantial amount of unsaturated fatty acids, few studies have investigated the fatty acid composition of bell pepper seed oils using Gas Chromatography/Mass Spectrometry (GC/MS). Emerging evidence indicates that the predominant fatty acids in bell pepper seed oils are linoleic acid (C18:2) (69.4%), palmitic acid (C16:0) (13.10%), oleic acid (C18:1) (12.51%), and stearic acid (C18:0) (4.82%) (Cocan et al. 2021; Embaby et al. 2022). Apart from the fatty acid composition in seed oil, a former study by Embaby and Mokhtar (2011) also studied the amino acid composition of defatted bell pepper seed flour. Interestingly, the defatted bell pepper seed flour is reported to be a good source of essential amino acids (such as leucine, lysine, phenylalanine, threonine, and histidine) and non-essential amino acids (such as cysteine, arginine, glutamic acid, and aspartic acid).

Table 2 Macronutrient composition of bell pepper by-products

By-products	Country of origin	Proximate composition				Reference
		Carbohydrate	Crude protein	Crude fat	Moisture	
Defatted seed flour[a]	Egypt	56.28	19.28	19.57	70.95	Embaby and Mokhtar (2011)
Powdered seeds[a]	Iraq	44.2	29.1	15.1	7.2	Eskander et al. (2023)
Seeds[b] Placentas[b]	Nigeria	0.30 3.35	14.1 3.35	1.03 2.86	50.2 88.2	Emmanuel Ilesanmi (2014)

[a] Proximate composition is expressed in % dry weight
[b] Proximate composition is expressed in g/100 g fresh weight

The current chapter reveals that only one study has investigated the proximate composition of bell pepper placentas. Emmanuel Ilesanmi (2014) reported that bell pepper placentas consist of 3.35 g of carbohydrates, 3.35 g of crude protein, 2.86 g of crude fat, and 88.2 g of moisture per 100 g of fresh weight. In addition, the placentas of bell peppers also contain a trace amount of micronutrients such as calcium (9.00 mg/100 g fresh weight), magnesium (2.00 mg/100 g fresh weight), zinc (63.0 mg/100 g fresh weight), iron (0.84 mg/100 g fresh weight), potassium (3.65 mg/100 g fresh weight), and sodium (32.0 mg/100 g fresh weight). Research on bell pepper by-products has predominantly focused on the seeds, with limited attention given to leaves, stems, placentas, and peels. Exploration of the chemical composition of the rest of the by-products (leaves, stems, placentas, and peels) remains an area for further investigation in future studies.

3 Antioxidant Properties and Bioactive Phytochemicals Extracted from Bell Pepper By-Products

The current chapter reveals that studies on the antioxidant capacity, antioxidant components, and profiling of bioactive phytochemicals in bell pepper by-products are scarce in the literature. Table 3 depicts the antioxidant capacity and antioxidant components of bell pepper by-products. It is observed that three studies centered on bell pepper seeds (Silva et al. 2013; Sora et al. 2015; Leng et al. 2022), while one study focuses on the leaf (Rácz et al. 2023). Interestingly, the most commonly used assays to quantify the antioxidant capacity of bell pepper by-products are the 2,2-diphenyl-1-picrylhydrazyl (DPPH), 2,2′-azino-bis (3-ethylbenzothiazoline-6-sulfonic acid) (ABTS), and Ferric Reducing Antioxidant Power (FRAP) assays.

Although a direct comparison of the antioxidant capacity of bell pepper seeds across these studies is impossible to carry out due to variations in the units of measurement (Table 3), emerging evidence suggests that bell pepper seeds are an excellent source of antioxidants. A recent study by Leng et al. (2022) compared the antioxidant properties of seeds derived from green, yellow, and red Australian-grown bell peppers. The yellow bell pepper's seeds exhibited the highest ability to scavenge the DPPH radicals and reduce the ferric ions (Fe^{3+}) to ferrous ions (Fe^{2+}) than the seeds derived from the green and red bell peppers. Even though the seeds derived from yellow and red bell peppers had the same antioxidant capacity as measured by the ABTS assay (0.56 mg AAE/g fresh weight), they exhibited a more pronounced ability to reduce the ABTS radical cation ($ABTS^{+}$) compared to the green bell pepper seeds. Concerning the antioxidant components of bell pepper seeds, it is noted that the total tannin content was comparable across all the seed extracts (0.11 mg CE/g fresh weight to 0.13 mg CE/g fresh weight). Nevertheless, red pepper seeds exhibited a higher total phenolic content (0.53 mg GAE/g fresh weight), green pepper seeds showed a higher total flavonoid content (4.98 μg QE/g fresh weight), and yellow pepper seeds had a higher total antioxidant capacity (0.88 mg AAE/g fresh weight) compared to their counterparts.

Table 3 Antioxidant capacity and antioxidant components of bell pepper by-products

By-product extracts/ Country of origin	Findings	Reference
Aqueous powdered seed extracts (1 g powdered seeds:100 ml water)/ Spain	1. DPPH· assay: IC_{25} = 0.413 mg/mL (sweet Italian powdered seeds) and > 18.2 mg/ml (Reus long pairal powdered seeds). 2. Superoxide radical assay: 45% (sweet Italian powdered seeds) and 47% (Reus long pairal powdered seeds). 3. Nitric oxide radical assay: IC_{25} = 0.105 mg/mL (sweet Italian powdered seeds) and 0.108 mg/mL (Reus long pairal powdered seeds).	Silva et al. (2013)
40% ethanolic powdered seed extract/ Brazil	1. DPPH assay: 11.32 μmol TE/g powdered seeds. 2. ABTS assay: 89.25 μmol TE/g powdered seeds. 3. FRAP assay: 9.94 μmol TE/g powdered seeds. 4. TPC: 409.45 mg GAE/100 g powdered seeds.	Sora et al. (2015)
80% ethanolic lyophilized powdered seed extracts/Australia	1. DPPH·assay: 0.12 mg AAE/g fresh weight (green bell pepper powdered seeds), 0.15 mg AAE/g fresh weight (yellow bell pepper powdered seeds), and 0.11 mg AAE/g fresh weight (red bell pepper powdered seeds). 2. ABTS assay: 0.38 mg AAE/g fresh weight (green bell pepper powdered seeds), 0.56 mg AAE/g fresh weight (yellow bell pepper powdered seeds), and 0.56 mg AAE/g fresh weight (red bell pepper powdered seeds). 3. FRAP assay: 0.02 mg AAE/g fresh weight (green bell pepper powdered seeds), 0.04 mg AAE/g fresh weight (yellow bell pepper powdered seeds), and 0.02 mg AAE/g fresh weight (red bell pepper powdered seeds). 4. TPC: 0.43 mg GAE/g fresh weight (green bell pepper powdered seeds), 0.52 mg GAE/g fresh weight (yellow bell pepper powdered seeds), and 0.53 mg GAE/g fresh weight (red bell pepper powdered seeds). 5. TFC: 4.98 μg QE/g fresh weight (green bell pepper powdered seeds), 3.66 μg QE/g fresh weight (yellow bell pepper powdered seeds), and 0.54 μg QE/g fresh weight (red bell pepper powdered seeds). 6. TTC: 0.11 mg CE/g fresh weight (green bell pepper powdered seeds), 0.13 mg CE/g fresh weight (yellow bell pepper powdered seeds), and 0.12 mg CE/g fresh weight (red bell pepper powdered seeds). 7. TAC: 0.76 mg AAE/g fresh weight (green bell pepper powdered seeds), 0.88 mg AAE/g fresh weight (yellow bell pepper powdered seeds), and 0.58 mg AAE/g fresh weight (red bell pepper powdered seeds).	Leng et al. (2022)

(continued)

Table 3 (continued)

By-product extracts/ Country of origin	Findings	Reference
70% ethanolic leaf extracts/Hungary	1. FRAP assay: 0.677 mM AAE/mg fresh weight (Darina), 0.698 mM AAE/mg fresh weight (Édesalma), and 0.876 mM AAE/mg fresh weight (Rekord). 2. TPC: 0.859 mM GAE/mg fresh weight (Darina), 0.951 mM GAE/mg fresh weight (Édesalma), and 1.018 mM GAE/mg fresh weight (Rekord).	Rácz et al. (2023)

Notes: *DPPH·* 2,2-diphenyl-1-picrylhydrazyl; *ABTS* 2,2′-azino-bis (3-ethylbenzothiazoline-6-sulfonic acid); *FRAP* Ferric Reducing Antioxidant Power; *TPC* Total phenolic content; *TFC* Total flavonoid content; *TTC* Total tannin content; *TAC* Total antioxidant capacity; *TE* Trolox equivalents; *GAE* Gallic acid equivalents; *AAE* Ascorbic acid equivalents; *QE* Quercetin equivalents; *CE* Catechin equivalents

Sora et al. (2015) reported that the total phenolic content of 40% ethanolic powdered seed extract was 409.45 mg GAE/100 g (or 4.09 mg GAE/g). This finding is approximately ten times higher than the total phenolic content of seeds derived from different colors of bell peppers (0.43–0.53 mg GAE/g fresh weight) reported by Leng et al. (2022). Such discrepancies could be explained by using different study protocols in assessing the total phenolic content or the diversity of bell pepper cultivars used for analysis (Razola-Díaz et al. 2022; Murugan and Tan 2023). In addition to seeds, a recent study by Rácz et al. (2023) examined the antioxidant properties of bell pepper leaves from three different Hungarian-grown bell pepper cultivars (Darina, Édesalma, and Rekord). The findings revealed that the antioxidant properties of bell pepper leaves, measured by the FRAP assay and total phenolic content, were in the descending order of Rekord> Édesalma> Darina.

Several sophisticated instruments have also been used to characterize and quantify the individual bioactive phytochemicals in bell pepper by-products. Although the literature has consistently reported that capsaicin and dihydrocapsaicin are the primary bioactive compounds that are responsible for the pungency and heat in peppers (Peña-Alvarez et al. 2009; Uarrota et al. 2021), Sora et al. (2015) claimed that capsaicin was not detected in the pulp or seed of bell peppers. The lack of capsaicin in the bell pepper may be attributed to the mutation in the putative aminotransferase (p-AMT) gene, which is involved in the biosynthesis of capsaicin (Nakaniwa et al. 2024). On the other hand, dihydrocapsaicin was reported to be 12.27 mg/100 g powdered seeds (Sora et al. 2015).

Leng et al. (2022) characterized the individual phenolic acids in bell pepper seeds derived from different colors using Liquid Chromatography coupled with Electrospray Ionization-Quadrupole-Time-of-Flight-Mass Spectrometry (LC-ESI-QTOF-MS/MS) and High-Performance Liquid Chromatography Photodiode Array (HPLC-PDA). Surprisingly, the findings indicated that 64.1% of phenolic acids and 50% of flavonoids of bell peppers are distributed in the seeds. Further analysis revealed that bell pepper seeds contain remarkable sources of phenolic acids (including chlorogenic acid, p-hydroxybenzoic acid, and hydroxybenzoic acids) and flavonoids (including quercetin, quercetin-3-rhamnoside, quercetin-3-glucoside, and epicatechin). The identification of individual phenolic acids and flavonoids in bell pepper leaves was also carried out by León-Chan et al. (2017) using Ultra-High Performance Liquid Chromatography (UHPLC). Luteolin-7-O-glucoside (flavonoid), apigenin-7-O-glucoside (flavonoid), and chlorogenic acid (phenolic acid) were reported as the primary bioactive phytochemicals in bell pepper leaves.

The current chapter reveals that no prior studies have investigated the antioxidant properties or bioactive phytochemicals in bell peppers' peels, placentas, and stems. Despite those previously mentioned, bell pepper peels are anticipated to share nearly identical bioactive phytochemical profiles with their flesh. This similarity can be attributed to their close physical connection, with the flesh positioned just underneath the peel. Since it has been previously reported that colored bell peppers are significant sources of vitamin C, β-carotene, lycopene, capsanthin, quercetin, and luteolin (Sun et al. 2007; León-Chan et al. 2017), it is plausible that these bioactive phytochemicals are also present in abundant amounts in the peels.

4 Biological and Functional Properties of Bioactive Phytochemicals Extracted from Bell Pepper By-Products

Emerging evidence on the health-promoting effects of bell pepper by-products is primarily derived from studies using cell lines and animal models, with a predominant focus on seeds and leaves. Investigations into the health-promoting effects of placentas, stem, and peels are yet to be undertaken.

4.1 Biological Properties of Bioactive Phytochemicals Extracted from Bell Pepper Leaves

Hazekawa et al. (2017) studied the anti-inflammatory properties of bell pepper leaves using mouse spleen cells. These cells were stimulated with concanavalin A to induce the secretion of pro-inflammatory cytokines, including interferon-gamma (IFN-γ), interleukin-6 (IL-6), and tumor necrosis factor-alpha (TNF-α). The findings indicated that bell pepper leaves' aqueous extract suppressed the production of pro-inflammatory cytokines without causing cytotoxicity. Even though the bioactive phytochemicals responsible for this effect have yet to be confirmed, the anti-inflammatory potential of bell pepper leaves is apparent.

In addition to the anti-inflammatory properties, Alsuhaibani et al. (2017) conducted a study examining the effects of chlorophyll extracted from bell pepper leaves on the body weight and blood glucose level of male Swiss albino rats. The findings demonstrated that chlorophyll extracted from bell pepper leaves significantly reduced body weight and decreased blood glucose levels in the rats. Overall, these findings indicate that bell pepper leaves have the potential to be used as anti-obesity and hypoglycemic agents.

4.2 Biological Properties of Bioactive Phytochemicals Extracted from Bell Pepper Seeds

A recent study by Younes and Mustafa (2024) investigated the oxidative stress mitigating, anti-cancer, anti-inflammatory, and anti-diabetic properties of phytocoumarins extracted from green sweet bell pepper seeds. Reactive oxygen species (ROS) were induced in SH-SY5Y human neuroblastoma cells to examine the oxidative stress mitigating effect through exposure to hydrogen peroxide. The findings suggested that phytocoumarins in green sweet bell pepper seeds alleviated oxidative stress by reducing the ROS levels in SH-SY5Y cells. In light of their reported ability to reduce oxidative stress, phytocoumarins in green sweet bell pepper seeds also demonstrated excellent anti-cancer effects against MCF-7 (breast carcinoma) and HeLa (human cervix carcinoma) cancer cell lines, as well as AR42J (rat pancreatic tumoral cells) cell lines. Moreover, these bioactive phytochemicals also suppressed the activities of inflammation-mediated enzymes, including cyclooxygenase 1 (COX1), cyclooxygenase 2 (COX2), and 5-lipoxygenase (5-LOX), thereby suggesting their potency as anti-inflammatory agents. The same study also tested the anti-diabetic properties of phytocoumarins in green sweet bell pepper seeds. Owing to their moderate ability to inhibit the glucose homeostasis enzymes α-amylase and α-glucosidase, these bioactive phytochemicals could potentially emerge as potent anti-diabetic agents.

Despite those previously mentioned above, a former study by Raybaudi-Massilia et al. (2017) indicated that the aqueous and methanolic extracts of red sweet pepper seeds exhibited no cytotoxic activity toward human cancer cell lines from the breast (MCF-7 and SKBr3), prostate (PC3), and cervix (HeLa). Moreover, these extracts showed no antimicrobial activity against bacteria that cause foodborne illnesses, including *Listeria monocytogenes*, *Salmonella enterica serovar Enteritidis*, and *Staphylococcus aureus*. The ability of bell pepper seeds to inhibit the enzymatic activity of acetylcholinesterase was tested by Silva et al. (2013). The findings revealed that the aqueous extract of bell pepper seeds failed to suppress the activity of the acetylcholinesterase enzyme, which is known to be associated with Alzheimer's disease. In short, the health-promoting effects of bell pepper seeds remain inconclusive due to a lack of studies.

4.3 Functional Properties of Bioactive Phytochemicals Extracted from Bell Pepper Seeds

Since the literature has consistently reported that bell pepper seeds are an excellent source of protein (Table 2), Li et al. (2018) compared the functional properties of bioactive protein isolates extracted from the bell pepper seeds using ultrasound-assisted extraction and alkaline dissolving and acid-precipitating methods. Despite applying two different extraction methods, both bioactive protein isolates exhibited higher solubility at alkaline pH. Such characteristics suggest that these bioactive protein isolates can be incorporated into products with neutral or alkaline pH to improve their functionality and nutritional value. The bioactive protein isolates were also reported to bear an excellent oil-holding capacity in addition to their effective emulsifying and foaming properties. Generally, the reported functional properties enable the bioactive protein isolates extracted from bell pepper seeds to be highly versatile for various food applications.

5 Food and Non-food Applications of Extracts and Bioactive Phytochemicals from Bell Pepper By-Products

It has been documented that linoleic sunflower oil is an exceptional source of linoleic acid (PUFA, 69%) and oleic acid (MUFA, 20%) (National Sunflower Association 2024). The high proportion of unsaturated fatty acids in sunflower oil makes it an excellent choice for cooking or salad dressing. In view that bell pepper seed oil has a nearly identical fatty acid composition (69.4% linoleic acid and 12.51% oleic acid) to linoleic sunflower oil, future studies could explore its potential as a cooking oil or salad dressing (Embaby and Mokhtar 2011). Interestingly, a recent study by Cocan et al. (2021) showed that bell pepper seed oil improved the thermo-oxidative stability of sunflower oil. This promising effect is likely attributed to the bioactive phytochemicals in bell pepper seed oil. In general, these findings suggest that bell pepper seed oil has the potential to become a natural additive for cooking vegetable oils.

Even though studies related to the food and non-food applications of bell pepper by-products are still limited, emerging evidence suggests that fruit and vegetable by-products that are rich in bioactive phytochemicals such as phenolic acid, flavonoids, carotenoids, and anthocyanins can be repurposed for the formulation of nutraceuticals, food colorants, or cosmetic products (Ahmad and Khan 2019; Sharma et al. 2021; Benucci et al. 2022). It is also worth noting that nitrogen, phosphorus, and carbon are essential nutrients for plant growth. Coincidentally, these nutrients are abundant in fruit and vegetable by-products such as bell pepper seeds (Ramamoorthy et al. 2024). This implies that bell pepper by-products can be potentially turned into organic fertilizer for agricultural use. Apart from those previously mentioned, fruit and vegetable by-products can also be utilized in the production of biodegradable food packaging materials such as bioplastic film, biodegradable food containers, or biodegradable cutlery (Otoni et al. 2021; Gong et al. 2024).

6 Valorization of Bell Pepper By-products for Non-health Purposes

A recent review by Chhandama et al. (2022) affirmed that food waste generated throughout the food supply chain could be valorized and converted into valuable biofuels (biogas, bioethanol, or biodiesel) or bioelectricity. Segundo et al. (2024) have recently developed a particular single-chamber device called a microbial fuel cell to convert pepper waste into usable electrical energy with the assistance of energy-generating bacteria. Given the successful generation of bioelectricity, the study mentioned above suggested that pepper waste could be a potential renewable energy source. Since the literature consistently reported that pepper seeds are an excellent source of lipids, Lee et al. (2017) carried out a study to investigate the possibility of transforming these lipids into biodiesel. The findings revealed that almost all (94.1%) of the lipids in pepper seeds can be converted into biodiesel without undergoing additional lipid extraction.

7 Conclusion

While it is widely recognized that *Capsicum annuum* comprises both pungent hot chilies and non-pungent bell peppers (Hernández-Pérez et al. 2020), the current chapter uncovers that the majority of existing studies did not specify the varieties of Capsicum studied. This lack of specification can create ambiguity in understanding the findings related to bell peppers. Another concern worth highlighting is that despite the abundance of research on bell peppers, studies are scarce focusing on the by-products (peels, seeds, placentas, stems, and leaves) derived from them. In conclusion, emerging evidence indicates that bell pepper by-products are exceptional sources of nutrients and bioactive phytochemicals. Additional research is required to gain deeper insights into the properties and potential food and non-food applications of bell pepper by-products discussed in this chapter.

References

Ahmad F, Khan ST (2019) Potential industrial use of compounds from by-products of fruits and vegetables. In: Malik A, Erginkaya Z, Erten H (eds) Health and safety aspects of food processing technologies. Springer International Publishing, Cham, pp 273–307

Alsuhaibani AM, Alkehayez NM, Alshawi AH, Al-Faris NA (2017) Effects of chlorophyll on body functioning and blood glucose levels. Asian J Clin Nutri 9:64–70. https://doi.org/10.3923/ajcn.2017.64.70

Anaya-Esparza LM, de la Mora ZV, Vázquez-Paulino O et al (2021) Bell peppers (*Capsicum annum L.*) losses and wastes: source for food and pharmaceutical applications. Molecules 26:5341

Baenas N, Belović M, Ilic N et al (2019) Industrial use of pepper (*Capsicum annum L.*) derived products: technological benefits and biological advantages. Food Chem 274:872–885. https://doi.org/10.1016/j.foodchem.2018.09.047

Benucci I, Lombardelli C, Mazzocchi C, Esti M (2022) Natural colorants from vegetable food waste: recovery, regulatory aspects, and stability-a review. Compr Rev Food Sci Food Saf 21:2715–2737. https://doi.org/10.1111/1541-4337.12951

CABI (2019) *Capsicum annuum* (bell pepper). CABI Compendium. https://doi.org/10.1079/cabicompendium.15784

Chhandama MVL, Chetia AC, Satyan KB et al (2022) Valorisation of food waste to sustainable energy and other value-added products: a review. Bioresour Technol Rep 17:100945

Cocan I, Negrea M, Cozma A et al (2021) Chili and sweet pepper seed oil used as a natural antioxidant to improve the thermo-oxidative stability of sunflower oil. Agronomy 11:2579. https://doi.org/10.3390/agronomy11122579

Colombari LF, Chamma L, da Silva GF et al (2022) Maturation and post-harvest resting of fruits affect the macronutrients and protein content in sweet pepper seeds. Plan Theory 11:2084. https://doi.org/10.3390/plants11162084

Cvetković T, Ranilović J, Jokić S (2022) Quality of pepper seed by-products: a review. Foods 11:748

Embaby HE, Miyakawa T, Hachimura S et al (2022) Physical and chemical properties of nabak (Zizyphus spina-christi) seed kernel and sweet pepper (*Capsicum annuum L.*) seed oils. J Sci Food Agric 102:2660–2666. https://doi.org/10.1002/jsfa.11605

Embaby HES, Mokhtar SM (2011) Chemical composition and nutritive value of lantana and sweet pepper seeds and Nabak seed kernels. J Food Sci 76:C736–C741. https://doi.org/10.1111/j.1750-3841.2011.02166.x

Emmanuel Ilesanmi A (2014) The contribution of flesh, placenta and seeds to the nutritional attributes of a variety of *Capsicum annum* (Bell pepper). Elixir Food Sci 68:22587–22594

Eskander MZ, Hassan KHA, Alshara AS et al (2023) Oil extraction and fatty acid characterization of sweet peppers seeds *capsicum annum (L.)* by gas chromatography-mass spectrometry (GC-MC) and its use in beef burger patties preservation. J Microbiol Biotechnol Food Sci 12:e9520. https://doi.org/10.55251/jmbfs.9520

Food and Agriculture Organization of the United Nations (2024) Statistics: chillies and peppers, green (*Capsicum* spp. and *Pimenta* spp.). In: Food and agriculture organization of the United Nations. http://data.un.org/Data.aspx?d=FAO&f=itemCode%3a401. Accessed 27 May 2024

Gong L, Passari AK, Yin C et al (2024) Sustainable utilization of fruit and vegetable waste bioresources for bioplastics production. Crit Rev Biotechnol 44:236–254

Hazekawa M, Hideshima Y, Ono K et al (2017) Anti-inflammatory effects of water extract from bell pepper (*Capsicum annuum L. var. grossum*) leaves in vitro. Exp Ther Med 14:4349–4355. https://doi.org/10.3892/etm.2017.5106

Hernández-Pérez T, del Gómez-García MR, Valverde ME, Paredes-López O (2020) *Capsicum annuum* (hot pepper): an ancient Latin-American crop with outstanding bioactive compounds and nutraceutical potential. A review. Compr Rev Food Sci Food Saf 19:2972–2993. https://doi.org/10.1111/1541-4337.12634

Lee J, Kim J, Ok YS, Kwon EE (2017) Rapid biodiesel synthesis from waste pepper seeds without lipid isolation step. Bioresour Technol 239:17–20. https://doi.org/10.1016/j.biortech.2017.05.011

Leng Z, Zhong B, Wu H et al (2022) Identification of phenolic compounds in Australian-grown bell peppers by liquid chromatography coupled with electrospray ionization-quadrupole-time-of-flight-mass spectrometry and estimation of their antioxidant potential. ACS Omega 7:4563–4576. https://doi.org/10.1021/acsomega.1c06532

León-Chan RG, López-Meyer M, Osuna-Enciso T et al (2017) Low temperature and ultraviolet-B radiation affect chlorophyll content and induce the accumulation of UV-B-absorbing and antioxidant compounds in bell pepper (*Capsicum annuum*) plants. Environ Exp Bot 139:143–151. https://doi.org/10.1016/j.envexpbot.2017.05.006

Li M, Wen X, Peng Y et al (2018) Functional properties of protein isolates from bell pepper (*Capsicum annuum L. var. annuum*) seeds. LWT-Food Sci Technol 97:802–810. https://doi.org/10.1016/j.lwt.2018.07.069

Murugan R, Tan ST (2023) Unveiling the antioxidant properties of fruit oils extracted from the by-products of *Pulasan* and *Salak*. Vegetos 36:694–700. https://doi.org/10.1007/s42535-022-00406-2

Nakaniwa R, Misawa Y, Nakasato S et al (2024) Biochemical aspects of putative aminotransferase responsible for converting vanillin to vanillylamine in the capsaicinoid biosynthesis pathway in capsicum plants. J Agric Food Chem 72:559–565. https://doi.org/10.1021/acs.jafc.3c07369

National Sunflower Association (2024) Sunflower oil fatty acid profiles. https://www.sunflowernsa.com/health/sunflower-oil-fatty-acid-profiles. Accessed 8 Jun 2024

Nirvesha Nair VK, Tan ST (2021) Antioxidant activities, total phenolic content and colour parameters in the aqueous extracts of avocado, banana and papaya leaves. Jurnal Sains Kesihatan 19:137–142. https://doi.org/10.17576/JSKM-2021-1901-15

North Carolina Extension Gardener Plant Toolbox (2024) *Capsicum annuum* (Grossum Group). https://plants.ces.ncsu.edu/plants/capsicum-annuum-grossum-group. Accessed 28 May 2024

Otoni CG, Azeredo HMC, Mattos BD et al (2021) The food-materials nexus: next generation bioplastics and advanced materials from Agri-food residues. Adv Mater 33:2102520

Peña-Alvarez A, Ramírez-Maya E, Alvarado-Suárez LÁ (2009) Analysis of capsaicin and dihydrocapsaicin in peppers and pepper sauces by solid phase microextraction-gas chromatography-mass spectrometry. J Chromatogr A 1216:2843–2847. https://doi.org/10.1016/j.chroma.2008.10.053

Rácz A, Czégény G, Kutyáncsánin D et al (2023) Fight against cold: photosynthetic and antioxidant responses of different bell pepper cultivars (*Capsicum annuum L.*) to cold stress. Biol Futur 74:327–335. https://doi.org/10.1007/s42977-023-00182-3

Ramamoorthy K, Dhanraj R, Vijayakumar N et al (2024) Vegetable and fruit wastes: valuable source for organic fertilizer for effective growth of short-term crops: *Solanum lycopersicum* and *Capsicum annum*. Environ Res 251:118727. https://doi.org/10.1016/j.envres.2024.118727

Raybaudi-Massilia R, Suárez AI, Arvelo F et al (2017) Cytotoxic, antioxidant and antimicrobial properties of red sweet pepper (*Capsicum annuum L. var. Llanerón*) extracts: in vitro study. Int J Food Stud 6:222–231. https://doi.org/10.7455/ijfs/6.2.2017.a8

Razola-Díaz MDC, Gómez-Caravaca AM, de Andrés JL et al (2022) Evaluation of phenolic compounds and pigments content in yellow bell pepper wastes. Antioxidants 11:557. https://doi.org/10.3390/antiox11030557

Segundo RF, Magaly DLCN, Luis CC et al (2024) Obtaining sustainable electrical energy from pepper waste. Sustainability (Switzerland) 16:3448. https://doi.org/10.3390/su16083448

Sharma M, Usmani Z, Gupta VK, Bhat R (2021) Valorization of fruits and vegetable wastes and by-products to produce natural pigments. Crit Rev Biotechnol 41:535–563

Silva LR, Azevedo J, Pereira MJ et al (2013) Chemical assessment and antioxidant capacity of pepper (*Capsicum annuum L.*) seeds. Food Chem Toxicol 53:240–248. https://doi.org/10.1016/j.fct.2012.11.036

Sood T, Sood S, Sood VK et al (2023) Characterisation of bell pepper (*Capsicum annuum L. var. grossum Sendt.*) accessions for genetic diversity and population structure based on agro-morphological and microsatellite markers. Sci Hortic 321:112308. https://doi.org/10.1016/j.scienta.2023.112308

Sora GTS, Haminiuk CWI, da Silva MV et al (2015) A comparative study of the capsaicinoid and phenolic contents and in vitro antioxidant activities of the peppers of the genus *capsicum*: an application of chemometrics. J Food Sci Technol 52:8086–8094. https://doi.org/10.1007/s13197-015-1935-8

Sun T, Xu Z, Wu C-T et al (2007) Antioxidant activities of different colored sweet bell peppers (*Capsicum annuum L.*). J Food Sci 72:S98–S102. https://doi.org/10.1111/j.1750-3841.2006.00245.x

Tan SS, Tan ST, Tan CX (2020) Antioxidant, hypoglycemic and anti-hypertensive properties of extracts derived from peel, fruit and kernel of Salak. Br Food J 122:3029–3038. https://doi.org/10.1108/BFJ-03-2020-0233

The United States Department of Agriculture (2021) FoodData Central. In: USDA. https://fdc.nal.usda.gov/fdc-app.html#/food-search?query=BELL%20PEPPER&type=Foundation. Accessed 27 May 2024

Tridge (2024) Fresh bell pepper. https://www.tridge.com/intelligences/bell-pepper/import. Accessed 27 May 2024

Uarrota VG, Maraschin M, de Bairros Â, Pedreschi R (2021) Factors affecting the capsaicinoid profile of hot peppers and biological activity of their non-pungent analogs (Capsinoids) present in sweet peppers. Crit Rev Food Sci Nutr 61:649–665

Wong PY, Tan ST (2020) Comparison of total phenolic content and antioxidant activities in selected coloured plants. Br Food J 122:3193. https://doi.org/10.1108/BFJ-12-2019-0927

Yasin M, Li L, Donovan-Mak M et al (2023) Capsicum waste as a sustainable source of capsaicinoids for metabolic diseases. Foods 12: 907. https://doi.org/10.3390/foods12040907

Younes AH, Mustafa YF (2024) Novel coumarins from green sweet bell pepper seeds: their isolation, characterization, oxidative stress-mitigating, anticancer, anti-inflammatory, and antidiabetic properties. J Mol Struct 1312:138629. https://doi.org/10.1016/j.molstruc.2024.138629

Phytochemicals from Chayote (*Sechium edule*) By-products

Ashiq Hussain, Qurat Ul An, Adnan Ahmed, Shazia Yqaub, Afifa Asif, Khurram Kabir, Rizwan Nisar, Haya Fatima, and Faiza Iftikhar Gorsi

Abstract The valorization of by-products of fruits and vegetables for different food, pharma, and industrial purposes has gained much attention recently. Apart from the aerial fruits and vegetables, the bulb vegetables also have several consumption and processing patterns, due to which different parts of these vegetables are generated as waste streams. Implementing proper handling practices for these waste materials to recover valuable components is an area of research nowadays. *Sesame edule*, or Chayote, is widely accepted as a food valued for its nutritional and biological qualities. Chayote is an edible plant that is a member of the *Cucurbitaceae*, the family of gourds. The current chapter outlines the research on Chayote by-products' pharmacological, nutritional, and phytochemical characteristics and suggests areas for future investigation. It also covers the many applications of Chayote in nutrition, as a helpful food ingredient, and in non-food sectors. While the ripe green-colored fruit is the most often eaten part of the plant, people can also eat unripe fruits, tender leaf shoots, tuberous roots, and mature seeds, which are generally considered waste products but are a valuable source of nutrients. Because they have therapeutic properties, Chayote plant shoots are deemed necessary for nutrition due to a wide range of phytochemicals. Though they have not yet been thoroughly studied as a nutraceutical, Chayote peels and seeds are known for their bioactive qualities and high ash, fiber, protein, and mineral contents. From an economic and environmental perspective, one crucial part of waste management can be the valorization of the underutilized Chayote by-products. Therefore, extracting and applying valuable chemicals from Chayote by-products is a major global scientific problem.

A. Hussain (✉) · Q. U. An · S. Yqaub · A. Asif · R. Nisar · F. I. Gorsi
Institute of Food Science and Nutrition, University of Sargodha, Sargodha, Pakistan

A. Ahmed
Department of Botany, University of Sargodha, Sargodha, Pakistan

K. Kabir · H. Fatima
PMAS Arid Agriculture University, Rawalpindi, Punjab, Pakistan

M. F. Ramadan (ed.), *Bioactive Phytochemicals in By-products from Bulb, Flower and Fruit Vegetables*, https://doi.org/10.1007/978-3-031-77399-0_8

Keywords Circular economy · Minerals · Biological activities · Extract · Nutrients · Peel · Seed · Root · Shoot

1 An Introduction to Chayote (*Sechium edule*)

The World Health Organization suggests consuming at least 400 g of vegetables and fruits, or five pieces, daily, apart from starchy tubers, to maintain excellent health. Following USDA guidelines, a person should eat one cup (~237 g) of freshly grown vegetables that are either raw or cooked or a couple of cups of unprocessed leafy greens during the day (Managa et al. 2021). As consuming fruits and vegetables is healthy, their by-products are discarded as a waste stream.

Chayote, or *Sechium edule*, is an edible plant from the *Cucurbitaceae* family of gourds. It is a perennial herbaceous climbing plant with green, pear-like fruits, tuberous roots, and persistent tendrils. Sites with moderate temperatures (13–21 °C), comparatively high relative humidity (80–85%), evenly scattered yearly rainfall (1500–2500 mm), and 12 h of daylight are ideal for Chayote growth (Cadena-Iñiguez et al. 2007; Gavia-García et al. 2023). Chayote has quickly grown due to its high adaptability to various environmental conditions. Chayote is grown chiefly for its healthy fruits in most areas. It has been historically proven that Chayote originated in Mexico, and after that, it spread to other regions. Currently, this species is found on five continents: Africa, Asia, America, Australia, and Europe (Chang et al. 2021). Popular names for the edible *Sechium edule* fruit include Chayote, christophene, vegetable pear, mirliton, merleton choko (in Australia and New Zealand), starpreciantė, citrayota, citrayote (in Ecuador and Colombia), chuchu (Brazil), machucha, caiota, pipinela (Portugal), chow chow (India), cho cho or chow chow (Jamaica), Sayote (Philippines), güisquil (El Salvador and Guatemala), and pear squash or iskus (Nepal) (Vieira et al. 2019).

Although Chayote was not a valuable crop for the economy or a luxury good, it was nonetheless significant to the non-elite. In symbolic terms, Chayote is associated with femininity; it is the receptacle of life, and its hirsute cleft symbolizes the vulva; it is contrasted with its antithesis, the chili pepper (Capsicum). Ancient Yucatec art depicts the "*Cucurbita* tree," an old name for Chayote, as Cucurbit vines, a synecdoche for the First World Tree. According to a linguistic study, the indigenous people categorized Chayote as cucurbits, grouping their subtypes based on color and thorniness (Moreira 2015).

Ten species, including *Sechium edule*, are found in the genus *Sechium* P. Br., a Cucurbitaceae family participant. Eight of these are found in the wild, and two are grown. The *Sechium edule* is reported to have twelve varietal groups (var.) that are often used. These groups differ in size, shape, color, flavor, consistency, and thorn presence (Barrera-Guzmán et al. 2021). The light green, slightly basally cleft fruits of the variety *Virens levis* have a light green mesocarp that has a neutral flavor and fiber that sticks to the mesocarp. Their dimensions vary from 6.0 to 11.40 cm in width and 9.30 to 18.30 cm in length. The dark fruits of *Nigrum xalapensis* range in size from 5.4 to 18 cm in width and 5.5 to 26.6 cm in length, and they include a noticeable basal cleft. The fruits of *nigrum spinosum* are dark green in color, around

5.8–17.1 cm long and 5.0–12.2 cm wide, with a high density of spines, a very evident basal cleft, a light green to dark green mesocarp, and a neutral to slightly sweet flavor. They also have fiber connected to the mesocarp (Ramírez-Rodas et al. 2021; Castro Rodríguez et al. 2015).

Reunion Island's traditional industry of chayote farming has become increasingly prevalent in recent years, utilizing a lot of mineral fertilizers and pesticides. Professionals in the field of agriculture on Reunion Island have spent the last 10 years advocating for the transition from agrochemical to agro-ecological Chayote cultivation, which complies with organic farming laws. More than 60% of the chayote-producing regions adhere to organic farming practices as of 2017. Increased yields resulted from reduced fruit damage; the protection's cost is lower, and its duration is shorter. The knowledge gained from organic Chayote cultivation has helped other crops become organic and led to the implementation of these agro-ecological techniques in neighboring Indian Ocean nations (Deguine et al. 2020; Veigas et al. 2020). Although this is considered a neglected and underutilized crop, it has much potential to become a highly cultivated, consumed, and processed crop because of its exceptional medicinal potential. Therefore, this chapter was designed to provide comprehensive data on the chayote by-products as a good source of nutritional and phytochemical contents and applications in the food, feed, pharma, and other sectors.

2 Economic Value of Chayote

The *Cucurbitaceae* family of medicinal plants includes several eatable fruits and vegetables used as food and medicine for ages worldwide. Numerous traditional literature has shown the crucial significance that these plants have played in the global ethnopharmacological and conventional therapeutic systems. Many human illnesses have been treated with various plant parts (Hussain et al. 2021, 2022a, b, 2023).

Finding the top commodities is crucial to boosting regional economic development competitiveness. Chayote is one of the commodities with the potential to rise to prominence in the future (Aminda et al. 2024). Several species in the *Cucurbitaceae* family are suitable for human consumption, many of which include beneficial chemicals (Hussain et al. 2022c). The plump fruits weighing 250 and 400 g are produced by the perennial Chayote, which has tuberous roots and is typically eaten like vegetables. Originating in Mesoamerica, the primary producers of this species are Mexico and Costa Rica. In terms of trade and economy, the most valued kinds of Chayote are *nigrum spinosum*, which is grown in moderate regions and high valleys between 2000 and 2800 meters above sea level, and *virens levis*, which is grown in tropical and subtropical climates (Riviello-Flores et al. 2018). Despite originating between southern Mexico and Guatemala, this crop is now grown worldwide in tropical and subtropical climates. The fruit known as "Chayote" has a high commercial value in commerce (Díaz-de-Cerio et al. 2019).

Vegetables like Chayote have a high market worth, and pectin may be used in processed foods. The establishment mechanism of the complex coacervation of sodium caseinate and Chayote pectin in an aqueous solution was examined in a study by Ke et al. (2023a, b), and strange thermal behaviors of the complexes generated were noted. Mexican Chayote tubers were used to separate starch, and its yield, physicochemical, rheological, and molecular properties were all examined. For comparison, commercial potatoes were utilized. Chayote tubers had a 49% starch yield and an 89.1% purity. The Chayote tuber possesses unique physicochemical and molecular properties, making it a potential substitute for starch separation (Cadena-Iñiguez et al. 2007). With notable in vitro digestive characteristics, Chayote starch provides a nutritional content appropriate for human consumption. Chayote fruit starch has demonstrated intriguing functional properties appropriate for relevant food, energy, and cosmetics applications. It may also be used as a raw material in starch production (Abdelhai et al. 2023).

It has successfully created the delicious treat "Petha" using Chayote fruit. Meanwhile, the by-products left over in the form of peels and seeds may be utilized to extract bioactives. Research into process optimization for manufacturing Chayote-based "Petha" can aid in producing value-added goods from inexpensive vegetables like Chayote. Through the transfer of innovations to small and medium-sized industries like farmers and produce processing organizations, the new product process can be scaled up and utilized at small and medium scales, helping to enhance socioeconomic position through additional income (Taynath et al. 2020). Despite being an underutilized crop, chayote still has a lot of industry and commercial potential because every part of the plant provides many benefits for multiple sectors of the economy. These fractions must be introduced and used for various purposes by being made into powders, extracts, and separated bioactives to valorize Chayote by-products. Figure 1 presents different fractions of chayote by-products.

Fig. 1 Different fractions of Chayote plant

3 Chayote By-products and Their Chemical Composition

Food waste is a foremost problem in many states. Almost 1.3 billion tonnes of food processing waste are dumped into the environment each year, which influences the ecosystem primarily due to the chemical components of food, effects on global warming, eutrophication, and soil acidification. Fruits and vegetables have the highest rate of food waste since their stems, leaves, peels, and seeds are typically not used in cuisine. The majority of these by-products are wasted during the following stages: 10% during harvesting, 50% during processing, which includes storage and transportation, 40% in industries, and 40% in homes, kitchens, and restaurants (Scherhaufer et al. 2018; Barik and Paul 2017). Vegetable waste, including biological by-products and residues, comprises complex molecules, including cellulose, lignin, pentosans, hemicelluloses, and carbohydrates like hexoses and pentoses. Antioxidant compounds like flavonoids, phenolic acids, carotenoids, bioactive lipids and peptides, and phytochemicals are also present. Waste is a more inexpensive renewable source for the production of bioproducts, helping to reduce the environmental impacts of the agricultural commodities market while promoting economic growth and social well-being without depending on non-renewable fossil or mineral resources (Aguilar-Rivera and Olvera-Vargas 2023; Gómez-García et al. 2024).

The climate, geography, growing environment, stage of the plant, and processing practices all affect the nutritional makeup of Chayote by-products. While the ripe green-colored fruit is the part of the plant that is most often eaten, you can also eat unripe fruits, tender leaf shoots, and tuberous roots, which are generally thought of as waste products but are a valuable source of nutrients (Vieira et al. 2019). In many regions of Latin America and beyond, Chayote fruit and root have played a significant role in people's diets. Cultivated Chayote stems and tender leaves are used as regular vegetables. The fruit of the Chayote contains sufficient amounts of micro- and macronutrients, but its edible components are heavy in calories and carbohydrates, particularly in the young stems, roots, and seeds. Numerous significant amino acids are abundant in the fruit, particularly seeds (Ordoñez et al. 2006).

More non-starch polysaccharides, notably cellulose and pectin, are present in Chayote fractions. The Chayote non-starch polysaccharide is thought to include 15–20% pectin homogalacturonans and rhamnogalacturonan I, according to the characteristics results (Ke et al. 2023a, b). Starch made up the majority of the carbohydrates in the expanded Chayote root. Additionally, it was discovered that 0.7% of the vigorously growing distal 10–12 cm of the fragile apical shoots consisted of starch and 0.3% of soluble sugars on a wet-weight basis. The presence of carbohydrates in Chayote sprouts is probably responsible for their flavor and nutritional value. In tropical locations, people particularly appreciate Chayote sprouts as a cooked delicate dish and as a source of nutritious vegetable greens (Cadena-Iñiguez et al. 2007).

Because they have therapeutic properties, Chayote plant shoots are considered necessary for nutrition. Because Chayote shoots are high in proximate content, antioxidant activity, and minerals, it is advised to consume them for health advantages

Table 1 Chemical composition (per 100 g dw) of Chayote by-products

Composition	Root	Leaf	Shoot	Seed	Peel	Fruit
Moisture (per 100 g fw)	79.7	89.7	88.4	82.5	85.0	90–94.5
Fat (g)	0.2	0.4–2.3	1.5	6.0–8.2	1.3	0.1–0.3
Protein (g)	2.0	2.6–4.8	30.3	5.5	11.1	0.9–1.1
Fiber (g)	0.4	1.2–12.1	19.2	1.5–1.9	23.3	0.4–1
Ash (g)	1.8	0.7	12.5	1.5	6.2	0.5
Carbohydrates (g)	17.8	4.7	36.6	60	76.4	82–86
Energy (Cal)	79.0	60.0	320	255	367	26–31
Ascorbic acid (mg)	19.0	16.0	12.50	1.9	51.6	11–20
Niacin (mg)	0.9	1.1	–	–	–	0.4–0.5
Riboflavin (mg)	0.03	0.18	–	–	–	0.04
Thiamin (mg)	0.05	0.08	–	–	–	0.03
Vitamin A (μg)	–	615.0	–	–	–	5.0

Source: Saade (1996); Cadena-Iñiguez et al. (2007); Castro Rodríguez et al. (2015); Loizzo et al. (2016); Vieira et al. (2019); Molinari et al. (2021); Mashine et al. (2023)

(Shariff et al. 2023). Protein (4.0 g), fat (0.4 g), carbs (4.7 g), fiber (1.2 g), calcium (58 mg), phosphorus (108 mg), iron (2.5 mg), thiamin (615 μg), riboflavin (0.08 mg), niacin (0.18 mg), and ascorbic acid (1.1 mg) are all present in 100 g of fresh Chayote leaves (Harriman 1998). At different phases of development, arginine, agmatine, putrescine, and spermidine were discovered in the apical sections and leaves of Chayote. When flower buds initially appeared in immature leaves, the concentration of agmatine—the immediate decarboxylation result of L-arginine—rose significantly. Agmatine content was usually comparatively higher in young leaves than in older ones. No agmatine buildup was observed in other *Cucurbitaceae* species (Vansuyt and Zinsou 1986).

Though they have not yet been thoroughly studied as a nutraceutical, Chayote seeds are known for their bioactive qualities and high protein content. The overview of a study by Vieira et al. (2023a, b) on Chayote seed proteins reveals that these proteins have balanced amounts of all amino acids, which gives them antioxidant and antidiabetic effects. Reviewed literature has witnessed the presence of nutritional contents in each fraction of underutilized Chayote plant and fruit, recommending their use in food and pharma sectors. Table 1 summarizes the chemical composition of Chayote by-products.

4 Bioactive Compounds in the Extracts of Chayote By-products

Because of their high antioxidant potential, which is primarily credited to the presence of bioactive compounds like polyphenols and terpenoids and essential nutrients like vitamins and minerals, fruits and their various by-products have been linked to an improvement in human quality of life by preventing chronic degenerative disease or extending lifespan (De Jesús Lira-Ricárdez et al. 2024). An analysis found 26 chemicals categorized as flavonoids, phenolic acids, and other volatile substances in the Chayote fruit using 70% ethanolic extracts. With flavones making up 60.6% of all the chemicals measured in Chayote, diosmetin 7-O-rutinoside (23.8%) is the main flavonoid (Díaz-de-Cerio et al. 2019). According to Siciliano et al. (2004), flavonoids have been identified in various edible organs of the Chayote plant. Eight flavonoids, comprising 3-C-glycosyl and 5-O-glycosyl flavones, were identified and measured in the plant's fruits, leaves, stems, and roots. Apigenin and luteolin are the aglycone moieties, and glucose, apiose, and rhamnose are the sugar units. According to their findings, the dried component of the leaves had the highest overall quantity of flavonoids (35 mg/10 g), followed by the roots (30.5 mg/10 g) and the stems (19.3 mg/10 g).

The total flavonoid, β-carotene, and phenolic contents of methanol extracts made from Chayote shoots were high; however, the exact amounts varied depending on the extraction method used (Chang et al. 2021). The green leaves of the medicinal Chayote plant are used in traditional medicine and cuisine. An attempt was made to use ultrasound-assisted extraction to extract phenolic compounds and carotenoids from the Chayote leaves. The primary class of carotenoids, accounting for 42–85% of the entire carotenoid content, was determined by HPLC investigation of the extracts to be the xanthophyll class, followed by β-carotene and tocopherol. Furthermore, it was found that the Chayote leaf extracts included 26 different chemicals that were categorized as flavonoids, phenolic acids, and other polar compounds. Of all the quantified chemicals, flavonols made up 55%, while phenolic acids, mostly comprised of ferulic acid, chlorogenic acid, and catechin, comprised about 35% (Vieira et al. 2022).

The bioactives from Chayote peel were extracted more effectively when ultrasound assistance was used. According to the extracts' HPLC analysis, tocopherol esters were the predominant class of carotenoids, followed by retinol esters and a-tocopherol. Using HPLC analysis, 26 compounds were found in the Chayote peel extracts; of these, phenolic acids made up 71% and flavonols 21% of the total compounds measured (Vieira et al. 2023b). To encourage its sustainable usage in the culinary, cosmetic, and pharmaceutical sectors, more research on the phytochemical composition of Chayote peels and the implementation of green extraction practices to recover valued bioactive chemicals are needed.

The Chayote fruit stems, shoots, seeds, peels, young leaves, and root tubers are highly valued sources of carbohydrates, proteins, fibers, minerals, and vitamins, especially vitamins A and C (Sood and Rana 2017). Upon examining the high quality of proteins found in Chayote seeds, it was discovered that the isolated isolates and bioactive proteins were abundant in bioactive peptides that hold potential for further investigation in food systems (Vieira et al. 2023a). Each Chayote fraction is loaded with bioactive components involved in several biological and functional roles. Therefore, it is necessary to utilize the by-products of this vegetable for identification, isolation, characterization of bioactive compounds, and application in the pharmaceutical industry for the wellness of humans. Table 2 provides data on different bioactive compounds determined in various fractions of Chayote.

Table 2 Bioactive compounds in the extracts of Chayote by-products

Chayote by-product fractions	Extract types	Total phenolics	Total flavonoids	Total carotenoids	FRAP value	ABTS value	Reference
Peel	Ultrasonic-assisted ethanolic extracts	406 mg GAE/100 g dw	–	38 mg/100 g dw	82.83 mg AAE/100 g dw	319 mg AAE/100 g dw	Vieira et al. (2023b)
Leaves	Ethanolic extracts	–	35 mg/10 g	–	–	–	Siciliano et al. (2004)
Roots	Ethanolic extracts	–	30.5 mg/10 g	–	–	–	Siciliano et al. (2004)
Stem	Ethanolic extracts	–	19.3 mg/10 g	–	–	–	Siciliano et al. (2004)
Shoot	Methanolic extracts	5.16 mg GAE/g dw	5.02 mg QE/g dw	–	–	–	Chang et al. (2021)
Tubers	Ultrasound-assisted hydro-ethanolic extracts	188 mg/100 g dw		–	–	–	Díaz-de-Cerio et al. (2019)
Fruit	Ethanolic extracts	480 mg/100 g dw	16 mg/100 g dw	–	–	–	Riviello-Flores et al. (2018)
Peel	Methanolic extracts	66 mg CAE/g of extract	48.6 mg of QE/g extract	9.7 mg/100 g	17.4 µM Fe(II)/g	0.7 IC_{50} (mg/mL)	Loizzo et al. (2016)
Leaves	Methanolic extracts	89.3 mg CAE/g extract	67.2 mg of QE/g extract	4.5 mg/100 g	22.3 µM Fe(II)/g	0.7 IC_{50} (mg/mL)	Loizzo et al. (2016)
Pulp	Methanolic extracts	58.5 mg CAE/g of extract	23.3 mg of QE/g extract	8.3 mg/100 g	8.3 µM Fe(II)/g	0.1 IC_{50} (mg/mL)	Loizzo et al. (2016)

Where dw, dry weight; AAE, ascorbic acid equivalent; GAE, gallic acid equivalent; QE, quercetin equivalent; CAE, chlorogenic acid equivalent

5 Functional and Biological Properties of Extracts and Bioactive Compounds from Chayote By-products

There are reports of the ethnobotanical and ethnomedicinal usage of the Chayote crop and its by-products as an herbal treatment. However, they are not as widespread as other medicinal crops. It has not been fully utilized commercially, but it has good potential as a therapeutic crop and might provide farmers with a reliable source of revenue (Mashine et al. 2023). Even though the pharmacological qualities of Chayote's by-products are currently well known, very few reports have been written about the separation and identification of the drug's distinct chemical elements, and even fewer in vivo investigations have evaluated the drug's biological efficacy (Vieira et al. 2019). One of the underappreciated *Cucurbitaceae* species is Chayote, which is a multipurpose plant that may be used for both therapeutic and culinary purposes since it is rich in nutrients, including protein, minerals, and phenols, and because its extracts have antioxidant, anti-cancer-and-anti-cardiovascular--effects (Cheng et al. 2023; Gavia-García et al. 2023). Extracts from chayote squash were tested for their antibacterial effectiveness against harmful microorganisms. Methanol was used to extract the powdered dried Chayote squash, and then hexane, chloroform, ethyl acetate, and butanol were used in that extraction order. Every extract showed moderate to potent antibacterial properties. Because the solvent had varying polarity, diverse chemical compositions of the studied extracts led to different bioactivities of the extracts (Sulistiyani et al. 2022). According to a study, Chayote leaf extracts in both aqueous and ethanolic forms showed anti-microbial effectiveness against strains of multi-resistant *Enterococci* and *Staphylococci* (Ordoñez et al. 2003). Fruit, young leaves, stems, and tuberized sections of the roots are all consumed as vegetables, either raw or cooked, and they can also be used in various stews. Its leaves can be infused to help dissolve kidney stones, treat arteriosclerosis, and lower blood pressure. Fruit infusions can help reduce urine retention, and the pulp can occasionally heal rashes (Sood and Rana 2017). In several angiotensin-induced hypertension models, hydroalcoholic extract from Chayote roots, including different sub-portions, showed beneficial benefits (Lombardo-Earl et al. 2014).

High levels of flavonoids in Chayote leaves have been discovered to be an adequate substitute for lowering hyperuricemia in rats; these findings have been confirmed by relevant investigations (Putri et al. 2021). Chayote leaves have a high content of flavonoids, particularly luteolin and apigenin, which can alleviate the inflammation and oxidative stress conditions brought on by hyperuricemia, according to a study on the influence of the flavonoid portion of Chayote leaves on levels of uric acid, oxidative stress, and inflammatory indicators in hyperuricemia rats (Wahyuningtyas et al. 2022). Infusions made from the Chayote leaves are used to treat hypertension, arteriosclerosis, and kidney stones. Therefore, extracting the antioxidant constituents from Chayote leaves and using them could be a valuable waste management solution. Using the leaves valuable chemicals in food products would also benefit the pharmaceutical and cosmetic industries (Vieira et al. 2019).

It has been reported that ethanolic extracts of Chayote leaves and aqueous seed extracts show a substantial lowering effect using a linoleate model and a strong inhibitory action by β-carotene bleaching. The primary flavonoids accountable for this protective action were flavanol (Ordoñez et al. 2006).

Numerous individuals worldwide utilize alternative medicine, utilizing the potential of herbs to treat a range of ailments at the level of traditional medicine. Because they are oxidants, drugs can change the radiolabeling morphology of red blood cells. The effects of Chayote bark extract on red blood cell shape and blood element radiolabeling were assessed in a study. The presence of flavonoids, several proteinase classes, and caspase-like proteases are sensitive to radiolabeling effects, DNA molecule damages, and cell membrane alterations explained the Chayote bark extract's action (Dire et al. 2010). In another study, twenty-four growing indigenous pigs from Mizoram were divided into four groups of six animals at random to investigate the impact of feeding chayote fruit and leaves meal on carcass features and sensory quality of pork. A Chayote meal replaced the pig grower ration in various amounts for 90 days. It was determined that a Chayote-based meal could safely substitute the typical grower ration in pigs' diets by up to 40% without negatively impacting the pork's sensory quality or carcass characteristics. It has been demonstrated that flavonoids found in Chayote shoot lower blood lipid and cholesterol levels, avoiding atherosclerosis and fatty liver (Wu et al. 2014). It was shown that the extract from the Chayote shoots increased the activity of the enzyme AMP-activated protein kinase and inhibited several lipogenic-related enzymes, such as sterol regulatory element-binding proteins and HMG-CoA reductase proteins, which are crucial regulators of hepatic lipid metabolism (Yang et al. 2015).

Bioactive proteins and protein isolates from Chayote seeds have been observed to be involved in different pharmacological actions involving antidiabetic and antioxidant potential (Veigas et al. 2020; Vieira et al. 2023a, b). Previously, using *Escherichia coli* as the host, a protein known as Sechiumin was isolated from the aqueous extracts of the seeds from the Chayote plant and purified, characterized, cloned, and produced. It is purported to be a possible chemotherapeutic drug with ribosomal inactivation capabilities (Wu et al. 1998). Since enzymes work well in mild environments like water, low temperatures, and pressure fluctuations, they are typically employed as industrial food catalysts and catalysts for energy production. β-Glucosidase was extracted from the epicarp and mesocarp pulp of Chayote fruit using molecular exclusion and ionic-interchange chromatography techniques. The hydrolysis of terminal non-reducing glycosyl residues in oligosaccharides and glycosides is done by β-glucosidase (Mateos et al. 2015). Within the kingdom of plants, acid phosphatases are extensively distributed and play a significant role in the metabolism of carbohydrates in plant tissue. One job is to catalyze the hydrolysis of phosphate esters, which can be attached to substances like nucleic acids, carbohydrates, or lipids. The relative phosphatase activity of the green form of Chayote's skin, pulp, and seed was examined. The epidermis had the greatest activity levels of all the phosphatases, whereas the seed had the lowest (Burnette and Flick Jr 1977). Different in vitro and in vivo controlled studies have confirmed that the extracts and isolated bioactive components from Chayote by-products have been involved in various biological and functional activities, and a summary of this data is shown in Table 3.

Table 3 Functional and biological properties of extracts and bioactive compounds from Chayote by-products

Bioactive compound	Chayote fraction	Biological and functional properties	Reference
Flavonoids	Shoots	Antihypercholesterolemic	Wu et al. (2014)
Sechiumin	Seeds	Chemotherapeutic drug with ribosomal inactivation capabilities	Wu et al. (1998)
Bioactive proteins and protein isolates	Seeds	Antidiabetic and antioxidant	Vieira et al. (2023a, b)
Flavonoids	Leaves	Potential of lowering hyperuricemia	Putri et al. (2021)
Luteolin and apigenin	Leaves	Alleviate inflammation and oxidative stress conditions	Wahyuningtyas et al. (2022)
β-glucosidase	Epicarp and mesocarp	Bioconversion of oligosaccharides	Mateos et al. (2015)
Acid phosphatases	Peel, flesh, and seeds	Hydrolysis of phosphate esters	Burnette and Flick Jr (1977)
Extracts	Leaves	Blood pressure modulator	Nahdi et al. (2016)
Extracts	Leaves	Cardioprotective	Orozco-Martínez et al. (2020)
Extracts	Buds	Antihypertensive and antisedative	Di Stasi et al. (2002)
Aqueous and ethanolic extracts	Leaves	Antibacterial activities	Ordoñez et al. (2003)
Hydroalcoholic extracts	Roots	Antihypertensive	Lombardo-Earl et al. (2014)
Polyphenols	Shoots	Hepatoprotective	Yang et al. (2015)
Ethanolic extracts	Leaves and seeds	Antioxidant activities	Ordoñez et al. (2006)
Methanolic extracts	Peel, leaves, and fruit	Antidiabetic activities	Loizzo et al. (2016)

6 Applications of Extracts and Bioactive Substances from By-products of Chayote in Food and Non-food Contexts

The Chayote plant is consumed almost entirely, and one of its main advantages over other cucurbits is its capacity to tolerate cold temperatures to some degree. Various techniques are employed to prepare different portions of this crop, such as boiling, baking, stuffing, mashing, frying, scalloping, pickling, or utilizing it as a green vegetable in soup preparation. It is a very nutritious crop used in food production because of its bland flavor (Mashine et al. 2023). The edible parts of the chayote are not limited to the fruit; you may also consume the stem, root, leaves, and seeds. The plant's leaves and branches are widely used in stir fries and salads, particularly in Asia, while the tubers are eaten like sweet potatoes, potatoes, and other related root vegetables. Due to its softness, the fruit is especially suitable for hospital diets and adds stability to baby foods, liquids, pasta dishes, and sauces (Loizzo et al. 2016; Raleng et al. 2022).

Products made from chayote different fractions are a rich source of dietary fiber, essential minerals, vitamins, and amino acids (Taynath et al. 2020). The inclusion of Chayote leaf in product development has great promise because of its rich nutritional content and biological properties, which may contribute to better livelihoods, less waste, and more sustainable agriculture (Managa et al. 2021). Chayote peel flour has been utilized to develop several value-added nutritional food formulations, including muffins, cookies, cereal bars, and sausages (Vieira et al. 2019).

Plant polysaccharides have outstanding water-holding, thickening, and biological qualities, which make them frequently used in food and health goods. Fruit peels contain Chayote polysaccharides used in the culinary, pharmaceutical, and other industries. To increase its potential uses, Chayote can be a valuable source of polysaccharides for the food and medicinal industries (Ke et al. 2023a, b). Another non-food application of Chayote was found as the development of silver nanoparticles. Silver nanoparticles were biosynthesized via aqueous extract developed from Chayote fruit as a capping and reducing agent. The disclosed method for producing silver nanoparticles was faster, more energy-efficient, and less environmentally harmful than many traditional approaches. In addition, the biosynthesis process required only the metal precursor and no other compounds (Chelli et al. 2017).

One of the best-value functional food components is pectin. Gaining more insight into the characteristics of Chayote pectin could encourage the application of Chayote fruit as a rich source of pectin. It was discovered that pectin obtained from Chayote fruit peels using ultrasonic-aided extraction had a low esterification degree, a high molecular mass, and an appropriate foaming capacity (Ke et al. 2020). Pectin is abundant in Chayote, a vegetable with great economic importance on a global scale. Pectin is among the several non-starch polysaccharides found in Chayote. However, Chayote's high moisture content and short shelf-life result in financial losses. Chayote pectin was more emulsified, had a higher viscosity, and could contain more oil and water than sour orange peel pectin (Ke et al. 2024).

Among some contemporary uses of Chayote, omitting the dietary purpose is in the cosmetic sector, as a source of ingredients in moisturizers, skin cleansers, sun-protective lotions, medicated toothpaste and mouthwash, healthy shaving creams, active deodorants, and shampoos (Huerta-Reyes et al. 2022). The biological features of chayote extracts warrant particular attention, particularly the antibacterial and antioxidant activities, which indicate that chayote extract may be effectively administered topically to avert and treat infections and skin damage. Whether applying modern or conventional medications topically, the active ingredient must be liberated from the carrier (vehicle) before making contact with the skin's surface. This allows the agent to enter the stratum corneum and penetrate deeper layers of the skin. Phytopharmaceuticals have the potential to be extracted from the formulation (Ordoñez et al. 2009).

Furthermore, to stop the long-term harmful effects of ultraviolet (UV) light, such as skin cancer and photoaging, photoprotection is crucial. The development of skin

Table 4 Food and non-food applications of Chayote by-products

Product	Chayote fraction used	Advantages	References
Cereal bar	Peel flour	Improved nutritional contents	Cristo et al. (2015)
Muffins	Peel flour	Improved nutritional and sensorial features	Bousfield et al. (2017)
Kikiam (Sausage-like Chinese dish)	Peel flour	Improved nutritional contents	Martelino et al. (2018)
Sfiha (bakery product)	Peel flour	Improved nutritional and sensorial features	Molinari et al. (2021)
Cookies	Peel powder	Improved nutritional and sensorial features	Mukherjee et al. (2022)
Petha	Tubers	Improved nutritional and sensorial features	Taynath et al. (2020)
Biscuits	Tubers	Improved minerals and vitamins	Sakung et al. (2020)
Petha	Tubers	Improved sensorial features	Taynath et al. (2020)
Fermented smoothie	Leaves	Increased total phenol and carotenoid contents	Managa et al. (2021)
Chaguro	Tubers	Medicinal food products with hypoglycemic and hypolipidemic effects	Sudargo et al. (2021)
Cosmetics	Essential oils from shoots	Skin protection	Huerta-Reyes et al. (2022)
Cosmetic composition	Leaves and fruit	Anti-aging, whitening, anti-wrinkle, anti-inflammation, and moisturizing	Yadav et al. (2015)
Skincare extracts	Extracts	Photoprotection against UV radiations	Metral et al. (2018)
Hydrogel-containing Chayote fluid extracts	Fluid extracts	Antibacterial, antimycotic, and antioxidant agent	Ordoñez et al. (2009)
Silver nanoparticles	Fruit extracts	Antifungal agents	Chelli et al. (2017)

photoprotective agents that are more effective than synthetic ones has been accelerated by the use of naturally occurring bioactive phytochemicals. According to a study, extracts from the Chayote fruit shielded keratinocytes against UV-induced cytotoxicity lowered intracellular ROS levels, and lessened DNA damage caused by oxidative stress following UVA exposure. Thus, a combination of antioxidant activity, decreased DNA damage, and increased DNA repair capabilities may account for the benefits of Chayote extracts against UVA (Metral et al. 2018). Although Chayote is a neglected and underutilized crop, a lot of food and non-food application of different fractions can be found in the literature, which has been summarized in Table 4.

7 Valorization of By-products for Non-health Purposes, i.e. Source of Bioenergy

Worldwide, an enormous volume of waste-activated sludge and agricultural wastes are produced annually. Anaerobic digestion co-digestion has been suggested as a replacement for utilizing materials from such organic wastes by producing biochemicals and bioenergy to build a circular economy. A viable and sustainable solution to address both the problem of environmental degradation and meet the world's growing energy needs is bioenergy produced from garbage. Because Chayote fruits have many terpenes, their by-products are also used to produce bioenergy (Pan et al. 2021). It was possible to distinguish between the leftovers from a restaurant's pre- and post-preparation processes thanks to the chemical examination of the food waste. Waste high in lignin or carbohydrates can be detected and used to greater use. Using this characterization as a basis, an analysis of the properties and chemical makeup of the wastes in a Brazilian case study was conducted to determine the potential for energy production concerning methane generation capability. The findings showed that the waste produced by the food sector has a significant potential for energy production and that the chemical makeup of the biomass will dictate the optimal application for it. Samples with Chayote residues had a 35% non-carbohydrate (lignin) content, suggesting that it might be used in conjunction with other energy sources to produce energy (Pereira et al. 2021). It has been discovered that the vegetable residue of Chayote can act as a natural coagulant to eliminate turbidity and chemical oxygen demand from wastewater in the textile industry (Almeida et al. 2017).

By using waste from fruit and vegetables as a feedstock in the circular economy, we can reduce the amount of conventional food commodities used for human consumption and animal feed, as well as the amount of deforestation caused by the production, pretreatment, adequate transportation, handling, and transformation of these materials. We can also reduce the detrimental effects on air, water, soil, and ecosystems. The expense of establishing workable production procedures also provides a solution for the waste disposal issue. Because it might be challenging to dispose of this organic waste in a landfill, it poses a significant threat to the ecology and causes contamination (Aguilar-Rivera and Olvera-Vargas 2023; Gómez-García et al. 2024). Chayote by-products of waste could be utilized by biorefineries for producing food, livestock feed, biofertilizers, dietary fibers, biofuel, bioproducts, and the value chain, along with socioeconomic and environmental implications, following the circular bioeconomy with the approach to life cycle assessment toward sustainable production (de Souza Medina et al. 2023) (Fig. 2).

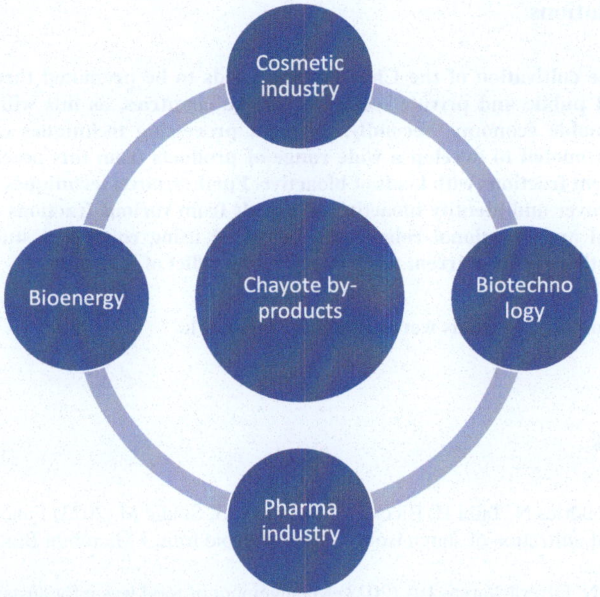

Fig. 2 Non-food applications of Chayote by-products

8 Conclusion

It has been demonstrated throughout this chapter that *Sechium edule* is a significant phytogenetic reserve for the food industry and agriculture sector, primarily because of the variety of biologic types that make up an infraspecific complex. We anticipate that this diversity will continue to grow due to the infraspecific complex's coexistence with human groups that have incorporated it into their culture. In addition to the edible fruit, the peels and seeds of the Chayote fruit and the plant's leaves, branches, and roots may be used for both food and non-food uses to promote human health. It is commonly known that Chayote waste products and their edible elements are highly concentrated in various health-promoting compounds that could be used in the food, cosmetic, bioenergy, and pharmaceutical industries. Though little thought has been paid to the underutilized or overlooked crops, Chayote leftovers are a great source of vital bioactive chemicals, and using them in waste management could be crucial from an environmental and economic perspective. It is recommended that more research be done on the chemical profile of *Sechium edule* shoots, roots, flowers, seeds, leaves, peel, and pulp to determine which bioactive components are present in these sections and how they relate to various pharmacological actions.

Recommendations

First of all, the cultivation of the Chayote crop needs to be promoted through the active involvement of public and private sectors of all the countries, as this will help build the world's sustainable economy. Secondly, different processing techniques could be implemented and promoted to develop a wide range of products from this precious crop while using its different fractions with loads of bioactive. Further, novel techniques could be implemented to discover and identify bioactive chemicals from various fractions of chayote, and their biological and functional roles could be tested using controlled studies.Conflict of InterestThe authors of the current work declare no conflict of interest.

Funding Statement No funds were availed for this work.

References

Abdelhai M, Boudries N, Taibi H, Blecker C, Sinnaeve G, Sindic M (2023) Functional properties and potential utilization of starch isolated from chayote fruit. J Microbiol Biotechnol Food Sci 13(1):9215

Aguilar-Rivera N, Olvera-Vargas LA (2023) Management of food waste for sustainable economic development and circularity. In: Filho WL, Azul AM, Doni F, Salvia AL (eds) Handbook of sustainability science in the future: policies, technologies and education by 2050. Springer, Cham, pp 1–23

Almeida CA, De Souza MTF, Freitas TKFS, Ambrósio E, Geraldino HCL, Garcia JC (2017) Vegetable residue of Chayote (*Sechium edule* SW.) as natural coagulant for treatment of textile wastewater. Int J Energy Water Resour 1(1):37–46

Aminda FR, Sari AK, Anggrasari H (2024) Development of leading commodities for vegetable and fruit crops in Brebes Regency, Central Java. West Sci Interdiscip Stud 2(01):155–165

Barik S, Paul KK (2017) Potential reuse of kitchen food waste. J Environ Chem Eng 5(1):196–204

Barrera-Guzmán LA, Cadena-Iñiguez J, Legaria-Solano JP, Sahagún-Castellanos J (2021) Phylogenetics of the genus Sechium P. Brown: A review. Span J Agric Res 19(1):e07R01

Bousfield IC, da Rocha DR, Costa P, dos Santos Oliveira PHP (2017) Desenvolvimento e avaliação sensorial da farinha de chuchu (*Sechium edule*) para produção de um bolo tipo muffin. Nutrição Brasil 16(5):292–300

Burnette FS, Flick Jr GJ (1977) Acid phosphatase activities of the skin, flesh, and seed in *Sechium edule*, Sw, the Chayote. J Food Prot 40(6):373–374

Cadena-Iñiguez J, Arévalo-Galarza L, Avendaño-Arrazate CH, Soto-Hernández M, Ruiz-Posadas LDM, Santiago-Osorio E et al (2007) Production, genetics, postharvest management and pharmacological characteristics of *Sechium edule* (Jacq.) Sw. Fresh Produce 1(1):41–53

Castro Rodríguez JM, Toledo Díaz AM, Rodríguez Galdón B, Perdomo Molina A, Rodríguez-Rodríguez EM, Díaz Romero C (2015) Morphological and chemical composition characterization of chayotas (*Sechium edule*) grown in the Canary Islands (Spain). Arch Latinoam Nutr 65(4):243–253

Chang KA, Ley SL, Lee MY, Yaw HY, Lee SW, Chew LY et al (2021) Determination of nutritional constituents, antioxidant properties, and α-amylase inhibitory activity of *Sechium edule* (chayote) shoot from different extraction solvents and cooking methods. LWT 151:112177

Chelli VR, Bag SS, Golder AK (2017) A biosynthesis route to nearly spherical AgNPs using chayote fruit extract. Environ Prog Sustain Energy 36(1):192–199

Cheng S, Liu Y, Su L, Liu X, Chu Q, He Z et al (2023) Physiological, anatomical and quality indexes of root tuber formation and development in chayote (*Sechium edule*). BMC Plant Biol 23(1):413

Cristo TD, Rodrigues BM, Santos ND, Candido CJ, Santos ED, Novello D (2015) Cereal bar with added chayote peel flour: physico-chemical and sensory characteristics among children. Semina: Ciências Biológicas e da Saúde (Londrina) 36(2):85–96

Deguine JP, Atiama-Nurbel T, Vanhuffel L, Cresson C (2020) Recent advances in organic cultivation of chayote (*Sechium edule*) in Reunion Island. Org Agric 10(2):135–143

De Jesús Lira-Ricárdez J, Jasso MEM, Cabello LO (2024) Technological and analytical aspects of bioactive compounds and nutraceuticals from fruit sources. In: Bioactive compounds and nutraceuticals from plant sources. Apple Academic Press, pp 79–120

de Souza Medina T, D'Almeida CTDS, Nascimento TPD, de Abreu JP, de Souza VR, Kalili DC et al (2023) Food service kitchen scraps as a source of bioactive phytochemicals: disposal survey, optimized extraction, metabolomic screening and chemometric evaluation. Metabolites 13(3):386

Díaz-de-Cerio E, Verardo V, Fernández-Gutiérrez A, Gómez-Caravaca AM (2019) New insight into phenolic composition of chayote (*Sechium edule* (Jacq.) Sw.). Food Chem 295:514–519

Dire GF, Fernandes JFO, Gomes ML, Albuquerque AC, Camacho ACLF, Cardoso MEO et al (2010) Comparative analysis of the biological effects related to a natural extract processed from the bark of chayote (*Sechium edule*). Global Veterinaria 4(3):225–229

Di Stasi LC, Oliveira GP, Carvalhaes MA, Queiroz-Junior M, Tien OS, Kakinami SH, Reis MS (2002) Medicinal plants popularly used in the Brazilian Tropical Atlantic Forest. Fitoterapia 73(1):69–91

Gavia-García G, Rosado-Pérez J, Arista-Ugalde TL, Aguiñiga-Sánchez I, Santiago-Osorio E, Mendoza-Núñez VM (2023) The consumption of *Sechium edule* (chayote) has antioxidant effect and prevents telomere attrition in older adults with metabolic syndrome. Redox Rep 28(1):2207323

Gómez-García R, Vilas-Boas AA, Campos DA, Pintado MM, Aguilar CN (eds) (2024) Food byproducts management and their utilization. CRC, New York

Harriman NA (1998) Promoting the conservation and use of underutilized and neglected crops. 8. Chayote. *Sechium edule* (Jacq.) Sw: Rafael Lira Saade. 1996. International Plant Genetic Resources Institute (IPGRI), Rome, Italy. 58 pp (paperback). Gratis. ISBN 92-9043-298-5

Hernandez-Uribe JP, Agama-Acevedo E, Gonzalez-Soto RA, Bello-Pérez LA, Vargas-Torres A (2011) Isolation and characterization of Mexican chayote tuber (*Sechium edule* Sw.) starch. Starch-Stärke 63(1):32–41

Huerta-Reyes M, Tavera-Hernández R, Alvarado-Sansininea JJ, Jiménez-Estrada M (2022) Selected species of the Cucurbitaceae family used in Mexico for the treatment of diabetes mellitus. Molecules 27(11):3440

Hussain A, Kausar T, Din A, Murtaza MA, Jamil MA, Noreen S et al (2021) Determination of total phenolic, flavonoid, carotenoid, and mineral contents in peel, flesh, and seeds of pumpkin (*Cucurbita maxima*). J Food Process Preserv 45(6):e15542

Hussain A, Kausar T, Sehar S, Sarwar A, Ashraf AH, Jamil MA et al (2022a) A comprehensive review of functional ingredients, especially bioactive compounds present in pumpkin peel, flesh and seeds, and their health benefits. Food Chem Adv 1:100067

Hussain A, Kausar T, Jamil MA, Noreen S, Iftikhar K, Rafique A et al (2022b) In vitro role of pumpkin parts as pharma-foods: antihyperglycemic and antihyperlipidemic activities of pumpkin peel, flesh, and seed powders, in alloxan-induced diabetic rats. Int J Food Sci. https://doi.org/10.1155/2022/4804408

Hussain A, Kausar T, Sehar S, Sarwar A, Ashraf AH, Jamil MA et al (2022c) Utilization of pumpkin, pumpkin powders, extracts, isolates, purified bioactives and pumpkin based functional food products: A key strategy to improve health in current post COVID 19 period: An updated review. Appl Food Res 2(2):100241

Hussain A, Kausar T, Sehar S, Sarwar A, Quddoos MY, Aslam J et al (2023) A review on biochemical constituents of pumpkin and their role as pharma foods; a key strategy to improve health in post COVID 19 period. Food Prod Process Nutr 5(1):22

James L, Samanta AK (2018) Effect of feeding chayote (*Sechium edule*) fruits and leaves on carcass characteristics of indigenous pig of Mizoram (Zovawk). Indian Vet J 95(4):20–22

Ke J, Jiang G, Shen G, Wu H, Liu Y, Zhang Z (2020) Optimization, characterization and rheological behavior study of pectin extracted from chayote (*Sechium edule*) using ultrasound assisted method. Int J Biol Macromol 147:688–698

Ke J, Chang Y, Nie C, Yang R, Ma J, Zhao T et al (2023a) Formation mechanism of complex coacervation of chayote (*Sechium edule*) pectin-sodium caseinate in aqueous solution. Food Hydrocoll 143:108902

Ke J, Zhang Y, Wang X, Sun J, Wang S, Ma Y et al (2023b) Structural characterization of cell-wall polysaccharides purified from chayote (*Sechium edule*) fruit. Food Chem X 19:100797

Ke J, Wang Y, Sun J, Wang X, Wang S, Ma Y et al (2024) Influence mechanism of chayote (*Sechium edule*) pectin on the stability of sodium caseinate emulsion. LWT 195:115831

Loizzo MR, Bonesi M, Menichini F, Tenuta MC, Leporini M, Tundis R (2016) Antioxidant and carbohydrate-hydrolysing enzymes potential of *Sechium edule* (Jacq.) Swartz (Cucurbitaceae) peel, leaves and pulp fresh and processed. Plant Foods Hum Nutr 71:381–387

Lombardo-Earl G, Roman-Ramos R, Zamilpa A, Herrera-Ruiz M, Rosas-Salgado G, Tortoriello J, Jiménez-Ferrer E (2014) Extracts and fractions from edible roots of *Sechium edule* (Jacq) Sw with antihypertensive activity. Evid Based Complement Alternat Med. https://doi.org/10.1155/2014/594326

Managa MG, Akinola SA, Remize F, Garcia C, Sivakumar D (2021) Physicochemical parameters and bioaccessibility of lactic acid bacteria fermented chayote Leaf (*Sechium edule*) and pineapple (*Ananas comosus*) smoothies. Front Nutr 8:649189

Martelino JA, Avila RLD, Saddi JKB, Virrey RTG (2018) Utilization of discarded vegetable peels in making Kikiam as quality street foods. Asia Pac Higher Educ Res J 5(2):1–4

Mashine L, Rime J, Datta S, Angousana S, Guragai A, Anal PM (2023) A review on chayote (*Sechium edule* (Jacq.) Swartz): the viviparous cucurbit. Pharma Innov J 12(6):3771–3775

Mateos SE, Cervantes CAM, Zenteno E, Slomianny MC, Alpuche J, Hernández-Cruz P et al (2015) Purification and partial characterization of β-glucosidase in chayote (*Sechium edule*). Molecules 20(10):19372

Metral E, Rachidi W, Damour O, Demarne F, Bechetoille N (2018) Long-term genoprotection effect of *Sechium edule* fruit extract against UVA irradiation in Keratinocytes. Photochem Photobiol 94(2):343–350

Molinari LC, do Amaral LA, de Menezes MB, dos Santos EF, Schiessel DL, Novello D (2021) Esfiha with added chayote peel flour: physical-chemical characterization and sensory analysis among children. Int J Innov Educ Res 9(6):117–128

Moreira FA (2015) Chayote: pre-Columbian origins and dispersal. Hortic Rev 43:89–144

Mukherjee P, Kudave RR, Uppaluri RV (2022) Formulation and characterization of squash enriched cookies. In: North-East research conclave. Springer, Singapore, pp 241–262

Nahdi MS, Martiwi INA, Arsyah DC (2016) The ethnobotany of medicinal plants in supporting the family health in Turgo, Yogyakarta, Indonesia. Biodiversitas 17(2):900–906

Ordoñez AAL, Gómez JD, Cudmani NM, Vattuone MA, Isla MI (2003) Antimicrobial activity of nine extracts of *Sechium edule* (Jacq.) Swartz. Microb Ecol Health Dis 15(1):33–39

Ordoñez AAL, Gomez JD, Vattuone MA (2006) Antioxidant activities of *Sechium edule* (Jacq.) Swartz extracts. Food Chem 97(3):452–458

Ordoñez AAL, Ordoñez RM, Zampini IC, Isla MI (2009) Design and quality control of a pharmaceutical formulation containing natural products with antibacterial, antifungal and antioxidant properties. Int J Pharm 378(1–2):51–58

Orozco-Martínez J, Lira-Saade R, Jiménez-Estrada M, Ávila-Acevedo JG, Serrano-Parrales R, Hernández-Delgado T (2020) Medicinal plants of Oaxaca, Mexico: Ethnobotany and antibacterial activity. Bol Latinoam Caribe Plant Med Aromat 19:221–235

Pan SY, Tsai CY, Liu CW, Wang SW, Kim H, Fan C (2021) Anaerobic co-digestion of agricultural wastes toward circular bioeconomy. Iscience 24(7)

Pereira BS, Castrisana RN, de Freitas C, Contiero J, Brienzo M (2021) Chemical composition determines the bioenergy potential of food waste from pre-and post-production. J Mater Cycles Waste Manag 23:1365–1373

Putri DP, Wahyuningtyas AP, Al-Baarri ANM, Maharani N (2021) The effect of chayote leaves (*Sechium edule*)'s flavonoid fraction on the reduction of the serum uric acid levels through the inhibition of xanthine oxidase activity. Slovak J Food Sci 15:1049–1055

Raleng A, Singh NJ, Sarangi PK, Manojkumar P, Wahengbam A (2022) Standardization of frying time-temperature strategy for enhancing the quality and storability of chayote chips. Appl Food Res 2(2):100167

Ramírez-Rodas Y, Arévalo-Galarza L, Cadena-Iñiguez J, Delgado-Alvarado A, Ruiz-Posadas L, Soto-Hernández M (2021) Postharvest storage of three chayote (*Sechium edule* (Jacq.) Sw.) varieties. Scientia Agropecuaria 12(2):239–247

Riviello-Flores MDLL, Arévalo-Galarza MDL, Cadena-Iñiguez J, Soto-Hernández RM, Ruiz-Posadas LDM, Gómez-Merino FC (2018) Nutraceutic characteristics of the extracts and juice of chayote (*Sechium edule* (Jacq.) Sw.) fruits. Beverages 4(2):37

Saade RL (1996) Chayote, Sechium edule (Jacq.) Sw (vol 8). Bioversity International. Edition: Promoting the conservation and use of neglected and underutilized crops. No. 8. Volume editors: Heller J, Engels JMM, Hammer K. IPK and IPGRI, Rome, Italy. ISBN: 92-9043-298-5

Sakung JM, Nurmayanti Y, Fitra H (2020) Nutritional evaluation of chayote flour-based biscuits (*Sechium edule*). Indian J Public Health Res Dev 11(3):2034–2038

Scherhaufer S, Moates G, Hartikainen H, Waldron K, Obersteiner G (2018) Environmental impacts of food waste in Europe. Waste Manag 77:98–113

Shariff AHM, Hainusa NA, Huda N, Zakaria M, Ullah S, Huyop F, Wahab RA (2023) Antioxidant activity, total phenolic content, and nutrient composition of Chayote shoot (*Sechium edule*, Jacq. Swartz) from Kundasang, Sabah: antioxidant activity, total phenolic content, and nutrient composition of Chayote shoot (*Sechium edule*, Jacq. Swartz) from Kundasang, Sabah. J Trop Life Sci 13(1):147–156

Siciliano T, De Tommasi N, Morelli I, Braca A (2004). Study of flavonoids of Sechium edule (Jacq) Swartz (Cucurbitaceae) different edible organs by liquid chromatography photodiode array mass spectrometry. Journal of Agricultural and Food Chemistry, 52(21), 6510–6515. https://doi.org/10.1021/jf040214q

Sood R, Rana MK (2017) Chayote (Chow Chow). In: Vegetable crop science. CRC, Boca Raton, pp 565–572

Sudargo T, Aulia B, Prameswari AA, Isnansetyo A, Puspita ID, Budiyanti SA (2021) Effect of administration of chaguro made of chayote (*Sechium edule*) and tuna (Thunnus sp.) on rats induced with streptozotocin-nicotinamide and a high-fat diet. Curr Res Nutr Food Sci J 9(1):258–266

Sulistiyani NI, Ainurrofiq A, Suryanti V (2022). Antibacterial Activity of Chayote (Sechium edule Swartz) Squash Extracts and Their Phytochemical Constituents. Journal of Biodiversity and Biotechnology, 2(1), 26–32

Taynath SJ, Singh BK, Jena S, Said PP (2020) Optimization of process variables for petha manufacture from chayote (*Sechium edule* Sw.). J Food Process Preserv 44(11):e14872

Vansuyt G, Zinsou C (1986) Accumulation of agmatine in chayote (*Sechium edule*) leaves during development. Physiol Plant 67(4):592–597

Veigas GJ, Bhattacharjee A, Hegde K, Shabaraya AR (2020) A brief review on *Sechium edule*. Int J Pharm Sci Rev Res 65:165–168

Vieira EF, Pinho O, Ferreira IM, Delerue-Matos C (2019) Chayote (*Sechium edule*): A review of nutritional composition, bioactivities and potential applications. Food Chem 275:557–568

Vieira EF, Souza S, Moreira MM, Cruz R, Silva ABD, Casal S, Delerue-Matos C (2022) Valorization of phenolic and carotenoid compounds of *Sechium edule* (Jacq. Swartz) leaves: Comparison between conventional, ultrasound-and microwave-assisted extraction approaches. Molecules 27(21):7193

Vieira EF, Fontoura AQ, Delerue-Matos C (2023a) Chayote (*Sechium edule* (Jacq.) Swartz) seed as an unexploited protein source: bio-functional and nutritional quality of protein isolates. Foods 12(15):2949

Vieira EM, Moreira M, Cruz R, da Silva A, Casal S, Matos C (2023b) An insight into chayote (*Sechium edule*) peels valorization: phytochemical characterization and bioactive potential. In: The 4th international electronic conference on foods session food analysis, biotechnology, and engineering. MDPI

Wahyuningtyas AP, Putri DP, Maharani N, Al-Baarri ANM (2022) Flavonoid fraction from chayote (*Sechium edule* (Jacq.) Sw) leaves reduced malondialdehyde (MDA) and tumor necrosis factor-α (TNF-α) in hyperuricemic rats. Nutr Food Sci 52(2):366–378

Wu TH, Chow LP, Lin JY (1998) Sechiumin, a ribosome-inactivating protein from the edible gourd, *Sechium edule* Swartz: Purification, characterization, molecular cloning and expression. Eur J Biochem 255(2):400–408

Wu CH, Ou TT, Chang CH, Chang XZ, Yang MY, Wang CJ (2014) The polyphenol extract from *Sechium edule* shoots inhibits lipogenesis and stimulates lipolysis *via* activation of AMPK signals in HepG2 cells. J Agric Food Chem 62(3):750–759

Yadav A, Avinash RK, Misra ADD, Kalita H, Gopi R, Maurya BN, Ngachan SV (2015) Genetic diversity of chayote (*Sechium edule* (Jacq.) Sw.) in Sikkim hills. Int J Pharm 378:51–58

Yang MY, Chan KC, Lee YJ, Chang XZ, Wu CH, Wang CJ (2015) *Sechium edule* shoot extracts and active components improve obesity and a fatty liver that involved reducing hepatic lipogenesis and adipogenesis in high-fat-diet-fed rats. J Agric Food Chem 63(18):4587–4596

Phytochemicals from Cucumber (*Cucumis sativus*) By-products

M. Z. Bauleth, J. Sheehama, Natascha Cheikhyoussef, Ahmad Cheikhyoussef, M. Makhaba, R. Sharma, C. N. Cupido, and Ahmed A. Hussein

Abstract Cucumber (*Cucumis sativus* L.) is a widely cultivated plant belonging to the Cucurbitaceae family. This plant is a popular vegetable crop used in several international cuisines and traditional medicine, such as the Mediterranean diet style and Indian culture, owing to its refreshing taste, high water content, and high nutritional value. It is mainly consumed in its fresh form and is known to constitute primary metabolites such as carbohydrates, proteins, and fats. The Cucumber by-products, namely flowers, leaves, peels, seeds, and juice wastes, have recently gained scientific interest due to bioactive phytochemical classes such as phenolic compounds, flavonoids, carotenoids, and cucurbitacins. Phytochemicals are secondary metabolites produced by plants. These products are biologically active, naturally occurring chemicals in various parts of a plant, providing health benefits for humans beyond those attrib-

The original version of the chapter has been revised. A correction to this chapter can be found at https://doi.org/10.1007/978-3-031-77399-0_14

M. Z. Bauleth
Department of Human, Biological & Translational Medical Sciences, Faculty of Health Sciences and Veterinary Medicine, University of Namibia, Windhoek, Namibia

J. Sheehama
Department of Human, Biological & Translational Medical Sciences, Faculty of Health Sciences and Veterinary Medicine, University of Namibia, Windhoek, Namibia

Hifikepunye Pohamba and Oshakati Campuses, University of Namibia, Oshakati, Namibia

N. Cheikhyoussef
Ministry of Higher Education, Technology, and Innovation (MHETI), Windhoek, Namibia

A. Cheikhyoussef
Multidisciplinary Research Services, Centre for Research Services, University of Namibia, Windhoek, Namibia

M. Makhaba · R. Sharma · A. A. Hussein (✉)
Chemistry Department, Cape Peninsula University of Technology, Cape Town, South Africa
e-mail: MOHAMMEDAm@cput.ac.za

C. N. Cupido
Department of Botany, University of Fort Hare, Alice, South Africa

M. F. Ramadan (ed.), *Bioactive Phytochemicals in By-products from Bulb, Flower and Fruit Vegetables*, https://doi.org/10.1007/978-3-031-77399-0_9

uted to macronutrients and micronutrients. They exhibit antioxidant, anti-inflammatory, antimicrobial, anticancer, and antidiabetic activities and have been reported to be used in the prevention and treatment of cardiovascular diseases. They further show potential in promoting skin health and improving digestive health. In addition to their potential health benefits, cucumber by-products can also be utilized in the food and cosmetic industries. There are suggestions to incorporate cucumber peel extracts into beverages and snacks for nutritional value enhancement and antioxidant activity. Cucumber seed oil, rich in essential fatty acids, can be utilized to formulate skincare products for its moisturizing and anti-aging effects. In conclusion, cucumber by-products are rich sources of bioactive phytochemicals with various health benefits. These compounds have been linked to human health by protecting against degenerative diseases. There is a research window of opportunity to explore these bioactive compounds' full potential and develop innovative ways to utilize cucumber by-products in various applications, including functional foods and cosmetics.

Keywords *Cucumis sativus* L. · Cucumber · By-products · Biological activities · Phytochemicals

Abbreviations

A549	Adenocarcinomic human alveolar basal epithelial cells
COX-2	Cyclooxygenase-2 enzyme
D. Mel-2	Schneider's Drosophila Line 2
DOCA	Deoxycorticosterone acetate
DPPH	1,1-diphenyl-2-picrylhydrazyl
DU145	Human prostate cancer cell line
ED_{50}	Median effective dose
FRAP	Ferric reducing antioxidant power
H1975	Human non-small cell lung carcinoma
H_2O_2	Hydrogen peroxide
HaCaT	Human epidermal keratinocytes
HCT116	Human colorectal carcinoma cell line
HD	Hydro-distillation
HDAC5	Histone deacetylase 5
HDF	Human dermal fibroblasts
HDF	Normal human dermal fibroblast cells
HepG2	Hepatoblastoma cell line
HPLC-MS/MS	High-performance liquid chromatography-mass spectrometry
IC_{50}	Fifty percent inhibition concentration
IL-6	Interleukin 6
LPS	Lipopolysaccharide
MDA-MB-231	Triple negative breast cancer cells
MIC	Minimum inhibition concentration
MMP-1	Matrix metalloproteinase-1
MTT	3-(4,5-Dimethylthiazol-2-yl)-2,5-Diphenyltetrazolium Bromide
NMDA	*N*-methyl-D-aspartate receptor

PKM2	Pyruvate kinase M2
RAW 264.7	Macrophage cell
SREBP-1c	Sterol regulatory element-binding protein 1c
SREBP-2	Sterol regulatory element-binding protein 2
STZ	Streptozotocin
SV40-MES-13	Mesangial cell line
TGF-β	Transforming growth factor beta
THP-1	Human monocytic cell line
TLR4	Toll-like receptor 4
TNF-α	Tumor necrosis factor
UV	Ultraviolet
VEGF	Vascular endothelial growth factor
VMHD	Vacuum microwave hydro-distillation
XTT-NR-SRB	Tetrazolium hydroxide-neutral red-sulforhodamine B

1 Introduction and Economic Values of Each Discussed By-product

The UN's call for sustainable agricultural growth has sparked increased interest in agricultural waste due to plants' crucial role in human survival and the economy. In general, plants are rich in nutrients and phytochemicals with pharmacological benefits.

Cucurbitaceae is a family of flowering plants called cucurbits and includes over 120 genera and 965 species (Omokhua-Uyi and Van Staden 2020). Several species of this family are commercially cultivated for various reasons worldwide (Mukherjee et al. 2022). Among them is *Cucumis sativus* L (common name cucumber). Cucumber is a widely consumed vegetable popular in Asia, Europe, and the USA (Mukherjee et al. 2013; Uthpala et al. 2020). Traditionally, *C. sativus* is used in more than one form for various applications: in Ayurveda, the fruits, leaves, and seeds are mainly for treating skin ailments (Mukherjee et al. 2013).

In Chinese medicine, the leaves, stems, and roots are used as bitter tonics (Khan et al. 2022). China is the top producer in the world, with ~75 million tons/year (Atlas Big 2024). Cucumber is a commonly consumed vegetable used in several international cuisines and traditional medicine, such as the Mediterranean diet style and Indian culture, due to its refreshing taste, high water content, and high nutritional value (Uthpala et al. 2020). Besides being a popular and essential food item in traditional medicine, cucumbers also have potential health benefits due to bioactive phytochemicals in their by-products. Cucumber by-products are known to contain bioactive phytochemicals that have potential health benefits. These bioactive compounds have garnered increasing interest in nutrition and medicine due to their antioxidant, anti-inflammatory, and anticancer properties. There is a recent surge in closed agronomic activities, which has led to an increase in plant waste, including plant by-products such as seeds, peels, stems, leaves, flowers, and roots (Gao et al.

2022; Olennikov and Kashchenko 2023a). It is no exception for the cucumber, the main crop for greenhouse crop rotation, according to Olennikov and Kashchenko (2023a); hence, maximizing the use of these by-products is necessary. Vilas-Boas et al. (2021) stated that agro-food wastes should be considered renewable sources of added-value bioactive compounds. The extraction and use of bioactive phytochemicals in cucumber by-products can be of significant economic value. An important economic benefit is the potential to create new revenue streams for cucumber producers by utilizing the by-products that previously used to be discarded. Companies may produce value-added products such as functional foods, dietary supplements, and cosmetics by extracting and isolating bioactive compounds from cucumber peels and seeds. Another possibility is to use by-products like leaves to manufacture plant-based biopesticides to reduce the excessive use of synthetic pesticides. However, it is crucial to investigate the phytotoxicity of these potential products as their metabolites can have an allelopathic effect, which may cause reduction or stimulate growth and development in other plants (Mežaka et al. 2023). Given the increasing demand for natural and plant-based ingredients in the food, cosmetic, and pharmaceutical industries, bioactive phytochemicals from cucumber by-products can be natural alternatives to synthetic additives and preservatives. Incorporating these compounds into functional foods or dietary supplements may improve health outcomes and reduce the burden of chronic diseases and healthcare costs. This chapter aims to highlight some of the latest highlights on cucumber by-products' bioactive phytochemicals, the functional properties of by-product extracts, their bioactive compounds, and their potential food and non-food applications.

2 Chemical Composition and Bioactive Compounds of Extracts from By-products

Cucumber comprises 95% water and contains many primary metabolites such as carbohydrates, proteins, fats, and vitamins A and C (Uthpala et al. 2020). The cucumber by-products, namely peels, seeds, leaves, and juice wastes, have recently gained scientific interest due to the presence of phytochemical compounds. Numerous studies have documented the chemical constituents of *C. sativus* and associated the biological activities with certain phytochemicals. Table 1, Figs. 1 and 2 illustrate the structures of the metabolites that were isolated in *C. sativus*.

2.1 Cucumber Leaves

Different metabolites have been isolated and identified from *C. sativus* in various parts. Leaves are a rich source of other flavonoids and cucurbitacins.

Table 1 Flavonoids identified in cucumber leaves

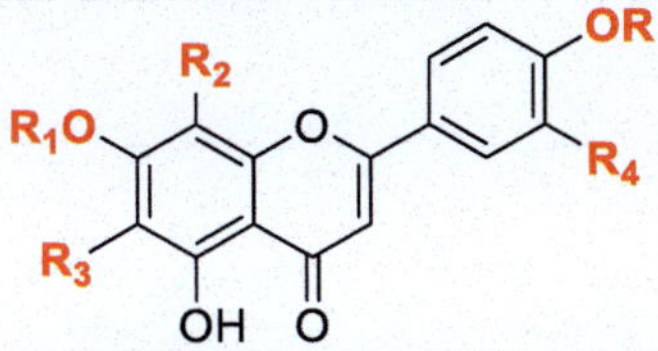

	Name	R1	R2	R3	R4	R	Ref
1	Apigenin 7-*O*-(6″-*O*-*p*-coumaroyl) glucoside	6-Cou-Glc					6
2	Apigenin 7-*O*-(6″-*O*-feruloyl)glucoside	6-Feru-Glc					3
3	Isoorientin			C-Glc	OH		4, 5
4	Isoscoparin 2″-*O*-(6‴-*O*-feruloyl) glucoside			*C*-Glc(2-1)-“6-Feru”-Glc	OCH_3	Glc	3,6
5	Isoscoparin 2″-*O*-(6‴-*O*-feruloyl) glucoside-4′-*O*-glucoside			*C*-Glc(2-1)-“6-Feru”-Glc	OCH_3	Glc	6
6	Isoscoparin 2″-*O*-(6‴-*O*-*p*-coumaroyl) glucoside			*C*-Glc(2-1)-“6-Coum”-Glc	OCH_3	Glc	6
7	Isovitexin			C-Glc			1,3,4,5,6
8	Isovitexin-2″-*O*-*p*-coumarate			*C*-“2-Coum”-Glc			3
9	Isovitexin-6″-*O*-*p*-coumarate			*C*-“6-Coum”-Glc			3
10	Isovitexin-2″-*O*-*p*-rhamnoside			*C*-Glc(2-1)Rha			1,3
11	Isovitexin 2′′-*O*-(6′′′-*O*-caffeyl)-glucoside			*C*-Glc(2-1)-“6-Caff”-Glc			2
12	Isovitexin 2′′-*O*-glucoside (meloside A)			C-Glc(2-1)Glc			1,2,3,5
13	Isovitexin-2′′-*O*-glucoside-6″-O-*p*-coumarate			*C*-“6-Coum”Glc(2-1) Glc			2,3,7
14	Isovitexin-2′′-*O*-glucoside-6″-O-ferulate			*C*-“6-Feru”Glc(2-1)Glc			3,6,7
15	Isovitexin 2″-*O*-(6‴-*O*-feruloyl) glucoside			C-Glc(2-1)-“6-Feru”-Glc			3,6,7
16	Isovitexin 2″-*O*-(6‴-*O*-*p*-coumaroyl) glucoside			*C*-Glc(2-1)-6-Coum-Glc			2,3,6,7

(continued)

Table 1 (continued)

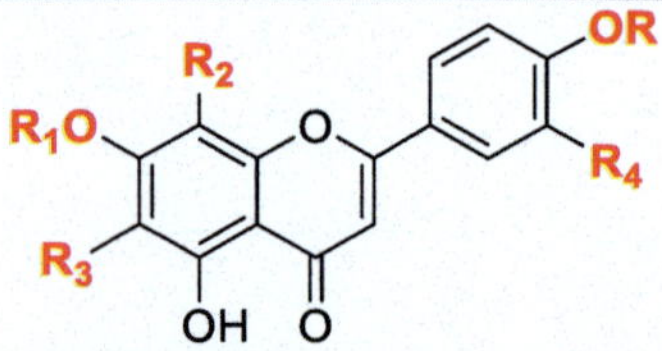

	Name	R1	R2	R3	R4	R	Ref
17	Isovitexin 4′-*O*-glucoside-2″-*O*-(6‴-*O*-*p*-coumaroyl) glucoside			*C*-Glc(2-1)-"6-Coum"-Glc		Glc	2,3,6,7
18	Isovitexin 4′,2′′-di-*O*-glucoside (8)			C-Glc(2-1)Glc		Glc	1,2,3
19	Isovitexin 4′-*O*-(2′′′-*O*-caffeyl)-glucoside-2′′-*O*-glucoside (cucumoside C)			C-Glc(2-1)Glc		2-"Caff"-Glc	2
20	Isovitexin 4′-*O*-glucoside-2′′-*O*-(2′′′-*O*-caffeyl)-glucoside (cucumoside B),			*C*-Glc(2-1)-"2-Caff"-Glc		Glc	2
21	Isovitexin 4′-*O*-glucoside-2′′-*O*-(2′′′-*O*-feruloyl)-glucoside (cucumoside F),			*C*-Glc(2-1)-"2-Feru"-Glc		Glc	2
22	Isovitexin 4′-*O*-glucoside-2′′-*O*-(2′′′-*O*-*p*-coumaroyl)-glucoside (cucumoside G)			*C*-Glc(2-1)-"2-Coum"-Glc		Glc	2
23	Isovitexin 4′-*O*-glucoside-2′′-*O*-(2′′′-*O*-sinapoyl)-glucoside (cucumoside E),			*C*-Glc(2-1)-"2-Sina"-Glc		Glc	2
24	Isovitexin 4′-*O*-glucoside-2′′-*O*-(6′′′-*O*-caffeyl)-glucoside (cucumoside A)			*C*-Glc(2-1)-"6-Caff"-Glc		Glc	2
25	Isovitexin 4′-*O*-glucoside-2′′-*O*-(6′′′-*O*-feruloyl)-glucoside			*C*-Glc(2-1)-"6-Feru"-Glc			2,3,6,7

(continued)

Table 1 (continued)

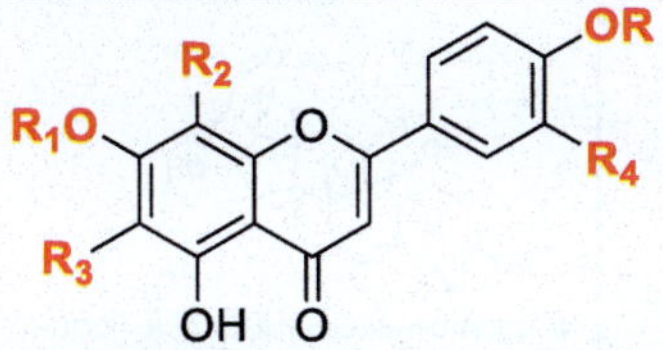

	Name	R1	R2	R3	R4	R	Ref
26	Isovitexin 4′-*O*-glucoside-2′′-*O*-(6′′′′-*O*-sinapoyl)-glucoside (cucumoside D)			*C*-Glc(2-1)-“6-Sina”-Glc		Glc	2
27	Isovitexin-2″-*O*-(6″-*O*-feruloyl)-glucoside-6″-*O*-ferulate			*C*-“6-Feru”-Glc(2-1)-“6-Feru”-Glc			3,7
28	Isovitexin-7-*O*-(6″-*O*-glucosyl)-glucoside	6(Glc)-Glc		*C*-Glc			3
29	Orientin		*C*-Glc		OH		4
30	Saponarin	Glc		*C*-Glc			1,3
31	Saponarin-4′-*O*-glucoside (Isovitexin-7,4′-*O*-glucoside)	Glc		*C*-Glc		Glc	1,3,4,6
32	Vicenin 2		*C*-Glc	*C*-Glc			4,6
33	Vitexin		C-Glc				3,4
34	Isovitexin-7,2″-di-*O*-glucoside	Glc		*C*-Glc(2-1)Glc			1
35	Isovitexin-4′-*O*-glucoside			*C*-Glc		Glc	3

(1) Olennikov and Kashchenko (2023b), (2) Olennikov and Kashchenko (2024), (3) Olennikov and Kashchenko (2023a), (4) McNally et al. (2003), (5) Krauze-Baranowska and Cisowski (2001), (6) Abou-Zaid et al. (2001), and (7) Olennikov and Kashchenko (2023c)
Coum coumaroyl, *Feru* feruloyl, *Sina* sinapoyl, *caff* caffeyl; *Glc* glucose

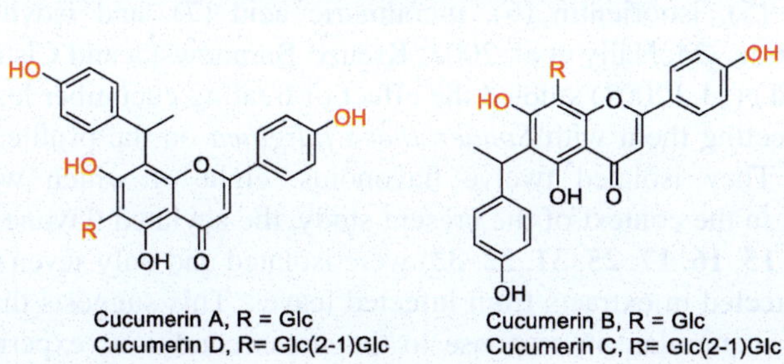

Fig. 1 Cucumerinoids reported from *C. sativus* leaves

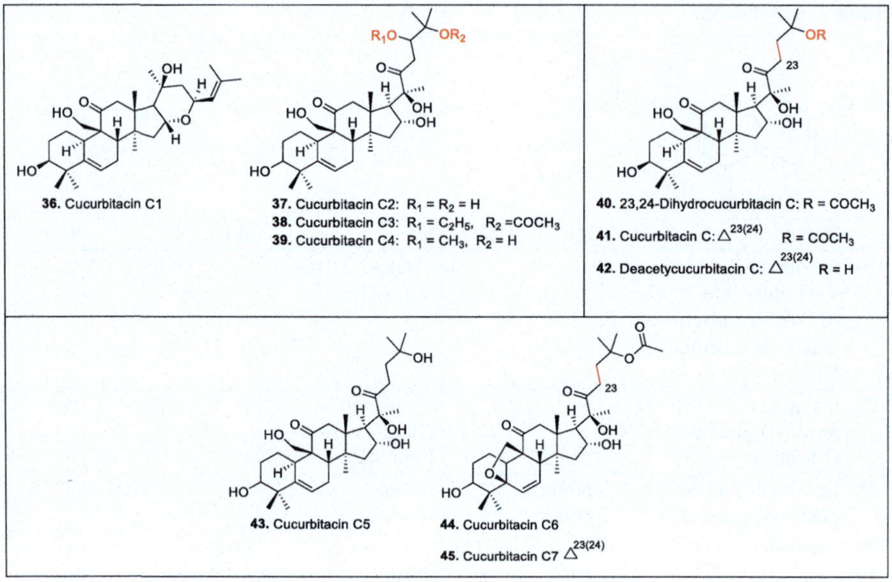

Fig. 2 Cucurbitacins isolated from *C. sativus* leaves

Flavonoids

The leaves of cucumbers are a rich source of flavonoids, including a specific type related to the isovitexin nucleus. To date, 37 compounds have been isolated and identified from the leaves (see Table 1), most of which are *C* and/or *O*-glycosylated (mainly with glucose) and acylated with phenolic acids such as coumaric, ferulic, sinapic, and caffeic acids. One notable metabolite has a unique structure featuring an ethylbenzene side chain connected to the flavonoid backbone's C-6 or C-8 position. These metabolites, known as curcumerins, are rare, with only four structures identified so far (curcumerins A-D, Fig. 1). These compounds have a C6-C2-C6-C3-C6 backbone and are analogs of vitexin and isovitexin.

Earlier studies on the leaves yielded cucumerin A and B, vitexin (3), isovitexin (4), orientin (5), isoorientin (6), *p*-coumaric acid (7), and isovitexin 2″-β-O-glucopyranoside, (McNally et al. 2003; Krauze-Baranowska and Cisowski 2001),

Abou-Zaid et al. (2001) studied the effect of treating cucumber leaves with silicone and infecting them with *Sphaerotheca fuliginea* on the profile of secondary metabolites. They isolated twelve flavonoids, eight of which were acylated. Interestingly, in the context of the present study, the acylated flavone *C*-glucosides (**1**, **4**, **5**, **6**, **7**, **15**, **16**, **17**, **25**, **31**, **32**, **33**) were isolated and only seven of them (**4–6**, **15–7**, **25**) detected in extracts from infected leaves. This suggests that these compounds may accumulate in response to the stress conditions experienced by the plant. However, other groups reported some of these compounds from non-infected leaves. Compounds **4**, **15**, **16,** and **25** were reported by Olennikov and Kashchenko (2023a, b, c, 2024) and McNally et al. (2003). This may indicate the quantitative variations of the detected compounds in cucumber leaves and the enhancement of their biosynthesis when infected with *S. fuliginea*.

Olennikov and Kashchenko (2024) reported fourteen acylated isovitexin derivatives (**11–13**, **16**, **18–26**), and seven of them (**19–24**, **26**) were given the names of cucumoside A-G. The same group, Olennikov and Kashchenko (2023c), identified seven acylated isovitexin derivatives (**13–17**, **25**, and **27**) from cucumber's green leaves. Another study by Olennikov and Kashchenko (2023b) identified curcumerins C and D with seven more flavonoids (**7**, **10**, **12**, **18**, **30**, **31**, and **34**).

Extensive studies using LC-MS by Olennikov and Kashchenko (2023a) on the green waste biomass of cucumber green leaves resulted in the tentative identification of 38 flavonoids, 20 of which (curcumins C and D, **2**, **4**, **8**, **9**, **10**, **12**, **13**, **14**, **15**, **16**, **18**, **25**, **27**, **28**, **30**, **31**, **33**, **35**) were identified based on direct comparison with authentic samples. The identified compounds revealed the potential of the green leaves as a source of more flavonoids with the same skeletons.

Cucurbitacins

Cucurbitacins are triterpenoid compounds reported in many plants, especially Cucurbitaceae, including the *Cucumis* genus. Different categories of cucurbitacins have been identified to date, primarily discovered in the Cucurbitaceae family, and they exhibit significant pharmacological properties. Among others, cucurbitacin C is the only one detected in the cucumber plant and contributes to its bitter taste (Rice et al. 1981; Horie et al. 2007; Min et al. 2023). Cucumber leaves have high cucurbitacin C content, while the fruit, root, and seeds have low content. Cucurbitacin C structure is similar to cucurbitacin B, E, and hemolysin A but has distinct structural differences at the C-2, C-3, and C-19 positions.

Starting with 110 kg of cucumber bitter leaves, in addition to cucurbitacin C, nine analogs of cucurbitacin C (**C1–C7**, deacetylcucurbitacin C, and 23,24-dihydrocucurbitacin C, Fig. 2) have been isolated (**36–45**) (Qing et al. 2014, 2022).

The same group (Chen et al. 2022) developed a rapid detection method using high-performance liquid chromatography/quadrupole-time-of-flight mass spectrometry (HPLC-Q-TOF-MS) combined with in-source fragmentation (ISF) to screen cucurbitacins and cucurbitacin glycosides in leaves, fruits, flowers, and roots of cucumber plants. The authors used seven chemical markers from their previous study, and their fragmentation pathways were compared with the extracts of the cucumber in different parts; 40 cucurbitacins (aglycons and glucosides) were identified tentatively. All the compounds have cucurbitacin C-related structures. The analysis indicated the high content of cucurbitacins in leaves more than in other parts and concluded the direct correlation of the biological activities of the leaves to the high content of cucurbitacins.

Other Metabolites

A study by Insanu et al. (2022) indicated that the ethanol extract of the leaf and stem of *C. sativus* contains alkaloids, glycosides, steroids, saponins, and tannins. In the same study, the chloroform extract of the leaf and stem of *C. sativus* contained alkaloids, glycosides, steroids, flavonoids, saponins, and tannins. Methanol leaf extract of *C. sativus* contained a high amount of vitamin A (23.00 ± 0.01 mg/kg), vitamin B1 (0.03 ± 0.01 mg/kg), vitamin B2 (0.03 ± 0.02 mg/kg), vitamin B6 (0.44 ± 0.01 mg/kg), vitamin C (6.11 ± 0.02 mg/kg), vitamin E (0.09 ± 0.01 mg/kg), niacin

(0.41 ± 0.01 mg/kg), and folate (15.00 ± 0.02 mg/kg). The study further revealed that chemical compounds of *C. sativus*'s essential oil were 3-nonenal, nonanal, (*E*,*Z*)-2,6-nonadienal, (*Z*)-3-nonenol, (*E*)-2-nonenal, (*Z*)-6-nonenol, *n*-decanal, (*E*,*E*)-2,4-nonadienal, (*E*,*E*)-2,4 decadienal, (*E*,*Z*)-2,4-decadienal, (*E*)-β-damascenone, 1-tetradecene, tetradecane, pentadecanal, α-humulene, β-ionone, caryophyllene oxide, tridecanal, tetradecanal, 9,12,15-octadecatrienal, and 9,17,octadecadienal (Insanu et al. 2022).

Three megastigmane, cucumegastigmanes I (**46**) and II (**47**), dehydrovomifoliol (**48**), and five other compounds were isolated from the leaves of *C. sativus*. lutein, (–)-(1*R*,2*S*,5*R*,6*S*)-2,6-di-(4-hydroxyphenyl)-3,7-dioxabicyclo(3,3,0)octane (**49**), adenosine (**50**), indole-3-aldehyde (**51**), and indole-3-carboxylic acid (**52**) (Kai et al. 2007).

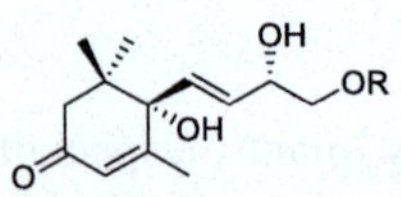

46. Cucumegastigmanes I, R=H
47. Cucumegastigmanes II, R=Glc

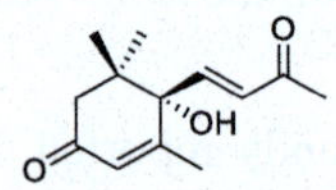

48. Dehydrovomifoliol

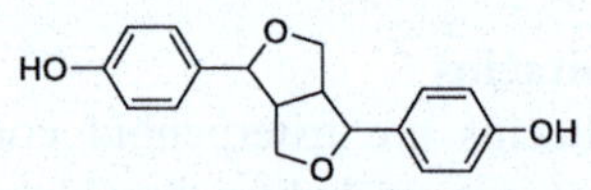

49. Ligballinol (2,6-di-(4'-hydroxyphenyl)-3,7-dioxabicyclo (3,3,0) octane)

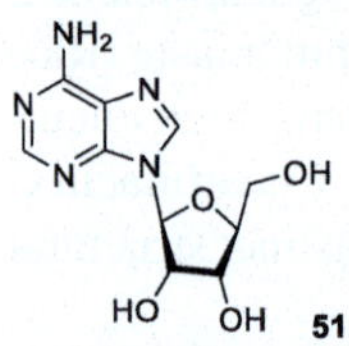

50. Adenosine

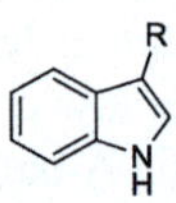

51. indole-3-aldehyde (16); R=CHO
52. indole-3-carboxylic acid (17) R= COOH

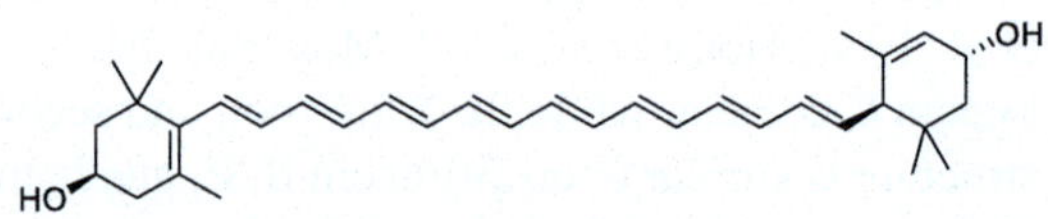

53. Lutein

Three antimicrobial sphingolipids were separated by bioassay-guided isolation from the chloroform fraction of the crude methanol extract of cucumber stems and identified as (2*S*,3*S*,4*R*,10*E*)-2-((2′*R*)-2-hydroxytetra-cosanoylamino)-1,3,4-octadecanetriol-10-ene (**54**), 1-O-glucopyranosyl(2*S*,3*S*,4*R*,10*E*)-2-((2′R)-2-hydroxytetracosanoylamino)-1,3,4-octadecanetriol-10-ene (**55**), and soya-cerebroside I (**56**) (Tang et al. 2010).

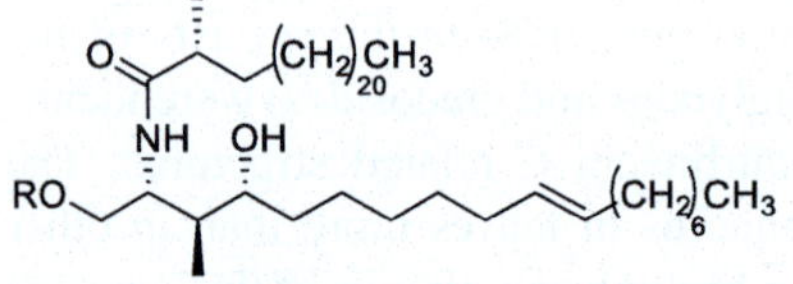

54. R=H
55. R=Glc

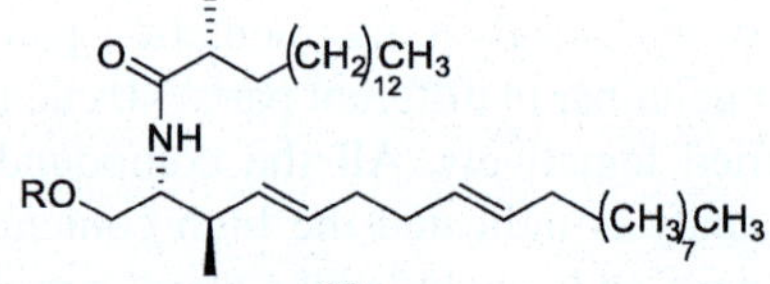

56

2.2 *Cucumber Root*

Allelochemical constituents from the hydrophobic root exudate at different growth stages were identified, including benzoic, *p*-hydroxybenzoic, 2,5-dihydroxybenzoic, 3-phenyl propionic, cinnamic, *p*-hydroxycinnamic, myristic, palmitic, and stearic acids, as well as *p*-thiocyanate phenol and 2-hydroxybenzothiazole. Except for 2-hydroxybenzothiazole, all these were toxic to lettuce growth (Yu and Matsui 1994).

Zhou et al. (2012) reported 15 compounds of triterpenoids and steroids from the roots. The isolated compounds were identified as stigmasta-8(14),22-diene-7α-methoxy-3β-ol, dehydroeburicoic acid, α-spinasterol-3-*O*-glucopyranoside, stigmasta-4,6,8(14),22-tetraen-3-one, ergosterol, ursolic acid, 3β-hydroxymultiflora-8-en-17-oic acid, karounidiol, 1*H*-indole-3-aldehyde, methyl 1*H*-indole-3-carboxylate, methyl 1*H*-indole-3-acetate, *p*-hydroxybenzaldehyde, isovanillin, 2-hydroxybenzyl alcohol, and 4-hydroxycinnamic acid.

2.3 *Cucumber Seeds and Oil*

Cucumber seeds are characterized by their high contents of crude proteins (42%) and fats (42.5%) and ash that is rich with phosphate (P_2O_5): 0.62%. Oil extracted from seeds is light yellow and clear and has the following physicochemical characteristics: acid value: 0.22, saponification value: 1930, soluble fatty acid: 0.4%, unsaponification matter: 0.91%, and specific gravity: 0.91 (Mukherjee et al. 2013).

The fatty acid profile for cucumber seeds' oil is dominated by linoleic acid (40.11%), oleic acid (38.70%), stearic acid (16.2%), and palmitic acid (0.63%); meanwhile, the seed cake is very rich in protein (72.53%), ash (9.7%), carbohydrates (8.64%), crude fiber (1%), and water content of 1.13% (Kapoor 1990). Some sterols reported in the cucumber seeds' oil include avenasterol, campesterol, clerosterol, codisterol, isofucosterol, stigmasterol, and sitosterol, while Hemphill et al. (1972) reported on the gibberellin hormone.

2.4 *Cucumber Flowers*

Earlier research by Krauze-Baranowska and Cisowski (2001) identified *O*-glycosylated derivatives such as quercetin 3-*O*-glucoside, kaempferol 3-*O*-glucoside, isorhamnetin 3-*O*-glucoside, and kaempferol 3-*O*-rhamnoside. However, a recent study (Olennikov 2023) employing HPLC–PDA–ESI–tQ–MS/MS on the flavonoid fraction from methanolic flower extract detected 47 compounds. This study identified C-, O-, and C-, O-glycosides of various flavonoid backbones, many of which were tentatively identified for the first time in *C. sativus*. The same study demonstrated that flower extracts from ten cucumber cultivars were quantified, revealing Isoorientin, isoorientin-4′-*O*-glucoside, and cucumerins B and D as predominant.

3 Biological and Functional Properties of Extracts and Bioactive Compounds from By-products

Total extracts and pure compounds from *C sativus* by-products have undergone different biological examinations. Various pharmacological effects have been reported, including hypolipidemic, anticancer, antibacterial, antioxidant, anti-inflammatory, and antidiabetic activities.

3.1 Antioxidant Activity

Free radicals (i.e., ROS and RNS) are generated from various stages of cellular metabolism. Consequently, high levels of free radicals in the body may lead to chronic conditions (Engwa 2018). The peels, seeds, leaves, and fruits of *C. sativus* contain exogenous antioxidants that can scavenge free radicals. The plants' methanolic leaf (Nasrin et al. 2015) and peel (Biswas et al. 2021) extracts have effectively scavenged free radicals in DPPH and H_2O_2 assays. Anjani et al. (2023) recently reported that the ethanolic peels had 54 and 71% DPPH and FRAP free radical scavenging capacity. The antioxidant properties of the aqueous and ethanol fruit extracts of *C. sativus* were studied in various assays at concentrations ranging from 0.001 to 1.00 mg/mL (Amani et al. 2024). Cucumber seeds can be a tonic and help treat burning sensation, constipation, and intermittent fevers (Warrier 1994). The methanolic extract of *C. sativus* seeds reported a significant ulcer potential due to the antioxidant activity at 300 mg/kg dose in pyloric ligation and water immersion stress-induced rat models, as reported by Gill et al. (2009).

On the other hand, the pure isolate vicenin-2 (**32**) displayed up to 30 and 70% DPPH and ABTS radical scavenging capacity, respectively (Marrassini et al. 2023). The compound was found to effectively scavenge superoxide anions and hydroxyl radicals at a concentration range of 25.0–100.0 µM (Patanè et al. 2023). Quercetin is a natural flavonoid whose pharmacological activities have been extensively studied. This compound, isolated from the peels of *C. sativus* (Anjani et al. 2023), was shown in a separate study to display antioxidant activity in H_2O_2-induced oxidative stress in red blood cells (Remigante et al. 2022).

3.2 Anti-inflammatory Activity

Anti-inflammatory drugs are key in regulating inflammation biomarkers that may be linked to several health disorders. Various by-products from *C. sativus* like the seeds (Wahid et al. 2021), leaves (Trejo-Moreno et al. 2018; Nasrin et al. 2015), and fruits (Amani et al. 2024; Agatemor et al. 2019) have been researched for their anti-inflammatory properties. Vicenin-2 (**31**), a flavonoid glycoside of the leaf parts of the plant, was confirmed to have anti-inflammatory activity by Marrassini et al. (2023). It was reported that at a concentration range of 0.1–10 µg/mL, the compound inhibits LPS-induced RAW264.7 macrophages. Saponarin (**30**), also obtained from the leaves, exerts anti-inflammatory activity in LPS-induced RAW264 cells (Min et al. 2021).

3.3 Anticancer Activity

It is reported that the anticancer activity of this plant is due to the presence of cucurbitacins. These triterpenoid derivatives exert their anticancer actions via different pathways. Recently, two cucurbitacin C-type derivatives (compounds **44** and **45**) isolated from the leaves of the plant were shown to inhibit cell proliferation on HepG2 (IC_{50} = 4.16–10.06 μM), A549 (IC_{50} = 10.13–18.66 μM), DU145 (IC_{50} = 1.82–3.61 μM), and HCT116 (IC_{50} = 10.66–11.35 μM) cells (Qing et al. 2022).

3.4 Antibacterial Activity

In Ayurvedic medicine, the boiled leaves of *C. sativus* with cumin seeds are helpful for the treatment of throat infections. Meanwhile, powdered leaves and sugar can be used as a diuretic and administered to treat remitted and inflammatory fevers (Kapoor 1990).

Sphingolipids (**54–56**) were isolated and separated by bioassay-guided isolation from the stems of cucumber extracts (chloroform fraction of the crude methanol extract). They reported a broad range antimicrobial activity spectrum against fungi (*Pythium aphanidermatum*, *Botryosphaeria dothidea*, *Fusarium oxysporum*, and *Botrytis cinerea*) and bacteria (*Xanthomonas vesicatoria*, *Pseudomonas lachrymans*, and *Bacillus subtilis*) (Tang et al. 2010).

The extract from *C. sativus* peels was found to inhibit the growth of *E. coli*, *P. aeruginosa*, and *S. mutans*, forming inhibitory zones of 11.36 ± 0.47 mm, 6.7 ± 0.36 mm, and 10.16 ± 0.11 mm, respectively. Scanning Electron Microscopy (SEM) results also confirmed the rupturing of bacterial cells or biofilms compared to control strains (Anjani et al. 2023).

Different extracts from the cucumber seeds were tested for their antibacterial activity. The results showed significant antimicrobial activity, suggesting that *C. sativus* should be used as an active nutrition to control microbial infections that lead to chronic diseases (Begum et al. 2019).

The antimicrobial activities of cucumber extract@Cu hybrid nanofowers (hNFs) against *E. coli* and *S. aureus* strains were determined by the micrillion method, and MIC values were determined in 125 μg/mL and 250 μg/mL, respectively (Erdem and Koca 2023.).

3.5 Antidiabetic Activity

Various in vitro and in vivo models are readily available to study the antidiabetic effects of plants and their metabolites. Saponarin (**30**) was isolated from the leaves of *C. sativus* plant (Abou-Zaid et al. 2001). This compound was reported in vivo by Simeonova et al. (2016) to ameliorate STZ-induced (40 mg/kg bw) glycemic effects in male diabetic rats. Another study (Seo et al. 2015) revealed that the compound regulates glucose uptake and glucogenesis through phosphorylation of CRTC2 and HDAC5 on HepG2 and TE671 cells. Chen et al. (2021) showed that isovitexin (**7**) effectively reduces glucose uptake in A549 and H1975 cells at concentrations ranging from 12.5 to 50 μg/mL. Choi et al. (2014) found the compound to be a potent inhibitor of aldose reductase enzymes (RLAR, IC_{50} = 0.49 μM, and HRAR IC_{50} = 0.13 μM), which are crucial in the polyol pathway.

3.6 *Hypolipidemic Activity*

The latest study by Olennikov (2023) reported new cucumber flavonoids that inhibit pancreatic lipase, and Olennikov and Kashchenko (2023c) indicated their possible hypolipidemic potential. Apigenin derivatives are the main groups of flavonoids from cucumber leaves that have a hypolipidemic effect (Hu et al. 2022). Apigenin can improve hyperlipidemia in high-fat diet-fed rats (Aldayel 2022) and induce a reduction of lipid accumulation via palmitic acid activation of the AMPK/SREBP pathway in HepG2 cells (Lu et al. 2019). Wu et al. (2021) reported that apigenin has an amelioration effect on lipid metabolism disorders via sterol regulatory element-binding protein 1c (SREBP-1c) downregulation, sterol regulatory element-binding protein 2 (SREBP-2) and some other genes down-streaming Luteolin derivatives were identified as the compounds responsible for the hypolipidemic activity. The author suggested that cucumber flowers could be a valuable dietary supplement (Table 2).

Table 2 Bioactive compounds from cucumber by-products and their reported function and biological properties

S/N	Compound name	Source/ by-product	Pharmacological action/activity	Assay method/ results	References
1	Cucurbitacin C6 (**44**)	Leaves	Anticancer	MTT assay against HepG2 (IC_{50} = 4.16–10.06 μM), A549 (IC_{50} = 10.13–18.66 μM), DU145 (IC_{50} = 1.82–3.61 μM), and HCT116 (IC_{50} = 10.66–11.35 μM)	Qing et al. (2022)
2	Cucurbitacin C7 (**45**)	Leaves	Anticancer	MTT assay against HepG2 (IC_{50} = 4.16–10.06 μM), A549 (IC_{50} = 10.13–18.66 μM), DU145 (IC_{50} = 1.82–3.61 μM), and HCT116 (IC_{50} = 10.66–11.35 μM)	Qing et al. (2022)

(continued)

Table 2 (continued)

S/N	Compound name	Source/ by-product	Pharmacological action/activity	Assay method/ results	References
3	Vicenin-2 (**32**)	Leaves	Anticancer, wound healing, antioxidant, and anti-inflammatory	Inhibiting the STAT-3 signaling pathway downregulates apoptotic and TNF-α/IL-6/IL-β cytokine levels linked with carcinogenesis in in vivo studies on mice Promotes wound healing effects on STZ-induced type 1 diabetic rat via the anti-inflammatory VEGF and TGF-β signaling pathways DPPH (up to 30%) and ABTS (up to 70%) radical scavenging capacity Inhibits LPS-induced RAW264.7 macrophages at 0.1–10 μg/mL concentrations The cinnamoyl active site inhibits the oxidative stress-induced damage to HAS Modulates the expression of MAPKs and MMPs signaling pathways linked with UV oxidative stress on HDF	Marrassini et al. (2023) Zhang et al. (2020) Huang et al. (2020) Tan et al. (2019, 2020) Duan et al. (2019)
4	Saponarin (**30**)	Leaves	Antidiabetic, anti-inflammatory, and antioxidant activity	Ameliorates STZ-induced glycemic effect and the release of nitric oxide and reactive oxygen species on diabetic rats Regulates glucose uptake and glucogenesis through phosphorylation of CRTC2 and HDAC5 on HepG2 and TE671 cells Prevents the expression of TNF-α/IL-1β/COX-2/IL-33/IL-6/TSLP/LL-37/p38 cytokines in LPS-induced RAW264.7, RBL-2H3, and HaCaT cells Displays moderate DPPH and ABTS scavenging capacity	Min et al. (2021) Simeonova et al. (2016) Lee et al. (2016a, b) Seo et al. (2015)

(continued)

Table 2 (continued)

S/N	Compound name	Source/ by-product	Pharmacological action/activity	Assay method/ results	References
5	Isovitexin (**7**)	Leaves	Anticancer, anti-inflammatory, antioxidant, and antidiabetic	Regulation of miR-98-5p/ Bcl-2/Bcl-xL and MAPKs/ NF-κB signaling pathways on HaCaT cells Induces apoptosis and regulates PI3K/Akt/mTOR/ PKM2-mediated pathways on HCT116, SW480, HT-29, LoVo, A549, and H1975 Inhibits LPS-induced inflammation and production of reactive oxygen species on SV40-MES-13 cells Alleviates inflammation via TLR4/myD88/NF-κB, RAS–MAPK, and Nrf2/ HO-1 pathways in mice DPPH scavenging capacity (IC_{50} = 0.5 μg/mL) At 12.5–50 μg/mL concentration, the compound attenuates glucose uptake in A549 and H1975 cells Inhibits production of RLAR (IC50 = 0.49 μM) and HRAR (IC50 = 0.13 μM)	Lv et al. (2023); Tseng et al. (2023); Chen et al. (2021); Sun et al. (2021); Zhang et al. (2021); Hu et al. (2021); Zhu et al. (2021); Liu et al. (2020); Choi et al. (2014)
6	Rutin	Seeds	Analgesic, anti-inflammatory, and anticancer	It mediates either by inhibiting COX enzymes or NMDA receptors in the CNS	Wahid et al. (2022); Imani et al. (2021)
7	Vitexin (**33**)	Seeds	Anti-inflammatory and antioxidant	Inhibits LPS-induced RAW264.7 macrophages at 0.1–10 μg/mL concentrations. DPPH (up to 30%) and ABTS (up to 70%) radical scavenging capacity	Marrassini et al. (2023); Wahid et al. (2022)

(continued)

Table 2 (continued)

S/N	Compound name	Source/ by-product	Pharmacological action/activity	Assay method/ results	References
8	Chlorogenic acid	Seeds	Antibacterial and anti-inflammatory	It affects the intracellular metabolic imbalance of the TCA cycle and glycolysis of *Bacillus subtilis* (MIC = 2.50 mg/mL) Attenuates pro-inflammatory IL-1β and TNF-α cytokines on LPS-stimulated RAW 264.7 cells Stimulates anti-inflammatory effects in mice via the TLR4/MyD88/NF-κB signaling pathway	Wahid et al. (2022); Hamid et al. (2022); Wu et al. (2020); Chen et al. (2018); Hwang et al. (2014)
9	Gallic acid	Seeds	Antidiabetic and antioxidant	Dose-dependent inhibition of α-amylase (IC_{50} = 1.09 μM) and α-glucosidase (IC_{50} = 1.22 μM) enzymes DPPH radical scavenging with ED_{50} = 5.73 μM	Wahid et al. (2022); Hamid et al. (2022) Adefegha et al. (2015)
10	Protocatechuic acid	Seeds and peels	Antidiabetic, antioxidant, antihypertensive, and antityrosinase	Dose-dependent inhibition of α-amylase (IC_{50} = 1.12 μM) and α-glucosidase (IC_{50} = 1.76 μM) enzymes DPPH and lipid peroxidation radical scavenging with ED_{50} = 8.29 μM Inhibits MMMP-1 secretion on UV-induced HDF cells Shows reversal effects of DOCA-salt-induced hypertension against Male Wistar rats Displays an IC_{50} value of 246.42 μg/mL in tyrosinase inhibition assay	Wahid et al. (2022) Shin et al. (2020) Girsang et al. (2020) Safaeian et al. (2018) Adefegha et al. (2015)

(continued)

Table 2 (continued)

S/N	Compound name	Source/ by-product	Pharmacological action/activity	Assay method/ results	References
11	Kaempferol	Peels	Anti-inflammatory, anticancer, and antioxidant	Inhibits NF-κB-DNA binding and reduces NF-κB p65 expression MTT assay on HepG2 (IC_{50} = 30.92 μM), CT26 (IC_{50} = 88.02 μM), and B16F1 (IC_{50} = 70.67 μM) Shows a scavenging capacity of DPPH (IC50 = 47.93 μM) and ABTS (IC50 = 0.337 mM)	Hamid et al. (2022); Wang et al. (2018) Kadioglu et al. (2015)
12	Caffeic acid	Peels	Antibacterial, antioxidant, anticancer, and wound healing	Displays inhibitions with an MIC of 56–1024 μg/mL on *Staphylococcus aureus* strains DPPH radical scavenging capacity (IC_{50} = 32.2 μM) Attenuates XTT-NR-SRB cell viability on MDA-MB-231 cells (IC_{50} = 135.85–150.94 μM after 24 h)	Hamid et al. (2022) Kępa et al. (2018) Sidoryk et al. (2018)
13	Vanillic acid	Peels	Antioxidant	Reduces H_2O_2-induced oxidative stress on the D. Mel-2 cell line at an IC_{50} of 250 μg/mL	Hamid et al. (2022); Taqvi et al. (2021)

4 Food and Non-food Applications of Extracts and Bioactive Compounds from By-products

Ahmad Cheikhyoussef

4.1 Food Applications

Cucumber is widely consumed either fresh in salads as a cooling food or fermented as pickles, and consuming skin peels and seeds expands its nutritional benefits (Hedges and Lister 2005). In traditional medicine, the leaves, stems, roots, seeds, and peels of *C. sativus* have been used as antioxidant, analgesic, anti-diarrheal, anti-gonorrheal, anti-inflammatory, de-toxicant, constipation and indigestion relief, hypertension-reducing agent, and diabetes mellitus and serum lipids regulator (Gill et al. 2010; Dixit and Kar 2010; Dhiman et al. 2012).

4.2 Non-food Applications

There is a need for an urgent cucumber waste disposal transformation to create new products with different biological activities, including aerobic composition (Kostov et al. 1995) and anaerobic co-digestion of cucumber by-products and sewage sludge (Lowe et al. 2020). The cucumber peels have 25.1% lactic acid and are used in the treatments of warts, xerosis, and chemical peeling of the skin together with other cucumber pulp tissues and juice (Wehr et al. 1986; Murad et al. 1995). The cucumber peel extract was reported to ameliorate diabetes mellitus in terms of the observed changes in serum lipids and hepatic lipid peroxidation (LPO) in diabetic mice, whereby a dose of 500 mg/kg b. wt/day for 15 days can reverse the enzyme associated with such health condition (Dixit and Kar 2010). The stems have been used in Traditional Chinese Medicine (TCM) thanks to their anti-inflammatory activity. As per the ancient book "*Ben Cao Gang Mu*" during the Ming Dynasty of China, cucumber can expand the blood vessels and lower blood pressure (Li 1988).

Venkatachalam et al. (2015) synthesized metallic silver nanoparticles using callus (CAgNPs) (21–23 nm) and leaf extracts (LEAgNPs) (11–19 nm) of *C. sativus* and evaluated its wound healing activity in Wistar albino rat model. They attributed the healing property to the presence of cucurbitacin in the synthesized AgNPs with similar findings on cucurbitacin triterpenoids, which are well known for their bitterness and toxicity in AgNPs (Mukherjee et al. 2013).

5 Valorization of By-products for Non-health Purposes, i.e. Source of Bioenergy

Reduction of food waste provides critical environmental and economic benefits. Valorization of food by-products into edible materials is one of the most exciting strategies in this field. However, food by-products and wastes can contain chemical contaminants or potential pathogens that may endanger consumers' health. Therefore, assuring the quality and safety of these by-products is of utmost importance in taking advantage of this valorization strategy. Based on the Food and Agriculture Organization (FAO) of the United Nations definition, food loss is the decrease in the quantity or quality of food resulting from decisions and actions by food suppliers in the chain, excluding retail, food service providers and consumers (FAO 2019). It reflects the inefficiency and unsustainability of agrifood systems and negatively impacts the economy, food security, nutrition, and the environment (FAO 2023).

Out of 14,000 articles found via a scientometric analysis of the scientific databases using the keywords "*Cucumis sativus*" and "cucumber," approximately 900 studies are dedicated to the disposal of waste from the production of cucumber and its by-products (Olennikov and Kashchenko 2023a) with more advancements on successful cucumber cultivation and increasing productivity have been reported, however, few studies focused on waste disposal from *Cucumis sativus* production. The exploration of cucumber waste as a potential biorefinery feedstock has been reported recently (Gao et al. 2022). Initially, the essential oils were isolated from cucumber waste via vacuum microwave hydro-distillation (VMHD). A yield of 0.49 wt % was obtained, and the quality of the extractives was compared concerning traditional hydro-distillation (HD), considering that VMHD is faster and more efficient in terms of energy use than HD (Gao et al. 2022).

6 The Use of Cucumber By-products as Natural Biosorbent

The peel of *Cucumis sativus* (cucumber) has been investigated as an eco-friendly and cost-effective adsorbent for removing harmful dyes from wastewater. Studies have shown that activated cucumber peel can effectively adsorb dyes like crystal violet, rhodamine B, and Acid Blue 113, among others (Table 3). These dyes pose significant environmental and health risks, including carcinogenic and mutagenic effects. The biosorption process using cucumber peel has been optimized by adjusting parameters such as dye concentration, pH, contact time, and temperature, demonstrating rapid and efficient dye removal. The kinetics generally follow pseudo-second-order models, and Langmuir and Freundlich models often describe the adsorption isotherms. This approach offers a sustainable alternative for treating dye-contaminated water, utilizing agricultural waste materials to create valuable biosorbents.

Table 3 Biosorbent prepared from cucumber by-products for removal of harmful dyes from wastewater

Adsorbent	Adsorbate	Applicable kinetic models	Reference
Cucumis sativus oven-dried and treated with H_2SO_4 for 12 h in different geometrical sizes (75–125, 125–250, and 250–500 μm)	Crystal violet and rhodamine B	Pseudo-second order	Smitha et al. (2017)
Biosorbent from overripe *C. sativus* peel, without chemical treatment, after drying at 800 °C	Acid Blue 113 dye	Pseudo-second order.	Lee et al. (2016a)
Biosorbent from *C. sativus* peel waste	Crystal violet dye	Pseudo-second order	Shakoor and Nasar (2019)
Magnetic nanoparticles (Fe_3O_4 NPs) were prepared using aqueous peel extracts of *C. sativus*.	Antibiotics, piperacillin, tazobactam, sulfamethoxazole, tetracycline, trimethoprim, ampicillin, and erythromycin	NA	Stan et al. (2017)

Adsorbent	Adsorbate	Applicable kinetic models	Reference
Biosorbent from cucumber peels after heating at 110 °C for 24 h, particle size range 250–500 μm	Methylene blue and Orange G	Pseudo-first-order, pseudo-second-order, Elovich and intra-particle diffusion models	Stavrinou et al. (2018)
Biosorbent from cucumber peels	Methylene blue	Pseudo-first order	Akkaya and Güzel (2014)
Biosorbent from cucumber peels after	Acid violet 17		Santhi and Smitha (2015)
Biochar from cucumber peels	Acid orange 7	Pseudo-second order	Kapoor and Sivamani (2023)
Biosorbent from cucumber peels	Crystal violet and safranin		Luikham et al. (2018)
Biosorbent from cucumber peels after drying at 110 ± 2 °C, different particle sizes	Acid black-7	Pseudo-second order	Smitha et al. (2015)
Biosorbent from cucumber peels after oven-drying at 80 °C for 24 h and treated with HCl. particle size of 500 μm	Cd(II) ions	Pseudo-second order	Pandey et al. (2014)

References

Abou-Zaid MM, Lombardo DA, Kite GC, Grayer RJ, Veitch NC (2001) Acylated flavone C-glycosides from Cucumis sativus. Phytochemistry 58(1):167–172

Adefegha SA, Oboh G, Ejakpovi II, Oyeleye SI (2015) Antioxidant and antidiabetic effects of gallic and protocatechuic acids: a structure–function perspective. Comp Clin Pathol 24:1579–1585

Agatemor UMM, Nwodo OFC, Ozah IR (2019) Inhibition of phospholipase A2 and prostaglandin synthase activities as possible mechanisms for the anti-inflammatory effect of Cucumis sativus fruit homogenate. Acta Sci Pharm Sci 3:68–73

Akkaya G, Guzel F (2014) Application of some domestic wastes as new low-cost biosorbents for removal of methylene blue: kinetic and equilibrium studies Chemical Engineering Communications, 201(4):557–578

Aldayel ST (2022) Apigenin attenuates high-fat diet-induced nephropathy in rats by hypoglycemic and hypolipidemic effects, and concomitant activation of the Nrf2/antioxidant axis. J Funct Foods 99:105295

Amani T, Surenthar M, Shanmugam R (2024) Anti-inflammatory and antioxidant activity of *Cucumis sativus* and *Citrus macroptera* herbal formulation: an in-vitro study. Cureus 16(1)

Anjani, Srivastava N, Mathur J (2023) Isolation, purification and characterization of quercetin from *Cucumis sativus* peels; its antimicrobial, antioxidant and cytotoxicity evaluations. 3 Biotech 13(2):46

Atlas Big. World cucumber production by country. Available online: https://www.atlasbig.com/en-us/countries-cucumber-production. Accessed 30 Apr 2024

Begum HA, Asad F, Sadiq A, Mulk S, Ali K (2019) 44. Antioxidant, antimicrobial activity and phytochemical analysis of the seeds extract of *Cucumis sativus* Linn. Pure Appl Biol (PAB) 8(1):433–441

Biswas S, Ghosh P, Dutta A, Biswas M, Chatterjee S (2021) Comparative analysis of nutritional constituents, antioxidant and antimicrobial activities of some common vegetable wastes. Curr Res Nutr Food Sci J 9(1):62–74

Chen D, Pan D, Tang S, Tan Z, Zhang Y, Fu Y, Lü G, Huang Q (2018) Administration of chlorogenic acid alleviates spinal cord injury via TLR4/NF κB and p38 signalling pathway anti inflammatory activity. Mol Med Rep 17(1):1340–1346

Chen RL, Wang Z, Huang P, Sun CH, Yu WY, Zhang HH, Yu CH, He JQ (2021) Isovitexin potentiated the antitumor activity of cisplatin by inhibiting the glucose metabolism of lung cancer cells and reduced cisplatin-induced immunotoxicity in mice. Int Immunopharmacol 94:107357

Chen SY, Zhou QYJ, Chen L, Li JY, Xie T, Zhang SH (2022) Screening and identifying cucurbitacins and cucurbitacin glycosides in Cucumis sativus using high-performance liquid chromatography/quadrupole-time-of-flight mass spectrometry combined with in-source fragmentation and alkali adduct ions. Rapid Commun Mass Spectrom 36(14):e9323

Choi JS, Islam MN, Ali MY, Kim EJ, Kim YM, Jung HA (2014) Effects of C-glycosylation on anti-diabetic, anti-Alzheimer's disease and anti-inflammatory potential of apigenin. Food Chem Toxicol 64:27–33

Dhiman K, Gupta A, Sharma DK, Gill NS, Goyal A (2012) A review on the medicinally important plants of the family Cucurbitaceae. Asian J Clin Nutr 4(1):16–26

Dixit Y, Kar A (2010) Protective role of three vegetable peels in alloxan induced diabetes mellitus in male mice. Plant Foods Hum Nutr 65(3):284–289

Duan X, Wu T, Liu T, Yang H, Ding X, Chen Y, Mu Y (2019) Vicenin-2 ameliorates oxidative damage and photoaging via modulation of MAPKs and MMPs signaling in UVB radiation exposed human skin cells. J Photochem Photobiol B Biol 190:76–85

Engwa GA (2018) Free radicals and the role of plant phytochemicals as antioxidants against oxidative stress-related diseases. Phytochemicals: source of antioxidants and role in disease prevention. BoD Books Demand 7:49–74

Erdem T, Koca FD (2023) Catalytic and antimicrobial activities of cucumber leaf extract-based organic@ inorganic hybrid nanoflowers through Fenton-like reaction. Rendiconti Lincei. Scienze Fisiche e Naturali 34(4):1255–1261

FAO (2019) The state of food and agriculture 2019. Moving forward on food loss and waste reduction. Rome. 182 pp

FAO (2023) Guidelines for action on food loss and waste reduction in the Near East and North Africa. Cairo. https://doi.org/10.4060/cc7330en. Accessed 07/05/2024

Gao Y, Briers HK, Matharu AS, Fan J (2022) Exploration of cucumber waste as a potential biorefinery feedstock. Processes 10(12):2694

Gill NS, Garg M, Bansal R, Sood S, Muthuraman A, Bali M, Sharma PD (2009) Evaluation of antioxidant and antiulcer potential of *Cucumis sativum* L. seed extract in rats. Asian J Clin Nutr 1:131–138

Gill NS, Sood S, Muthuraman A, Garg M, Kumar R, Bali M, Dev Sharma P (2010) Antioxidant, anti-inflammatory and analgesic potential of *Cucumis sativus* seed extract. Lat Am J Pharm 29:927–932

Girsang E, Lister INE, Ginting CN, Bethasari M, Amalia A, Widowati W (2020) Comparison of antiaging and antioxidant activities of protocatechuic and ferulic acids. Mol Cell Biomed Sci 4(2):68–75

Hamid K, Muhammad H, Waseem Iqbal M, Bhatti S (2022) Response surface methodology for the extraction of polyphenol contents and HPLC profiling of *Cucumis sativus* peels. Jilin Univ Eng Technol Ed 41:41–52

Hedges LJ, Lister CE (2005) Nutritional attributes of salad vegetables. Crop & food research confidential report, 1473, pp 1–29

Hemphill Jr DD, Baker LR, Sell HM (1972) Isolation and identification of the gibberellins of *Cucumis sativus* and *Cucumis melo*. Planta 103(3):241–248

Horie H, Ito H, Ippoushi K, Azuma K, Sakata Y, Igarashi I (2007) Cucurbitacin C—bitter principle in cucumber plants. Jpn Agric Res Q JARQ 41(1):65–68

Hu X, Li R, Sun M, Kong Y, Zhu H, Wang F, Wan Q (2021) Isovitexin depresses osteoarthritis progression via the Nrf2/NF-κB pathway: An in vitro study. J Inflamm Res 1403–1414

Hu H, Weng J, Cui C, Tang F, Yu M, Zhou Y, Shao F, Zhu Y (2022) The hypolipidemic effect of hawthorn leaf flavonoids through modulating lipid metabolism and gut microbiota in hyperlipidemic rats. Evid-Based Complement Alternat Med eCAM 2022:3033311

Huang G, Li S, Zhang Y, Zhou X, Chen W (2020) Vicenin-2 is a novel inhibitor of STAT3 signaling pathway in human hepatocellular carcinoma. J Funct Foods 69:103921

Hwang SJ, Kim YW, Park Y, Lee HJ, Kim KW (2014) Anti-inflammatory effects of chlorogenic acid in lipopolysaccharide-stimulated RAW 264.7 cells. Inflamm Res 63:81–90

Imani A, Maleki N, Bohlouli S, Kouhsoltani M, Sharifi S, Maleki Dizaj S (2021) Molecular mechanisms of anticancer effect of rutin. Phytother Res 35(5):2500–2513

Insanu M, Rizaldy D, Silviani V, Fidrianny I (2022) Chemical compounds and pharmacological activities of *Cucumis* genus. Biointerface Res Appl Chem 12(1):1324–1334

Kadioglu O, Nass J, Saeed ME, Schuler B, Efferth T (2015) Kaempferol is an anti-inflammatory compound with activity towards NF-κB pathway proteins. Anticancer Res 35(5):2645–2650

Kai H, Baba M, Okuyama T (2007) Two new megastigmanes from the leaves of *Cucumis sativus*. Chem Pharm Bull 55(1):133–136

Kapoor LD (1990) CRC handbook of ayurvedic medicinal plants. CRC, Boca Raton

Kapoor RT, Sivamani S (2023) Biosorption behavior of acid orange 7 dye onto *Cucumis sativus* peel biochar and its effect on growth of *Triticum aestivum*. Environ Prog Sustain Energy 42(5):14106

Kępa M, Miklasińska-Majdanik M, Wojtyczka RD, Idzik D, Korzeniowski K, Smoleń-Dzirba J, Wąsik TJ (2018) Antimicrobial potential of caffeic acid against *Staphylococcus aureus* clinical strains. Biomed Res Int 2018(1):7413504

Khan A, Mishra A, Hasan SM, Usmani A, Ubaid M, Khan N, Saidurrahman M (2022) Biological and medicinal application of *Cucumis sativus* Linn. Review of current status with future possibilities. J Complement Integr Med 19(4):843–854

Kostov O, Tzvetkov Y, Kaloianova N, Van Cleemput O (1995) Cucumber cultivation on some wastes during their aerobic composting. Bioresour Technol 53(3):237–242

Krauze-Baranowska M, Cisowski W (2001) Flavonoids from some species of the genus Cucumis. Biochem Syst Ecol 29(3):321–324

Lee LY, Gan S, Tan MSY, Lim SS, Lee XJ, Lam YF (2016a) Effective removal of Acid Blue 113 dye using overripe Cucumis sativus peel as an eco-friendly biosorbent from agricultural residue. J Clean Prod 113:194–203

Lee JH, Park MJ, Ryu HW, Yuk HJ, Choi SW, Lee KS, Kim SL, Seo WD (2016b) Elucidation of phenolic antioxidants in barley seedlings (Hordeum vulgare L.) by UPLC-PDA-ESI/MS and screening for their contents at different harvest times. J Funct Foods 26:667–680

Li SZ (1988) Xinhua compendium of material medica, vol 2. Shanghai Science & Technology Publishing House, Shanghai, p 310

Liu S, Zhang X, Wang J (2020) Isovitexin protects against cisplatin-induced kidney injury in mice through inhibiting inflammatory and oxidative responses. Int Immunopharmacol 83:106437

Lowe TB, Hatch BT, Antle C, Nartker S, Ammerman ML (2020) One- and two-stage anaerobic co-digestion of cucumber waste and sewage sludge. Environ Technol 41(24):3157–3165

Lu J, Meng Z, Cheng B, Liu M, Tao S, Guan S (2019) Apigenin reduces the excessive accumulation of lipids induced by palmitic acid via the AMPK signaling pathway in HepG2 cells. Exp Ther Med 18(4):2965–2971

Luikham S, Malve S, Gawali P, Ghosh S (2018) A novel strategy towards agro-waste mediated dye biosorption for water treatment. World J Pharm Res 7(4):197–208

Lv X, Guan H, Liu H, Hao R, Zhang W, Li F, Guo J, Jiang Y, Li D (2023) The anti-apoptosis effect of isovitexin on human keratinocytes by regulating miR-98-5p/Bcl-2/Bcl-xL and MAPKs/NF-κB signaling pathways. Food Chem Adv 2:100238

Marrassini C, Saint Martin EM, Alonso MDR, Anesini CA (2023) Vicenin-2 and vitexin participate in the in vitro modulation of the anti-inflammatory and antioxidant activities exerted by two *Urtica circularis* extracts. Latin Am Caribb Bull Med Aromatic Plants 23(1):48–58

McNally DJ, Wurms KV, Labbé C, Quideau S, Bélanger RR (2003) Complex C-glycosyl flavonoid phytoalexins from *Cucumis sativus*. J Nat Prod 66(9):1280–1283

Mežaka I, Kronberga A, Berga M, Kaļāne L, Pastare L, Skudriņš G, Nakurte I (2023) Biochemical and physiological responses of *Cucumis sativus* L. to application of potential bioinsecticides—Aqueous Carum carvi L. seed distillation by-product based extracts. Agriculture 13(5):1019

Min SY, Park CH, Yu HW, Park YJ (2021) Anti-inflammatory and anti-allergic effects of saponarin and its impact on signaling pathways of RAW 264.7, RBL-2H3, and HaCaT Cells. Int J Mol Sci 22(16):8431

Min K, Song K, Lim S, Yi G, Lee EJ (2023) Cucurbitacin and volatile compound profiling reveals independent domestication of cucumber (*Cucumis sativus* L.) fruit. Food Chem 405:135006

Mukherjee PK, Nema NK, Maity N, Sarkar BK (2013) Phytochemical and therapeutic potential of cucumber. Fitoterapia 84:227–236

Mukherjee PK, Singha S, Kar A, Chanda J, Banerjee S, Dasgupta B, Haldar PK, Sharma N (2022) Therapeutic importance of Cucurbitaceae: A medicinally important family. J Ethnopharmacol 282:114599

Murad H, Shamban AT, Premo PS (1995) The use of glycolic acid as a peeling agent. Dermatol Clin 13(2):285–307

Nasrin F, Bulbul IJ, Aktar F, Rashid MA (2015) Anti-inflammatory and antioxidant activities of Cucumis sativus leaves. Bangladesh Pharm J 18(2):169–173

Olennikov DN (2023) Separation, characterization and mammal pancreatic lipase inhibitory potential of cucumber flower flavonoids. Separations 10(4):255

Olennikov DN, Kashchenko NI (2023a) Green waste from cucumber (*Cucumis sativus* L.) cultivation as a source of bioactive flavonoids with hypolipidemic potential. Agronomy 13(9):2410

Olennikov DN, Kashchenko NI (2023b) New flavonoids from *Cucumis sativus*. Chem Nat Compd 59(4):651–654

Olennikov DN, Kashchenko NI (2023c) Acylated flavonoids from *Cucumis sativus* inhibit the activity of human pancreatic lipase. Appl Biochem Microbiol 59(4):530–538

Olennikov DN, Kashchenko NI (2024) New acylated *C, O*-glycosylflavones from *Cucumis sativus*. Chem Nat Compd 60(2):235–240

Omokhua-Uyi AG, Van Staden J (2020) Phytomedicinal relevance of South African Cucurbitaceae species and their safety assessment: A review. J Ethnopharmacol 259:112967

Pandey R, Ansari NG, Prasad RL, Murthy RC (2014) Removal of Cd (II) ions from simulated wastewater by HCl modified Cucumis sativus peel: Equilibrium and kinetic study. Air Soil Water Res 7:ASWR-S16488

Patanè GT, Lombardo L, Putaggio S, Tellone E, Ficarra S, Barreca D, Laganà G, De Luca L, Calderaro A (2023) Anti-aggregative and protective effects of vicenin-2 on heat and oxidative stress-induced damage on protein structures. Int J Mol Sci 24(24):17222

Qing ZX, Zhou Y, Liu XB, Cheng P, Zeng JG (2014) 23, 24-Dihydrocucurbitacin C: a new compound regarded as the next metabolite of cucurbitacin C. Nat Prod Res 28(15):1165–1170

Qing Z, Shi Y, Han L, Li P, Zha Z, Liu C, Liu X, Huang P, Liu Y, Tang Q, Zeng K (2022) Identification of seven undescribed cucurbitacins in *Cucumis sativus* (cucumber) and their cytotoxic activity. Phytochemistry 197:113123

Remigante A, Spinelli S, Straface E, Gambardella L, Caruso D, Falliti G, Dossena S, Marino A, Morabito R (2022) Antioxidant activity of quercetin in a H2O2-induced oxidative stress model in red blood cells: functional role of band 3 protein. Int J Mol Sci 23(19):10991

Rice CA, Rymal KS, Chambliss OL, Johnson FA (1981) Chromatographic and mass spectral analysis of cucurbitacins of three Cucumis sativus cultivars. J Agric Food Chem 29(1):194–196

Safaeian L, Emami R, Hajhashemi V, Haghighatian Z (2018) Antihypertensive and antioxidant effects of protocatechuic acid in deoxycorticosterone acetate-salt hypertensive rats. Biomed Pharmacother 100:147–155

Santhi T, Smitha T (2015) Adsorption of acid violet-17 from synthetic aqueous solution on two kinds of *Cucumis sativus* L. Peel adsorbent. Int J Res Chem Environ 5(2):35–43

Seo WD, Lee JH, Jia Y, Wu C, Lee SJ (2015) Saponarin activates AMPK in a calcium-dependent manner and suppresses gluconeogenesis and increases glucose uptake via phosphorylation of CRTC2 and HDAC5. Bioorg Med Chem Lett 25(22):5237–5242

Shakoor S, Nasar A (2019) Utilization of *Cucumis sativus* peel as an eco-friendly biosorbent for the confiscation of crystal violet dye from artificially contaminated wastewater. Anal Chem Lett 9(1):1–19

Shin S, Cho SH, Park D, Jung E (2020) Anti-skin aging properties of protocatechuic acid in vitro and in vivo. J Cosmet Dermatol 19(4):977–984

Sidoryk K, Jaromin A, Filipczak N, Cmoch P, Cybulski M (2018) Synthesis and antioxidant activity of caffeic acid derivatives. Molecules 23(9):2199

Simeonova R, Vitcheva V, Krasteva I, Zdraveva P, Konstantinov S, Ionkova I (2016) Antidiabetic and antioxidant effects of saponarin from *Gypsophila trichotoma* on streptozotocin-induced diabetic normotensive and hypertensive rats. Phytomedicine 23(5):483–490

Smitha T, Santhi T, Makeswari M (2015) Adsorption of acid black-7 from synthetic aqueous solution onto Cucumis sativus peel. J Chem Pharm Res 7(4):1617–1625

Smitha T, Santhi T, Prasad AL, Manonmani S (2017) Cucumis sativus used as adsorbent for the removal of dyes from aqueous solution. Arab J Chem 10:S244–S251

Stan M, Lung I, Soran ML, Leostean C, Popa A, Stefan M, Lazar MD, Opris O, Silipas TD, Porav AS (2017) Removal of antibiotics from aqueous solutions by green synthesized magnetite nanoparticles with selected agro-waste extracts. Process Saf Environ Prot 107:357–372

Stavrinou A, Aggelopoulos CA, Tsakiroglou CD (2018) Exploring the adsorption mechanisms of cationic and anionic dyes onto agricultural waste peels of banana, cucumber and potato: Adsorption kinetics and equilibrium isotherms as a tool. J Environ Chem Eng 6(6):6958–6970

Sun X, Li P, Qu X, Liu W (2021) Isovitexin alleviates acute gouty arthritis in rats by inhibiting inflammation via the TLR4/MyD88/NF-κB pathway. Pharm Biol 59(1):1324–1331

Tan WS, Arulselvan P, Ng SF, Mat Taib CN, Sarian MN, Fakurazi S (2019) Improvement of diabetic wound healing by topical application of Vicenin-2 hydrocolloid film on Sprague Dawley rats. BMC Complement Altern Med 19:1–16

Tan WS, Arulselvan P, Ng SF, Taib CNM, Sarian MN, Fakurazi S (2020) Healing effect of vicenin-2 (VCN-2) on human dermal fibroblast (HDF) and development VCN-2 hydrocolloid film based on alginate as potential wound dressing. Biomed Res Int 2020(1):4730858

Tang J, Meng X, Liu H, Zhao J, Zhou L, Qiu M, Zhang X, Yu Z, Yang F (2010) Antimicrobial activity of sphingolipids isolated from the stems of cucumber (*Cucumis sativus* L.). Molecules 15(12):9288–9297

Taqvi S, Bhat EA, Sajjad N, Sabir JS, Qureshi A, Rather IA, Rehman S (2021) Protective effect of vanillic acid in hydrogen peroxide-induced oxidative stress in D. Mel-2 cell line. Saudi J Biol Sci 28(3):1795–1800

Trejo-Moreno C, Méndez-Martínez M, Zamilpa A, Jiménez-Ferrer E, Perez-Garcia MD, Medina-Campos ON, Pedraza-Chaverri J, Santana MA, Esquivel-Guadarrama FR, Castillo A, Cervantes-Torres J (2018) *Cucumis sativus* aqueous fraction inhibits angiotensin II-induced inflammation and oxidative stress in vitro. Nutrients 10(3):276

Tseng CY, Yu PR, Hsu CC, Lin HH, Chen JH (2023) The effect of isovitexin on lipopolysaccharide-induced renal injury and inflammation by induction of protective autophagy. Food Chem Toxicol 172:113581

Uthpala TG, Marapana RAU, Lakmini KP, Wettimuny DC (2020) Nutritional bioactive compounds and health benefits of fresh and processed cucumber (*Cucumis sativus* L.). Sumerianz J Biotechnol 3(9):75–82

Venkatachalam P, Sangeetha P, Geetha N, Sahi SV (2015) Phytofabrication of bioactive molecules encapsulated metallic silver nanoparticles from *Cucumis sativus* L. and its enhanced wound healing potential in rat model. J Nanomater 2015(1):753193

Vilas-Boas AA, Pintado M, Oliveira AL (2021) Natural bioactive compounds from food waste: Toxicity and safety concerns. Foods 10(7):1564

Wahid S, Alqahtani A, Khan RA (2021) Analgesic and anti-inflammatory effects and safety profile of Cucurbita maxima and Cucumis sativus seeds. Saudi Journal of Biological Sciences, 28(8):4334–4341

Wahid M, Saqib F, Chicea L, Ahmedah HT, Sajer BH, Marc RA, Pop OL, Moga M, Gavris C (2022) Metabolomics analysis delineates the therapeutic effects of hydroethanolic extract of *Cucumis sativus* L. seeds on hypertension and isoproterenol-induced myocardial infarction. Biomed Pharmacother 148:112704

Wang J, Fang X, Ge L, Cao F, Zhao L, Wang Z, Xiao W (2018) Antitumor, antioxidant and anti-inflammatory activities of kaempferol and its corresponding glycosides and the enzymatic preparation of kaempferol. PLoS One 13(5):e0197563

Warrier PK (1994) Indian medicinal plants: a compendium of 500 species. Press Orientlongman, Chennai

Wehr R, Krochmal ML, Bagatelle F, Ragsdale WA (1986) Controlled two-centre study of lactate 12% lotion and a petrolatum-based cream in patients with xerosis. Cutis 37:205–207

Wu Y, Liang S, Zhang M, Wang Z, Wang Z, Ren X (2020) The effect of chlorogenic acid on Bacillus subtilis based on metabolomics. Molecules 25(18):4038

Wu L, Guo T, Deng R, Liu L, Yu Y (2021) Apigenin ameliorates insulin resistance and lipid accumulation by endoplasmic reticulum stress and SREBP-1c/SREBP-2 pathway in palmitate-induced HepG2 cells and high-fat diet-fed mice. The Journal of pharmacology and experimental therapeutics, 377(1):146–156

Yu JQ, Matsui Y (1994) Phytotoxic substances in root exudates of cucumber (Cucumis sativus L.). J Chem Ecol 20:21–31

Zhang C, Chen Y, Zhang M, Xu C, Gong G, Veeraraghavan VP, Bolla SR, Li Y (2020) Vicenin-2 treatment attenuated the Diethylnitrosamine-induced liver carcinoma and oxidative stress through increased apoptotic protein expression in experimental rats. J Environ Pathol Toxicol Oncol 39(2):113–123

Zhang Y, Qi Z, Wang W, Wang L, Cao F, Zhao L, Fang X (2021) Isovitexin inhibits ginkgolic acids-induced inflammation through downregulating SHP2 activation. Front Pharmacol 12:630320

Zhou XJ, Li XS, Shen Y, Pei G, Wang JF, Cheng YX (2012) Steroids and triterpenoids from Cucumis sativus roots. Chem Nat Compd 48:419–422

Zhu H, Zhao N, Jiang M (2021) Isovitexin attenuates tumor growth in human colon cancer cells through the modulation of apoptosis and epithelial-mesenchymal transition via PI3K/Akt/mTOR signaling pathway. Biochem Cell Biol 99(6):741–749

Phytochemicals from Eggplant (*Solanum melongena* L.) By-Products

Reham Hassan Mekky , Mostafa M. Hegazy, Sherouk Hussein Sweilam, Mohamed S. Abd El Hafeez, Ghadir Ali Sayed, Mahmoud A. Mansour, and María del Mar Contreras

Abstract Eggplant, a fundamental ingredient in vegetarian cuisine, has been utilized since ancient eras. It is a highly commercially viable crop in America, Africa, India, Indochina, European countries, and South-East Asia, with China being the leading producer. Eggplants possess significant nutritional content and can be beneficial in treating chronic conditions such as anemia, arthritis, atherosclerosis, asthma, bronchitis, malignancies, cholera, coronary heart disease, hypertension, gastrointestinal difficulties, and the prevention of metabolic syndrome. Their composition includes alkaloids, carotenoids, phenolic compounds, flavonoids, tannins, saponins, steroids, fibers, minerals, vitamins, and antioxidants. In their natural state, eggplants possess a bitter flavor and are extensively utilized as a food source for different culinary applications. There is significant interest in the phytochemical makeup of egg-

R. H. Mekky (✉) · M. S. Abd El Hafeez · M. A. Mansour
Department of Pharmacognosy, Faculty of Pharmacy, Egyptian Russian University, Badr City, Cairo, Egypt
e-mail: reham-mekky@eru.edu.eg

M. M. Hegazy
Department of Pharmacognosy and Medicinal Plants, Faculty of Pharmacy, Al-Azhar University, Cairo, Egypt

Department of Pharmacognosy, Faculty of Pharmacy, Sinai University - Arish Branch, Arish, Egypt

S. H. Sweilam
Department of Pharmacognosy, Faculty of Pharmacy, Egyptian Russian University, Badr City, Cairo, Egypt

Department of Pharmacognosy, College of Pharmacy, Prince Sattam Bin Abdulaziz University, Al-Kharj, Saudi Arabia

G. A. Sayed
Department of Biochemistry, Faculty of Pharmacy, Egyptian Russian University, Badr City, Cairo, Egypt

M. del Mar Contreras
Department of Chemical, Environmental and Materials Engineering, Universidad de Jaén, Jaén, Spain

M. F. Ramadan (ed.), *Bioactive Phytochemicals in By-products from Bulb, Flower and Fruit Vegetables*, https://doi.org/10.1007/978-3-031-77399-0_10

plant by-products, which have been used in Africa for over 2000 years for culinary and medicinal purposes. These qualities have been demonstrated to be effective in treating diabetes and hypertension. Nevertheless, there is a lack of data supporting the utilization of these by-products as food, leading to a substantial depletion of vital bioactive food. In addition, research projects examine how these substances can enhance the value of different food items, including jelly crystals, custard, cookies, bread, gluten-free spaghetti, yogurt-like goods, and gummy candies.

Keywords Eggplant by-products · Calyx · Peels · Agri-food residue · Valorization

List of Abbreviations

Aae	Ascorbic acid equivalent
Abts	2,2′-azino-bis(3-ethylbenzothiazoline-6-sulfonic acid)
Akt	Protein kinase b
Cox-2	Cyclooxygenase-2
Dad	Diode array detector
Dpph	1,1-diphenyl-2-picrylhydrazyl
Erk	Extracellular signal-regulated kinase
Gc	Gas chromatography
Hld	High-density lipoproteins
Hmp	High methoxyl pectin
Hplc	High performance liquid chromatography
Ic50	Half maximal inhibitory concentration
Inos	Inducible nitric oxide synthase
Jnk	C-jun n-terminal kinases
Ldl	Low-density lipoproteins
Mapk	Mitogen-activated protein kinase
Mapk	Mitogen-activated protein kinase
Mbc	Minimum bactericidal concentration
Mic	Minimum inhibitory concentration
Ms	Mass spectrometry
No	Nitric oxide
P-bad	Phosphorylated bad pathway
Pge2	Prostaglandin e2
Pi3k	Phosphoinositide 3-kinase
Ros	Reactive oxygen species
Rp	Reducing power
Sle	Solid–liquid extraction
Sod	Superoxide dismutase
TBHQ	Tertiary butyl hydroquinone
Uae	Ultrasound-assisted extraction
Uhrf1	Ubiquitin-like with PHD and ring finger domains 1
Vldl	Very low-density lipoproteins

1 Introduction

Medicinal plants belonging to the genus *Solanum* represent nearly one-half of the Solanaceae family (Sękara et al. 2007). Despite the fact that the green parts and unripe fruits of *Solanum* genus plants are considered toxic to humans, they are among the most valuable and eco-edible vegetables (Shaikh et al. 2022). The plants of the *Solanum* genus can be divided into 13 clades (Taher et al. 2017), where the edible crops are considered one of the essential crops in different developing countries, i.e., potato (*S. tuberosum* L.), tomato (*S. lycopersicum* L), and eggplants (African/Middle Eastern species, *S. incanum* L., and *S. macrocarpon* L.) (Chapman 2019; Sękara et al. 2007).

Significantly, eggplant is an essential component in vegetarian food because it has a unique texture and potent activities, so it is known as a king of vegetables or black magic (Shaikh et al. 2022). It is relevant from the past to today. For example, in the third century B.C., in India, southern Asia, and South Africa, eggplants were known as Brinjal, Varttaka, Bhantaki, and Nattingan (Chapman 2019; Sękara et al. 2007; Shaikh et al. 2022), and in the fourth and ninth century in China and Africa, respectively, the eggplant began to be characterized, in the Arabic world known as Melongena from the fifteenth to sixteenth century (Sękara et al. 2007; Shaikh et al. 2022). Other names are aubergine (Chapman 2019) and guinea squash in the United Kingdom and France (Rasouli et al. 2023). In the fifteenth century, it was recently investigated in Poland and by Avicenna as a medicinal food (Sękara et al. 2007). Eggplant is one of the most economically edible crops in America, Africa, India, Indochina, European countries, and South-East Asia (Mibei et al. 2017; Rasouli et al. 2023; Sękara et al. 2007). There are many native species of eggplant distributed in Africa, including *S. aethiopicum* L., *S. macrocarpon* L., and *S. incanum* L. (Chapman 2019; Rasouli et al. 2023; Taher et al. 2017). Different varieties of eggplant are classified based on the fruit shape into three main types: dwarf, egg-shaped, and long slender type (Yarmohammadi et al. 2021). Approximately fifty million and a half tons of eggplant vegetables are produced annually in tropical and subtropical weather in only five countries; China ranks first among them with 58% of the total output, followed by India, then Iran and Egypt, and fifth by Turkey (Mainali 2014; Salamatullah et al. 2021a). Nowadays, most countries cultivate different species of eggplants with a significant variability in the shape, size, and taste of their fruits, leaves, and flowers, resulting from hybridization, mutation, or human selection (Rasouli et al. 2023; Sękara et al. 2007). Earlier studies established that various parts of eggplants, such as leaves and fruits, are used for cooking. Literature reports provide evidence of their high nutritional value and potential to treat chronic disorders, including anemia (Yarmohammadi et al. 2021), arthritis (Yarmohammadi et al. 2021), atherosclerosis (Singh et al. 2019), asthma and bronchitis (Yarmohammadi et al. 2021), cancers "coronary and prostate" (Mibei et al. 2017; Salamatullah et al. 2021a), cholera (Singh et al. 2019), coronary heart disease (Mibei et al. 2017; Salamatullah et al. 2021a), hypertension (Salamatullah et al. 2021a), gastrointestinal issues (Chapman 2019) such as aiding digestion (Shaikh et al. 2022), cholesterol control (Yarmohammadi et al. 2021), diabetes mellitus (Mibei et al. 2017; Salamatullah et al. 2021a; Yarmohammadi et al. 2021), neuralgia (Singh et al. 2019), nasal ulcers (Singh et al. 2019), and prevent the occurrence of metabolic syndrome—a condition characterized by dyslipidemia, insulin resistance, and obesity (Condurache et al. 2021; Yarmohammadi et al. 2021).

The phytochemical isolations significantly revealed the presence of micro- and nutritional primary and secondary metabolites such as:

- Alkaloids (Mibei et al. 2017)
- Carotenoids: alpha-carotene, lutein, and lycopene (Mibei et al. 2017; Sękara et al. 2007; Shaikh et al. 2022)
- Phenolic compounds (Mibei et al. 2017), such as anthocyanins in the reddish-purple pigment called nasunin (70–90%), or delphinidin and its glycosides, are the most abundant pigments in peel part (Chapman 2019; Condurache et al. 2021; Salamatullah et al. 2021a; Shaikh et al. 2022; Yarmohammadi et al. 2021), and in the pulp part delphinidin is the predominant pigment (Yarmohammadi et al. 2021). They also contain phenolic acids, such as hydroxycinnamic acid (Chapman 2019) and chlorogenic acid (Salamatullah et al. 2021a); flavonoids like kaempferol (Condurache et al. 2021; Shaikh et al. 2022) and myricetin (Shaikh et al. 2022); and tannins (Mibei et al. 2017), which validate their medicinal purposes (Chapman 2019; Mibei et al. 2017; Rasouli et al. 2023)
- Saponins (Mibei et al. 2017; Sękara et al. 2007)
- Steroids (Mibei et al. 2017)
- Fibers: cellulose, hemicellulose, lignin, and pectin (Salamatullah et al. 2021a; Shaikh et al. 2022)
- Minerals (mainly, iron, calcium, copper, potassium, and magnesium) (Salamatullah et al. 2021a; Shaikh et al. 2022)
- Vitamins: vitamin C (Mainali 2014; Salamatullah et al. 2021a; Shaikh et al. 2022), vitamins B1 and B6 (Shaikh et al. 2022), vitamin A (Shaikh et al. 2022), and folates (Shaikh et al. 2022)

The raw fruits of the eggplants are considered the edible part, and they are widely used as food in different ways (boiling with water, frying, or grilling) with a bitter taste (Salamatullah et al. 2021a; Sękara et al. 2007). However, there is high interest in phenolic compounds from eggplant, especially from the by-products, to widen the utility of this plant. For example, the pulp was used alone or with the by-product peel part as food, which is the richest part with the antioxidant "nasunin" (Shaikh et al. 2022), so the disposal of eggplant peels as waste represents a massive loss of bioactive material (Karimi et al. 2021). Understandably, the correlation between the whole fruit fresh or its skin with its biological activities: antimicrobial, antimutagenicity, antioxidant, and antiviral activities; antidiabetic and antihypertensive activities are due to the presence of high percentage of chlorogenic acid (70–90%) and anthocyanins (Delphinidin), respectively (Condurache et al. 2021; Salamatullah et al. 2021a; Shaikh et al. 2022; Yarmohammadi et al. 2021). The antioxidant activity of the eggplant's anthocyanins is decreased or increased by cooking, cultivar, and environmental factors (Salamatullah et al. 2021b). Traditional or recent extraction methods can be used to exhibit the phenolic compounds from the raw materials "by-products" (Condurache et al. 2021). The antioxidant bioactive components are distributed in many parts of eggplant by-products such as calyx, leaves, pulp, peels, seeds, and stems (Salamatullah et al. 2021a; Shaikh et al. 2022; Yarmohammadi et al. 2021). Indeed, agri-food residues remain a subject of interest for researchers since they serve as a valuable source of bioactive metabolites. These metabolites can play a crucial role in promoting sustainability in both the food and non-food industries (Abouzed et al. 2018; Hegazy et al. 2023; Mekky et al. 2019, 2022, 2023a, b).

The carotenoids are also interesting: lycopene is essential in its antioxidant biological potentiality (diabetes mellitus and prostate cancer (Mibei et al. 2017; Shaikh et al. 2022)). Dietary fibers present in high concentrations in eggplant fruit increase the absorption of blood cholesterol through the liver to treat obesity and improve gastrointestinal activity in humans (Shaikh et al. 2022). Moreover, peel and calyx are by-products of eggplant and are an incredibly high source of fibers; populations didn't consume their benefits and discard them (Karimi et al. 2021). Feeding eggplant to pregnant women improves infant and bone health, prevents anemia, and is safe for pregnant women because it contains folates, calcium, iron, and copper elements (Shaikh et al. 2022).

Noticeably, the by-product leaves of the eggplants are used as green vegetables in a raw or cooked way (Sękara et al. 2007). In Africa, both the fruit and the leaves of eggplant local species have been used for more than 2000 years for food and treatment purposes (Chapman 2019). In 2007, nipple seeds of eggplants from Cheras (Malaysia) showed the highest antioxidant activity among different varieties (Shaikh et al. 2022). Otherwise, the calyx by-product and the leaves showed the highest total phenolic content in another study, whereas the peel exhibited the maximum anthocyanin content (Shaikh et al. 2022). Moreover, the leaves seem to be a good source of kaempferol (Condurache et al. 2021). It has recently been shown that by-products (roots, skin, pulp, leaves, or purple eggplant extracts) exhibited antidiabetic properties by decreasing alpha-glucosidase activity or blood sugar levels. Another study revealed that the calyx and unripe fruits showed antihypertensive activity by reducing blood pressure levels (Yarmohammadi et al. 2021). There is no evidence of the use of fruit-clasping calyxes, flowers, seeds, or unripe fruits as food; instead they are discarded as waste products (Karimi et al. 2021; Sękara et al. 2007) and discarded of these raw materials is considered a massive loss of a great valuable bioactive food (Karimi et al. 2021).

Owing to the diversity and relevance of bioactive compounds in eggplant by-products, the focus of this chapter book is to provide a valuable review of these compounds but focused on the species: *S. incanum* L., *S. melongena* L., *S. macrocarpon* L., and *S. aethiopicum* L. Their biological properties are also reported to provide a vast compendium about this plant.

2 Bioactive Compounds Reported in Eggplants

Several classes of secondary metabolites were detected in eggplant by-products, which indicate their high medicinal value. Antioxidant and anticancer activities of eggplant by-products may be attributed to them (Contreras-Angulo et al. 2022).

2.1 Phenolic Compounds

Eggplants contain a considerable quantity of phenolic chemicals. Phenolic compounds are considered the most abundant compounds extracted from eggplant by-products. For example, 30–60% of these phenolic compounds are in by-products, particularly peels (Mauro et al. 2020). Producing these compounds during plant growth includes protection from UV radiation and defense mechanisms against certain diseases (Gürbüz et al. 2018). The amount and composition of eggplant phenolics and the percentage of concentrated phenolic compounds could vary depending on the agronomic condition of the plant (plant species, cultivar, developmental stage) and environmental factors (such as climate and season factors, light and water availability, and pH).

This work has reviewed and classified phenolic compounds into flavonoids and phenolic acids, considering each metabolite name, molecular formula, part used, occurrence, geographical origins, extraction, and analytical method. They are listed in Tables 1 and 2, and their chemical structure is shown in Figs. 1 and 2.

Flavonoids

Most reported studies focused on flavonoids, especially anthocyanin, as they possess valuable biological activities. Anthocyanins can be used as natural food colorants and extracted from various flower petals, leaves, fruits, and vegetables. The antioxidant activity of anthocyanins depends on their glycosylation and acylation patterns, the number of hydroxyl and methyl groups in their structure, and catechol moiety and oxonium ion in their B and C rings, respectively. It is reported that among six common anthocyanidins found in nature, delphinidin possesses the highest antioxidant potential (Karimi et al. 2021).

Phenolic Acids

Phenolic acids have significant antioxidant activity due to their interaction with reactive oxygen and nitrogen species. The reduction of reactive species is the main biological activity of phenolic acids. The predominant phenolic compound found in all eggplant varieties is chlorogenic acid. Phenolic compounds are known to display a wide range of biological activities, such as anti-inflammatory, antidiabetic, antimicrobial, or anti-carcinogenic activities (Abd El Hafeez et al. 2022; Abd El Hafeez 2024; Horincar et al. 2020a).

Table 1 Reported flavonoids in *Solanum melongena* L. eggplant by-products

No	Compound	Molecular formula	Part	Occurrence	Geographical origin	Extraction	Method of analysis	Reference
1	[(2R,3R,4S,5S,6R)-4,5-dihydroxy-2-[5-hydroxy-2-(4-hydroxyphenyl)-4-oxo-7-[(2S,3R,4S,5S,6R)-3,4,5-trihydroxy-6-(hydroxymethyl)oxan-2-yl]oxychromen-3-yl]oxy-6-(hydroxymethyl)oxan-3-yl] (E)-3-(4-hydroxyphenyl)prop-2-enoate	$C_{36}H_{36}O_{18}$	Leaves	American eggplants	Mexico	Solid–liquid extraction	HPLC–MS	Contreras-Angulo et al. (2022)
2	Apigenin	$C_{15}H_{10}O_5$	Peels	Local market	Iraq	Solid–liquid extraction	HPLC	Kareem and Naji (2022)
3	Catechin	$C_{15}H_{14}O_6$	Peels	Local market	Iraq	Solid–liquid extraction	HPLC	Kareem and Naji (2022)
4	Cyanidin-3-*O*-rutinoside	$C_{27}H_{31}O_{15}$	Peels	Dragaica	Romania	Ultrasound-assisted extraction	HPLC	Condurache et al. (2021)
				Commercial market in Compiègne	France			Ferarsa et al. (2018)

(continued)

Table 1 (continued)

No	Compound	Molecular formula	Part	Occurrence	Geographical origin	Extraction	Method of analysis	Reference
5	Delphinidin-3-*O*-glucoside	$C_{21}H_{21}O_{12}$	Peels	Dragaica	Romania	Ultrasound-assisted extraction	HPLC	Condurache et al. (2021)
				Local name of *Muktakeshi*	India	Solid–liquid extraction	HPLC–MS	Chatterjee et al. (2013)
6	Delphinidin-3-*O*-rutinoside	$C_{27}H_{31}O_{16}$	Peels	Dragaica	Romania	Ultrasound-assisted extraction	HPLC	Condurache et al. (2021)
				Commercial market in Compiègne	France	Ultrasound-assisted extraction	HPLC	Ferarsa et al. (2018)
				Local supermarkets from Orihuela	Spain	Solid–liquid extraction	HPLC	Hernández-Herrero and Frutos (2011)
				Local name of *Muktakeshi*	India	Solid–liquid extraction	HPLC–MS	Chatterjee et al. (2013)
				Local market	Portugal	Ultrasound-assisted extraction	HPLC-DAD	Silva et al. (2021)
7	Delphinidin-3,5-diglucoside	$C_{27}H_{30}O_{17}$	Peels	Local name of *Muktakeshi*	India	Solid–liquid extraction	HPLC–MS	Chatterjee et al. (2013)
8	Isorhamnetin-3-*O*-glucoside	$C_{22}H_{22}O_{12}$	Stems Leaves Roots	American eggplants	Mexico	Solid–liquid extraction	LC-MS	Contreras-Angulo et al. (2022)
9	Kaempferol	$C_{15}H_{10}O_6$	Peels	Local market	Iraq	Solid–liquid extraction	HPLC	Kareem and Naji (2022)
10	Kaempferol-3,7-diglucoside	$C_{27}H_{30}O_{16}$	Peels	Local market	Egypt	Solid–liquid extraction	HPLC-MS	Mohamed et al. (2019)

No	Compound	Molecular formula	Part	Occurrence	Geographical origin	Extraction	Method of analysis	Reference
11	Kaempferol-3-rutinoside	$C_{27}H_{30}O_{15}$	Peels			Solid–liquid extraction	LC-DAD-MS	Manousaki et al. (2016)
12	Kaempferol-3-galactoside	$C_{21}H_{20}O_{11}$	Peels	Local market	Egypt	Solid–liquid extraction	HPLC-MS	Mohamed et al. (2019)
13	Kaempferol-3-glucoside Astragalin	$C_{21}H_{20}O_{11}$	Peels	Local market	Egypt	Solid–liquid extraction	HPLC-MS	Mohamed et al. (2019)
		$C_{21}H_{20}O_{11}$	Stems Leaves Roots	American eggplants	Mexico	Solid–liquid extraction	LC-MS	Contreras-Angulo et al. (2022)
14	Kaempferol-7-neohesperidoside	$C_{27}H_{30}O_{15}$	Stems Leaves Roots	American eggplants	Mexico	Solid–liquid extraction	LC-MS	Contreras-Angulo et al. (2022)
15	Malvidin-3-rutinoside-5-glucoside	$C_{35}H_{45}O_{21}+$	Peels	Local market	France	Ultrasound-assisted extraction	HPLC	Ferarsa et al. (2018)
16	Myricetin-3-neohesperoidoside	$C_{27}H_{30}O_{17}$	Peels	Local market	Egypt	Solid–liquid extraction	HPLC-MS	Mohamed et al. (2019)
17	Myricetin-3-*O*-galactoside	$C_{21}H_{20}O_{13}$	Peels	Local market	Egypt	Solid–liquid extraction	HPLC-MS	Mohamed et al. (2019)
18	Petunidin 3-*O*-rutinoside	$C_{28}H_{33}O_{16}+$	Peels	Dragaica	Romania	Ultrasound-assisted extraction	HPLC	Condurache et al. (2021)
				Local market	France	Ultrasound-assisted extraction	HPLC	Ferarsa et al. (2018)
19	Quercetin	$C_{15}H_{10}O_7$	Peels	Local market	Pakistan	Ultrasound-assisted extraction	HPLC	Kainat et al. (2023)
				Local market	Iraq	Solid–liquid extraction	HPLC	Kareem and Naji (2022)
20	Quercetin-3-rutinoside	$C_{27}H_{30}O_{16}$	Peels	Local market	Egypt	Solid–liquid extraction	HPLC-MS	Mohamed et al. (2019)

(continued)

No	Compound	Molecular formula	Part	Occurrence	Geographical origin	Extraction	Method of analysis	Reference
21	Quercetin-3-diglucoside	$C_{27}H_{30}O_{17}$	Peels	Local market	Egypt	Solid–liquid extraction	HPLC-MS	Mohamed et al. (2019)
22	Quercetin-3-galactoside	$C_{21}H_{20}O_{12}$	Peels	Local market	Egypt	Solid–liquid extraction	HPLC-MS	Mohamed et al. (2019)
23	Quercetin-3-*O*-glucoside	$C_{21}H_{20}O_{12}$	Peels	Local market	Egypt	Solid–liquid extraction	HPLC-MS	Mohamed et al. (2019)
24	Quercetin-3-*O*-rhamnoside	$C_{21}H_{20}O_{11}$	Peels	Local market	Egypt	Solid–liquid extraction	HPLC-MS	Mohamed et al. (2019)
25	Rutin	$C_{27}H_{30}O_{16}$	Peels	Local market	Iraq	Solid–liquid extraction	HPLC	Kareem and Naji (2022)
			Stems Leaves	American eggplants	Mexico	Solid–liquid extraction	HPLC-MS	Contreras-Angulo et al. (2022)

DAD diode array detection; *HPLC* high-performance liquid chromatography; *LC* liquid chromatography; MS mass spectrometry

Table 2 Reported phenolics in *Solanum melongena L.* eggplant by-products

	Compound	Molecular formula	Part used	Occurrence	Geographical origin	Extraction after defatting and/or oil pressing	Method of analysis	Reference
26	Gallic acid	$C_7H_6O_5$	Peel	Local market of Lahore district	Pakistan	Ultrasound-assisted extraction	HPLC	Kainat et al. (2023)
27	Benzoic acid	$C_7H_6O_2$	Peel	Local market of Lahore district	Pakistan	Ultrasound-assisted extraction	HPLC	Kainat et al. (2023)
28	Chlorogenic acid	$C_{16}H_{18}O_9$	Peel	Local market of Lahore district	Pakistan	Ultrasound-assisted extraction	HPLC	Kainat et al. (2023)
				Local market	China	Solid–liquid extraction	HPLC	Manousaki et al. (2016)
29	Syringic acid	$C_9H_{10}O_5$	Peel	Local market of Lahore district	Pakistan	Ultrasound-assisted extraction	HPLC	Kainat et al. (2023)
30	*p*-Coumaric acid	$C_9H_8O_3$	Peel	Local market of Lahore district	Pakistan	Ultrasound-assisted extraction	HPLC	Kainat et al. (2023)
				Local market	China	Solid–liquid extraction	HPLC	Manousaki et al. (2016)
31	Sinapic acid	$C_{11}H_{12}O_5$	Peel	Local market of Lahore district	Pakistan	Ultrasound-assisted extraction	HPLC	Kainat et al. (2023)
32	*p*-Coumaroylquinic acid	$C_{16}H_{18}O_8$	Peel	Local market	China	Solid–liquid extraction	HPLC	Manousaki et al. (2016)
33	di-Caffeoylquinic acid	$C_{25}H_{24}O_{12}$	Peel	Local market	China	Solid–liquid extraction	HPLC	Manousaki et al. (2016)

HPLC high-performance liquid chromatography

2.2 *Terpenes and Terpenoids*

Table 3 describes the main terpene and terpenoids in eggplant leaves and other volatile compounds. Mainly, they consisted of monoterpenes, sesquiterpenes, and oxygenated derivatives (Bouhajeb et al. 2020). The presence of the sesquiterpenoid 2-naphthaleneacetic acid, decahydro-8-hydroxy-4a,8-dimethyl-alpha-methylene was reported in a root extract (Table 4) (Contreras-Angulo et al. 2022). Their chemical structures are found in Figs. 3 and 4, respectively.

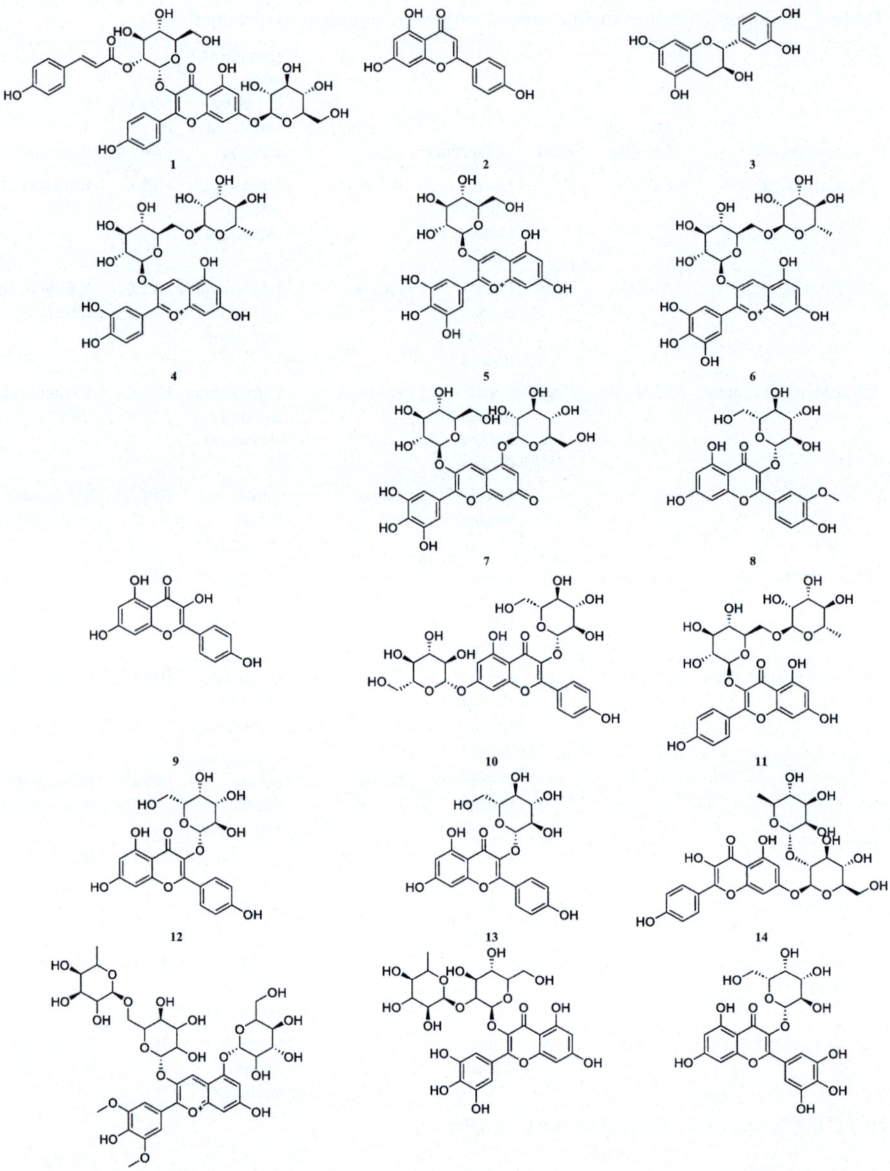

Fig. 1 Flavonoid compounds previously reported in by-products of *Solanum melongena* L. eggplant

2.3 Other Compounds

Other compound types have been observed in eggplant by-products. They were grouped into steroid components (Table 4), carbohydrates, amino acids, peptides,

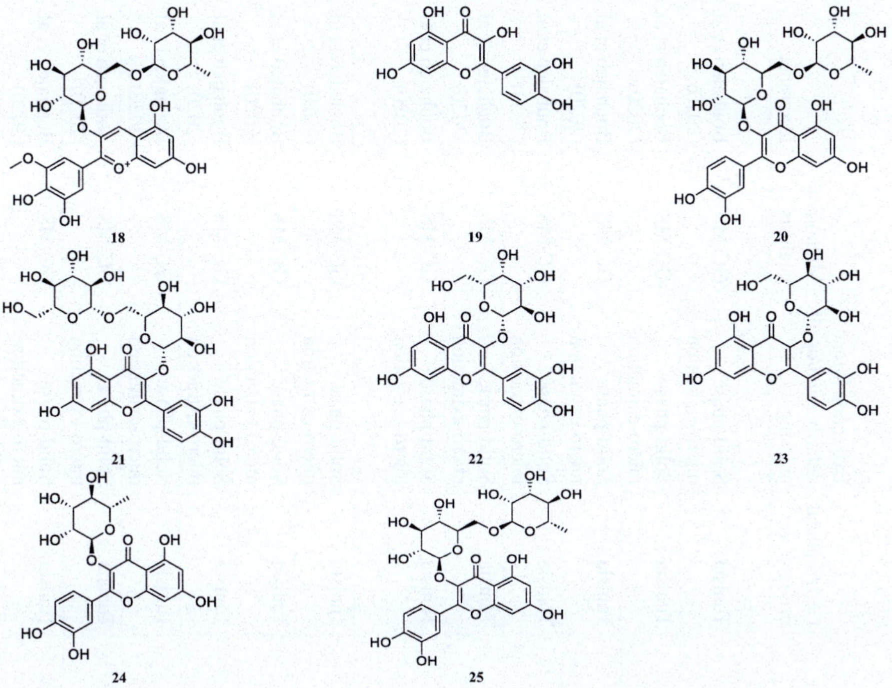

Fig. 10.1 (continued)

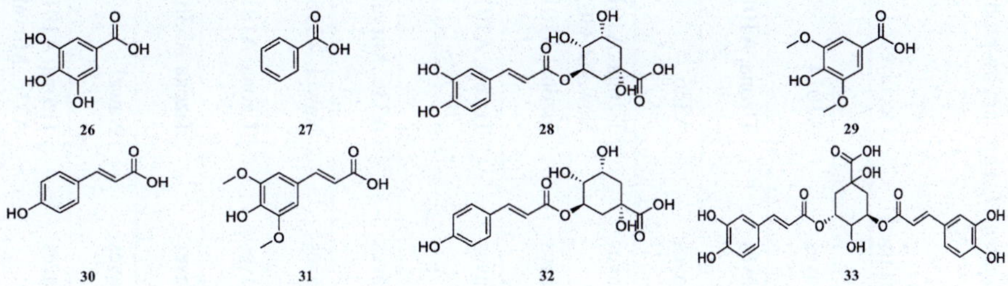

Fig. 2 Phenolic compounds previously reported in by-products of *Solanum melongena*

nucleotides, organic acids, and other derived compounds (Tables 4, 5, and 6; Figs. 4, 5, and 6).

Moreover, eggplant peels contain pectins with 60.2% of galacturonic acid (GalA). Arabinose is the predominant neutral monosaccharide with a content of 27.8%, followed by rhamnose (7.4%), galactose (2.3%), xylose (0.7%), and fructose (0.7%). This would suggest that galactans and arabinans are the most prevalent side chains (Kazemi et al. 2019a). It is reported that the pectin in eggplant calyxes contains 66.8–69.7% GalA. Among the neutral monosaccharides in the structure of EPP, galactose (18.1%) is the most abundant, followed by rhamnose (8.8%), arabinose (2.8%), and fructose (0.7%). This would suggest that galactans and arabinans, respectively, are the most prevalent side chains (Kazemi et al. 2019a)

Table 3 Reported volatile oil in *Solanum melongena L.* eggplant by-products

	Compound	Molecular formula	Part used	Occurrence	Geographical origin	Extraction after defatting and/or oil pressing	Method of analysis	Reference
(A) Monoterpenes								
34	*α*-Pinene	$C_{10}H_{16}$	Leaves	A1014	Tunisia	Solid-phase micro-extraction	GC-MS	Bouhajeb et al. (2020)
35	*β*-Pinene	$C_{10}H_{16}$	Leaves	Tizona, A1014	Tunisia	Solid-phase micro-extraction	GC-MS	Bouhajeb et al. (2020)
36	α-Terpinene	$C_{10}H_{16}$	Leaves	A336	Tunisia	Solid-phase micro-extraction	GC-MS	Bouhajeb et al. (2020)
37	Limonene	$C_{10}H_{16}$	Leaves	Tizona ,A1014, Nawel,A336,Mirval	Tunisia	Solid-phase micro-extraction	GC-MS	Bouhajeb et al. (2020)
38	γ-Terpinene	$C_{10}H_{16}$	Leaves	Tizona, A1014	Tunisia	Solid-phase micro-extraction	GC-MS	Bouhajeb et al. (2020)
39	Terpinolene	$C_{10}H_{16}$	Leaves	Tizona, A1014	Tunisia	Solid-phase micro-extraction	GC-MS	Bouhajeb et al. (2020)
(B) Oxygenate monoterpenes								
40	1,8-Cineole	$C_{10}H_{18}$	Leaves	A336, Mirval	Tunisia	Solid-phase micro-extraction	GC-MS	Bouhajeb et al. (2020)
41	trans-Sabinene hydrate	$C_{10}H_{18}O$	Leaves	Nawel, A336	Tunisia	Solid-phase micro-extraction	GC-MS	Bouhajeb et al. (2020)
42	Linalool	$C_{10}H_{18}O$	Leaves	Tizona, A1014	Tunisia	Solid-phase micro-extraction	GC-MS	Bouhajeb et al. (2020)
43	Camphor	$C_{10}H_{16}O$	Leaves	Tizona,A336,Mirval	Tunisia	Solid-phase micro-extraction	GC-MS	Bouhajeb et al. (2020)
44	cis-Dihydro carvone	$C_{10}H_{16}O$	Leaves	Nawel	Tunisia	Solid-phase micro-extraction	GC-MS	Bouhajeb et al. (2020)
45	Carvone	$C_{10}H_{14}O$	Leaves	Tizona, A1014, Nawel, A336, Mirval	Tunisia	Solid-phase micro-extraction	GC-MS	Bouhajeb et al. (2020)

Table 3 (continued)

	Compound	Molecular formula	Part used	Occurrence	Geographical origin	Extraction after defatting and/or oil pressing	Method of analysis	Reference
46	Dihydrolinalyl acetate	$C_{12}H_{22}O_2$	Leaves	Tizona, Mirval	Tunisia	Solid-phase micro-extraction	GC-MS	Bouhajeb et al. (2020)
(C) Sesquiterpene hydrocarbons								
47	α-Copaene	$C_{15}H_{24}$	Leaves	Tizona	Tunisia	Solid-phase micro-extraction	GC-MS	Bouhajeb et al. (2020)
48	Safranal	$C_{10}H_{14}O$	Leaves	Tizona, A1014, Nawel, A336, Mirval	Tunisia	Solid-phase micro-extraction	GC-MS	Bouhajeb et al. (2020)
49	*α*-Cyclocitral	$C_{10}H_{16}O$	Leaves	A1014, Nawel, A336	Tunisia	Solid-phase micro-extraction	GC-MS	Bouhajeb et al. (2020)
50	*β*-Cyclocitral	$C_{10}H_{16}O$	Leaves	Tizona, A1014, Nawel, A336, Mirval	Tunisia	Solid-phase micro-extraction	GC-MS	Bouhajeb et al. (2020)
51	β-Cyclohomocitral	$C_{11}H_{18}O$	Leaves	Tizona, Nawel, A336, Mirval	Tunisia	Solid-phase micro-extraction	GC-MS	Bouhajeb et al. (2020)
52	(E)-Geranylacetone	$C_{13}H_{22}O$	Leaves	Tizona, A1014, Nawel, A336	Tunisia	Solid-phase micro-extraction	GC-MS	Bouhajeb et al. (2020)
(D) Nitrogen derivatives								
53	Methylpyrazine	$C_5H_6N_2$	Leaves	A336, Mirval	Tunisia	Solid-phase micro-extraction	GC-MS	Bouhajeb et al. (2020)
54	2,6-Lutidine	C_7H_9N	Leaves	A336	Tunisia	Solid-phase micro-extraction	GC-MS	Bouhajeb et al. (2020)
55	2,5-Dimethylpyrazine	$C_6H_8N_2$	Leaves	A336, Mirval	Tunisia	Solid-phase micro-extraction	GC-MS	Bouhajeb et al. (2020)
56	2,6-Dimethylpyrazine	$C_6H_8N_2$	Leaves	A336, Mirval	Tunisia	Solid-phase micro-extraction	GC-MS	Bouhajeb et al. (2020)
57	2-Ethyl-3-methyl pyrazine	$C_7H_{10}N_2$	Leaves	A336	Tunisia	Solid-phase micro-extraction	GC-MS	Bouhajeb et al. (2020)

(continued)

Table 3 (continued)

	Compound	Molecular formula	Part used	Occurrence	Geographical origin	Extraction after defatting and/or oil pressing	Method of analysis	Reference
(E) Non-terpene derivatives								
58	(E)-2-Penten-1-ol	$C_5H_{10}O$	Leaves	A1014	Tunisia	Solid-phase micro-extraction	GC-MS	Bouhajeb et al. (2020)
59	1,3-Butanediol	$C_4H_{10}O_2$	Leaves	Tizona	Tunisia	Solid-phase micro-extraction	GC-MS	Bouhajeb et al. (2020)
60	Hexanal	$C_6H_{12}O$	Leaves	Tizona, A1014, Nawel, A336, Mirval	Tunisia	Solid-phase micro-extraction	GC-MS	Bouhajeb et al. (2020)
61	Isovaleric acid	$C_5H_{10}O_2$	Leaves	Tizona, Mirval	Tunisia	Solid-phase micro-extraction	GC-MS	Bouhajeb et al. (2020)
62	Ethyl 2-methylbutyrate	$C_7H_{14}O_2$	Leaves	Tizona, Mirval	Tunisia	Solid-phase micro-extraction	GC-MS	Bouhajeb et al. (2020)
63	(E)-2-Hexenal	$C_6H_{10}O$	Leaves	Tizona, Nawel, A336	Tunisia	Solid-phase micro-extraction	GC-MS	Bouhajeb et al. (2020)
64	(Z)-3-Hexen-1-ol	$C_6H_{12}O$	Leaves	Tizona, A014, Mirval	Tunisia	Solid-phase micro-extraction	GC-MS	Bouhajeb et al. (2020)
65	1-Hexanol	$C_6H_{14}O$	Leaves	A1014, Nawel, A336, Mirval	Tunisia	Solid-phase micro-extraction	GC-MS	Bouhajeb et al. (2020)
66	Heptanal	$C_7H_{14}O$	Leaves	Tizona, A1014, Nawel, A336, Mirval	Tunisia	Solid-phase micro-extraction	GC-MS	Bouhajeb et al. (2020)
67	(Z)-3-Hexenyl formate	$C_7H_{12}O_2$	Leaves	Tizona	Tunisia	Solid-phase micro-extraction	GC-MS	Bouhajeb et al. (2020)
68	3-Methyl butyrolactone	$C_5H_8O_2$	Leaves	A1014	Tunisia	Solid-phase micro-extraction	GC-MS	Bouhajeb et al. (2020)
69	Benzaldehyde	C_7H_6O	Leaves	Tizona, A1014, Nawel, A336	Tunisia	Solid-phase micro-extraction	GC-MS	Bouhajeb et al. (2020)

Table 3 (continued)

	Compound	Molecular formula	Part used	Occurrence	Geographical origin	Extraction after defatting and/or oil pressing	Method of analysis	Reference
70	6-Methyl-5-hepten-2-one	$C_8H_{14}O$	Leaves	Tizona, A1014, Nawel, A336, Mirval	Tunisia	Solid-phase micro-extraction	GC-MS	Bouhajeb et al. (2020)
71	2-Pentylfuran	$C_9H_{14}O$	Leaves	Tizona, A1014, Nawel, A336, Mirval	Tunisia	Solid-phase micro-extraction	GC-MS	Bouhajeb et al. (2020)
72	2-(1-Pentenyl) furan	$C_9H_{12}O$	Leaves	A336	Tunisia	Solid-phase micro-extraction	GC-MS	Bouhajeb et al. (2020)
73	Octanal	$C_8H_{16}O$	Leaves	Tizona, A1014, Nawel	Tunisia	Solid-phase micro-extraction	GC-MS	Bouhajeb et al. (2020)
74	2,2,6-Trimethyl-cyclohexan-1-one	$C_9H_{16}O$	Leaves	Tizona, A1014, Nawel, A336, Mirval	Tunisia	Solid-phase micro-extraction	GC-MS	Bouhajeb et al. (2020)
75	3-Octen-2-one	$C_8H_{14}O$	Leaves	Tizona, A1014, Nawel	Tunisia	Solid-phase micro-extraction	GC-MS	Bouhajeb et al. (2020)
76	(E)-2-Octenal	$C_8H_{14}O$	Leaves	A336	Tunisia	Solid-phase micro-extraction	GC-MS	Bouhajeb et al. (2020)
77	3,5-Octadien-2-one (isomer 1)	$C_8H_{12}O$	Leaves	Tizona, A1014, Nawel, A336, Mirval	Tunisia	Solid-phase micro-extraction	GC-MS	Bouhajeb et al. (2020)
78	3,5-Octadien-2-one (isomer 2)	$C_8H_{12}O$	Leaves	Tizona	Tunisia	Solid-phase micro-extraction	GC-MS	Bouhajeb et al. (2020)
79	*n*-Undecane	$C_{11}H_{24}$	Leaves	Tizona, A336	Tunisia	Solid-phase micro-extraction	GC-MS	Bouhajeb et al. (2020)
80	Nonanal	$C_9H_{18}O$	Leaves	Tizona, A1014, Nawel, A336, Mirval	Tunisia	Solid-phase micro-extraction	GC-MS	Bouhajeb et al. (2020)
81	2,6-Dimethyl-cyclohexanol	$C_8H_{16}O$	Leaves	Tizona, A1014, Nawel, A336, Mirval	Tunisia	Solid-phase micro-extraction	GC-MS	Bouhajeb et al. (2020)
82	(Z)-2-Nonenal	$C_9H_{16}O$	Leaves	A336, Mirval	Tunisia	Solid-phase micro-extraction	GC-MS	Bouhajeb et al. (2020)

(continued)

	Compound	Molecular formula	Part used	Occurrence	Geographical origin	Extraction after defatting and/or oil pressing	Method of analysis	Reference
83	8-Methylnonanal	$C_{10}H_{20}O$	Leaves	Tizona, A1014, Nawel	Tunisia	Solid-phase micro-extraction	GC-MS	Bouhajeb et al. (2020)
84	Naphthalene	$C_{10}H_8$	Leaves	Tizona, A1014, Nawel	Tunisia	Solid-phase micro-extraction	GC-MS	Bouhajeb et al. (2020)
85	*n*-Dodecane	$C_{12}H_{26}$	Leaves	A336	Tunisia	Solid-phase micro-extraction	GC-MS	Bouhajeb et al. (2020)
86	(E)-2-Dodecene	$C_{12}H_{24}$	Leaves	A1014, Nawel	Tunisia	Solid-phase micro-extraction	GC-MS	Bouhajeb et al. (2020)
87	Decanal	$C_{10}H_{20}O$	Leaves	Tizona, A1014, Nawel, A336	Tunisia	Solid-phase micro-extraction	GC-MS	Bouhajeb et al. (2020)
88	*n*-Tridecane	$C_{13}H_{28}$	Leaves	A336, Mirval	Tunisia	Solid-phase micro-extraction	GC-MS	Bouhajeb et al. (2020)
89	1-Tetradecene	$C_{14}H_{28}$	Leaves	Nawel, Mirval	Tunisia	Solid-phase micro-extraction	GC-MS	Bouhajeb et al. (2020)
90	*n*-Tetradecane	$C_{14}H_{30}$	Leaves	Tizona, A1014, Nawel, A336, Mirval	Tunisia	Solid-phase micro-extraction	GC-MS	Bouhajeb et al. (2020)
91	2-Methyltetra-decane	$C_{15}H_{32}$	Leaves	Tizona	Tunisia	Solid-phase micro-extraction	GC-MS	Bouhajeb et al. (2020)
92	*n*-Pentadecane	$C_{15}H_{32}$	Leaves	Tizona, Nawel, A336	Tunisia	Solid-phase micro-extraction	GC-MS	Bouhajeb et al. (2020)

GC gas chromatography; *MS* mass spectrometry

Table 4 Reported steroids and other compounds in *Solanum melongena L.* eggplant by-products

No	Compound	Molecular formula	Class	Part used	Occurrence	Geographical origin	Extraction after defatting and/or oil pressing	Method of analysis	Reference
93	Khasianine	$C_{39}H_{63}NO_{11}$	Steroidal glycosides	Roots	American eggplants	Mexico	Solid–liquid extraction	LC-MS	Contreras-Angulo et al. (2022)
94	2-Naphthaleneacetic acid, decahydro-8-hydroxy-4a,8-dimethyl-alpha-methylene	$C_{15}H_{24}O_3$	Sesquiterpenoids	Roots	American eggplants	Mexico	Solid–liquid extraction	LC-MS	Contreras-Angulo et al. (2022)
95	Spirost-5-en-3-ol, (3beta,25R)	$C_{27}H_{42}O_3$	Steroid sapogenins	Roots	American eggplants	Mexico	Solid–liquid extraction	LC-MS	Contreras-Angulo et al. (2022)
96	Solamargine	$C_{45}H_{73}NO_{15}$	Steroidal glycosides	Roots	American eggplants	Mexico	Solid–liquid extraction	LC-MS	Contreras-Angulo et al. (2022)
97	Solasonine	$C_{45}H_{73}NO_{16}$	Steroidal glycosides	Roots	American eggplants	Mexico	Solid–liquid extraction	LC-MS	Contreras-Angulo et al. (2022)
98	Solasodine	$C_{27}H_{43}NO_2$	Steroidal alkaloids	Fruits Roots	American eggplants	Mexico	Solid–liquid extraction	LC-MS	Contreras-Angulo et al. (2022)
99	(3beta,22R,25R)-26-(beta-D-Glucopyranosyloxy)-22-hydroxyfurost-5-en-3-yl 6-deoxy-alpha-L-mannopyranosyl-(1→2)-[beta-D-glucopyranosyl-(1→4)]-beta-D-glucopyranoside	$C_{51}H_{84}O_{23}$	Steroidal glycosides	Fruits Roots	American eggplants	Mexico	Solid–liquid extraction	LC-MS	Contreras-Angulo et al. (2022)

(continued)

Table 4 continued.

No	Compound	Molecular formula	Class	Part used	Occurrence	Geographical origin	Extraction after defatting and/or oil pressing	Method of analysis	Reference
100	1-(9Z,12Z-octadecadienoyl)-glycero-3-phospho-(1″-myo-inositol)	$C_{27}H_{49}O_{12}P$	Glycerophospholipids/ carbohydrates	Stems Leaves Roots	American eggplants	Mexico	Solid–liquid extraction	LC-MS	Contreras-Angulo et al. (2022)
101	2,3,4,5-tetrahydroxy-6-(hydroxymethyl) cyclohexyltetracosanoate	$C_{31}H_{60}O_7$	Organic acids/ carbohydrates	Leaves	American eggplants	Mexico	Solid–liquid extraction	LC-MS	Contreras-Angulo et al. (2022)
102	6-{[3-(dimethylamino)-2-hydroxy-5-methylcyclohexyl] oxy}-16-ethyl-4-hydroxy-15-(hydroxymethyl)-5,7,9,13-tetramethyl-1-oxacyclohexadec-11-ene-2,10-dione	$C_{31}H_{55}NO_7$	Steroidal glycosides	Fruits Roots	American eggplants	Mexico	Solid–liquid extraction	LC-MS	Contreras-Angulo et al. (2022)

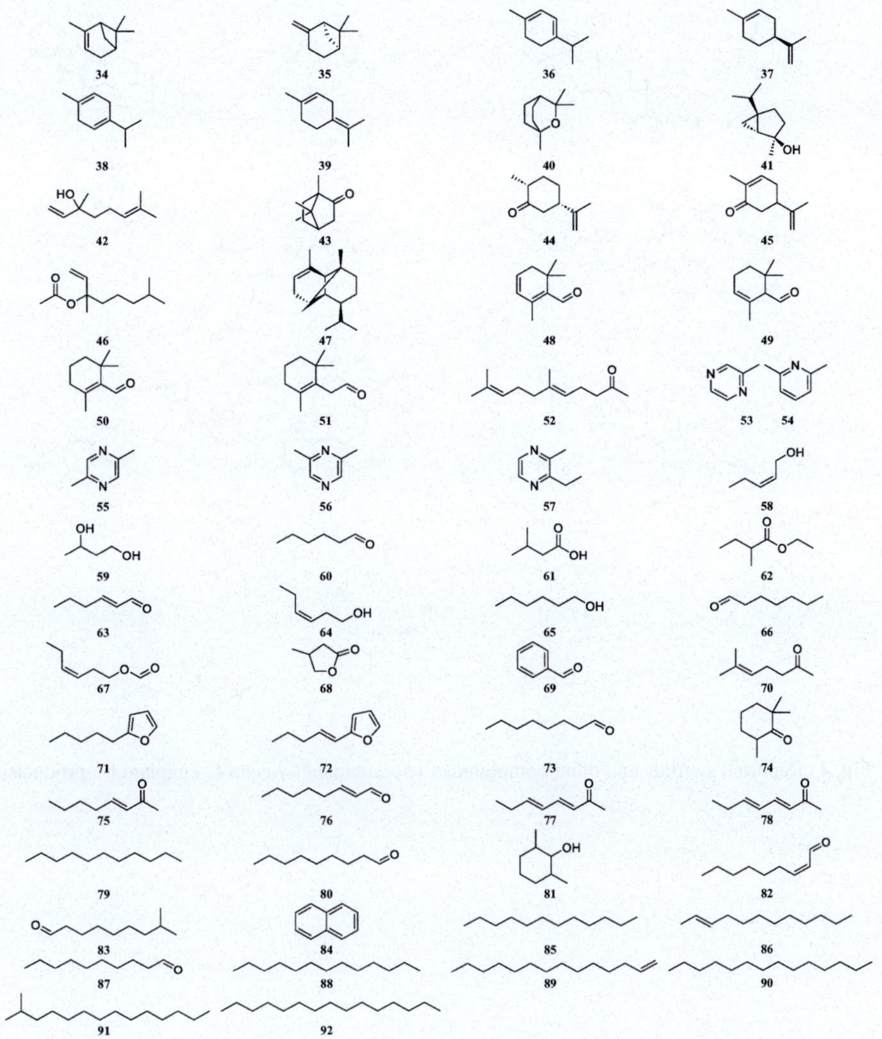

Fig. 3 Volatile oil compounds previously reported in by-products of *Solanum melongena* L. eggplant

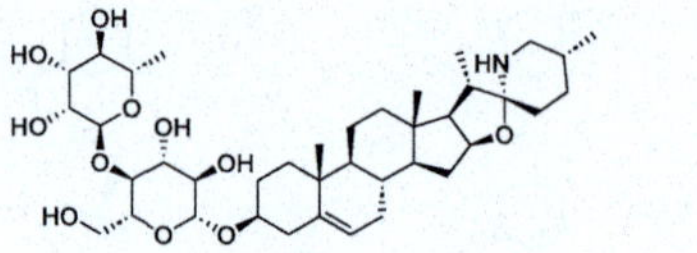

93

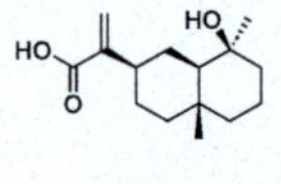

94

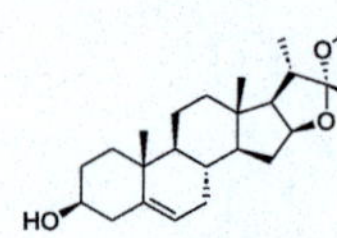

95

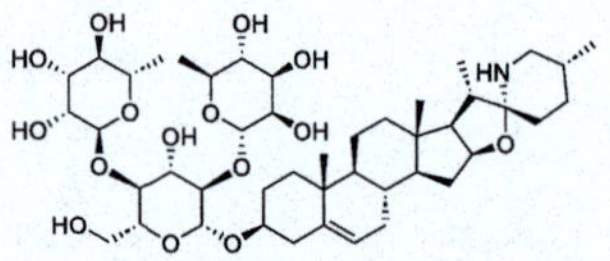

96

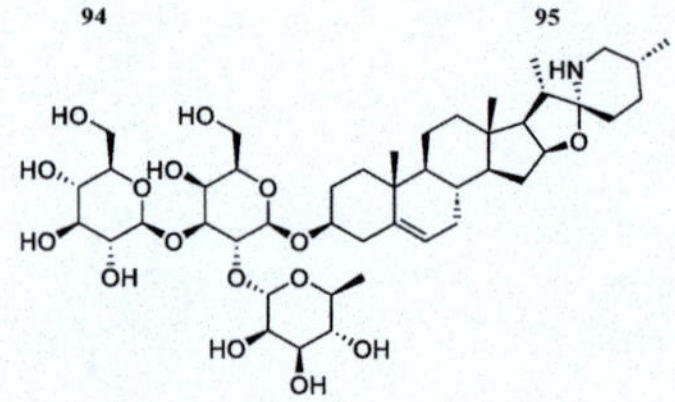

97

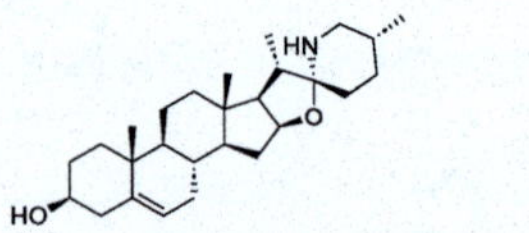

98

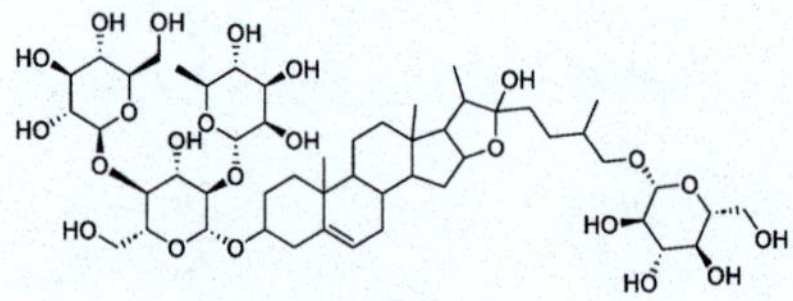

99

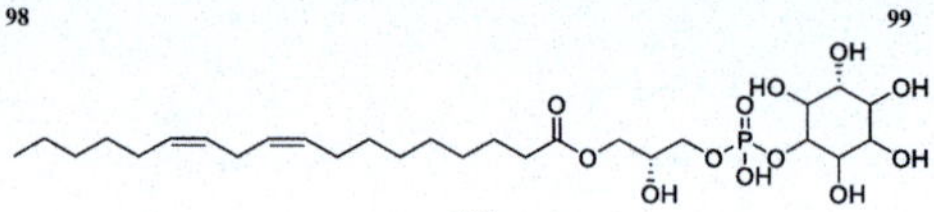

100

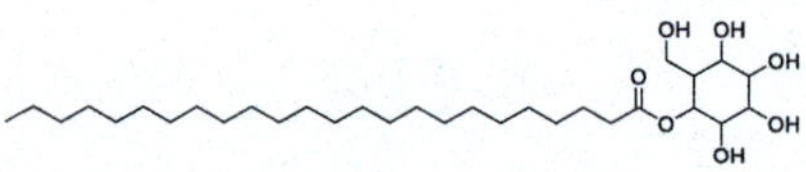

101

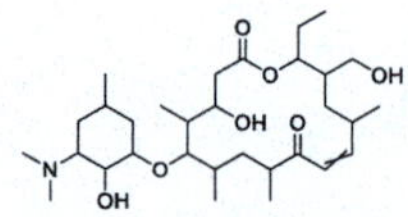

102

Fig. 4 Reported steroids and other compounds in *Solanum melongena L.* eggplant by-products

Table 5 Reported carbohydrates and carbohydrate conjugates in *Solanum melongena* L. eggplant by-products

No	Compound	Molecular formula	Part used	Occurrence	Geographical origin	Extraction after defatting and/or oil pressing	Method of analysis	Reference
103	[(2S)-2-nonanoyloxy-3 [(2R,3S,4S,5R,6S)-3,4,5-trihydroxy-6-[[(2S,3R,4S,5R,6S)-3,4,5-trihydroxy-6-(hydroxymethyl)oxan-2-yl] oxymethyl]oxan-2-yl] oxypropyl] tetradecanoate	$C_{38}H_{70}O_{15}$	Roots	American eggplants	Mexico	Solid–liquid extraction	LC-MS	Contreras-Angulo et al. (2022)
104	2″,3″,5″-Tri-*O*-acetylnebularine	$C_{16}H_{18}N_4O_7$	Fruit Roots	American eggplants	Mexico	Solid–liquid extraction	LC-MS	Contreras-Angulo et al. (2022)
105	[6-[[6-[3,4-dihydroxy-2,5 bis(hydroxymethyl) oxolan-2-yl]oxy-3,4,5-trihydroxyoxan-2-yl] methoxy]-3,4,5-trihydroxyoxan-2-yl]methyl (E)-2-methylbut-2-enoate	$C_{23}H_{38}O_{17}$	Roots	American eggplants	Mexico	Solid–liquid extraction	LC-MS	Contreras-Angulo et al. (2022)
106	[2-(2,5-Dioxo-hexahydro-imidazo[4,5-d]imidazol-1-yl)-acetylamino]-acetic acid	$C_8H_{11}N_5O_5$	Roots	American eggplants	Mexico	Solid–liquid extraction	LC-MS	Contreras-Angulo et al. (2022)
107	[(2S)-2-tridecanoyloxy-3-[(2R,3S,4S,5R,6S)-3,4,5-trihydroxy-6-(hydroxymethyl)oxan-2-yl] oxypropyl] tetradecanoate	$C_{36}H_{68}O_{10}$	Stems Leaves Roots	American eggplants	Mexico	Solid–liquid extraction	LC-MS	Contreras-Angulo et al. (2022)

(continued)

Table 5 continued.

No	Compound	Molecular formula	Part used	Occurrence	Geographical origin	Extraction after defatting and/or oil pressing	Method of analysis	Reference
108	Decyl-lA-D-maltoside	$C_{22}H_{42}O_{11}$	Leaves Roots	American eggplants	Mexico	Solid–liquid extraction	LC-MS	Contreras-Angulo et al. (2022)
109	Solphenazine F	$C_{24}H_{28}N_2O_{10}$	Roots	American eggplants	Mexico	Solid–liquid extraction	LC-MS	Contreras-Angulo et al. (2022)
110	Oligomycin	$C_{45}H_{74}O_{11}$	Leaves	American eggplants	Mexico	Solid–liquid extraction	LC-MS	Contreras-Angulo et al. (2022)
111	1-(9Z-tetradecenoyl)-glycero-3-phospho-(1″-myo-inositol)	$C_{23}H_{43}O_{12}P$	Leaves Roots	American eggplants	Mexico	Solid–liquid extraction	LC-MS	Contreras-Angulo et al. (2022)
112	6-{[4-(dimethylamino)-3-hydroxy-6-methyloxan-2-yl]oxy}-16-ethyl-4,10-dihydroxy-7-(2-hydroxyethyl)-5,9,13,15-tetramethyl-1-oxacyclohexadeca-11,13-dien-2-one	$C_{31}H_{55}NO_8$	Fruits Leaves Roots	American eggplants	Mexico	Solid–liquid extraction	LC-MS	Contreras-Angulo et al. (2022)

Table 6 Reported fatty acids, amino acids, peptides, and other nitrogen compounds in *Solanum melongena* L. eggplant by-products

No	Compound	Molecular formula	Class	Part used	Occurrence	Geographical origin	Extraction after defatting and/or oil pressing	Method of analysis	Reference
113	Butyrolactone	$C_4H_6O_2$	Fatty acid	Peels	Local market of Lahore district	Pakistan	Ultrasound-assisted extraction	GS-MS	Kainat et al. (2023)
114	4H-Pyran-4-one, 2,3-dihydro-3, 5-dihydroxy-6-methyl-	$C_6H_8O_4$	Fatty acid	Peels	Local market of Lahore district	Pakistan	Ultrasound-assisted extraction	GS-MS	Kainat et al. (2023)
115	*n*-Hexadecanoic acid	$C_{16}H_{32}O_2$	Fatty acid	Peels	Local market of Lahore district	Pakistan	Ultrasound-assisted extraction	GS-MS	Kainat et al. (2023)
116	Oleic acid	$C_{18}H_{34}O_2$	Fatty acid	Peels	Local market of Lahore district	Pakistan	Ultrasound-assisted extraction	GS-MS	Kainat et al. (2023)
117	Jasmonic acid	$C_{12}H_{18}O_3$	Fatty acid	Stems Leaves Roots	American eggplants	Mexico	Solid–liquid extraction	LC-MS	Contreras-Angulo et al. (2022)
118	17(18)-EpETE	$C_{20}H_{30}O_3$	Fatty acids	Stems Leaves Roots	American eggplants	Mexico	Solid–liquid extraction	LC-MS	Contreras-Angulo et al. (2022)
119	Carmabin B	$C_{40}H_{59}N_5O_7$	Amino acids, peptides, and analogs	Fruits Leaves Roots	American eggplants	Mexico	Solid–liquid extraction	LC-MS	Contreras-Angulo et al. (2022)
120	Phenacetyl KKR-acid	$C_{26}H_{44}N_8O_5$	Amino acids, peptides, and analogs	Leaves Roots	American eggplants	Mexico	Solid–liquid extraction	LC-MS	Contreras-Angulo et al. (2022)

(continued)

Table 6 (continued)

No	Compound	Molecular formula	Class	Part used	Occurrence	Geographical origin	Extraction after defatting and/or oil pressing	Method of analysis	Reference
121	Val-Ala-His-Ser-Gly-Ala	$C_{22}H_{37}N_9O_7$	Amino acids, peptides, and analogs	Roots	American eggplants	Mexico	Solid–liquid extraction	LC-MS	Contreras-Angulo et al. (2022)
122	His-Arg-Lys(Me)-Val-Leu-Arg	$C_{36}H_{67}N_{15}O_7$	Amino acids, peptides, and analogs	Leaves	American eggplants	Mexico	Solid–liquid extraction	LC-MS	Contreras-Angulo et al. (2022)
123	Ac-Ala-Asp-Val-Ala-Asp-al	$C_{21}H_{33}N_5O_{10}$	Amino acids, peptides, and analogs	Stems Leaves Roots	American eggplants	Mexico	Solid–liquid extraction	LC-MS	Contreras-Angulo et al. (2022)
124	(2S,4S,5R)-N-[3-[3-[(2,3-dihydroxybenzoyl)amino] propyl-[(2S,4S,5R)-2-(2,3-dihydroxyphenyl)-5-methyl-1,3-oxazolidine-4-Carbonyl] amino] propyl]-2-(2,3-dihydroxyphenyl)-5-methyl-1,3-oxazolidine-4-carboxamide	$C_{35}H_{43}N_5O_{11}$	Amino acids, peptides, and analogs	Stems Leaves Roots	American eggplants	Mexico	Solid–liquid extraction	LC-MS	Contreras-Angulo et al. (2022)
125	Piplartine dimer A	$C_{34}H_{38}N_2O_{10}$	Amino acids, peptides, and analogs	Stems Fruits Leaves Roots	American eggplants	Mexico	Solid–liquid extraction	LC-MS	Contreras-Angulo et al. (2022)

Table 6 (continued)

No	Compound	Molecular formula	Class	Part used	Occurrence	Geographical origin	Extraction after defatting and/or oil pressing	Method of analysis	Reference
126	N-[4-[2-[2,4-bis(2-methylbutan-2-yl)phenoxy] butanoylamino]-2-hydroxyphenyl]-2,2,3,3,4,4,4-heptafluorobutanamide	$C_{30}H_{37}F_7N_2O_4$	Amino acids, peptides, and analogs	Leaves	American eggplants	Mexico	Solid–liquid extraction	LC-MS	Contreras-Angulo et al. (2022)
127	3-Aza-lipid X	$C_{34}H_{67}N_2O_{11}P$	Amino acids, peptides,	Leaves	American eggplants	Mexico	Solid–liquid extraction	LC-MS	Contreras-Angulo et al. (2022)
128	5-oxo-delta-Bilirubin	$C_{33}H_{34}N_4O_7$	Amino acids, peptides, and analogs	Stems Leaves	American eggplants	Mexico	Solid–liquid extraction	LC-MS	Contreras-Angulo et al. (2022)
129	Stearoyl Ara-C	$C_{27}H_{47}N_3O_6$	Amino acids, peptides, and analogs	Leaves Roots	American eggplants	Mexico	Solid–liquid extraction	LC-MS	Contreras-Angulo et al. (2022)
130	4‴ deoxy EM-A enol ether	$C_{37}H_{65}NO_{11}$	Amino acids, peptides, and analogs	Fruits Leaves Roots	American eggplants	Mexico	Solid–liquid extraction	LC-MS	Contreras-Angulo et al. (2022)
131	Ala-Arg-Thr-Lys-Gln-al	$C_{24}H_{46}N_{10}O_7$	Amino acids, peptides, and analogs	Stems Leaves Roots	American eggplants	Mexico	Solid–liquid extraction	LC-MS	Contreras-Angulo et al. (2022)

(continued)

Table 6 (continued)

No	Compound	Molecular formula	Class	Part used	Occurrence	Geographical origin	Extraction after defatting and/or oil pressing	Method of analysis	Reference
132	Rodaplutin	$C_{24}H_{39}N_9O_7$	Amino acids, peptides, and analogs	Leaves	American eggplants	Mexico	Solid–liquid extraction	LC-MS	Contreras-Angulo et al. (2022)
133	Antho-rwamide I	$C_{31}H_{46}N_{10}O_7$	Amino acids, peptides, and analogs	Stems Fruits Leaves Roots	American eggplants	Mexico	Solid–liquid extraction	LC-MS	Contreras-Angulo et al. (2022)
134	Methionyl puromycin	$C_{27}H_{38}N_8O_6S$	Amino acids, peptides, and analogs	Stems Fruits Leaves Roots	American eggplants	Mexico	Solid–liquid extraction	LC-MS	Contreras-Angulo et al. (2022)
135	2,4,6-Tri-(6-aminocaproic acid)-1,3,5-triazine	$C_{21}H_{36}N_6O_6$	Amino acids, peptides, and analogs	Stems Leaves Roots	American eggplants	Mexico	Solid–liquid extraction	LC-MS	Contreras-Angulo et al. (2022)
136	[(2S)-3-(4-butoxycarbonylpiperazin-1-yl)-2-[[6-[(3S)-3-methoxypyrrolidin-1-yl]-2-phenyl pyrimidine-4-carbonyl] amino]-3-oxopropyl] phosphonic acid	$C_{28}H_{39}N_6O_8P$	Amino acids, peptides, and analogs	Stems Fruits Leaves Roots	American eggplants	Mexico	Solid–liquid extraction	LC-MS	Contreras-Angulo et al. (2022)
137	Solanamide A	$C_{30}H_{46}N_4O_6$	Amino acids, peptides, and analogs	Stems Leaves Roots	American eggplants	Mexico	Solid–liquid extraction	LC-MS	Contreras-Angulo et al. (2022)

Table 6 (continued)

No	Compound	Molecular formula	Class	Part used	Occurrence	Geographical origin	Extraction after defatting and/or oil pressing	Method of analysis	Reference
138	4-tert-Butylcalix[5]arene	$C_{55}H_{70}O_5$	Amino acids, peptides, and analogs	Leaves Roots	American eggplants	Mexico	Solid–liquid extraction	LC-MS	Contreras-Angulo et al. (2022)
139	5-(4-ethylphenyl)-4-[hydroxy-(3-methyl-4-propan-2-yloxyphenyl) methylidene]-1-(3-morpholin-4-ylpropyl) pyrrolidine-2,3-dione	$C_{30}H_{38}N_2O_5$	Amino acids, peptides, and analogs	Roots	American eggplants	Mexico	Solid–liquid extraction	LC-MS	Contreras-Angulo et al. (2022)
140	N-[1-amino-3-(1H-indol-3-yl)-1-oxopropan-2-yl]-3-(cyclohexylcarbamoylamino)-1-(pyrazine-2-carbonyl) pyrrolidine-2-carboxamide	$C_{28}H_{34}N_8O_4$	Amino acids, peptides, and analogs	Roots	American eggplants	Mexico	Solid–liquid extraction	LC-MS	Contreras-Angulo et al. (2022)
141	Ala-Gly-Leu-Leu-Gly-Gly	$C_{21}H_{38}N_6O_7$	Amino acids, peptides, and analogs	Leaves Roots	American eggplants	Mexico	Solid–liquid extraction	LC-MS	Contreras-Angulo et al. (2022)
142	2-[[(2S)-2-[[(2S)-2-[[(2S)-2-[[2-hydroxy-3-(methylamino) decanoyl]amino]-3-methylbutanoyl]-methylamino]-3-methylbutanoyl]-methylamino]-3-(4-	$C_{41}H_{63}N_5O_9$	Amino acids, peptides, and analogs	Fruits Leaves Roots	American eggplants	Mexico	Solid–liquid extraction	LC-MS	Contreras-Angulo et al. (2022)

(continued)

Table 6 continued.

No	Compound	Molecular formula	Class	Part used	Occurrence	Geographical origin	Extraction after defatting and/or oil pressing	Method of analysis	Reference
143	(2S)-N-[(2S,3S)-1-[[(2S)-1-[[(2S)-5-amino-1,5-dioxopentan-2-yl] amino]-1-oxopropan-2-yl] amino]-3-methyl-1-oxopentan-2-yl]-1-[(2S)-2-[[(2S)-2-aminopropanoyl] amino]-3-methylbutanoyl] pyrrolidine-2-carboxamide	$C_{27}H_{47}N_7O_7$	Amino acids, peptides, and analogs	Fruits Leaves Roots	American eggplants	Mexico	Solid–liquid extraction	LC-MS	Contreras-Angulo et al. (2022)
144	Cyclo(grgdspa)	$C_{24}H_{38}N_{10}O_{12}$	Amino acids, peptides, and analogs	Stems Fruits Leaves Roots	American eggplants	Mexico	Solid–liquid extraction	LC-MS	Contreras-Angulo et al. (2022)
145	Thr-Val-Glu-Glu-Val-Asp	$C_{28}H_{46}N_6O_{14}$	Amino acids, peptides, and analogs	Stems Leaves Roots	American eggplants	Mexico	Solid–liquid extraction	LC-MS	Contreras-Angulo et al. (2022)
146	5-(3-ethoxy-4-hydroxyphenyl)-3-hydroxy-1-(2-methoxyethyl)-4-{[3-methyl-4-(meth ylethoxy)phenyl] carbonyl}-3-pyrrolin-2-one	$C_{26}H_{31}NO_7$	Amino acids, peptides, and analogs	Stems Leaves Roots	American eggplants	Mexico	Solid–liquid extraction	LC-MS	Contreras-Angulo et al. (2022)
147	Ethyl 4-sulfonamidonaphthyl sulfone	$C_{12}H_{13}NO_4S_2$	Carbohydrates and carbohydrate conjugates	Stems Leaves	American eggplants	Mexico	Solid–liquid extraction	LC-MS	Contreras-Angulo et al. (2022)

No	Compound	Molecular formula	Class	Part used	Occurrence	Geographical origin	Extraction after defatting and/or oil pressing	Method of analysis	Reference
148	(2S)-N-[(2S)-1-[[(2S)-1-[[(2S)-6-amino-1-oxohexan-2-yl] amino]-3-(4-hydroxyphenyl)-1-oxopropan-2-yl]amino]-3-methyl-1-oxobutan-2-yl]-2-[[(2S)-2-aminopropanoyl] amino]-4-methylpentanamide	$C_{29}H_{48}N_6O_6$	Depsipeptides	Stems Fruits Leaves Roots	American eggplants	Mexico	Solid–liquid extraction	LC-MS	Contreras-Angulo et al. (2022)
149	4,4″-(4,4″-Bipyridine-1,1″-diyl)bis(2,6-diaminopyrimidine 1-oxide)	$C_{18}H_{28}N_{10}O_2$	Amino acids, peptides, and analogs	Leaves Roots	American eggplants	Mexico	Solid–liquid extraction	LC-MS	Contreras-Angulo et al. (2022)
150	Ala-Ser-Leu-Ser(PO3H2)-Ala-Pro-Gly-al	$C_{25}H_{44}N_7O_{12}P$	Amino acids, peptides, and analogs	Stems Leaves Roots	American eggplants	Mexico	Solid–liquid extraction	LC-MS	Contreras-Angulo et al. (2022)
151	N-[(Benzyloxy)carbonyl]-L-Leucyl-N-[(3s)-5-Methyl-1-(Methylsulfonyl) hexan-3-Yl]-L-Leucinamide	$C_{28}H_{47}N_3O_6S$	Amino acids, peptides, and analogs	Leaves Roots	American eggplants	Mexico	Solid–liquid extraction	LC-MS	Contreras-Angulo et al. (2022)
152	Gly-Asn-Asn-Ser-Tyr-Ser	$C_{25}H_{36}N_8O_{12}$	Amino acids, peptides, and analogs	Stems Fruits Leaves Roots	American eggplants	Mexico	Solid–liquid extraction	LC-MS	Contreras-Angulo et al. (2022)
153	L-Prolinamide, N-(1-oxo-2-propylpentyl)-L-alpha-aspartyl-N-(4-((aminoiminomethyl) amino)-1-formyl butyl)-, (S)-	$C_{23}H_{40}N_6O_6$	Amino acids, peptides, and analogs	Stems Fruits Leaves Roots	American eggplants	Mexico	Solid–liquid extraction	LC-MS	Contreras-Angulo et al. (2022)

(continued)

Table 6 continued.

No	Compound	Molecular formula	Class	Part used	Occurrence	Geographical origin	Extraction after defatting and/or oil pressing	Method of analysis	Reference
154	(3r,3as,4r,6ar)-4-(2-Methoxyethoxy)hexahydrofuro [2,3-B]furan-3-Yl {(2s,3r)-3-Hydroxy-4-[{[(2z)-2-(Methylimino)-2,3-Dihydro-1,3-Benzoxazol-6-Yl]sulfonyl}(2-Methylpropyl)amino]-1-Phenylbutan-2-Yl}carbamate	$C_{32}H_{44}N_4O_{10}S$	Amino acids, peptides, and analogs	Stems Leaves	American eggplants	Mexico	Solid–liquid extraction	LC-MS	Contreras-Angulo et al. (2022)
155	Dibekacin 2″-phosphate	$C_{18}H_{41}N_5O_{11}P^{+3}$	Amino acids, peptides, and analogs	Fruits Leaves	American eggplants	Mexico	Solid–liquid extraction	LC-MS	Contreras-Angulo et al. (2022)
156	[4-[(1-hydroxy-2-oxoazepan-3-yl)amino]-4-oxobutan-2-yl] 2-[[2-(2,3-dihydroxyphenyl)-4-methyl-5H-1,3-oxazole-4-carbonyl] amino]-6-[hydroxy-[(Z)-tetradec-5-enoyl]amino] hexanoate	$C_{41}H_{63}N_5O_{11}$	Amino acids, peptides, and analogs	Fruits Roots	American eggplants	Mexico	Solid–liquid extraction	LC-MS	Contreras-Angulo et al. (2022)
157	Lexamine S 13	$C_{23}H_{48}N_2O$	Amino acids, peptides, and analogs	Leaves Roots	American eggplants	Mexico	Solid–liquid extraction	LC-MS	Contreras-Angulo et al. (2022)
158	Proferrioxamine G1t	$C_{23}H_{46}N_6O_7$	Amino acids, peptides, and analogs	Stems Leaves Roots	American eggplants	Mexico	Solid–liquid extraction	LC-MS	Contreras-Angulo et al. (2022)

No	Compound	Molecular formula	Class	Part used	Occurrence	Geographical origin	Extraction after defatting and/or oil pressing	Method of analysis	Reference
159	Dimethylaminopropyl palmitamide	$C_{21}H_{44}N_2O$	Amino acids, peptides, and analogs	Roots	American eggplants	Mexico	Solid–liquid extraction	LC-MS	Contreras-Angulo et al. (2022)
160	Ala-Gly-Tyr-Ala-Asn-Ala-Ala-al	$C_{27}H_{40}N_8O_9$	Amino acids, peptides, and analogs	Leaves	American eggplants	Mexico	Solid–liquid extraction	LC-MS	Contreras-Angulo et al. (2022)
161	[2-({4-[(2-amino-4-oxo-4,7-dihydro-3H-pyrrolo[2,3-d] pyrimidin-5-yl)methyl]benzene-1-	$C_{23}H_{21}N_5O_5$	Amino acids, peptides, and analogs	Leaves	American eggplants	Mexico	Solid–liquid extraction	LC-MS	Contreras-Angulo et al. (2022)
162	[(2R,3S,4S,5R,6S)-3,4,5-trihydroxy-6-[(1S,2R)-2-hydroxy-2-(2-hydroxy-2-methylpropyl)-4,4,6,6-tetramethyl-3,5-dioxocyclohexyl] oxyoxan-2-yl]methyl 3,4,5-trihydroxybenzoate	$C_{27}H_{38}O_{14}$	Amino acids, peptides, and analogs	Leaves Roots	American eggplants	Mexico	Solid–liquid extraction	LC-MS	Contreras-Angulo et al. (2022)
163	Itraconazole	$C_{35}H_{38}F_2N_8O_4$	Amino acids, peptides, and analogs	Stems Leaves Roots	American eggplants	Mexico	Solid–liquid extraction	LC-MS	Contreras-Angulo et al. (2022)
164	5-methyltetrahydrofolyl polyglutamate	$C_{30}H_{39}N_9O_{12}$	Amino acids, peptides, and analogs	Stems Fruits Leaves Roots	American eggplants	Mexico	Solid–liquid extraction	LC-MS	Contreras-Angulo et al. (2022)

(continued)

Table 6 continued.

No	Compound	Molecular formula	Class	Part used	Occurrence	Geographical origin	Extraction after defatting and/or oil pressing	Method of analysis	Reference
165	Ala-Thr-Leu-Leu-Asp-Ala-al	$C_{26}H_{46}N_6O_9$	Amino acids, peptides, and analogs	Fruits Roots	American eggplants	Mexico	Solid–liquid extraction	LC-MS	Contreras-Angulo et al. (2022)
166	1-[(3R,9R,10S)-9-[(dimethylamino)methyl]-12-[(2S)-1-hydroxypropan-2-yl]-3,10-dimethyl-13-oxo-2,8-dioxa-12-azabicyclo[12.4.0]octadeca-1(14),15,17-trien-16-yl]-3-(3,5-dimethyl-4-isoxazolyl)urea	$C_{29}H_{45}N_5O_6$	Amino acids, peptides, and analogs	Roots	American eggplants	Mexico	Solid–liquid extraction	LC-MS	Contreras-Angulo et al. (2022)
167	S-(3Alpha,7Alpha,12Alpha-Trihydroxy-5Beta-Cholestan-26-Oyl)-4′-phosphopantetheine	$C_{38}H_{67}N_2O_{11}P$ S	Amino acids, peptides, and analogs	Leaves Roots	American eggplants	Mexico	Solid–liquid extraction	LC-MS	Contreras-Angulo et al. (2022)

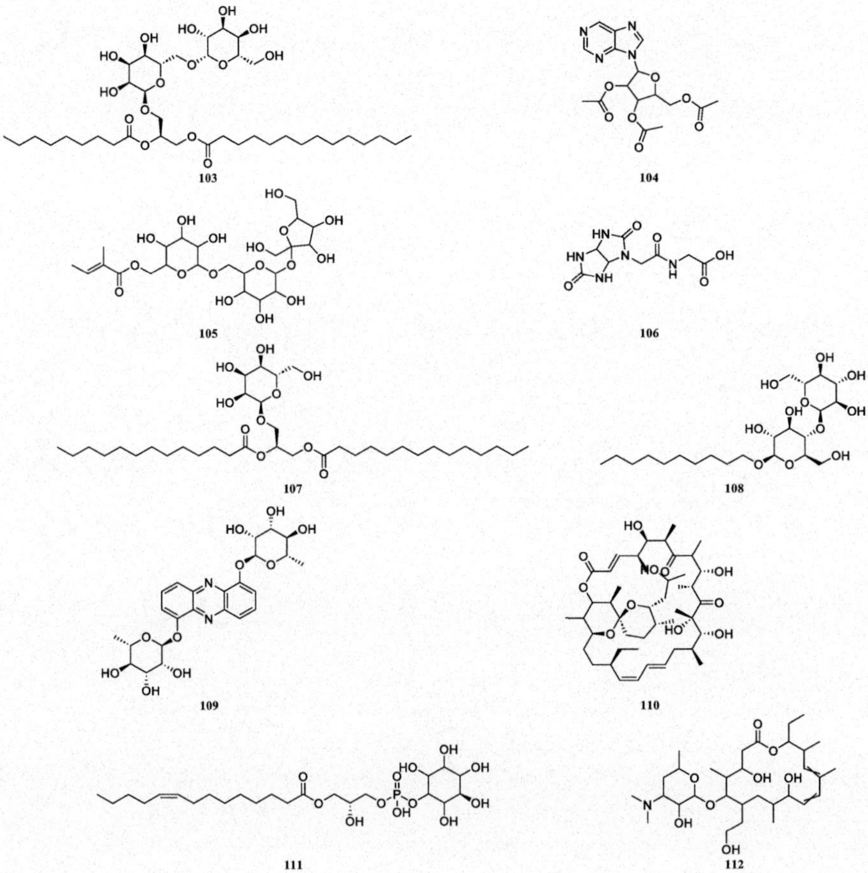

Fig. 5 Reported carbohydrates and carbohydrate conjugates in *Solanum melongena* L. eggplant by-products

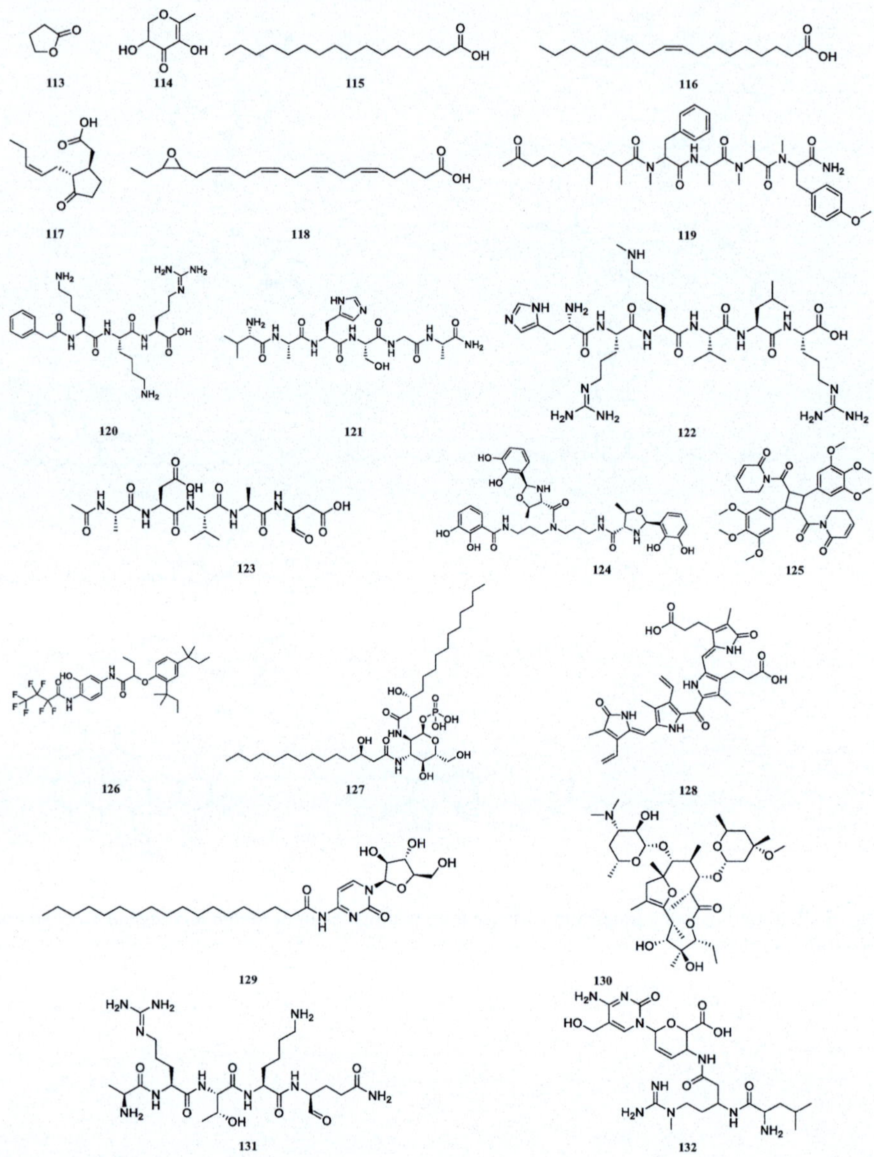

Fig. 6 Reported fatty acids and amino acids in *Solanum melongena* L. eggplant by-products

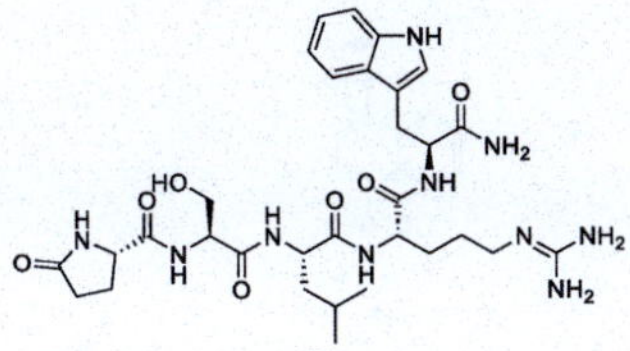

133

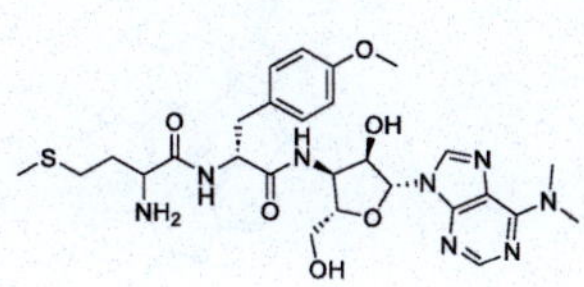

134

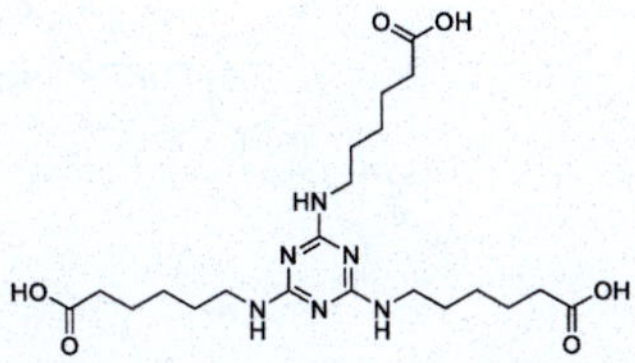

135

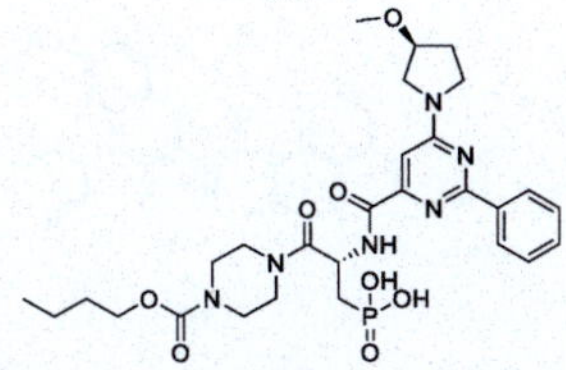

136

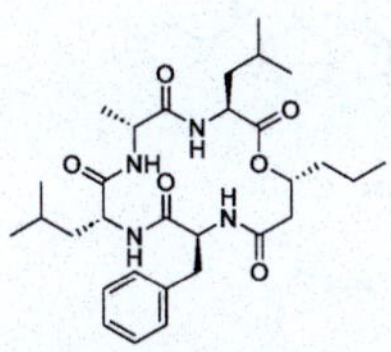

137

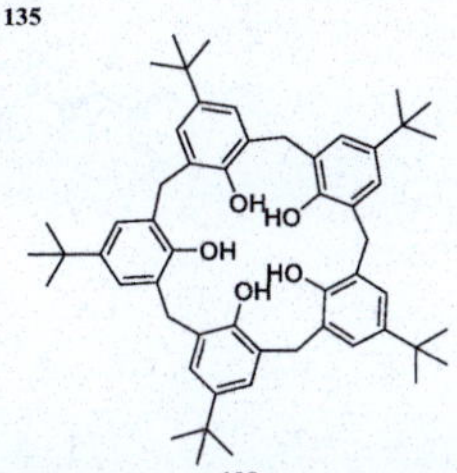

138

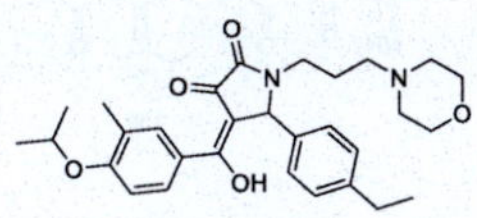

139

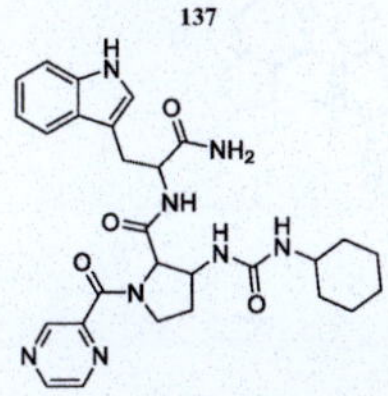

140

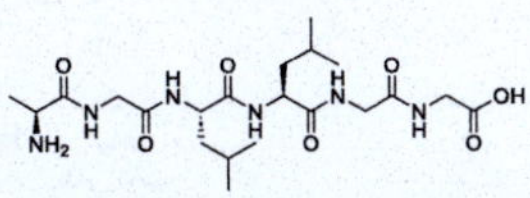

141

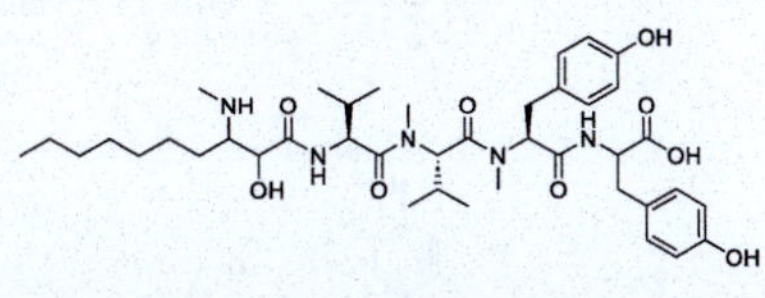

142

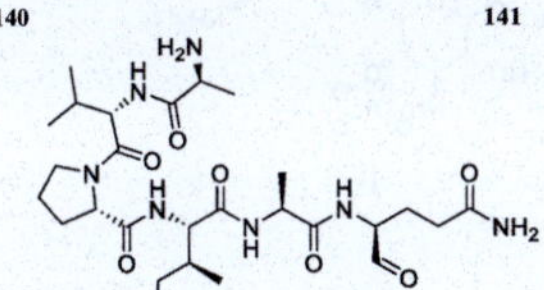

143

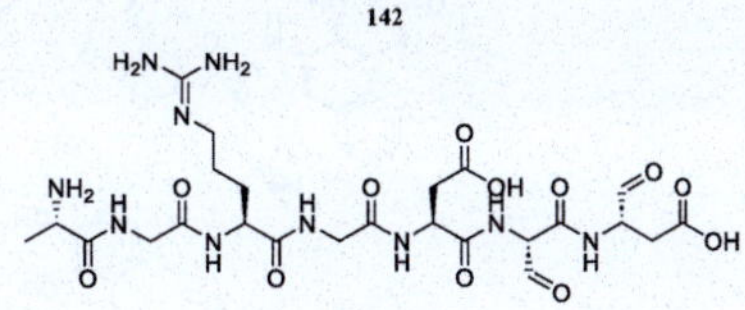

144

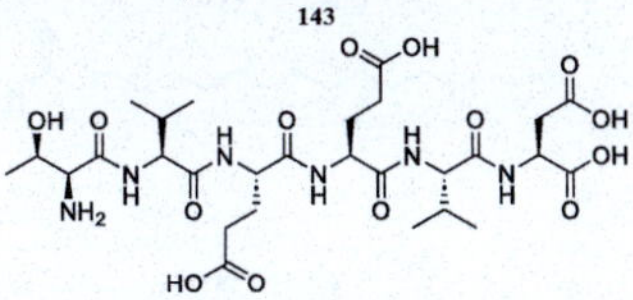

145

Fig. 10.6 (continued)

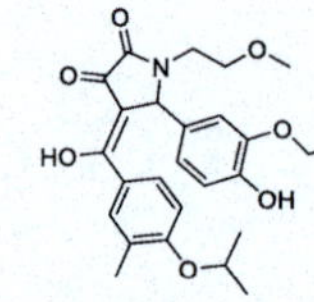

146

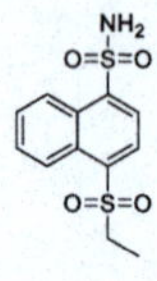

147

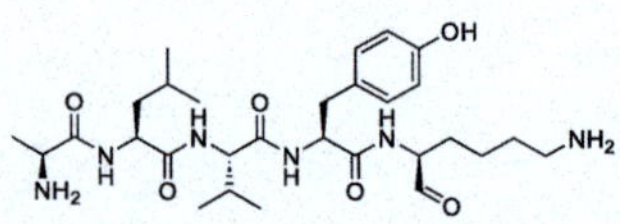

148

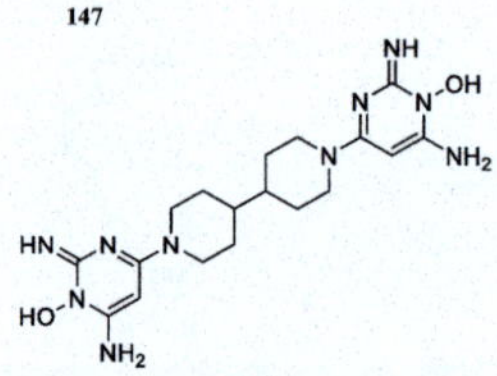

149

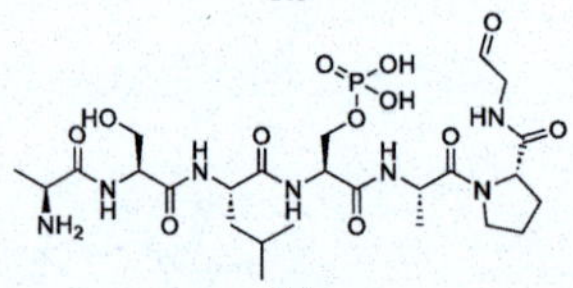

150

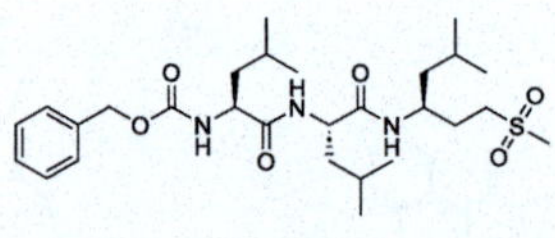

151

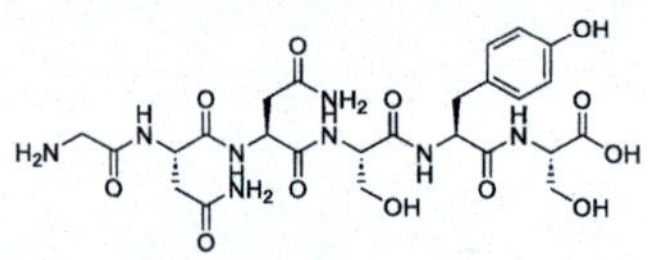

152

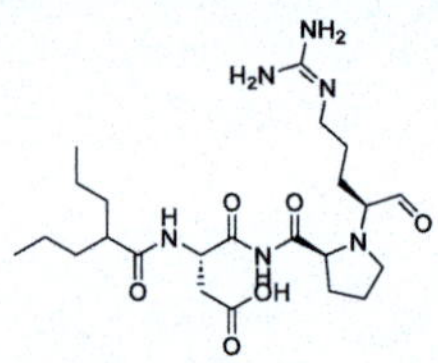

153

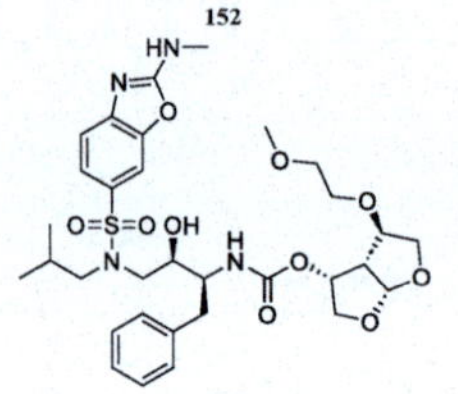

154

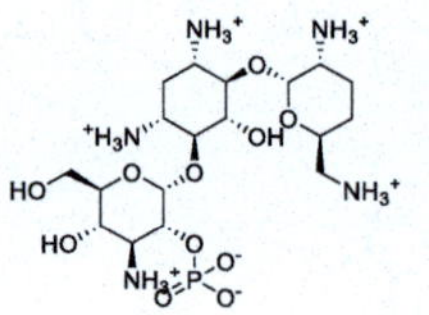

155

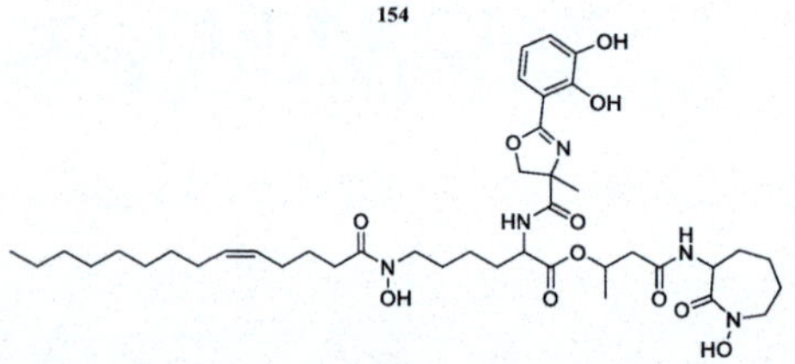

156

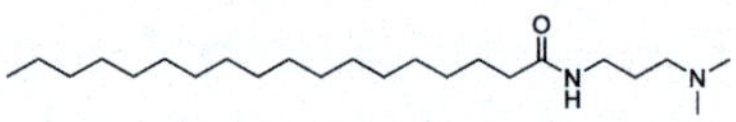

157

Fig. 10.6 (continued)

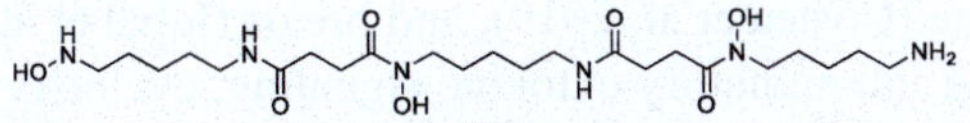

158

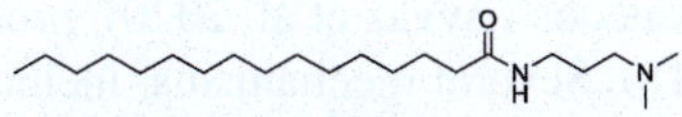

159

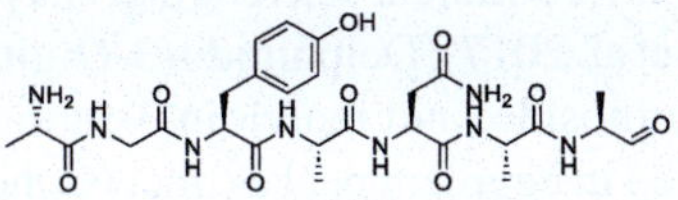

160

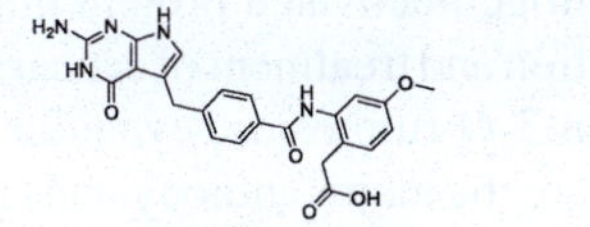

161

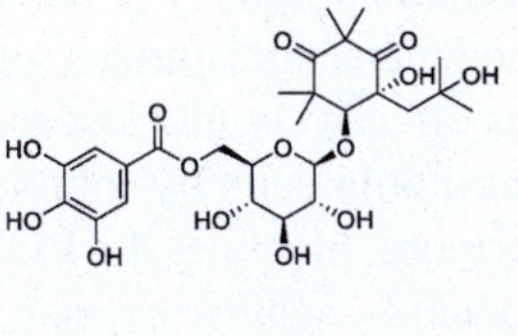

162

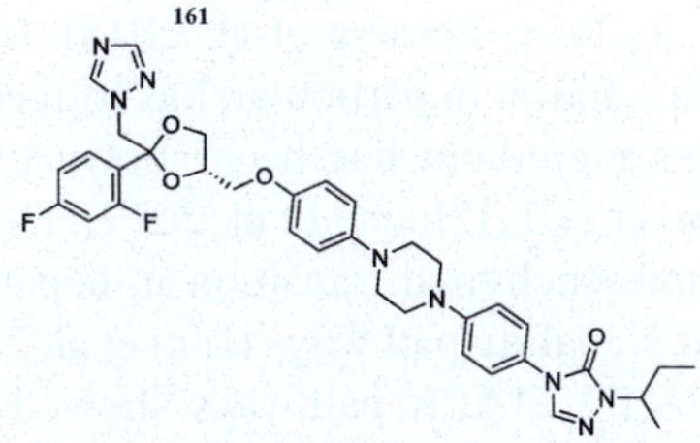

163

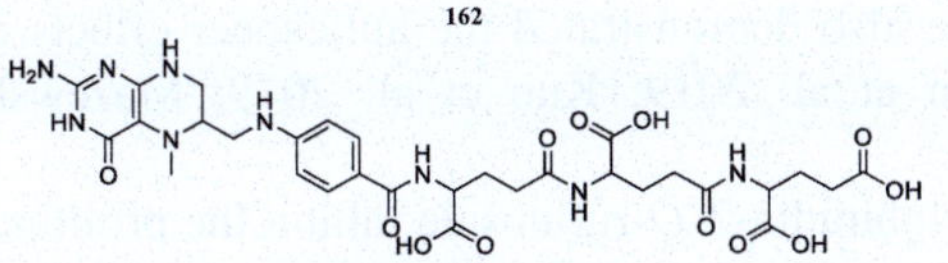

164

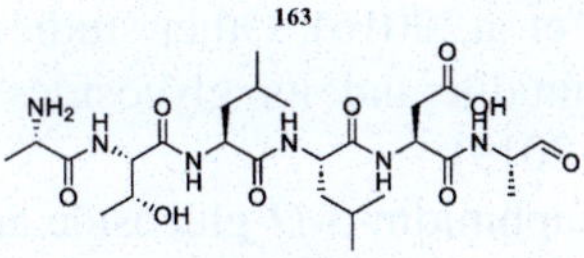

165

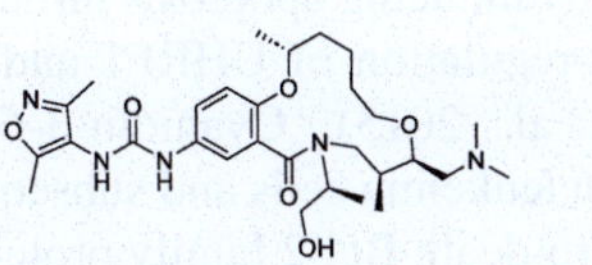

166

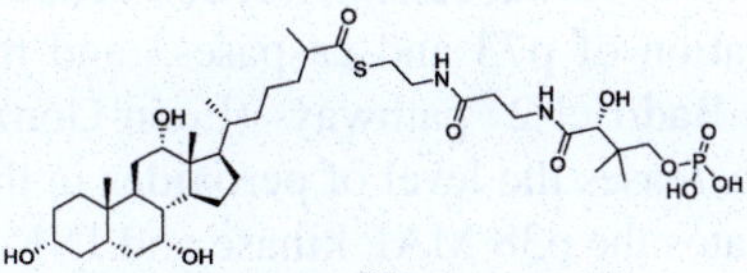

167

Fig. 10.6 (continued)

3 Biological Activity of Eggplants By-Product

3.1 Anticancer Activity

The eggplant peel extract effectively reduced cancer cell migration in a dose-dependent manner. Using the MTT assay, the cytotoxic effect of this extract on human cancer cell lines, including one breast cancer cell line, MDA-MB-231, two colon cancer cell lines, DLD-1 and HCT-116, and one prostate cancer cell line, PC3, was assessed in the presence of varying concentrations of the EPP extract. All cancer cell lines demonstrated a substantial decrease in survival, with IC_{50} values of 1.52 mg/ml (MDA-MB-231), 1.71 mg/ml (DLD-1), 1.99 mg/ml (HCT-116), and 1.88 mg/ml (PC3) (Kainat et al. 2023).

In three human tumor cell lines, HeLa (cervical carcinoma), NCI-H460 (non-small cell lung cancer), and HepG2 (hepatocellular carcinoma), the eggplant peel hydroethanolic extract demonstrated an inhibitory capability (GI50 < 400 μg/ml) (Silva et al. 2021).

Further research on anthocyanins and cancer treatment has been done extensively (Wu et al. 2020). Anthocyanins have been shown in studies to be helpful in a variety of malignancies, including gastrointestinal (Dharmawansa et al. 2020), cutaneous (Ayvaz et al. 2020), prostate (Costea et al. 2019), and breast (Iqbal et al. 2018). Several mechanisms, including inflammatory cytokine signaling, cyclooxygenase, mitogen-activated protein kinase, and free-radical scavenging, may impact the prevention and treatment of cancer (Khoo et al. 2017). Delphinidin 3-*O*-glucoside, delphinidin 3-*O*-rutinoside, cyanidin 3-*O*-rutinoside, and malvidin-3-rutinoside-5-glucoside are the major anthocyanins identified in eggplant peel extract (Condurache et al. 2021; Ferarsa et al. 2018). Concerning HER-2-positive breast cancer cells, delphinidin, in particular, has triggered apoptosis and autophagy (Chen et al. 2018). This ingredient has been shown to have an anti-metastatic impact on colorectal cancer cells (Huang et al. 2019). Delphinidin has effectively blocked the epithelial-to-mesenchymal transition in hepatocellular carcinoma cells by inhibiting numerous signaling pathways (Lim et al. 2019). Furthermore, blocking the PI3K/AKT and ERK1/2 MAPK pathways showed protective benefits against ovarian cell cancer (Lim et al. 2016). Other studies have also demonstrated the anticancer effects of delphinidin and its glycosides (Han et al. 2019; Kuo et al. 2019; Mazewski et al. 2019).

Delphinidin-3-*O*-glucoside and delphinidin-3-*O*-rutinoside inhibit the proliferation of Jurkat cells in a redox-sensitive way by inducing apoptosis through the activation of p73 and caspase-3 and the down-regulation of UHRF1 and the p-Akt/p-Bad/Bcl-2 pathway (León-González et al. 2015). Cyanidin-3-*O*-rutinoside increases the level of peroxides in the human leukemia cells and subsequently activates the p38 MAP kinase and JNK so the pro-death Bcl-2 family proteins, leading to mitochondrial release of apoptogenic factors and cell death. In contrast, cyanidin-3-*O*-rutinoside caused little reactive oxygen species (ROS) generation in the normal human peripheral blood mononuclear cells and had very low toxicity toward these cells (Feng et al. 2007). Malvidin suppressed JAK/STAT-3 phosphorylation and nuclear translocation, thereby inducing cell cycle arrest and mitochondrial-mediated apoptosis (Baba et al. 2017). The mechanism of different anthocyanins found in EPPs in the activation of apoptosis in cancer cells is demonstrated in Fig. 7.

Eggplant calyxes also exhibit anticancer properties. This bioactivity has been attributed to *N-trans-p*-coumaroyltyramine (168), N-*trans-p*-coumaroyloctopamine (169), *N-trans-p*-coumaroylnoradrenline (170), *N-trans*-feruloyloctopamine (171), and the phenylpropanoid neochlorogenic acid (172). The cytotoxic effects of these five representative compounds were observed on A549, HCT116, and MCF7 cells (Song et al. 2021).

3.2 Antimicrobial Activity

Anthocyanins limit microbial populations by several means, including destabilizing the cytoplasm membrane, permeabilizing the plasma membrane, and damaging the cell wall and intercellular matrix (Cisowska et al. 2011). It might interfere with microbial metabolism by making some substrates inaccessible or altering intra- and extra-cellular microbial enzymes (Niño-Medina et al. 2017). Some anthocyanins

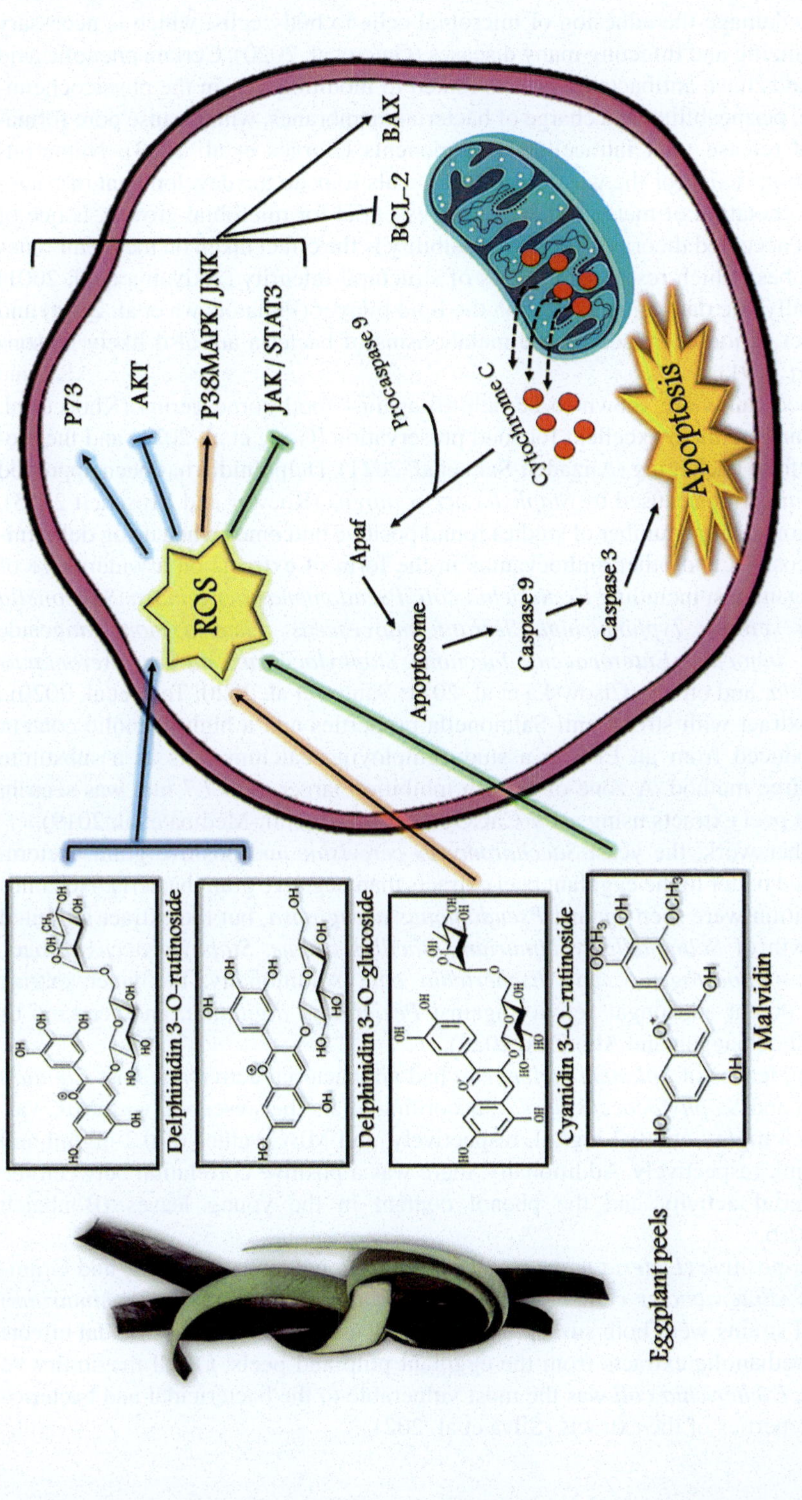

Fig. 7 Eggplant peels extract anthocyanins induce apoptosis in cancer cells and the mechanisms of the process

can also damage the adhesion of microbial cells to body cells, which is necessary for colonizing and infecting many diseases (Gato et al. 2020). Certain phenolic acid compounds have antibacterial effects linked to modifications in the physicochemical state, permeability, and charge of bacterial membranes, which cause pore formation and release vital intracellular components (Borges et al. 2013). Numerous explanations exist for the way phenol chemicals impede the development of bacteria. The chelation of metals, such as iron, essential for microbial growth, is one of the most accepted theories. Another possibility is the effect taken on the membranes of microbes, which results in the loss of structural integrity (Akiyama et al. 2001) and, finally, the damaging impact on the lipid bilayer (Funatogawa et al. 2004) and the direct action exercised on the metabolisms of bacteria are also likely reasons (Scalbert 1991).

Anthocyanins are shown to be helpful against food-borne germs (Khoo et al. 2017), making them excellent for food preservation (Gong et al. 2021) and the creation of food packaging (Alizadeh-Sani et al. 2021). Delphinidin has been approved to treat infections caused by *Staphylococcus aureus* (Roewer and Broscheit 2015). In addition, a large number of studies found positive outcomes when using delphinidin glycosides and other anthocyanins in the form of extracts on a wide range of microorganisms, including *Escherichia coli, Pseudomonas aeruginosa, Salmonella enterica serovar Typhimurium, Serratia marcescens, Listeria monocytogenes, Listeria innocua, Enterococcus faecium, Staphylococcus aureus, Aeromonas hydrophila,* and others (Cisowska et al. 2011; Santos et al. 2020; Tena et al. 2020).

An extract with strong anti-Salmonella properties and a high phenolic content was produced from an EPP in a study employing calcium salts as a substitute solvent-free method. A zone of growth inhibition larger than 7.7 mm was seen in eggplant peel extracts using a 1% $CaCO_3$ solution (Rochín-Medina et al. 2019).

In other work, the yeast *Saccharomyces cerevisiae* and positive gram bacteria responded better to the eggplant peel extracts than negative gram bacteria. No zones of inhibition were seen against *Pseudomonas aeruginosa*, but the extract inhibited the growth of *Salmonella typhimurium, Bacillus cereus, Staphylococcus aureus, Listeria monocytogenes,* and *Escherichia coli.* Additionally, the latter extract showed potent antifungal activity against *Penicillium digitatum*, another kind of yeast (Aboelhaggag and Abo-Zaid 2023).

Young leaves of *Solanum melongena* had antibacterial activity against *Candida albicans* and *Staphylococcus aureus*, according to another research, with MIC values of 0.8 mg/ml and 0.2 mg/ml, respectively, and MBC values of 0.2 mg/ml and 3.1 mg/ml, respectively. Additionally, there was a positive correlation between the antibacterial activity and the phenol content in the young leaves (Bouhajeb et al. 2020).

Gram-positive (*Listeria monocytogenes* and Staphylococcus aureus) and Gram-negative (*Enterobacter cloacae*, *Escherichia coli*, and *Salmonella Typhimurium*) bacterial strains were both susceptible to the bacteriostatic and bactericidal effects of hydroethanolic extracts from the eggplant pulp and peels. Of all the strains of bacteria, *Escherichia coli* was the most vulnerable to the bactericidal and bacteriostatic properties of the extracts (Silva et al. 2021).

3.3 Hypoglycemic Activity

In the small intestine, α-amylase mostly breaks down starch to reduce sugars such as maltose, maltotriose, and amylodextrin. α-glucosidase then proceeds to hydrolyze these reducing sugars, producing glucose. Thus, α-amylase plays a crucial role in the hydrolysis of starch. Therefore, the enzyme activity should be modulated to prevent and cure postprandial hyperglycemia and related metabolic problems (Sun et al. 2020; Sun et al. 2016). For example, prior research found that purple-colored extracts from peels had a 66.37% α-amylase inhibitory effect, while white-colored ones had 45.93% and black-colored ones had 37.73%. This difference can be attributed to different compositions, e.g., the former extract was rich in chlorogenic acid in free form (Colak et al. 2022). However, other compounds can contribute since in vitro research found that delphinidin 3-*O*-rutinoside inhibited lipoxygenase, lipase, and α-amylase (Condurache et al. 2021).

3.4 Anti-hyperlipidemic Activity

Metabolic syndrome is a collection of metabolic abnormalities that include insulin resistance, atherogenic dyslipidemia, central obesity, and hypertension. Its etiology includes numerous hereditary and acquired elements that fall under the general category of persistent low-grade inflammation and insulin resistance (Fahed et al. 2022). Lipoxygenase (LOX) catalyzes the conversion of polyunsaturated fatty acids into hydroperoxides, which causes several inflammatory diseases, including multiple sclerosis (Sroka et al. 2017). Due to its role in catalyzing the hydrolysis of triglycerides into fatty acids, high levels of pancreatic lipase in the human body are responsible for pancreatic diseases and obesity. Phenolic components from eggplant peels have been linked to reducing metabolic syndrome risk by blocking the corresponding enzymes (Condurache et al. 2021). Inhibiting the activities of the cyclooxygenase enzyme linked to several illnesses is another benefit of cyanidin 3-*O*-glucoside, delphinidin 3-*O*-glucoside, and petunidin 3-*O*-glucoside (Lee et al. 2017).

3.5 Other Bioactivities

Warts, burns, and various inflammatory conditions, including stomatitis, arthritis, and gastritis, can all be effectively treated with eggplant stalk. The anti-inflammatory properties of ethanolic extracts of aubergine stalks were shown by Im et al. (2016) by a considerable reduction in pro-inflammatory cytokine production, NO, iNOS, COX-2, and PGE2.

The effects of eggplant peel anthocyanins were investigated using a rat aortic ring angiogenesis model. This method is widely recognized for its utility in assessing the efficacy of antiangiogenic agents within a complex system involving interactions between endothelial cells, fibroblasts, pericytes, and smooth muscle cells. After 2–3 days, fibroblastic fusiform cells migrated from the ends of the aortic rings and spread throughout the collagen gel. After 5–6 days, microvessels began to emerge from the ends of the aortic rings and underwent elongation. Notably, anthocyanins at a concentration of 200 μM strongly inhibited the outgrowth of microvessels, with an even more significant inhibitory effect observed at a higher concentration of 10 μM (Matsubara et al. 2005).

Eggplant calyx extracts contain higher concentrations of minerals and vitamins compared to the edible portion of the plant. This makes it a valuable food for diabetics and cardiovascular patients, as well as for maintaining the health of bones and muscles and reducing inflammation while supporting the body's immune system. Additionally, some studies have shown that when the body gets the right amounts of phosphorus, potassium, and vitamin C, it can effectively reduce pain. Eggplant calyxes contain vital elements thought to be lacking in billions of individuals and can substitute for these deficiencies (Al Nachar 2019).

It was also discovered that an eggplant calyxes extract has hepatoprotective potential in diabetic rats. It activated detoxification enzymes and antioxidants while lowering ROS, inflammatory cytokines, cell apoptosis, and the expression of the p53 and Bax genes. It restored the histological and functional properties of the hepatocyte (Roshankhah et al. 2020).

3.6 *Antioxidant Activity*

The phenolic contents of eggplant make it one of the top ten vegetables regarding oxygen radical scavenging capabilities (Cao et al. 1996). In this context, several studies addressed the antioxidant activity of different extracts of eggplant by-products from various geographical origins employing 1,1-diphenyl-2-picrylhydrazyl (DPPH) radical-scavenging activity, reducing power, SOD-like activity, scavenging effect of H_2O_2, and metal chelating activity. Table 7 demonstrates the antioxidant activities of some eggplant by-products from different geographical origins.

Table 7 Antioxidant activity of eggplant by-products from different geographical origins

Sample	Origin	Extract type	DPPH RSA (mg/ml)	RP (mg/ml)	SOD-like activity (mg/ml)	Scavenging effect of H_2O_2 (minute)	ABTS (mg/ml)	Phosphomo-lybdenum (mg/ml)	Metal chelating activity (% inhibition)	Superoxide anion scavenging capacity (nmol/ml)	Hydroxyl radical scavenging capacity (nmol/ml)	Ref.
Calyxes	Korea	70% Ethanol	IC_{50} 3.45 ± 0.02	IC_{50} 2.42 ± 0.26	IC_{50} 0.51 ± 0.02							Jung et al. (2011)
		Water	IC_{50} 0.49 ± 0.16	IC_{50} 0.03 ± 5.49	IC_{50} 0.39 ± 0.01							
Leaves	Korea	70% Ethanol	IC_{50} 4.94 ± 0.03	IC_{50} 2.92 ± 0.06	IC_{50} 0.58 ± 0.02							Jung et al. (2011)
		Water	IC_{50} 6.89 ± 0.20	IC_{50} 162.20 ± 23.34	IC_{50} 0.30 ± 0.07							
Peels	Korea	70% Ethanol	IC_{50} 0.98 ± 0.33	IC_{50} 1.10 ± 0.13	IC_{50} 0.21 ± 0.02							Jung et al. (2011)
		Water	IC_{50} 1.69 ± 0.13	IC_{50} 1.66 ± 0.09	IC_{50} 0.22 ± 0.00							
Pulp	Korea	70% Ethanol	IC_{50} 12.30 ± 0.27	IC_{50} 3.90 ± 0.08	IC_{50} 0.40 ± 0.01							Jung et al. (2011)
		Water	IC_{50} 6.47 ± 0.18	IC_{50} 2.83 ± 0.02	IC_{50} 0.69 ± 0.00							
Stems	Korea	70% Ethanol	IC_{50} 13.13 ± 0.06	IC_{50} 3.10 ± 0.04	IC_{50} 0.38 ± 0.01							Jung et al. (2011)
		Water	IC_{50} 26.20 ± 1.96	IC_{50} 329.80 ± 13.80	IC_{50} 0.16 ± 0.00							
Peels	Algeria	70% Methanol	IC_{50} 4 ± 0.20	39 ± 2.5 mg QE/100 g DE		IT_{50} 36.97 ± 2.19			14.24 ± 1.26			Boulekbache-Makhlouf et al. (2013)
Peels	Algeria	70% Ethanol	IC_{50} 4.05 ± 0.08	30 ± 0.6 mg QE/100 g DE		IT_{50} 46.21 ± 1.50			3.71 ± 0.21			Boulekbache-Makhlouf et al. (2013)
Peels	Algeria	70% Acetone	IC_{50} 3.97 ± 0.95	33 ± 1.6 mg QE/100 g DE		IT_{50} 26.1 ± 0.53			18.53 ± 0.47			Boulekbache-Makhlouf et al. (2013)
Peels	Iran	73.5% Ethanol microwave-assisted extraction (MAE)	IC_{50} 0.52									Doulabi et al. (2020)

(continued)

Table 7 continued

Sample	Origin	Extract type	DPPH RSA (mg/ml)	RP (mg/ml)	SOD-like activity (mg/ml)	Scavenging effect of H_2O_2 (minute)	ABTS (mg/ml)	Phosphomo-lybdenum (mg/ml)	Metal chelating activity (% inhibition)	Superoxide anion scavenging capacity (nmol/ml)	Hydroxyl radical scavenging capacity (nmol/ml)	Ref.
Peels	Egypt	Ethanolic extract	66.5% at 25 µl concentration									El-Gammal (2012)
Stalks	Korea	95% Ethanol	IC_{50} 59.7									Im et al. (2016)
Peels	Italy	1% $CaCO_3$ Extract	IC_{50} 89									Rochín-Medina et al. (2019)
Peels	Pakistan	70% Ethanol	IC_{50} 95.23 ± 2.39									Kainat et al. (2023)
Peels	Italy	Semi-purified nasunin extract in chlorogenic acid								EC_{50} 9.73	EC_{50} 5.40	Braga et al. (2016)
Peels	Italy	Semi-purified delphinidin-3-rutinoside extract in chlorogenic acid								EC_{50} 20.61	EC_{50} 10.51	Braga et al. (2016)
Peels	India	Extracted with a solvent mixture of water, ethanol, and citric acid in the ratio 50:48:2 (v/v) Temperature (37.32 °C) Enzyme conc. (5%) Extraction time (1 h)	79.92%						29.90 mmol AAE/100 g AAE: Ascorbic acid equivalent			Amulya and ul Islam (2023)

Sample	Origin	Extract type	DPPH RSA (mg/ml)	RP (mg/ml)	SOD-like activity (mg/ml)	Scavenging effect of H_2O_2 (minute)	ABTS (mg/ml)	Phosphomo-lybdenum (mg/ml)	Metal chelating activity (% inhibition)	Superoxide anion scavenging capacity (nmol/ml)	Hydroxyl radical scavenging capacity (nmol/ml)	Ref.
Tizona young leaves	Tunisia	Ethanolic extract	IC_{50} 1.4 ± 0.1	IC_{50} 0.03 ± 0.00			IC_{50} 0.2 ± 0.0	IC_{50} 0.02 ± 0.00				Bouhajeb et al. (2020)
Tizona mature leaves	Tunisia	Ethanolic extract	IC_{50} 0.1 ± 0.0	IC_{50} 0.03 ± 0.00			IC_{50} 0.2 ± 0.0	IC_{50} 0.01 ± 0.00				Bouhajeb et al. (2020)
A1014 young leaves	Tunisia	Ethanolic extract	IC_{50} 0.2 ± 0.0	IC_{50} 0.02 ± 0.00			IC_{50} 0.3 ± 0.0	IC_{50} 0.02 ± 0.00				Bouhajeb et al. (2020)
A1014 mature leaves	Tunisia	Ethanolic extract	IC_{50} 0.2 ± 0.0	IC_{50} 0.01 ± 0.00			IC_{50} 0.2 ± 0.0	IC_{50} 0.02 ± 0.00				Bouhajeb et al. (2020)
Nawel young leaves	Tunisia	Ethanolic extract	IC_{50} 0.2 ± 0.0	IC_{50} 0.10 ± 0.00			IC_{50} 0.3 ± 0.0	IC_{50} 0.01 ± 0.00				Bouhajeb et al. (2020)
Nawel mature leaves	Tunisia	Ethanolic extract	IC_{50} 0.4 ± 0.1	IC_{50} 0.10 ± 0.00			IC_{50} 0.4 ± 0.0	IC_{50} 0.04 ± 0.00				Bouhajeb et al. (2020)
A336 young leaves	Tunisia	Ethanolic extract	IC_{50} 0.2 ± 0.0	IC_{50} 0.04 ± 0.00			IC_{50} 0.3 ± 0.0	IC_{50} 0.03 ± 0.00				Bouhajeb et al. (2020)
A336 mature leaves	Tunisia	Ethanolic extract	IC_{50} 0.3 ± 0.0	IC_{50} 0.02 ± 0.00			IC_{50} 0.3 ± 0.0	IC_{50} 0.02 ± 0.00				Bouhajeb et al. (2020)
Mirval young leaves	Tunisia	Ethanolic extract	IC_{50} 0.4 ± 0.0	IC_{50} 0.10 ± 0.00			IC_{50} 0.3 ± 0.0	IC_{50} 0.02 ± 0.00				Bouhajeb et al. (2020)
Mirval mature leaves	Tunisia	Ethanolic extract	IC_{50} 0.2 ± 0.0	IC_{50} 0.10 ± 0.00			IC_{50} 0.3 ± 0.0	IC_{50} 0.03 ± 0.00				Bouhajeb et al. (2020)

Table 7 continued

Sample	Origin	Extract type	DPPH RSA (mg/ml)	RP (mg/ml)	SOD-like activity (mg/ml)	Scavenging effect of H_2O_2 (minute)	ABTS (mg/ml)	Phosphomo-lybdenum (mg/ml)	Metal chelating activity (% inhibition)	Superoxide anion scavenging capacity (nmol/ml)	Hydroxyl radical scavenging capacity (nmol/ml)	Ref.
Roots	America	Methanolic extract	IC_{50} 4.25				IC_{50} 2.09					Contreras-Angulo et al. (2022)
Leafs	America	Methanolic extract	IC_{50} 0.46				IC_{50} 0.69					Contreras-Angulo et al. (2022)
Stems	America	Methanolic extract	IC_{50} 2.41				IC_{50} 2.63					Contreras-Angulo et al. (2022)
Leaves			IC_{50} 4.94									Jung et al. (2011)
Stems			IC_{50} 13.13									Jung et al. (2011)
Peels		Ethanolic extract					IC_{50} 1.00					Jo et al. (2012)

4 Valorization of Eggplant By-Products

Solanum melongena L. is a highly productive and cost-effective agricultural crop belonging to the solanaceae family. It is grown in a wide range of shapes, sizes, and colors and has its native origin in India (Horincar et al. 2020b). It is cultivated extensively in various regions across Asia, Europe, Africa, and America. The global interest in eggplant cultivation is increasing rapidly due to its rich nutritional content and versatility (Kazemi et al. 2019b). However, the industries producing such products also generate significant quantities of eggplant by-products, which are considered waste and end up in landfills. The peels and calyxes are the eggplant's main by-products (Mauro et al. 2020). The peel is particularly abundant in anthocyanins, while the peel and calyx are excellent sources of dietary fibers like pectin and cellulose (Niño-Medina et al. 2017).

4.1 Obtainment of Eggplant Peel Pigments

Anthocyanins are used as food pigments. Eggplant peels are rich in anthocyanin, but the amount of these compounds is affected by several factors, including the ripening stage, as the amount of anthocyanins decreases in overripe eggplant and process conditions. The ripped Black Bell cultivar has the highest amounts of peel anthocyanins among different eggplant cultivars. On the other hand, the ripped *Black Moon* cultivar has the lowest concentration of anthocyanins (Mauro et al. 2020). Therefore, selecting the cultivar and species is important to obtain more of these pigments. This is important to look for their color. For example, the amount and location of hydroxyl and methoxy groups in the anthocyanidin aglycone and the position, quantity, and structure of these conjugate sugars determine the fundamental differences between the different types of anthocyanins. Although there are more than 20 different kinds of anthocyanidins, the majority of plant anthocyanins come from six of them: cyanidin, peonidin, pelargonidin, petunidin, malvidin, and delphinidin (Tena et al. 2020). The dark purple color of the peel is due to the presence of anthocyanins of delphinidin derivatives. Delphinidin-3-*O*-rutinoside was the most significant anthocyanin. Other delphinidin identified includes delphinidin-3-(*p*-coumaroylrutinoside)-5-glucoside, delphinidin-3-rutinoside-5-glucoside, delphinidin-3-glucoside and delphinidin-3-rutinosyl-glucoside (Condurache et al. 2020).

Another important factor is related to the extraction conditions. For example, a study compared conventional solid–liquid extraction (SLE) with ultrasound-assisted extraction (UAE). High-performance liquid chromatography was utilized to analyze the chromatographic profile of the anthocyanins present in the eggplant peel extracts. The chromatographic analysis revealed the presence of five compounds: delphinidin 3-*O*-rutinoside-5-glucoside, delphinidin 3-*O*-glucoside, delphinidin 3-*O*-rutinoside, cyanidin 3-*O*-rutinoside, and petunidin 3-*O*-rutinoside. The SLE extract exhibited the following relative percentage concentrations of individual anthocyanins: delphinidin 3-O-rutinoside-5-glucoside (9.4%), delphinidin 3-*O*-glucoside (9.7%), delphinidin 3-*O*-rutinoside (73.6%), cyanidin 3-*O*-rutinoside (2.0%), and petunidin 3-*O*-rutinoside (0.4%). On the other hand, the UAE extract displayed the subsequent concentrations of anthocyanins: delphinidin 3-*O*-rutinoside-5-glucoside (7.3%), delphinidin 3-*O*-glucoside (12.2%), delphinidin 3-*O*-rutinoside (75.4%), cyanidin 3-*O*-rutinoside (1.5%), and petunidin 3-*O*-rutinoside (0.5%). (Condurache et al. 2021). Nonetheless, in both cases, delphinidin 3-*O*-rutinoside was the main pigment.

Another interesting pigment is delphinidin-3-(*p*-coumaroylrutinoside)-5-glucoside (nasunin), which isolated purple-colored crystals from eggplant. It showed a potent free radical scavenger with protective activity against lipid peroxidation (Noda et al. 2000). Therefore, besides being used as a colorant, the valuable biological activities of eggplant anthocyanins could make these compounds replace synthetic food additives with a good impact on health (Karimi et al. 2021) (the focus of Sect. 4.4).

Al-Shammary et al. (2015) and Kamali Moghaddam et al. (2019) have studied other exciting uses of this pigment. For example, the first authors examined ethanolic extract of eggplant peels for chromosome banding visualization, which appeared bright dark green. Al-Shammary et al. (2015) suggested using eggplant extract as a staining in cancer research for chromosome staining as a novel, easily prepared, and available alternative (Al-Shammary et al. 2015). The other work studied eggplant peel extract as a natural colorant for woolen yarn. After using different mordants with different mordanting procedures, the extract showed various shades of greenish-yellow to semi-blue. The semi-blue color was noticeable when tin chloride was used as a mordant, and chromaticity values showed a navy blue hue, which was better for sulfuric acid for mordanted woolen yarn (Kamali Moghaddam et al. 2019).

4.2 Obtainment of Eggplant Pectins

Pectin is a complex heteropolysaccharide interested in the food, pharmaceutical, and cosmetic sectors. As commented before, pectin from peels has around 60.2% GalA. The optimal pectin range from different extraction techniques applied to eggplant peels was 26.1–33.6%, respectively. When compared to other high-yielding sources, this amount is highly significant, e.g., citrus peel (~5–26%), tomato by-products (~10–35%), and passion fruit peel (~7–30%) (Marić et al. 2018). Pectin can also be obtained from the calyxes, and it can also be classified as HMP. The pectin extraction yield was about 18.4% (Marić et al. 2018), a lower yield than peels but in the range of the other sources.

Eggplant pectin falls under the category of HMP, which is widely employed as a gelling and thickening agent in the food industry. HMP can form gels in environments with high sugar concentrations and acidic pH, making it an ideal substance for use in foods with similar conditions, such as jams, jellies, and marmalades

(Christiaens et al. 2016). Additionally, HMP can form thin films under these conditions, making it valuable for producing antioxidant, antimicrobial, preservative, and pH indicator films (Kumar et al. 2020). Recent trends in pectin utilization have focused on its application in pharmaceutical science and medicine, either as an independent medicinal substance or for the development of drug delivery systems such as hydrogels (Amirian et al. 2021), liposomes, and other types of nanoparticles (Zhao et al. 2020).

4.3 Uses in Fermentation Approaches

Eggplant peel has the potential to be used as a carbon source in the fermentation process to produce the food additive pullulan. Pullulan is a linear exo-homopolysaccharide that is produced through submerged fermentation method by various strains of yeast-like microorganism, *Aureobasidium pullulans* (Talekar et al. 2018). However, there are certain possible disadvantages. For instance, the presence of pectin in plant cell walls may prevent hydrolytic enzymes from accessing cellulose molecules. Moreover, phenolic compounds found in high concentrations may have antimicrobial effects (Talekar et al. 2018)

An integrated valorization approach has been reported to produce the food additive pullulan to obtain many benefits from this by-product and convert disadvantages into opportunities. This method involves extracting the pectin and phenolic compounds and using the leftovers to produce pullulan through microbial fermentation and enzymatic hydrolysis as a zero-waste valorization method. Pullulan production yield was reported to be 16.8 g/l based on this approach. Additionally, the obtained pullulan showed $\alpha(1 \rightarrow 6)$ and $\alpha(1 \rightarrow 4)$ glucoside linkages, which are signs of recurring maltotriose units; it also had acceptable protein and ash concentrations (~1.8% and ~2.1%, respectively), and its chemical properties were comparable to those of pullulan sold in stores (Kazemi et al. 2019b).

Pullulan extracted from eggplant by-product is commonly utilized in conjunction with other polymers to create films and coatings for a variety of applications, including fruit preservation, food packaging (Luís et al. 2020), edible films, and drug delivery (Shah et al. 2020). Additionally, the fabrication of hydrogels from pullulan, sometimes in combination with other polymers, has emerged as a popular trend in recent years. Furthermore, pullulan's high-water solubility and unique rheological properties make it an excellent material for producing consistent and flawless micro and nanofibers through electrospinning (Luís et al. 2020).

Eggplant Peel Anthocyanins Eggplant peels were extracted and determined to contain an anthocyanin content of 158.00 parts per million (ppm). Aliquots of concentrated eggplant peel contain 200, 400, and 800 ppm, respectively. Tertiary Butyl Hydroquinone (TBHQ) was found to be at 200 ppm. Efforts were made to evaluate the antioxidation effects of these compounds in rats fed on a hyperlipidemic diet daily by stomach tube for 8 weeks. The results of the antioxidation experiments showed a significant decrease ($P < 0.05$) in serum total cholesterol levels, LDL cholesterol levels, LV cholesterol levels, and triglyceride levels, as well as an increase in HLD cholesterol levels and a decrease in VLDL cholesterol levels (Basuny et al. 2012).

4.4 Applications of Eggplant Waste Extracts and Bioactive Compounds

4.4.1 Food Applications

Annually, over ten million tons of eggplant peels and calyxes are generated as food industry processing waste, with little to no utilization (Karimi et al. 2021). However, the by-products of eggplant waste are a source of valuable bioactive compounds, as it was the focus of 3 (Agregán et al. 2021). Notably, anthocyanin derivatives of the delphinidin type present in peels have outstanding antioxidant, anticancer, and antimicrobial properties (Karimi et al. 2021). Therefore, in addition to serving as colorant agents for food products, anthocyanins and other phenolic compounds provide various functionalities in foods and beverages, such as improving shelf-life as preservatives (Agregán et al. 2021) and offering health-promoting properties (Karimi et al. 2021).

For example, the oxidation stability of cookies formulated with wheat flour substituted with 10 and 15% eggplant flour was improved since it showed a significant decrease in peroxide value after 3 weeks of storage compared to the control ones. This oxidation stability could be explained by increasing total phenolic content up to 2.6 times and DPPH free-radical scavenging activity up to 3.4 times. The antioxidant activity of eggplant extract could be a suitable natural preservative for foods and beverages (Agregán et al. 2021).

Anthocyanin utilization in food fortification is encountered as a major challenge since anthocyanin stability could be affected by temperature, enzymatic activity, oxygen, metal, pH, and exposure to light or UV (Horincar et al. 2020a). For instance, cooking methods using high temperatures by boiling, frying, grilling, and baking showed significant loss of eggplant anthocyanins by 43.7%, 60.3%, 84.4%, and 91.2%, respectively (Martini et al. 2021).

Zhang et al. (2020) reported the effect of different cooking methods of cookies prepared with anthocyanin dough and dietary fiber on anthocyanin's thermal stability. The results showed a significant decrease in anthocyanin concentrations due to thermal treatment. However, the dietary fiber showed a protective effect on anthocyanins. The thermal stability of anthocyanins is due to their entrapment into the fiber matrix, which is hypothesized through the formation of interactions, such as van der Waals forces, hydrogen bonds, hydrophobic interactions, and electrostatic interactions (Zhang et al. 2020).

Chatterjee et al. (2013) compared the thermal stability of eggplant waste extracted by the solvent mixture of water:ethanol, and supercritical CO_2. The extraction with CO_2e showed much more stability than solvent mixture extract, and CO_2 extract contains a higher amount of thermolabile delphinidin-3-*O*-rutinoside and lower amounts of trace metals. Therefore, depending on the purpose, the extracts were incorporated in custard and jelly crystals as a thermal and non-thermal food application model, respectively (Chatterjee et al. 2013). However, the thermal stability can be modulated by encapsulation. For example, it was reported that encapsulated eggplant extract has a higher stability time of 1500% than non-encapsulated extract (Karimi et al. 2021).

Pastry cream was improved by adding 10% of encapsulated eggplant extract in whey protein isolate and acacia gum to produce value-added food products. The pastry cream fortified with encapsulated eggplant extract showed an increase of total phenolic and flavonoid contents by 983% and 141%, respectively. Encapsulated eggplant extract did not affect sensory evaluation compared to control samples and improved the enriched samples' overall color and acceptance scores. Horincar et al. (2020a, b) concluded that eggplant extract is more suitable as a healthier additive for foods such as chocolate, candy, and jelly (Horincar et al. 2020a).

Another study was conducted by Sarabandi et al. (2019) comparing maltodextrin, gum Arabic, and a combination of these to find the best compound for encapsulation of eggplant extract for gummy candy fortification. The maltodextrin-encapsulated eggplant extract showed higher solubility, total phenolic content, and antioxidant activity. Moreover, the color and acceptability score were improved by the addition of 1.5% of encapsulated eggplant extract (Sarabandi et al. 2019).

In conclusion, anthocyanins, phenolic compounds, and eggplant extracts could be applied in many food products such as jelly crystals, custard, cookies, bread, gluten-free spaghetti, yogurt-like products, and gummy candy (Agregán et al. 2021; Karimi et al. 2021). Finally, anthocyanin stability could be solved by different methods such as encapsulation of extract or dietary fiber addition. Other applications could be for producing and stabilizing various emulsions (Gong et al. 2020).

4.4.2 Non-food Applications

Diethyl ether extract of eggplant showed antifungal activity against *Aspergillus flavus* (maize fungus) and 96.11% inhibition activity against aflatoxin B1 in comparison with 72.95% inhibition activity for topsin fungicide after 4 weeks of maize storage treatment. These results suggested that this extract can be applied to stored crops as an effective and safe agricultural fungicide to prevent stored food spoilage and treat plant diseases (Ismail et al. 2021).

References

Abd El Hafeez MS (2024) Alternative natural therapeutic plants and diabetes mellitus. ERU Res J. https://doi.org/10.21608/erurj.2024.221225.1051

Abd El Hafeez MS, El Gindi O, Hetta MH, Aly HF, Ahmed SA (2022) Quality control, anti-hyperglycemic, and anti-inflammatory assessment of Colvillea racemosa leaves using in vitro, in vivo investigations and its correlation with the phytoconstituents identified via LC-QTOF-MS and MS/MS. Plants 11(6):830

Aboelhaggag RM, Abo-Zaid EM (2023) Healthy sorbet with eggplant peel and beet stalk-leaves extracts as natural colorants. Egypt J Chem 66(9):145–157

Abouzed TK, Contreras MDM, Sadek KM, Shukry M, Abdelhady DH, Gouda WM, Abdo W, Nasr NE, Mekky RH, Segura-Carretero A, Kahilo KA-A, Abdel-Sattar E (2018) Red onion scales ameliorated streptozotocin-induced diabetes and diabetic nephropathy in Wistar rats in relation to their metabolite fingerprint. Diabetes Res Clin Pract 140:253–264. 10.1016/j.diabres.2018.03.042

Agregán R, Munekata PE, Feng X, Astray G, Gullón B, Lorenzo JM (2021) Recent advances in the extraction of polyphenols from eggplant and their application in foods. LWT 146:111381

Akiyama H, Fujii K, Yamasaki O, Oono T, Iwatsuki K (2001) Antibacterial action of several tannins against Staphylococcus aureus. J Antimicrob Chemother 48(4):487–491

Alizadeh-Sani M, Tavassoli M, McClements DJ, Hamishehkar H (2021) Multifunctional halochromic packaging materials: Saffron petal anthocyanin loaded-chitosan nanofiber/methyl cellulose matrices. Food Hydrocoll 111:106237

Al Nachar K (2019) Investigation and measurement of some mineral and vitamins in eggplant fruit calyx, and the possibility of being used as food supplements and alternative medicine. J Food Nutr 5:1–10

Al-Shammary HA, Salh R, Sabah H (2015) Eggplant peel ethanolic extract: a novel and alternative stain for chromosome banding. J Bioprocess Biotech 5(201):2

Amirian J, Zeng Y, Shekh MI, Sharma G, Stadler FJ, Song J, Du B, Zhu Y (2021) In-situ cross-linked hydrogel based on amidated pectin/oxidized chitosan as potential wound dressing for skin repairing. Carbohydr Polym 251:117005

Amulya P, ul Islam R (2023) Optimization of enzyme-assisted extraction of anthocyanins from eggplant (Solanum melongena L.) peel. Food Chem X 18:100643

Ayvaz Z, Danciu C, Ayvaz H, Stanila A, Nistor M, Socaciu C (2020) Anthocyanins, vibrant color pigments, and their role in skin cancer prevention. Biomedicines 8:336

Baba AB, Nivetha R, Chattopadhyay I, Nagini S (2017) Blueberry and malvidin inhibit cell cycle progression and induce mitochondrial-mediated apoptosis by abrogating the JAK/STAT-3 signalling pathway. Food Chem Toxicol 109:534–543

Basuny AM, Arafat SM, El-Marzooq MA (2012) Antioxidant and antihyperlipidemic activities of anthocyanins from eggplant peels. J Pharma Res Rev 2(3):50–57

Borges A, Ferreira C, Saavedra MJ, Simões M (2013) Antibacterial activity and mode of action of ferulic and gallic acids against pathogenic bacteria. Microb Drug Resist 19(4):256–265

Bouhajeb R, Selmi S, Nakbi A, Jlassi I, Montevecchi G, Flamini G, Zarrad I, Dabbou S (2020) Chemical composition analysis, antioxidant, and antibacterial activities of eggplant leaves. Chem Biodivers 17(12):e2000405

Boulekbache-Makhlouf L, Medouni L, Medouni-Adrar S, Arkoub L, Madani K (2013) Effect of solvents extraction on phenolic content and antioxidant activity of the byproduct of eggplant. Ind Crop Prod 49:668–674

Braga PC, Scalzo RL, Dal Sasso M, Lattuada N, Greco V, Fibiani M (2016) Characterization and antioxidant activity of semi-purified extracts and pure delphinidin-glycosides from eggplant peel (Solanum melongena L.). J Funct Foods 20:411–421

Cao G, Sofic E, Prior RL (1996) Antioxidant capacity of tea and common vegetables. J Agric Food Chem 44(11):3426–3431

Chapman MA (2019) Introduction: The importance of eggplant. In: Chapman M (ed) The eggplant genome. Compendium of plant genomes. Springer, Cham. https://doi.org/10.1007/978-3-319-99208-2_1

Chatterjee D, Jadhav NT, Bhattacharjee P (2013) Solvent and supercritical carbon dioxide extraction of color from eggplants: Characterization and food applications. LWT Food Sci Technol 51(1):319–324

Chen J, Zhu Y, Zhang W, Peng X, Zhou J, Li F, Han B, Liu X, Ou Y, Yu X (2018) Delphinidin induced protective autophagy via mTOR pathway suppression and AMPK pathway activation in HER-2 positive breast cancer cells. BMC Cancer 18(1):1–13

Christiaens S, Van Buggenhout S, Houben K, Jamsazzadeh Kermani Z, Moelants KR, Ngouémazong ED, Van Loey A, Hendrickx ME (2016) Process–structure–function relations of pectin in food. Crit Rev Food Sci Nutr 56(6):1021–1042

Cisowska A, Wojnicz D, Hendrich AB (2011) Anthocyanins as antimicrobial agents of natural plant origin. Nat Prod Commun 6(1):1934578X1100600136

Colak N, Kurt-Celebi A, Gruz J, Strnad M, Hayirlioglu-Ayaz S, Choung M-G, Esatbeyoglu T, Ayaz FA (2022) The phenolics and antioxidant properties of black and purple versus white eggplant cultivars. Molecules 27(8):2410

Condurache NN, Aprodu I, Grigore-Gurgu L, Petre BA, Enachi E, Râpeanu G, Bahrim GE, Stănciuc N (2020) Fluorescence spectroscopy and molecular modeling of anthocyanins binding to bovine lactoferrin peptides. Food Chem 318:126508

Condurache N-N, Croitoru C, Enachi E, Bahrim G-E, Stănciuc N, Râpeanu G (2021) Eggplant peels as a valuable source of anthocyanins: Extraction, thermal stability and biological activities. Plants 10(3):577. https://www.mdpi.com/2223-7747/10/3/577

Contreras-Angulo LA, Moreno-Ulloa A, Carballo-Castañeda RA, León-Felix J, Romero-Quintana JG, Aguilar-Medina M, Ramos-Payán R, Heredia JB (2022) Metabolomic analysis of phytochemical compounds from agricultural residues of eggplant (Solanum melongena L.). Molecules 27(20):7013

Costea T, Nagy P, Ganea C, Szöllősi J, Mocanu M-M (2019) Molecular mechanisms and bioavailability of polyphenols in prostate cancer. Int J Mol Sci 20(5):1062

Dharmawansa KS, Hoskin DW, Rupasinghe HV (2020) Chemopreventive effect of dietary anthocyanins against gastrointestinal cancers: A review of recent advances and perspectives. Int J Mol Sci 21(18):6555

Doulabi M, Golmakani MT, Ansari S (2020) Evaluation and optimization of microwave-assisted extraction of bioactive compounds from eggplant peel by-product. J Food Process Preserv 44(11):e14853

El-Gammal RE (2012) Utilization of some vegetables wastes and by-products as natural sources of antioxidants. J Food Dairy Sci 3(2):101–110

Fahed G, Aoun L, Bou Zerdan M, Allam S, Bou Zerdan M, Bouferraa Y, Assi HI (2022) Metabolic syndrome: updates on pathophysiology and management in 2021. Int J Mol Sci 23(2):786

Feng R, Ni H-M, Wang SY, Tourkova IL, Shurin MR, Harada H, Yin X-M (2007) Cyanidin-3-rutinoside, a natural polyphenol antioxidant, selectively kills leukemic cells by induction of oxidative stress. J Biol Chem 282(18):13468–13476

Ferarsa S, Zhang W, Moulai-Mostefa N, Ding L, Jaffrin MY, Grimi N (2018) Recovery of anthocyanins and other phenolic compounds from purple eggplant peels and pulps using ultrasonic-assisted extraction. Food Bioprod Process 109:19–28

Funatogawa K, Hayashi S, Shimomura H, Yoshida T, Hatano T, Ito H, Hirai Y (2004) Antibacterial activity of hydrolyzable tannins derived from medicinal plants against Helicobacter pylori. Microbiol Immunol 48(4):251–261

Gato E, Rosalowska A, Martinez-Guitian M, Lores M, Bou G, Perez A (2020) Anti-adhesive activity of a Vaccinium corymbosum polyphenolic extract targeting intestinal colonization by Klebsiella pneumoniae. Biomed Pharmacother 132:110885

Gong C, Lee MC, Godec M, Zhang Z, Abbaspourrad A (2020) Ultrasonic encapsulation of cinnamon flavor to impart heat stability for baking applications. Food Hydrocoll 99:105316

Gong S, Fei P, Sun Q, Guo L, Jiang L, Duo K, Bi X, Yun X (2021) Action mode of cranberry anthocyanin on physiological and morphological properties of Staphylococcus aureus and its application in cooked meat. Food Microbiol 94:103632

Gürbüz N, Uluişik S, Frary A, Frary A, Doğanlar S (2018) Health benefits and bioactive compounds of eggplant. Food Chem 268:602–610

Han B, Peng X, Cheng D, Zhu Y, Du J, Li J, Yu X (2019) Delphinidin suppresses breast carcinogenesis through the HOTAIR/micro RNA-34a axis. Cancer Sci 110(10):3089–3097

Hegazy MM, Mekky RH, Afifi WM, Mostafa AE, Abbass HS (2023) Composition and biological activities of Ficus carica latex. In: Ramadan MF (ed) Fig (Ficus carica): production, processing, and properties. Springer, Cham, pp 597–641. https://doi.org/10.1007/978-3-031-16493-4_27

Hernández-Herrero JA, Frutos MJ (2011) Degradation kinetics of pigment, colour and stability of the antioxidant capacity in juice model systems from six anthocyanin sources. Int J Food Sci Technol 46(12):2550–2557

Horincar G, Enachi E, Barbu V, Andronoiu DG, Râpeanu G, Stănciuc N, Aprodu I (2020a) Value-added pastry cream enriched with microencapsulated bioactive compounds from eggplant (Solanum melongena L.) peel. Antioxidants 9(4):351

Horincar G, Enachi E, Bolea C, Râpeanu G, Aprodu I (2020b) Value-added lager beer enriched with eggplant (Solanum melongena L.) peel extract. Molecules 25(3):731. https://www.mdpi.com/1420-3049/25/3/731

Huang C-C, Hung C-H, Hung T-W, Lin Y-C, Wang C-J, Kao S-H (2019) Dietary delphinidin inhibits human colorectal cancer metastasis associating with upregulation of miR-204-3p and suppression of the integrin/FAK axis. Sci Rep 9(1):18954

Im K, Lee JY, Byeon H, Hwang KW, Kang W, Whang WK, Min H (2016) In Vitro antioxidative and anti-inflammatory activities of the ethanol extract of eggplant (Solanum melongena) stalks in macrophage RAW 264.7 cells. Food Agric Immunol 27(6):758–771

Iqbal J, Abbasi BA, Batool R, Mahmood T, Ali B, Khalil AT, Kanwal S, Shah SA, Ahmad R (2018) Potential phytocompounds for developing breast cancer therapeutics: nature's healing touch. Eur J Pharmacol 827:125–148

Ismail IA, Qari SH, Shawer R, Elshaer MM, Dessoky ES, Youssef NH, Hamad NA, Abdelkhalek A, Elsamra IA, Behiry SI (2021) The application of pomegranate, sugar apple, and eggplant peel extracts suppresses *Aspergillus flavus* growth and aflatoxin B1 biosynthesis pathway. Horticulturae 7(12):558

Jo Y-N, Jeong H-R, Jeong J-H, Heo H-J (2012) The skin protecting effects of ethanolic extracts of eggplant peels. Korean J Food Sci Technol 44(1):94–99

Jung E-J, Bae M-S, Jo E-K, Jo Y-H, Lee S-C (2011) Antioxidant activity of different parts of eggplant. J Med Plant Res 5(18):4610–4615

Kainat F, Ali M, Akbar A, Masih R, Mehnaz S, Sadiq MB (2023) Ultrasonic extraction of phenolic compounds from eggplant peel and formulation of eggplant peel extract-enriched ice-cream. J Food Qual. https://doi.org/10.1155/2023/3267119

Kamali Moghaddam M, Ghanbari Adivi M, Tehrani Dehkordi M (2019) Effect of acids and different mordanting procedures on color characteristics of dyed wool fibers using eggplant peel (Solanum melongena L.). Prog Color Colorant Coat 12(4):219–230

Kareem AH, Naji EZ (2022) Study of the chemical properties of pectin extracted from residues of some plant sources. Ann Biol 38:252–257

Karimi A, Kazemi M, Samani SA, Simal-Gandara J (2021) Bioactive compounds from by-products of eggplant: Functional properties, potential applications and advances in valorization methods. Trends Food Sci Technol 112:518–531

Kazemi M, Khodaiyan F, Hosseini SS (2019a) Utilization of food processing wastes of eggplant as a high potential pectin source and characterization of extracted pectin. Food Chem 294:339–346

Kazemi M, Khodaiyan F, Hosseini SS, Najari Z (2019b) An integrated valorization of industrial waste of eggplant: Simultaneous recovery of pectin, phenolics and sequential production of pullulan. Waste Manag 100:101–111

Khoo HE, Azlan A, Tang ST, Lim SM (2017) Anthocyanidins and anthocyanins: Colored pigments as food, pharmaceutical ingredients, and the potential health benefits. Food Nutr Res 61(1):1361779

Kumar M, Tomar M, Saurabh V, Mahajan T, Punia S, del Mar Contreras M, Rudra SG, Kaur C, Kennedy JF (2020) Emerging trends in pectin extraction and its anti-microbial functionalization using natural bioactives for application in food packaging. Trends Food Sci Technol 105:223–237

Kuo H-CD, Wu R, Li S, Yang AY, Kong A-N (2019) Anthocyanin delphinidin prevents neoplastic transformation of mouse skin JB6 P+ cells: epigenetic re-activation of Nrf2-ARE pathway. AAPS J 21:1–10

Lee Y-M, Yoon Y, Yoon H, Park H-M, Song S, Yeum K-J (2017) Dietary anthocyanins against obesity and inflammation. Nutrients 9(10):1089

León-González AJ, Sharif T, Kayali A, Abbas M, Dandache I, Etienne-Selloum N, Kevers C, Pincemail J, Auger C, Chabert P (2015) Delphinidin-3-O-glucoside and delphinidin-3-O-rutinoside mediate the redox-sensitive caspase 3-related pro-apoptotic effect of blackcurrant juice on leukaemia Jurkat cells. J Funct Foods 17:847–856

Lim W, Jeong W, Song G (2016) Delphinidin suppresses proliferation and migration of human ovarian clear cell carcinoma cells through blocking AKT and ERK1/2 MAPK signaling pathways. Mol Cell Endocrinol 422:172–181

Lim WC, Kim H, Ko H (2019) Delphinidin inhibits epidermal growth factor-induced epithelial-to-mesenchymal transition in hepatocellular carcinoma cells. J Cell Biochem 120(6):9887–9899

Luís Â, Ramos A, Domingues F (2020) Pullulan films containing rockrose essential oil for potential food packaging applications. Antibiotics 9(10):681

Mainali RP (2014) Biology and management of eggplant fruit and shoot borer, Leucinodes orbonalis Guenee (Lepidoptera: Pyralidae): a review. Int J Appl Sci Biotechnol 2:18–28. https://doi.org/10.3126/ijasbt.v2i1.10001

Manousaki A, Jancheva M, Grigorakis S, Makris DP (2016) Extraction of antioxidant phenolics from agri-food waste biomass using a newly designed glycerol-based natural low-transition temperature mixture: a comparison with conventional eco-friendly solvents. Recycling 1(1):194–204. https://www.mdpi.com/2313-4321/1/1/194

Marić M, Grassino AN, Zhu Z, Barba FJ, Brnčić M, Brnčić SR (2018) An overview of the traditional and innovative approaches for pectin extraction from plant food wastes and by-products: Ultrasound-, microwaves-, and enzyme-assisted extraction. Trends Food Sci Technol 76:28–37

Martini S, Conte A, Cattivelli A, Tagliazucchi D (2021) Domestic cooking methods affect the stability and bioaccessibility of dark purple eggplant (Solanum melongena) phenolic compounds. Food Chem 341:128298

Matsubara K, Kaneyuki T, Miyake T, Mori M (2005) Antiangiogenic activity of nasunin, an antioxidant anthocyanin, in eggplant peels. J Agric Food Chem 53(16):6272–6275

Mauro RP, Agnello M, Rizzo V, Graziani G, Fogliano V, Leonardi C, Giuffrida F (2020) Recovery of eggplant field waste as a source of phytochemicals. Sci Hortic 261:109023

Mazewski C, Kim MS, Gonzalez de Mejia E (2019) Anthocyanins, delphinidin-3-O-glucoside and cyanidin-3-O-glucoside, inhibit immune checkpoints in human colorectal cancer cells in vitro and in silico. Sci Rep 9(1):11560

Mekky RH, Abdel-Sattar E, Segura-Carretero A, Contreras MDM (2019) Phenolic compounds from sesame cake and antioxidant activity: a new insight for agri-food residues' significance for sustainable development. Foods 8(10):432. https://www.mdpi.com/2304-8158/8/10/432

Mekky RH, Abdel-Sattar E, Segura-Carretero A, Contreras MDM (2022) A comparative study on the metabolites profiling of linseed cakes from Egyptian cultivars and antioxidant activity applying mass spectrometry-based analysis and chemometrics. Food Chem 395:133524. https://doi.org/10.1016/j.foodchem.2022.133524

Mekky RH, Hegazy MM, Afifi WM, Mostafa AE, Abbass HS, Abd-Elraouf M (2023a) Composition and functional properties of cardamom leaves. In: Ramadan MF (ed) Cardamom (Elettaria cardamomum): production, processing and properties. Springer, Cham, pp 157–178. https://doi.org/10.1007/978-3-031-35426-7_12

Mekky RH, Hegazy MM, Cádiz-Gurrea ML, Fernández-Ochoa Á, Segura Carretero A (2023b) Bioactive phytochemicals from sesame oil processing by-products. In: Ramadan Hassanien

MF (ed) Bioactive phytochemicals from vegetable oil and oilseed processing by-products. Springer, Cham, pp 155–194. https://doi.org/10.1007/978-3-030-91381-6_9

Mibei EK, Ambuko J, Giovannoni JJ, Onyango AN, Owino WO (2017) Carotenoid profiling of the leaves of selected African eggplant accessions subjected to drought stress. Food Sci Nutr 5(1):113–122. https://doi.org/10.1002/fsn3.370

Mohamed M, Zeitoun A, Abdalla AE (2019) Assessment of chemical composition and bioactive compounds in the peel, pulp and whole egyptian eggplant flour. J Adv Agric Res 24(1):14–37

Niño-Medina G, Urías-Orona V, Muy-Rangel MD, Heredia J (2017) Structure and content of phenolics in eggplant (Solanum melongena)-a review. S Afr J Bot 111:161–169

Noda Y, Kneyuki T, Igarashi K, Mori A, Packer L (2000) Antioxidant activity of nasunin, an anthocyanin in eggplant peels. Toxicology 148(2-3):119–123

Rasouli F, Babla M, Li L, Liang W, Yong M-T, Ahmed T, Tissue D, Huda S, Chen Z-H (2023) Physiological and yield performance is partially linked to water use efficiency of eggplant genotypes in a high-tech glasshouse. Horticulturae 9(1):19. https://www.mdpi.com/2311-7524/9/1/19

Rochín-Medina JJ, Sotelo-Castro JA, Salazar-Salas NY, López-Valenzuela JA, Ramírez K (2019) Antioxidant and anti-Salmonella activities of eggplant peel compounds obtained by solvent-free calcium-based extraction. CYTA J Food 17(1):873–881

Roewer N, Broscheit J (2015) Use of delphinidin against Staphylococcus aureus. In: Google Patents

Roshankhah S, Shabanizadeh A, Abdolmaleki A, Gholami MR, Salahshoor MR (2020) Evaluation of biomarkers in liver following Solanum melongena green calyx administration in diabetic rats. J Diab Metab Disord 19:1115–1127

Salamatullah AM, Ahmed MA, Alkaltham MS, Hayat K, Aloumi NS, Al-Dossari AM, Al-Harbi LN, Arzoo S (2021a) Effect of air-frying on the bioactive properties of eggplant (Solanum melongena L.). Processes 9(3):435. https://www.mdpi.com/2227-9717/9/3/435

Salamatullah AM, Alkaltham MS, Hayat K, Ahmed MA, Arzoo S, Husain FM, Alzahrani A (2021b) Bioactive and antimicrobial properties of eggplant (Solanum melongena L.) under microwave cooking. Sustainability 13(3):1519. https://www.mdpi.com/2071-1050/13/3/1519

Santos CA, Almeida FA, Quecán BX, Pereira PA, Gandra KM, Cunha LR, Pinto UM (2020) Bioactive properties of *Syzygium cumini* (L.) skeels pulp and seed phenolic extracts. Front Microbiol 11:990

Sarabandi K, Jafari SM, Mahoonak AS, Mohammadi A (2019) Application of gum Arabic and maltodextrin for encapsulation of eggplant peel extract as a natural antioxidant and color source. Int J Biol Macromol 140:59–68

Scalbert A (1991) Antimicrobial properties of tannins. Phytochemistry 30(12):3875–3883

Sękara A, Cebula S, Kunicki EJFH (2007) Cultivated eggplants—origin, breeding objectives and genetic resources, a review. Folia Hortic 19:97–114

Shah A, Ashames AA, Buabeid MA, Murtaza G (2020) Synthesis, in vitro characterization and antibacterial efficacy of moxifloxacin-loaded chitosan-pullulan-silver-nanocomposite films. J Drug Deliv Sci Technol 55:101366

Shaikh A, Iqbal M, Qadeer A (2022) A review on therapeutic potential and nutritional composition of eggplant. Eur Acad Res X:545–557

Silva GFP, Pereira E, Melgar B, Stojković D, Sokovic M, Calhelha RC, Pereira C, Abreu R, Ferreira IC, Barros L (2021) Eggplant fruit (Solanum melongena L.) and bio-residues as a source of nutrients, bioactive compounds, and food colorants, using innovative food technologies. Appl Sci 11(1):151

Singh S, Singh B, Alam T (2019) Evaluation of shelf-life, antioxidant activity and nutritional quality attributes in carnauba wax coated eggplant genotypes. J Food Sci Technol 56(11):4826–4833. https://doi.org/10.1007/s13197-019-03944-x

Song Y, Mei T, Liu Y, Kong S, Zhang J, Xie M, Ou S, Liang M, Wang Q (2021) Metabolites identification of chemical constituents from the eggplant (*Solanum melongena* L.) calyx in rats by UPLC/ESI/qTOF-MS analysis and their cytotoxic activities. Front Pharmacol 12:655008

Sroka Z, Sowa A, Dryś A (2017) Inhibition of lipoxygenase and peroxidase reaction by some flavonols and flavones: The structure-activity relationship. Nat Prod Commun 12(11):1934578X1701201111

Sun L, Warren FJ, Netzel G, Gidley MJ (2016) 3 or 3′-Galloyl substitution plays an important role in association of catechins and theaflavins with porcine pancreatic α-amylase: The kinetics of inhibition of α-amylase by tea polyphenols. J Funct Foods 26:144–156

Sun L, Wang Y, Miao M (2020) Inhibition of α-amylase by polyphenolic compounds: Substrate digestion, binding interactions and nutritional intervention. Trends Food Sci Technol 104:190–207

Taher D, Solberg SØ, Prohens J, Chou Y-y, Rakha M, Wu T-h (2017) World vegetable center eggplant collection: origin, composition, seed dissemination and utilization in breeding [review]. Front Plant Sci 8:1484. https://doi.org/10.3389/fpls.2017.01484

Talekar S, Patti AF, Vijayraghavan R, Arora A (2018) An integrated green biorefinery approach towards simultaneous recovery of pectin and polyphenols coupled with bioethanol production from waste pomegranate peels. Bioresour Technol 266:322–334

Tena N, Martín J, Asuero AG (2020) State of the art of anthocyanins: Antioxidant activity, sources, bioavailability, and therapeutic effect in human health. Antioxidants 9(5):451

Wu X, Li M, Xiao Z, Daglia M, Dragan S, Delmas D, Vong CT, Wang Y, Zhao Y, Shen J (2020) Dietary polyphenols for managing cancers: What have we ignored? Trends Food Sci Technol 101:150–164

Yarmohammadi F, Ghasemzadeh Rahbardar M, Hosseinzadeh H (2021) Effect of eggplant (Solanum melongena) on the metabolic syndrome: A review. Iran J Basic Med Sci 24(4):420–427. https://doi.org/10.22038/ijbms.2021.50276.11452

Zhang Y, Sun Y, Zhang H, Mai Q, Zhang B, Li H, Deng Z (2020) The degradation rules of anthocyanins from eggplant peel and antioxidant capacity in fortified model food system during the thermal treatments. Food Biosci 38:100701

Zhao X, Zhang X, Tie S, Hou S, Wang H, Song Y, Rai R, Tan M (2020) Facile synthesis of nano-nanocarriers from chitosan and pectin with improved stability and biocompatibility for anthocyanins delivery: An in vitro and in vivo study. Food Hydrocoll 109:106114

Phytochemicals from Pumpkin (*Cucurbita maxima*) By-Products

Vanessa Pellicorio, Marta Barberis, Raffaella Colombo, Chiara Milanese, and Adele Papetti

Abstract Cultivated plant by-products are a worldwide environmental issue that can be revalued with benefits to human health and the environment. Pumpkin (*Cucurbita maxima*) is one of the most widely consumed and cultivated vegetable crops. Pumpkin by-products (leaves, seeds, and skin/peel) are almost 25% of the total pumpkin processed because only the flesh fraction is edible. Notably, pumpkin wastes are health-promoting sources of phytochemicals, such as sugars, amino acids, phenols, glucosides, terpenes, flavonoids, lignans, plant steroids, and flavonoids. In addition, pumpkin by-products can be low-cost sources of potentially important bioactive compounds, i.e., tocopherols, carotenoids, fatty acids, and vitamins. Owing to the versatility of chemical compounds present in pumpkin wastes, they can be used in food supplements, showing antidiabetic, antioxidant, anticancer, and anti-inflammatory activities. Moreover, they can be used as ingredients for functional foods, including in baking, beverages, meat, and dairy products. In addition, they can be exploited for non-food and non-health purposes, becoming natural ingredients for cosmetics, natural dyes, biodegradable packaging components, wastewater purifying agents, or bioenergy sources. This chapter contributes to a more complete understanding of pumpkin by-products, considering chemical composition, bioactivity profile, and health and non-health applications of seeds and skin/peel.

Keywords Pumpkin · By-products · Circular economy · Environmental sustainability · Phytochemical analysis · Bioactive compounds · Health-promoting effect · Biofuel · Energy application

V. Pellicorio · M. Barberis · R. Colombo
Department of Drug Sciences, University of Pavia, Pavia, Italy

C. Milanese
Pavia Hydrogen Lab, Department of Chemistry & C.S.G.I, University of Pavia, Pavia, Italy

A. Papetti (✉)
Department of Drug Sciences & C.S.G.I., University of Pavia, Pavia, Italy
e-mail: adele.papetti@unipv.it

M. F. Ramadan (ed.), *Bioactive Phytochemicals in By-products from Bulb, Flower and Fruit Vegetables*, https://doi.org/10.1007/978-3-031-77399-0_11

Abbreviations

ABTS	2,2′-azino-bis(3-ethylbenzothiazoline-6-sulfonic acid)
Ala	Alanine
ALP	Alkaline phosphatase
ALT	Alanine transaminase
Arg	Arginine
Asp	Aspartic acid
AST	Aspartate amino transferase
BGL	Blood glucose level
CIP	Ciprofloxacin
CPEA	Cucurbita polyesteramide
CPEAU	Cucurbita polyesteramide urethane
DM	Degree of methylation
dm	Dry matter
DPPH	2,2-Diphenyl-1-picrylhydrazyl
DPP-IV	Dipeptidyl peptidase IV
EACC	Ehrlich ascites carcinoma cells
FESEM	Field emission scanning electron microscopy
FRAP	Ferric reducing antioxidant power
FTIR	Fourier-transform infrared spectroscopy
GAE	Gallic acid equivalent
GalA	Galacturonic acid
GI	Glycemic index
Glu	Glutamic acid
Gly	Glycine
GPx	Glutathione peroxide
GTT	Glucose tolerance test
HCAs	Heterocyclic aromatic amines
HDL	High-density lipoprotein
HECFA	*N,N′*-bis (2-hydroxyethyl) cucurbita oil fatty amide
HIFU	High-intensity focused ultrasound
His	Histidine
HOPSC	High oil pumpkin seed cake
IC50	Half-maximal inhibitory concentration
Ile	Isoleucine
IPB	Benign prostatic hyperplasia
IPSS	International Prostate Symptom Score
LDL	Low-density lipoprotein
Leu	Leucine
Lys	Lysine
MAE	Microwave-assisted extraction
NADEs	Natural deep eutectic solvents
NLCs	Lipid nanocarriers
OABSS	Overactive Bladder Symptom Score
PBPs	Pumpkin by-products

Phe	Phenylalanine
PLE	Pressurized liquid extraction
POB	Pumpkin seed oil by-product
PPG	Postprandial reduction in glucose
Pro	Proline
PSO	Cold-pressed pumpkin seed oil
QE	Quercetin equivalent
ROS	Reactive oxygen species
SCF	Supercritical fluids
Ser	Serine
SFE	Supercritical fluid extraction
SOD	Superoxide dismutase
SWE	Subcritical water extraction
TEM	Transmission electron microscopy
Thr	Threonine
Trp	Tryptophan
Tyr	Tyrosine
UAE	Ultrasound -assisted extraction
Val	Valine
VLDL	Very low-density lipoprotein
WHC	Water-holding capacity
ZnO NPs	Zinc oxide nanoparticles

1 Introduction

Pumpkin is a crop widely cultivated in Africa, South Asia, India, United States, and Europe; it belongs to *Cucurbita* genus, *Cucurbitaceae* family, and the most diffused varieties are *Cucurbita pepo*, *C. maxima*, and *C. moschata* (Peter et al. 2013).

Pumpkin fruit weight generally ranges from 1 to 10 kg when fully mature and is picked 4–6 weeks after flowering (Isutsa and Mallowa 2013). The pulp is characterized by good nutritional and sensory value, as it contains protein, pectin, dietary fiber, carotenoids (β-carotene), vitamins B1, B2, and C, and minerals such as Mg, P, K, Fe, and Se (Promsakha et al. 2017). In addition, many essential phytocompounds are present. It is generally consumed after cooking as vegetable food, or it is used for soup, bakery, puree, dessert, marmalade, beverage, meat, dairy products, ready-to-eat snacks, and flour-based preparations (Konopacka et al. 2010; Provesi et al. 2012; Quintana et al. 2018; Aljahani 2022). Because of its nutritional and health properties, the interest in using it not only as human food but also as feed and as a source of compounds useful in the cosmetic and pharmaceutical industry has increased in the last two decades (Sojak and Głowacki 2010).

Pumpkin production, including squashes and gourds, was estimated to be above 23 million tons cultivated worldwide in 2019 (FAO 2019). Therefore, the amount of by-products deriving from the pumpkin food sector is high and consists mainly of seeds, shells, peels, and skin, which are rich sources of different bioactive compounds, including macronutrients (protein, carbohydrates, sterols, and fatty acids), micronutrients (carotenoids and vitamins), and phenolic compounds (Batool et al. 2022; Aziz et al. 2023). Peels account for 3–16% of the by-products of pumpkin (PBPs) processing, and, therefore it could be considered the main by-product, followed by seeds, also rich in macro- and micronutrients, especially fatty acids, both polyunsaturated and monounsaturated (Saavedra et al. 2015; Bardaa et al. 2016; Hussain et al. 2022a). Peel and seeds are used as an ingredient source in the development of functional foods because of their antioxidant, anti-inflammatory, antimicrobial, antidiabetic, and anticancer effects (Kaur et al. 2020; Susmi et al. 2023). Effective PBPs need the suitable extraction of bioactive components by applying analytical procedures and technological processes that are time- and reagents-saving, environmentally friendly, and possibly comply with reuse policies to manage high waste disposal (Papargyropoulou et al. 2014).

This chapter will discuss the PBPs chemical composition and the different biological properties concerning the nutritional and non-nutritional composition. The functional properties will also be considered correlating to food and non-food applications, including cosmetic and dye fields. Finally, biofuels and energy applications will be considered.

2 Pumpkin By-Products Chemical Composition

The most important *Cucurbita maxima* by-products are seeds and peel. From seeds, an edible oil or a cake oil containing several bioactive compounds can be obtained, as well as from peel, which is a good source of nutraceuticals exploitable in the medical and nutraceutical fields (Pazinato Cuco et al. 2019).

Recently, many studies have been focused on pumpkin seeds as a relevant by-product for their potential health effects. Their chemical composition differs depending on the variety, the area of pumpkin production, the soil composition, and/or the harvesting period. The general proximate composition and the mineral content are reported in Table 1.

Seed proteins are generally extracted to obtain bioactive peptides. Different extraction methods can be used, such as Ultrasounds (UAE), High-Intensity Focused Ultrasound (HIFU), NAtural Deep Eutectic solvents (NADEs), supercritical fluids (SCF) like supercritical CO_2, Microwave-Assisted Extraction (MAE), or Pressurized Liquids Extraction (PLE). Despite these less conventional methods, Soxhlet extraction is common and considered advantageous because bioactives can be extracted even starting from small amounts of material. Simply changing the solvent allows the extraction of different compounds with varying solubility and high extraction efficiency. However, according to a study by Pazinato Cuco and coworkers, the use of supercritical CO_2 offers both a higher γ-tocopherols (283 mg/100 g oil) extraction yield than the Soxhlet method (105 mg/100 g oil) and a greater total phenolic content (851 mg gallic acid equivalent—GAE/100 g oil using CO_2 extraction vs.

Table 1 Pumpkin seed proximate chemical composition and mineral content (%, w/w dry material-dm) (Amoo et al. 2004; Rezig et al. 2012)

Moisture	3.08–8.46
Crude fiber	2.55–21.97
Crude protein	14.31–33.92
Total sugar	0.11–24.45
Crude fat	31.57–52.13
Total ash	3.60–3.97
Potassium	0.036–0.886
Sodium	0.030–0.357
Calcium	0.029–0.272
Phosphorus	0.224–0.824
Manganese	0.002–0.003
Iron	0.004–0.015
Zinc	0.004–0.025

615 mg GAE/100 g oil using Soxhlet) (Pazinato Cuco et al. 2019). Extraction with supercritical CO_2 occurs at 304.3 K and 7.9 mPa and is widely used, especially for the extraction of thermolabile compounds, due to the low temperature. CO_2 is considered non-toxic, non-flammable, non-explosive, and safe for food in these conditions. Methods such as UAE, NADEs, and SCF are especially advantageous when some more volatile components are present in seeds; hence, lower temperatures are needed to avoid any degradation, differently from MAE and Soxhlet, which require higher temperatures. Once obtained, proteins are purified and hydrolyzed, most often enzymatically, using an acidic or basic medium. Regarding the enzymatic process, alcalase, papain, chymotrypsin, and flavourzyme are generally employed. The choice of a specific enzyme is related to the cutting preference: in fact, alcalase hydrolyzes at hydrophobic amino acids level, while chymotrypsin is more specific for Tryptophan (Trp), Tyrosine (Tyr), Phenylalanine (Phe), and Leucine (Leu), and papain for Phe, Valine (Val), and Leu. Differently, flavourzyme has a broad specificity. Therefore, depending on the enzyme used, the obtained peptides may have different actions: antioxidant, anti-inflammatory, antimicrobial, anticancer, or antihypertensive (Colombo et al. 2024).

As reported in Table 1, the main components in the seed's chemical composition are fats, leading to the possibility of extracting a high amount of oil (Rezig et al. 2012) that attracts attention not only as an edible oil but also as a potential nutraceutical to be used as a food ingredient. In recent years, to overcome the conventional extraction issues, several other techniques have been proposed (Lemus-Mondaca et al. 2019), such as pressurized carbon dioxide (Pazinato Cuco et al. 2019), Soxhlet extraction (Tsaknis et al. 1997), subcritical water extraction (SWE) (Švarc-Gajić et al. 2022), or cold pressing extraction (Rezig et al. 2018) which are cost-saving methods with low impact on human health. The supercritical water technique has been recently investigated as it is very sustainable and food grade, free of organic solvents (Švarc-Gajić et al. 2022).

Regarding micronutrients, seeds are rich in phosphorus, sodium, magnesium, and calcium, which are important for bone, tooth, muscle, impulse conduction, nerve health, and maintaining homeostatic pressure. Conversely, they are deficient in manganese, zinc, and iron, which are important co-factors for many enzymes for glucose synthesis, wound healing, and red blood cell production (Amoo et al. 2004). In addition, pumpkin seeds are rich in polyphenols and carotenoids, bioactives that can be extracted using alcoholic, aqueous, or hydroalcoholic mixture solvents with different extraction yields. For example, polyphenols, flavonoids, and ß-carotene have been obtained with a yield of 8.27 mg GAE/g dm, 6.14 mg of Quercetin Equivalent (QE)/g dm, and 5.82 mg/100 g dm, respectively, using only methanol as extraction solvent (Jahan et al. 2023).

Seeds oil comprises 99.4–99.5% of fatty acids, mainly unsaturated, whose relative amounts are reported in Table 2. The most abundant unsaturated fatty acid is linoleic acid, which is important for different hormone synthesis and cell membrane synthesis, followed by oleic acid, which can reduce infection and cardiovascular disease risk. Saturated fatty acids account for only a small percentage of the total (Boujemaa et al. 2020; Tsaknis et al. 1997). Analyzing its composition, this oil can help prevent cardiovascular diseases and maintain body health when consumed appropriately.

Considering the chemical and physical parameters indicating the oil quality and stability (Table 3), the related peroxide value indicates generally good oil preservation from peroxidation and, hence, excellent resistance to degradation, even if the high iodine value suggests that the oil can be easily oxidized producing secondary products of oxidation. The acidic value (a good indicator of the oil's edibility) indicates that this oil can be used as a food ingredient. The saponification value is used to control the adulteration of the oil (Tsaknis et al. 1997; Amoo et al. 2004).

Seeds oil is also composed of different bioactive compounds, such as β-carotene (ranging from 1.5 g/100 g dm to 2.1 g/100 g dm), an excellent antioxidant that can counteract free radicals generation. It is also a primary source of vitamin A, which benefits vision and skin. Its content in tocopherols, naturally occurring in α, β, γ, and δ forms (well-known antioxidant compounds and precursors of vitamin E), is variable and ranges from 105.3 mg/100 g oil to 283.0 mg/100 g oil with the γ form being the most recurring one (Rezig et al. 2012; Boujemaa et al. 2020). In addition, a high amount of polyphenols (611.39–1053.18 mg GAE/100 g oil) is present, mainly ferulic, chlorogenic, vanillic, gallic, caffeic, syringic, and *p*-coumaric,

Table 2 Pumpkin seed oil fatty acid composition (Boujemaa et al. 2020; Tsaknis et al. 1997)

Fatty acid	Content (%)
Myristic acid (C14:0)	0.10–0.11
Palmitic acid (C16:0)	12.70–17.4
Stearic acid (C18:0)	6.00–6.56
Arachidic acid (C20:0)	0.30–0.47
Palmitoleic acid (C16:1)	0.08–0.11
Oleic acid (C18:1)	18.12–38.10
Linoleic acid (C18:2)	42.10–56.98
Linolenic acid (C18:3)	0.20–0.28

Table 3 Physico-chemical properties (mean values) of seed oil (Tsaknis et al. 1997; Amoo et al. 2004)

Peroxide value	2.80–9.20
Saponification value	126–201
Acidic value	0.36–0.97
Iodine value	18.66–107
Color	Light yellow
Specific gravity	1.45–1.46

4-hydroxybenzoic, and 4-hydroxyphenylacetic acid, catechins, rutin, resveratrol, myricetin, kaempferol, and quercetin-3-*O*-glucopyranoside. Generally, each polyphenol is present in equal concentration (on average between 6 mg/100 g dm and 20 mg/100 g dm) (Švarc-Gajić et al. 2022; Monica et al. 2022).

Seed pre-treatment using MAE or a roasting process increases the extraction yield of bioactive compounds and the stability of the obtained oil (Rezig et al. 2022). In addition, the pressing process at 40–50 °C produces cake oil which is not an edible product, but it is rich in squalene (1.684 mg/kg dm), phytosterols (3.9 mg/kg dm), tocopherols (220.3 mg/kg dm), and protochlorophylls which are intermediate in the biosynthesis of chlorophyll A (72.6 mg/kg dm) and phenols (90.1 mg/kg dm). A cryogenic grinding pre-treatment has been successfully applied to enhance the extractability of such molecules, leading to an amount of phytosterols and protochlorophylls being 2 and 2.5-fold higher, respectively (Balbino et al. 2019). Protein powder, flour, and fibrous extract can be also obtained by cake oil by applying the grinding process. The resulting powder is made up not only of proteins (61–63 g/100 g dm), but also of carbohydrates (11–12 g/100 g dm), fats (7–8 g/100 g dm), and fiber (9–10 g/100 g dm). It contains a good amount of both essential (Val 25.77 g/kg dm, isoleucine (Ile) 9.98 g/kg dm, Leu 37.59 g/kg dm, lysine (Lys) 14.29 g/kg dm, methionine 6.77 g/kg dm, threonine (Thr) 31.01 g/kg dm, Phe + Tyr 42.23 g/kg dm, and Trp 11.00 g/kg dm) and non-essential (alanine (Ala) 16.60 g/kg dm, arginine (Arg) 50.98 g/kg dm, aspartic acid (Asp) 24.94 g/kg dm, histidine (His) 4.64 g/kg dm, glycine (Gly) 39.13 g/kg dm, glutamic acid (Glu) 66.78 g/kg dm, proline (Pro) 31.00 g/kg dm, and serine (Ser) 14.36 g/kg dm) amino acids. The flour is poorer in proteins (39–40 g/100 g dm) but richer in carbohydrates, fats, and fiber (19–21 g, 9–10 g, and 12–13 g/100 g dm, respectively), and this composition makes it ideal for bakery products. The fibrous extract has an intermediate composition in proteins (44–46 g/100 g dm), sugars (15–16 g/100 g dm), and fats (7–8 g/100 g dm), but a higher content in fiber (30–32 g/100 g dm) and therefore it can be considered a healthy ingredient for human diet (Kalyna et al. 2021).

As regards pumpkin peel, it is an extremely rich by-product of natural beneficial substances essential for human nutrition and ensures the daily recommended intake of micronutrients. They are rich in sodium, which is necessary for balancing body fluids and activating important channels for nerve impulses; potassium, which is essential for muscle function and sodium balance; and calcium, which is fundamental for several biochemical pathways for opening important channels for nerve impulses and for bone health (Table 4).

Peel is also a good source of tocopherols, carotenoids, polyphenols, ascorbic acid, and pectins. Their concentration varies depending on the area, the type of soil in which pumpkins have been cultivated, and the harvesting period.

The total phenolic and flavonoid contents range from 354 mg GAE/100 g to 385 mg GAE/100 g and from 130 mg QE/100 g to 153 mg QE/100 g, respectively, while the ascorbic acid concentration is in the range 25.2–34.4 mg/100 g dm (Hagos et al. 2023; Baississe and Fahloul 2018), depending on the applied extraction procedure. SWE has to be preferred to supercritical fluid extraction (SFE) for carotenoid extraction (15.22 mg/100 g vs. 11.48 mg/100 g), differently from the total polyphenols, which are extracted in higher amounts using SFE (353.5 mg GAE/100 g vs. 213.6 mg GAE/100 g) (Salami et al. 2020). Heat and UAE techniques have also been successfully applied using as solvent alcoholic, aqueous, or hydroglycolic mixtures and thus obtaining different polyphenolic compounds, or the same polyphenols but in various concentrations, depending on both the used solvent and on the peel variety, as demonstrated in recent studies (Gaweł-Bęben et al. 2022; Leichtweis et al. 2023). For example, peels deriving from Halloween and Hokkaido varieties were extracted, and a higher concentration of polyphenols (mainly, apigenin, chlorogenic acid, coumaric acid, and kaempferol) was obtained for the Halloween variety using ethanol concerning aqueous or hydroglycolic solvents. Conversely, the highest yield was obtained using the aqueous medium for the Hokkaido variety, which also allows for extracting proteins in high amounts.

Pumpkin peel pectin consists of a chain of α-D-galacturonic acid (GalA) units linked at α-1,4 positions with rhamnose residues interspersed along the chain. These rhamnose residues create hairy branches with longer or shorter side chains containing rhamnose, D-galactose, L-arabinose, and D-xylose attached. Pectins are food hydrocolloids (Baississe and Fahloul 2018) that can be extracted using different strategies. The most common method involves using nitric, hydrochloric, or acetic acid. Higher yields are obtained using a hot shaking water bath or equipment such as a Soxhlet extractor (Frosi et al. 2023). For example, the composition of pectin extracted with hydrochloric acid is displayed in Table 5.

Generally, peel pectins are highly methylated (degree of methylation—DM > 50%) and rich in GalA (Baississe and Fahloul 2018), making them gelling agents or stabilizers, useful in the production of jam and marmalade.

Flour can be obtained from seeds and peels, replacing the normal flour in the food industry. According to some studies, the partial replacement of normal wheat

Table 4 Proximate composition (g/100 g dm) and mineral content (mg/100 g dm) of pumpkin peel (Jahan et al. 2023)

Moisture	7.95
Ash	4.47
Protein	15.45
Sodium	33.92
Potassium	378.81
Calcium	6.87
Iron	0.24

Table 5 Chemical composition of pumpkin peel pectin fraction

Total sugars (g/100 g)	65.66
GalA (g/100 g)	55.14
Methyl groups (g/100 g)	6.20
DM (%)	63.88
Protein (g/100 g)	0.44
Ash (g/100 g)	2.41
Polyphenols (mg/100 g)	5.86

flour with this one reduces the product elasticity due to the low gluten formation. However, the taste improves, and the amount of long-chain saturated fatty acids (C14:0, C16:0, C18:0) that could lead to diseases is reduced. Seed flour comprises 8.15% of water, 5.4% of ash, 10.42% of fat, 12.33% of protein, and 63.70% of carbohydrate. Differently, peel flour is made of 9.96% of water, 10.3% of ash, 5.59% of fat, 12.84% of protein, and 61.25% of carbohydrate (Indrianingsih et al. 2019).

A decrease in glucose and triglycerides in animal feed with this flour has also been proven (Lemus-Mondaca et al. 2019).

An interesting investigation has been performed on the peel and seed mixture, which leads to an oil richer in bioactives and thus shows higher beneficial activities than the individual by-product. For example, the yield of β-carotene is from 145.0 mg/100 g to 375.8 mg/100 g for the mixture oil and from 1.5 mg/100 g to 2.1 mg/100 g for seed oil, while phenolic compounds ranged between 1.2 mg GAE/100 g and 1090.0 mg GAE/100 g for the mixture vs. 611.39 mg GAE/100 g to 1053.18 mg GAE/100 g for seeds oil. As for tocopherols, the mixture oil contains α and γ-tocopherols, with the total tocopherols contents ranging from 143.7 mg/100 g oil to 660.1 mg/100 g oil, while seeds oil contains only the γ-tocopherol, that is present in a lower concentration (105.3 mg/100 g oil to 283.0 mg/100 g oil) (Pazinato Cuco et al. 2019).

3 Biological and Functional Properties of Pumpkin By-Products Bioactive Compounds

3.1 Biological Properties

3.1.1 Antioxidant Activity

Both seeds and peel extracts contain polyphenols and other compounds possessing antioxidant capacity. These molecules can counteract the formation of Reactive Oxygen Species (ROS), which are typically generated from oxygen metabolism and play a crucial role in several biological processes, such as cell signaling or defense against pathogens. Conditions such as ionizing radiations, UV irradiation, or the presence of pollutants, heavy metals, or some xenobiotics can lead to ROS overproduction. This results in an imbalance between ROS production and removal, leading to oxidative stress, with health implications such as cell and tissue damage, cancer, heart disease, diabetes, obesity, or neurological disorders (Pizzino et al. 2017). Generally, the antioxidant activity of compounds can be screened using various simple in vitro tests, including $DPPH^{\bullet}$, $ABTS^{\bullet+}$, and FRAP assays. The first two tests assess the compound's ability to act as a scavenger against the two stable radicals $DPPH^{\bullet}$ and $ABTS^{\bullet+}$. The third test is based on the sample's Fe^{3+} ion-reducing capacity.

Pumpkin seed and peel flour extracts obtained after by-product drying and ethanol maceration only slightly inhibited DPPH radicals, with values ranging from 17.99 to 33.80% when tested at 800 µg/ml (Indrianingsih et al. 2019). Antioxidant activity is not always strictly related to the presence of a higher amount of total phenolic compounds because it may depend on other factors, such as the compound chemical structure or other components like proteins, fats, and polysaccharides. However, by comparing the oils obtained using pentane as a solvent with those obtained with hexane or with a chloroform/methanol mixture as a solvent had a higher total phenolic content concerning those prepared using hexane or pentane, resulting in a higher radical scavenging activity (250 µmoles Trolox/kg vs. 110 and 100 µmoles Trolox/kg for hexane and pentane extracts, respectively). Good antioxidant activity was also achieved using the cold pressing extraction technique (150 µmoles Trolox/kg) (Rezig et al. 2018). Soxhlet extraction of seeds resulted in oil with good antiradical activity against $DPPH^{\bullet}$ (IC50 126.20 µg/ml) (Boujemaa et al. 2020). This activity was most likely due to the high concentration of fatty acids and carotenoids that cannot only have radical scavenging activity but also prevent their formation (Maldonade et al. 2020).

Regarding seed extracts, the highest phenolic and flavonoid content was obtained using 80% methanol (v/v) or 80% ethanol (v/v) differently from water. Hydroalcoholic solvent mixtures resulted in a greater antiradical activity against $DPPH^{\bullet}$ and $ABTS^{\bullet+}$, as well as greater iron-reducing capacity in the FRAP test (Al-Turky et al. 2022). This higher activity was recently confirmed by testing seeds and peel methanolic extracts. Their total antioxidant activity (determined by the presence of flavonoids, phenols, and carotenoids) was 50.19 and 60.72 mg Ascorbic Acid Equivalent (AAE)/g, respectively, and the antiradical DPPH capacity was 1090.07 and 337.87 ppm (IC50), respectively (Jahan et al. 2023).

Seed protein extracts showed significant activity against free oxygen radicals, mainly when hydrolyzed to generate biopeptides. For example, hydrolysates obtained using chymotrypsin had considerable activity, most likely due to the enzyme's capacity to release tryptophan, which could be able to absorb free radicals due to its chemical structure (indole ring and side chain) (Vargas et al. 2021).

3.1.2 Antimicrobial and Antifungal Activity

Many microorganisms are food contaminants, leading to deterioration, worsening food quality, and spoiling organoleptic characteristics. However, the main problem concerns the risk for consumers.

PBP extracts can effectively act against fungi and bacteria belonging to different strains. For example, hydro-ethanolic (80:20, v/v) seed extracts and fibrous filaments derived from pumpkin processing possess antimicrobial activity against several strains, such as *Yersinia enterocolitica* and *Staphylococcus aureus*. This bioactivity is important because *Y. enterocolitica* is responsible for many disorders like abdominal pain, diarrhea, and fever when ingested, and *S. aureus*, a coagulase-positive bacterium, is responsible for many diseases and septicemia in humans. The same extracts have been tested against the *Aspergillus* fungi responsible for food spoilage and mycotoxin production. The results indicated good inhibition, especially of *A. brasiliensis* and *A. fumigatus*, which are generally monitored in food packaging control (Leichtweis et al. 2022). The positive impact of pumpkin seed extract on infections caused by *S. aureus* was confirmed by Moniruzzaman et al. (2022), who explored the activity of *Cucurbita maxima* and *Momordica charantia* seed extracts. Although an inhibition zone was observed against all bacteria, the highest zone of inhibition was displayed against *S. aureus*. In silico study identified a high affinity to the active site of the staphylococcal BAP protein (7C7U) responsible for bacterial biofilm formation, providing an interesting perspective on the potential use of the extract as a source of antimicrobial agents and its applicability in the development of new molecules against antibiotic-resistant bacteria.

Antimicrobial effects have also been demonstrated against pathogens responsible for mastitis in bovines. Mastitis is a problem for milk producers, as it poses a risk to cattle and results in lower milk production and quality. To overcome this problem, antibiotics such as penicillin, cephalosporins, or broad-spectrum antibiotics are generally used, but the constant use of these drugs leads to an increase in antibiotic resistance, which is now a severe problem. To avoid this phenomenon, the use of natural extracts replacing drugs is becoming mandatory. Polysaccharides and carotenoids extracted from peel using 90% ethanol and acetone, respectively, exerted antimicrobial activity against *Escherichia coli*, coagulase-negative Staphylococci, several strains of *Acinetobacter*, *Klebsiella pneumoniae*, and some types of Bacilli (Tadee et al. 2020).

Among the hydroalcoholic extracts, pronounced antimicrobial properties were also found for 80:20 (v/v) methanol: water extracts obtained from powder and peel seeds, which inhibited fungi growth differently depending on the considered fungal strain. In fact, for peel extract, the highest zone of inhibition was registered for *Candida albicans* (8.69 mm inhibition zone) and the lowest for *Trichoderma* spp. strain (3.11 mm); no substantial differences were detected between *Fusarium oxysporum* and *Mucor miehei* (5.07 and 6.13 mm, respectively). The inhibition was more effective for seed extract. The inhibition average was 7.49 mm, with the highest inhibition zone obtained for *C. albicans* (9.34 mm), followed by *F. oxysporum* (7.80 mm), *M. miehei* (6.57 mm), and *Trichoderma* spp. (6.24 mm). The same extracts were also active against *S. aureus*, *B. subtilis*, *E. coli*, and *Salmonella typhi* with similar inhibition zones for all bacteria (6.60, 4.43, 7.03, and 13.41 mm for peel extract and 6.43, 3.70, 5.07, 14.40 mm for seed extract, respectively) (Hussain et al. 2021).

3.1.3 Anti-inflammatory Activity

Inflammation is the body's response to various stimuli of a harmful nature, such as infections and injury. It is defined as acute inflammation if an immediate and early response to a damaging stimulus is registered and chronic inflammation if a long-lasting inflammatory process involving active inflammation, tissue destruction, and attempted repair is present. It involves the vascular, somatosensory, autonomic, and immune systems. The mediators of inflammation are prostaglandins, chemokines, and proinflammatory cytokines (Matsuda et al. 2019). Inflammation is often caused by protein denaturation and can lead to significant health problems.

Peel and seed extracts, rich in flavonoids and phenols, exert an anti-inflammatory activity. In particular, peel methanolic extracts of two different Cucurbita maxima varieties (native and hybrid) showed intense anti-inflammatory activity preventing bovine serum albumin degradation with IC50 of 136.65 and 165.48 μg/ml, respectively. This activity was higher than that registered for salicylic acid used as a control in the study (IC50 35.41 μg/ml) (Susmi et al. 2023). These results confirmed those obtained in previous work for seed oil, but in this case, the oil obtained by the native variety had higher anti-inflammatory activity than that coming from the hybrid variety (Amin et al. 2020). The anti-inflammatory effects of ethanolic seed extracts were also confirmed in vivo in preclinical studies. After inducing edema and inflammation in rat and mice paw, it was observed that after 24 h following the extract administration, an inhibition of about 77.7% paw edema was registered, compared to the control (Wahid et al. 2021).

3.1.4 Antidiabetic Activity

Diabetes is a set of metabolic conditions characterized by total or partial insulin insufficiency. This leads to a state of hyperglycemia that can, in turn, lead, when protracted for long periods, to very important disorders of two types, namely microvascular complications affecting kidneys, retina, and peripheral nerves, and macrovascular complications such as myocardial infarction, stroke, or peripheral artery disease. Diabetes can be classified into type 1, in which a mechanism of autoimmune inflammation destroys ß-pancreatic cells; type 2, in which there is a dysfunction of the ß-cells, and different levels of insulin resistance can occur; and gestational type, an intolerance to carbohydrates that happens during pregnancy (Egan and Dinneen 2018). In glucose metabolism and the onset of diabetes, α-amylase, α-glucosidase, and dipeptidyl peptidase IV (DPP-IV) enzymes are of primary importance. α-amylase is present both in saliva, where it acts during chewing, and at the pancreatic level. Both isoforms cleave the bonds present in complex sugars, releasing glucose units. Therefore, inhibiting this enzyme leads to a lower release of glucose, resulting in lower blood glucose levels. α-glucosidase breaks down maltose molecules, which consist of two glucose units. The third enzyme breaks down incretins, which generally cause insulin secretion and interruption of glucagon release, with a general hypoglycemic effect.

The ethanolic extract of the seed was investigated in the concentration range of 6.2–500 mg/ml for its capacity to inhibit these three enzymes in vitro, showing antidiabetic action. When tested at 500 mg/ml, 85% of α-amylase, 91.16% of α-glucosidase, and 81.20% of DPP-IV were effectively inhibited with IC50 values ranging from 138 to 246 mg/ml. The best activity was registered against α-glucosidase followed by α-amylase and DPP-IV (Monica et al. 2022). These properties were also confirmed in vivo for an aqueous seed extract that could control blood glucose levels in normal, mild, and severe diabetic rat animal models in which diabetes has been induced (Kushawaha et al. 2017).

The activity of PBPs on blood glucose control in critically ill diabetic patients was examined, and the administration significantly decreased the blood glucose level from 214.9 to 178.4 mg/dl ($P < 0.001$) (Mahmoodpoor et al. 2018). The effect of an aqueous seed extract on normal, mildly diabetic, and severely diabetic rats was also studied. The extract dose of 200 mg/kg body weight was able to reduce the blood glucose level (BGL) in normal rats by 29.02% at 6 h fasting and 23.23% at 3 h during the glucose tolerance test (GTT), while in mild diabetic rats a reduction of 26.15% BGL was observed during GTT. Regarding the effect on severe diabetes, a significant decrease of 39.33% BGL and a 33.20% postprandial reduction in glucose (PPG) was observed after 28 days of treatment.

The hypoglycemic and hypolipidemic effects of pumpkin seed powder and oil on alloxan-induced diabetic rats were also investigated. The results indicated several positive effects, including a significant reduction in the levels of glucose, glycated hemoglobin, cholesterol, triglycerides, low-density lipoprotein (LDL), very low-density lipoprotein (VLDL), and lipid peroxidation (Abd-Elnoor 2019).

3.1.5 Anticancer Activity

Cancer is a worldwide multifactorial pathology whose development risk is affected by environmental, genetic, or lifestyle factors. The research focuses on new natural compounds to counteract the effects of chemotherapy or radiotherapy on health and quality of life and reduce the risk of development.

For this purpose, seeds and peel ethanolic extracts have been tested using two cancer cell lines, i.e., Ehrlich ascites carcinoma cells (EACC) and HeLa cell line. The seed extract could completely inhibit carcinoma in EACC when tested at 100 μg/ml; differently, lower activity was registered for the peel extract when tested at the same concentration. Excellent results were also obtained in the case of the HeLa cell lines for seed extract, showing 84% inhibition of the cancer cells at a concentration of 100 μg/ml. This high potential capacity of the extracts may be due to polyphenols, tocopherols, and carotenoids (Abou-Elella and Mourad 2015). In addition to phytocompounds, seed protein can also provide effective anticancer activity, as highlighted for albumin 2S, which had a cytotoxic effect on different cancer cell lines (breast cancer, ovarian teratocarcinoma, prostate cancer, and hepatocellular carcinoma). This effect is due to the extract's ability to induce cell apoptosis and damage the DNA of cancer cells (Tomar et al. 2014).

PBPs have also been investigated as potential reagents for the green synthesis of nanoparticles whose antitumor efficacy has been evaluated using a human breast cancer cell line (MDA-MB 231). Zinc oxide nanoparticles (ZnO NPs) were produced using pumpkin seed extract, and their morphology (size and structure) and stability have been confirmed by UV-Vis spectroscopy, field emission scanning electron microscopy (FESEM), transmission electron microscopy (TEM), and Fourier-transform infrared spectroscopy (FTIR) analyses. ZnO NPs exhibited significant activity against breast cancer cells with an IC50 of 10 μg/ml, inducing morphological changes, apoptosis, and production of ROS. In addition, ZnO NPs inhibited cell adhesion and migration, suggesting a potential action in inhibiting breast cancer progression and metastasis (Tabrez et al. 2022).

The oil extracted from seeds also exhibited anticancer potential action against liver (HepG-2), colorectal (Caco-2), and breast cancer (MCF-7) cells, as evident for two different oils deriving from Egyptian and European seed pumpkins. Both the oils had antiproliferative activity in the colorectal cancer line with similar values. Conversely, the Egyptian one had a higher antiproliferative capacity in liver cancer cells than the European one, which, in contrast, acted to a greater extent on breast cancer cell lines (Al-Okbi et al. 2016). The registered properties have been attributed to the oil composition rich in fatty acids and phytocompounds.

3.1.6 Other Health Applications

Many other biological properties have been detected in PBPs.

Recently, a pumpkin seeds alcoholic extract has been investigated for its hepatoprotective and antihyperlipidemic properties against paracetamol-induced liver damage and hyperlipidemia induced by the high-fat diet of rats. This extract could improve the activity of antioxidant enzymes and reduce total cholesterol, triglycerides, and LDL cholesterol levels while increasing high-density lipoprotein (HDL) cholesterol. The activity has been ascribed to the phytochemical compounds present in the extract, including flavonoids, phenols, and lignans. In particular, kaempferol, known for its antioxidant, antihypertensive, antidiabetic, and anti-inflammatory properties, could reduce ALT, AST, ALP, total protein, and bilirubin levels. These results suggested that the extract could prevent excessive lipid accumulation in liver tissue, making it a potentially helpful food ingredient for cholesterol-lowering and anti-hepatotoxicity therapy (Farooqi et al. 2022).

The potential use of pumpkin seeds to improve urinary tract disorders has also been investigated. Nishimura et al. (2014) developed a product derived from pumpkin seed oil and evaluated its effect on urinary disorders in volunteers with overactive bladder symptoms. The oil was orally administered to 45 subjects for 12 weeks, and the hematic results indicated a significant reduction of OABSS degree (Overactive Bladder Symptom Score) for the enrolled subjects. This study suggested pumpkin seed oil extract helps prevent and treat human urinary disorders and attributed the action to its phytosterol composition. However, it is also essential to consider the study limitations, including the lack of a randomized, double-blind, placebo-controlled study design and the study's small sample size; therefore, further research is needed. Conversely, Vahlensieck et al. (2022) conducted two randomized placebo-controlled trials to evaluate the ability of pumpkin seed soft extract (PSE) to treat mild-to-moderate lower urinary tract symptoms suggestive of benign prostatic hyperplasia (IPB). The first study demonstrated a significant improvement in the International Prostate Symptom Score (IPSS) compared to the placebo after 12 months of administration (improvement of 5.3 vs. 4.8 points in the PSE group vs. placebo group), while the second study was a three-arm study that tested the efficacy and safety of PSE capsules highlighting no differences between PSE and placebo. These data suggested an advantage of using PSE in treating patients with moderate lower urinary tract/IPB symptoms.

3.2 *Functional Properties*

3.2.1 Oxidative Stability

An oil generally undergoes an oxidation process, originating unpleasant organoleptic characteristics typical of rancidness.

Pumpkin seed oil has an oxidative stability of about 3 h 46 min, as confirmed by a study in which the oil was extracted using the cold pressing technique. However, the stability can increase to about 4 h and 32 min by applying MAE as a pretteatment for seeds before extraction, allowing higher phytosterol or tocopherol extraction yields and, accordingly, better oil antioxidant properties while using roasting. The stability did not change significantly (Rezig et al. 2022).

Using green and innovative extraction techniques, such as MAE and UAE, can also lead to higher stability of peel oil-containing pigments such as carotenoids (Sharma and Bhat 2021).

3.2.2 Color

Color is evaluated by CIE parameters representing the brightness axis in a three-dimensional diagram, i.e., L^*, a^*, b^*, and ΔE^*. L^* ranges from 0 (black) to 100 (white), a^* represents the axis from red (127) to green (–127), and b^* is the representative parameter from yellow (127) to blue (–127). ΔE^* indicates the geometric distance in the diagram between colors. The color of the peel extract obtained by conventional extraction tends to orange/green color ($L^* = 16.74$, $a^* = 15.43$, $b^* = 16.71$, and $\Delta E^* = 36.69$). The application of MAE or UAE can lead to a change in these values, resulting in a more yellow or orange color probably due to a change in the chemical structure of the peel components ($L^* = 1.84$, $a^* = 13.85$, $b^* = 3.78$, and $\Delta E^* = 43.65$ for MAE; $L^* = 3.90$, $a^* = 7.09$, $b^* = 5.67$, and $\Delta E^* = 60.23$ for UAE) (Sharma and Bhat 2021).

3.2.3 Solubility

The water solubility of pectin increases with the increase in the methoxyl groups' number, depending on their position. Pectin extracted from *Cucurbita maxima* peel is almost completely soluble in citric acid/sodium citrate buffer (89.26%) but has lower solubility in $CaCl_2$ (74.66%) and NaCl (83.68%) solutions (Baississe et al. 2022).

The solubility of seed proteins is strictly related to the extraction conditions. For example, the solubility in deionized water under acidic conditions is very low for protein fraction extracted with a chloroform/methanol mixture as well as for with pentane; conversely, it increases under more alkaline conditions except for proteins extracted with chloroform/methanol mixture, most likely due to the protein denaturation that occurs using these polar solvents. Therefore, to obtain a protein-rich oil, pentane or hexane should be used to avoid denaturation, and salts should be added to increase solubility (Rezig et al. 2013).

3.2.4 Viscosity

The intrinsic viscosity of pectin extracted from the peel is about 4.06 dl/g, but this value may change depending on the used solvent since it increases with the degree of dissociation and the viscosity increases (Baississe et al. 2022). The intrinsic viscosity of pectin obtained by MAE is generally lower than that obtained by conventional extraction, e.g., under acidic conditions (Yoo et al. 2012) because this parameter depends on pectin molecular size. Heating caused by MAE most likely alters the pectin structure, leading to a more efficient breakage of its structure and, thus, to smaller molecules, resulting in a lower intrinsic viscosity. Viscosity and molecular weight (which generally depends on the enzymatic degradation that can occur) of peel pectin are essential parameters because they can lead to the determination of hydrodynamically exciting factors, such as diffusion coefficient (*D*) and relaxation time (τs), which are $2.96 \pm 0.01 \times 10^{-5}$ cm^2/s and $3.82 \pm 0.00 \times 10^{-9}$ ms, respectively (Baississe et al. 2022). *D* generally indicates how the pectin structure is variable; a highly variable structure can be advantageous when different pectin functional properties are required; on the other hand, a more uniform structure with a lower *D* value can be helpful in the pharmaceutical field when it is important to control pectin behavior specifically. The relaxation time is an index representing the time pectin needs to return to its original state after being moved from equilibrium. If it is long, it may indicate more stable binding to other molecules, which can affect a product's properties, whether it is food or pharmaceutical.

Considering the viscosity of seed oil, it is in the range of 70 (purified) –72 (crude) mPa (Tsaknis et al. 1997).

3.2.5 Rheological Properties

The most important rheological properties attributed to peel pectins are thinning, thickening, and Newtonian behavior, which are influenced by parameters such as molecular weight, shape, and structure rigidity. Pectin solutions exhibit shear-thinning or pseudoplastic behavior; when the shear rate increases, the apparent viscosity decreases; when it increases, the deformation stresses increase. The yield stress values of pectin solutions range from 11.89 to 50.75 mPa and generally increase when pectin concentration in the solution increases. The zeta potential (ζ) indicates stability, with good range values from + 30 to –30 mV.

Peel pectin ζ is highly affected by concentration, ionic strength, and pH. The optimal value (–22.17 mV) has been obtained under the following conditions: pH 8.99, 0.5% concentration, and 0.85 M ionic strength. When it approaches its isoelectric point, pectin may form a flocculated agglomerate and present less stability. Its polyelectrolytic behavior increases as the ionic valence decreases and depends on the galacturonic acid units' number of ionic groups (carboxyl groups). Using these groups, pectins are net charged in solution, depending on pH, which affects their chemical and physical properties. For example, the charge can determine the interaction with other substances, such as minerals or proteins. Higher polyelectrolyte behavior is helpful in the food industry for forming gels, jams, or even stronger edible films.

Furthermore, there is a relationship between the pectin concentration and ζ potential, which increases with increasing particle concentration.

Pectins also have important emulsifying properties, which seem to be positively related to the presence of methyl and neutral oxy groups in their structure, as well as to the bond they can form with proteins, which can stabilize the emulsion by going on to form a three-dimensional lattice. The methylation degree also seems important for intrinsic viscosity, molecular weight, distribution coefficient, and relaxation time (Baississe et al. 2022).

Seeds oil is also able to form emulsions, as reported by a study performed in 2021 based on measures of ζ potential of nanoemulsions (obtained from a coarse, cloudy emulsion that became clear after lowering the initial temperature with cold water) at different storage temperatures. At 4 °C, the potential was 18.47 mV, but at 25 °C, the value rose to –18.50 mV; therefore, the emulsion is more stable at room temperature than at near-zero temperatures (Ordoñez Lozada et al. 2021).

4 Food and Non-food Applications of By-Products

Afterward, different applications of PBPs in the food, pharmaceutical, and cosmetic sectors will be addressed. This will clarify how the valorization of pumpkin waste can positively impact different fields, improving human and animal health or maintaining high-quality standards needed for the market.

4.1 Food Applications

4.1.1 Functional Food

In recent years, there has been a growing awareness that dietary factors contribute to various diseases, such as cardiovascular dysfunction, obesity, hypertension, diabetes, and even cancer. Many scientists believe that a proper diet can help reduce the risk of chronic diseases and limit their progression (Agarkova et al. 2019). Thanks to their composition in nutrients, non-nutrients, and biological activities, PBPs find application in the food industry as natural ingredients for functional foods.

In the bakery industry, PBPs enriched final products address nutritional deficiencies, prevent chronic diseases, and promote health. The research work by Gao et al. focused on developing cookies made of pumpkin seeds flour, konjac, and maltitol to produce a healthier product than traditional cookies high in sugar and fat. Nutritionally, the resulting cookies have a low glycemic index (GI), making them optimal for postprandial blood sugar stabilization (GI = 40.5%) (Gao et al. 2022). A recent investigation evaluated oilseed cakes, including pumpkin cakes, in food preparation for health purposes. The data from this study indicated that the high protein and mineral content, the low fat and moderate ash content make PBPs excellent functional ingredients to be considered in food preparations such as dietary supplements (Sobczak et al. 2020). In addition, pumpkin seed oil cake powder has been recently used to produce gluten-free bread to develop a celiac-friendly product with a desirable nutritional value higher than the currently available. This highlighted how PBPs have the potential to become alternative supplements to enhance the formulation of gluten-free products (Voučko et al. 2022).

Other examples, however, consider using pumpkin seeds and peel in food products to mitigate childhood malnutrition and adequately supplement breastfeeding by providing the energy and nutrient density necessary for immunity, well-being, growth, and development of infants and children. Their great potential utilization as functional ingredients is mainly in muffins and porridge preparation as regards the use of pumpkin seeds, while pumpkin skin is used in combination with seeds for making cookies due to the proper balance of phytochemicals (phenols, flavonoids, carotenoids, β-carotene) and nutritional composition (Białek et al. 2016; Marcel et al. 2022; Hussain et al. 2022a, b; Mdluli et al. 2024).

Recently, including pumpkin seed oil in mayonnaise has led to functional and technological improvements, as shown in the study by Abed and Luma Khairy (2023). The presence of phenol compounds contributes to the nutritional sensory function of the product, owing to their antioxidant, antibacterial, antiviral, anti-inflammatory, cardioprotective, antidiabetic, and anti-obesity properties. The polyphenols present in the highest amounts were vanillin, *p*-coumaric acid, vanillic acid, and tyrosol, finding themselves at 13.07%, 12.45%, 11.69%, 10.43%, and 8.22%, respectively. Furthermore, pumpkin seed oil has characterized the mayonnaise for an extended shelf life. Indeed, the authors' thermal stability study shows that mayonnaise fortified with pumpkin seed oil has a heat stability of 90.22%. This parameter was calculated by evaluating the product's weight before and after heat treatment (50 °C) for 48 h, considering emulsion separation, if necessary.

4.1.2 Food Additives

Nowadays, the possibility of replacing traditional food additives with plant-base ones is becoming a trend that must be supported. PBPs are gaining more relevance in various food fields, reducing their environmental impact and replacing or surpassing additives traditionally found in the food industry. As shown by the study conducted by Tekin-Cakmak et al. (2021), cold-pressed pumpkin seed oil by-product (POB) was used as a natural fat substitute and stabilizer in reduced-fat salad dressings. POB, rich in proteins, phenolic compounds, and polysaccharides, resulted in desirable shear-thinning flow behavior, increasing the pseudoplastic and visco-elastic solid character and increased oxidative stability. Indeed, the data show that polysaccharides and proteins played the role of stabilizers and emulsifying agents, creating homogeneous and stable networks over time.

Furthermore, storage stability was improved due to the small particle size of POB, which positively affected the size and distribution of oil droplets. In addition, the presence of phenolic compounds improved the oxidative stability of low-fat salad dressing samples thanks to the free radical scavenging activity of the POB phenolic fraction. Considering all the above, POB can be used in various food applications (i.e., low-fat salad dressings) by acting as a natural fat replacer, stabilizer, and oxidative agent.

Similarly, Uzlasir (2020) has investigated using pumpkin seed oil as a natural fat replacer to produce emulsified meat products. The investigation compared cold-pressed pumpkin seed oil (PSO) with meat and beef fat, typically used in sausage production. After a complete evaluation of physiochemical parameters such as water-holding capacity (WHC), jelly-fat separation, emulsion stability and viscosity, the authors concluded that pumpkin seed oil in certain levels could be a potential replacement for animal-based fats in the production of sausages. The data showed that using PSO at certain levels did not lead to a significant difference in the water-holding capacity parameter, although there were differences in gelatin-fat separation, probably due to the different nature of the matrices tested. The WHC is the primary determinant of stability and consumer acceptance, as it can cause taste, flavor, and juiciness losses. PBPs are not only used to replace animal fats in slaughtered products but are also chosen in the production of dry-cured hybrid meat sausages to replace animal protein.

An example is replacing pork with pumpkin seed protein extrudates, which resulted in sausages with higher protein and lower fat content. This investigation aimed to test the effect of plant-based proteins on the final product to identify a coherent and sliceable matrix that tastes good and is safe for the consumer. The meat was substituted in the final product with 12.5, 25, 37.5, and 50% textured pumpkin seed protein, after which drying behavior and moisture and water distribution were evaluated and compared with a control (all-meat formulation). The results suggested substituting meat above 25% is the threshold for avoiding adverse effects on final products' drying behavior, texture, and color. The authors explain that the lower use of animal protein in favor of textured pumpkin seed protein reduces nitrosomyoglobin formation, which is essential for the meat's red color. Comparing control extrudates with those containing textured pumpkin seed protein greater and less than 25% denotes a change in color from red to green.

Moreover, drying behavior and texture differ depending on the type and amount of protein added to the extrudates. Comparing the product containing less than 25% textured pumpkin seed protein with the control, there is a change from accelerated to slow drying, while the texture remains almost unchanged. A significantly different behavior occurs when comparing the control to the product with more than 25% pumpkin seed protein; however, slow/delayed drying occurs in this case, and texture is no more consistent than in the control. For these reasons, the authors suggested that to produce a meat hybrid with an appearance and technological characteristics similar to the control, it is recommended not to exceed 25% pumpkin seed protein. Despite these considerations, the study provided valuable insights into the favorable application of textured plant proteins as alternatives to meat proteins (Ebert et al. 2022).

Another application of pumpkin seed oil as a food additive is to reduce the effect of cancerogenic compounds exposure to prevent adverse effects on human health. The study considered the impact of pumpkin seed oil on heterocyclic aromatic amines (HCAs) formation in meatballs during two different cooking methods, i.e., pan frying and oven methods, and different temperatures (150, 200, and 250 °C). HCAs formation is directly related to the cooking temperature, therefore at 250 °C high amounts of HCAs were recorded for both cooking methods, but with values higher for pan frying (23.5 ± 0.25 ng/g at 250 °C) than oven-baking (9.32 ± 1.65 ng/g at 250 °C). Considering the same temperature condition, the addition of 1% and 2% pumpkin seed oil recorded an inhibition of HCAs formation of 76% and 76.25% for pan-frying samples and 31.11% and 19.95% for oven-cooking samples, respectively. Therefore, pumpkin seed oil is a promising natural antioxidant inhibiting HCA formation, making it a possible food additive in several cooked meat products to reduce the formation of carcinogenic compounds (Erdoğan and Özdestan-Ocak 2022).

The addition of natural additives derived from by-products must not only meet safety standards and bring improvements in functional and technological properties, but it is also necessary for the final product to have a high acceptability index. For this purpose, pumpkin peel by-products have been investigated for their inclusion in beef burgers, evaluating the functional, technological, and sensory characteristics. A sensory study was conducted on 60 consumers evaluating the sensory acceptability index, which showed that more than 70% of the volunteers considered the burgers enriched with PBPs to have good sensory acceptance. Subsequently, further analyses were performed, highlighting that the effect of the additive was positive both on the chemical composition and on cooking and appearance characteristics (Hartmann et al. 2020). Meat products have not been the only ones to see the inclusion of pumpkin by-products as food additives. The application of these natural by-products has also been explored in other areas, using them as thickening and gelling agents in yogurt (Kanareikina et al. 2019), oxidative stabilizers and fat replacements in mayonnaise (Rezig et al. 2022), and even as food carriers to improve stabilization during storage (Čakarević et al. 2021). In the work of Kanareikina et al. (2019), pumpkin seeds, in addition to giving a pleasant taste and color to the product, were used to prepare a product with promising viscosity properties. The authors explain the gelling activity of pumpkin seeds by the presence of pectins and calcium. Pectins belong to the polygalacturon group and exert structural functions in plants. In an acidic environment, such as yogurt, the dissociation of free carboxyl groups is suppressed, preventing the chains' electrostatic repulsion. However, the presence of calcium allows the formation of calcium-pectin bridges, which results in obtaining the desired viscosity and gelling activity. In the studies by Čakarević et al. (2021) and Rezig et al. (2022), pumpkin seed oil stabilizes two different food matrices. In the first research, pumpkin oil cake was examined as a source of new protein material to be encapsulated to bring technological and sensory improvements to cookies. This research showed how this by-product provided stable encapsulation for 3 months and protected betalain, a compound added to the cookies, from degradation. Meanwhile, in the second work, it was found that pumpkin seed oil could increase oxidative stability through pre-treatment, which resulted in the formation of Maillard reaction products, known as potent natural antioxidants, which could be responsible for the increased stability.

4.2 Feed Additives

PBPs are part of the millions of tons of plant residues produced yearly that could be used in livestock and poultry feed. Given their functional and technological properties, they can be considered excellent additives and substitutes for traditional forage. Their application could help decrease global pollution-related issues but, most importantly, bring quality benefits to meat, milk, and eggs, which greatly value human nutrition (Valdez-Arjona and Ramírez-Mella 2019). For example, one strategy to achieve these objectives was to replace soybeans with pumpkin seed cake in the diet of dairy goats. The study involved 28 French Alpine dairy goats divided into three groups: one fed with soybean and considered as a control, one fed partially replacing the control forage with linseed, and the last one wholly fed with pumpkin seed cakes. During the monitoring time (20th, 48th, and 75th days after starting the experiment), parameters such as milk protein, fat, lactose, and urea content and the number of somatic cells in the milk did not change compared with the control sample. In addition, from the analysis of the milk fat acid profile, the trend was comparable to what emerged from the analysis of the control. Therefore, no substantial difference was found in the three groups differently fed and in milk quality and composition. C6:0, C8:0, and C10:0 medium-chain fatty acids are responsible for organoleptic properties, and their content (g/kg fatty acids) was similar in the group of goats fed with soybean and the group fed with PBPs. Therefore, PBPs can be considered excellent substitutes for traditional forages while maintaining high standard yields for human nutrition (Klir et al. 2017). Li et al. (2022) obtained similar results when incorporating soybean with high-oil pumpkin seed cake (HOPSC) into the diet of lactating Holstein dairy cows. Also, in this case, replacing traditional forage with HOPSC did not change the milk fatty acid composition. In addition, the study evaluated the lactation performance of the bacterial community in the ruminant lumen, considering also the occurrence of fermentation. This unconventional feeding resulted in increased fat content and fatty acid composition within the safe range (5%) in a different diet, but no changes in the product performance and rumen fermentation parameters were recorded. As regards the bacterial community, genetic analysis indicated a change in the rumen bacterial abundance but not in their profile, probably due to the presence of unsaturated fatty acids, which create a protective effect on the feed particles, thus hindering the fermentation and resulting in rumen microbial abundance changes. In conclusion, although preliminary, similar results for the control (soybean meal feeding) and HOPSC indicated that HOPSC is a promising alternative for high protein quality feeding.

Pumpkin seed cake could also be a valuable source of protein and fat for slaughter lamb production, as shown in a study investigating its use as a partial replacement for traditional Carcass traits. Hemato-chemical parameters and the fatty acid profile have been monitored. As previously demonstrated by the other studies concerning the evaluation of the effect of PBPs on milk quality, no significant differences between the meat produced by pumpkin-fed and conventionally-fed lambs have been registered. Partial substitution (10 and 15%) of soybean meal had no significant effect on either lamb health status or the quality of the final products. In contrast, some changes in hematological analysis were recorded when 15% pumpkin seed cake was used in lamb feeding. Increased serum glucose and albumin were registered, while lower concentrations of non-esterified fatty acids and β-hydroxybutyrate were recorded. However, these differences were insignificant, and the animal energy status remained adequate. Other changes were recorded by evaluating the activities of glutathione peroxide (GPx) and superoxide dismutase (SOD) activities that did not significantly increase, probably due to the presence of polyunsaturated fatty acids and phenolic acids in the feed. The authors concluded that the stable lamb antioxidant status and desirable energy conditions were adequate (Antunović et al. 2018).

Similar studies have been performed on other slaughtered animals, such as chickens (Steiner et al. 2020) and pigs (Stahn et al. 2023), showing that the incorporation of PBPs in non-traditional feed can become an efficient feed additive without having adverse effects on animal health or meat quality.

Recently, PBPs have gained a role not only as a replacement additive but also as an additive that can benefit animals and human health. The choice of feed supplemented with PBPs can bring reproductive improvements while also going on to naturally control pathogen/parasitic organisms, stimulate nutrition or increase resistance to pathological infections, and reduce abdominal fat and serum levels of harmful lipids while increasing serum levels of beneficial lipids in poultry (Achilonu et al. 2018; Lotfi et al. 2021).

4.3 Cosmetic Uses

The uses of pumpkin seed oils in the cosmetic industry are also noteworthy. In recent years, the interest in adding ingredients derived from natural wastes to cosmetic formulations has been a much sought-after trend, both for commercial and environmental reasons and for the bioactive components contained within, which are needed to perform the required beneficial activities. In the wide range of cosmetic products, pumpkin seed oils have found their application for skin treatment and protection, as potential sunscreens, with anti-inflammatory and antioxidant activities, and as makeup removal agents (Lacatusu et al. 2018; Setsiripakdee et al. 2019; Laaraj et al. 2022). For example, pumpkin seed oil was incorporated into lipid nanocarriers (NLCs) to formulate innovative sunscreens. The results of that study pointed out that NLCs could sustain the release of antioxidants and were UVB/UVA filters with excellent physical stability. In addition, NLCs had high antioxidant activity, with a free radical inhibition capacity of up to 98%. These formulations had SPF values between 46.5 and 50.5 and effective UVA protection, with UVA-PF values between 13.9 and 15.5 (Lacatusu et al. 2018). The goal of evaluating the potential use of pumpkin seeds as sunscreens was also pursued in Laaraj's research. This work compared different oils considering their photoabsorbent, antioxidant, and anti-inflammatory effects. Pumpkin seed extract had antioxidant and anti-inflammatory effects (39% spleen cell inhibition of proliferation), and pumpkin seed oil also had significant absorption in the UVB zone, blocking 22% of the most intense UVB rays, but had no significant absorption in the UVA zone (Laaraj et al. 2022). A different research investigated using pumpkin seed oil as a solvent for makeup removal. The high content of fatty acids and tocopherols was responsible for the oil's emollient and moisturizing properties, further enhancing the formulation. Efficacy studies showed that the oil was able to remove 79.92% of foundation, 41.02% of liquid eyeliner, and 23.54% of pen eyeliner, also resulting in better skin hydration than conventional makeup removers (Setsiripakdee et al. 2019).

The summary of the PBPs applications is reported in Table 6.

5 Other Industrial Applications

PBPs also have a variety of non-food/health applications, mainly as corrosion inhibitors or wastewater agents, natural textile dye or in the bioenergy field (Karthikeyan and Vidya 2020; Omari et al. 2022; Alacabey 2022; Umoru et al. 2023; Paiva et al. 2024). The most recent applications, covering such application areas, will be discussed in the following paragraphs.

Table 6 Food and non-food applications of *Cucurbita maxima* by-products

Application	Product type	Function	References
Functional food	Cookies	Postprandial blood sugar stabilization	Gao et al. (2022)
		Pro-health purposes	Sobczak et al. (2020)
	Gluten-free bread		Voučko et al. (2022)
	Porridge	Supplementing breastmilk	Marcel et al. (2021)
	Biscuit	Therapeutic purposes	Hussain et al. (2022a, b)
	Muffin	Enhance nutrition	Mdluli et al. (2024)
	Mayonnaise	Nutritional purposes	Abed and Luma Khairy (2023)
	Salad dressing	High protein and carbohydrate content	Tekin-Cakmak et al. (2021)
Food additives	Salad dressing	Fat replacer, stabilizer, oxidative agent	Tekin-Cakmak et al. (2021)
	Bologna-type beef sausage	Fat replacer	Uzlaşir et al. (2020)
	Hybrid pock sausages	Protein replacer	Ebert et al. (2022)
	Meatballs	Antioxidant agent	Doğăn et al. (2022)
	Bovine burger	General food additive	Hartmann et al. (2020)
	Yogurt	Thickening and gelling agents	Kanareikina et al. (2019)
	Mayonnaise	Fat replacer	Rezig et al. (2022)
		Food carrier and stabilizing agent	Čakarevič et al. (2022)
Feed additives	Dairy goat feed	Replacing feed	Klir et al. (2017)
	Holstein dairy cows	Replacing feed	Li et al. (2022)
	Slaughter lamb	Replacing feed	Antunović et al. (2018)
	Slaughter chickens	Replacing feed	Steiner et al. (2020) Nikolaev et al. (2021)
	Slaughter pig	Replacing feed	Stahn et al. (2021)
Other health applications	Cosmetic	Sunscreens with anti-inflammatory and antioxidant activities	Lacatusu et al. (2018); Laaraj et al. (2022)
	Cosmetic	Makeup removal	Setsiripakdee et al. (2019)

5.1 Corrosion Inhibitor

Corrosion is the chemical and electrochemical process that occurs between metals and alloys, and it involves the formation of a stable compound by losing the metal itself. This reaction leads to several economic and environmental issues (i.e., halts in process, fires, liberation of toxic products to the atmosphere). Corrosion inhibitors are generally used to counteract this high-energy process, which unfortunately presents risks as they are toxic, costly, and harmful to man and the environment (Benítez et al. 2014). *Cucurbita maxima* is a perfect source of eco-friendly, low-cost, and biodegradable corrosion inhibitors in the chemical, petroleum, naval, civilian construction, transport method, and communications sectors. Many researchers have analyzed the applicability of PBPs for inhibiting corrosion on mild steels, aluminum, stainless steel, and alloys. The choice of using such by-products as green corrosion inhibitors is due to the optimal chemical composition. The chemical structure of heteroatoms such as N, S, P, O, multiple bonds, and aromatic rings is essential for creating a protective layer and, subsequently, blocking the active corrosion site. A recent interesting application of pumpkin seeds as a corrosion green inhibitor for mild steel started from the by-product prepared with sulfur, nitrogen, and phosphorus carbon quantum dots. Carbon quantum dots are a recent technique that gives very satisfactory results and interesting features from an environmental point of view. The use of this methodology makes it possible to exploit green raw materials, depending on their origins, due to their richness in O, N, P, and S. To obtain this green corrosion inhibitor, roasted and crushed pumpkin seeds have been treated with H_3PO_4 in a silicone bath at 200 °C for 8 h. After intermediate steps consisting in filtration, dialysis, and drying, a 100 ppm stock solution was prepared and added to the mild steel plates in HCl 1 M for 2 and 24 h at room temperature for the gravimetric test. The performance of the pumpkin seeds inhibitor (corrosion rate and corrosion inhibition efficiency) has been measured by the variation in the mass of the plates before and after each gravimetric test. The corrosion process was evaluated with and without pumpkin seed inhibitor at different times and concentrations (10, 20, 40, 60, and 80 ppm in 1 M of HCl). The corrosion rate decreased at 2 and 24 h, reaching 91% efficiency in 2 h with 60 ppm and 95% in 24 h with only 10 ppm (Paiva et al. 2024).

Pumpkin seeds have also been considered the eco-friendly starting material from which oil was obtained and used for the synthesized cucurbita polyesteramide urethane (CPEAU), a mild steel corrosion inhibitor. Recently, organic polymers have gained more interest, making them excellent candidates as corrosion inhibitors and in this study, a three-step CPEAU synthesis involving the synthesis of *N,N′*-bis (2-hydroxyethyl) Cucurbita Oil Fatty Amide (HECFA), the synthesis of Cucurbita Polyesteramide (CPEA), and syntheses of Cucurbita CPEAU was developed. The potentiodynamic polarization measurements have been performed to understand the behavior of mild steel with 100 or 250 ppm and without CPEAU. The results demonstrated a 93.23% inhibition capacity of the corrosive process, allowing the authors to call CPEAU an excellent corrosion inhibitor for mild steel in 1.0 M HCl (Okai et al. 2021). Pumpkin seeds have also been used as an eco-friendly corrosion inhibitor for aluminum alloy 7075-T6 in 3.5% NaCl solution. Aluminum and its alloys are highly used in different sectors, especially in the aeronautics industry. The powder obtained after seed washing and drying has been pelletized at about 1% (w/w) in potassium bromide, and electrochemical measurements have been carried out to define the corrosion inhibition efficiency by dissolving different amounts of potential inhibitor (250, 500, 1000, and 1500 ppm) in 3.5% of NaCl solution. Results indicated that the increase in pumpkin seed concentration is correlated with a decrease in the corrosion current density, a parameter linked to the corrosion inhibition efficiency. More in detail, increasing the inhibitor concentration from 250 to 1000 ppm, the inhibition efficiency ranged from 51.34 to 92.7%, reaching values higher than that registered for date palm fruit juice (63%) and *Phoenix dactylifera* L. cationic dye (78%). When used at 1500 ppm, the capacity decreased, probably due to the desorption of the inhibitor molecules and the lack of the inhibitor over a relatively small area of the alloy surface (Radi et al. 2021).

Similar results were also obtained for pumpkin peel (PCM) used as a corrosion inhibitor of mild steel in different acid solutions (HCl 1 N and H_2SO_4 1 N). The increase in inhibition efficiency was inversely related to the corrosion rate. When PCM was used at the highest concentration, the inhibition efficiency of the metal was about 93% and 98% in HCl 1 N and H_2SO_4 1 N, respectively, due to the different corrosive activity of the two acids (Anbarasi and Vasudha 2014; Anbarasi 2016). Pumpkin peel was also an extremely qualified inhibitor for stainless steel in a 1 M H_2SO_4 medium. Umoru et al. (2023) investigated the corrosion inhibition efficiency on Grade 304 austenitic stainless steel, considering the weight loss method and electrochemical measurements. Preliminarily, the effects of concentration (0.3, 0.5, and 0.7 g/l), combination with halide ions (KI and KCl), and corrosive system temperature (303, 333, and 363 K) have been investigated. Considering the electrochemical experiment that has been carried out to understand the thermodynamic effects behind the corrosion process, an increase in the inhibitor concentration led to an increase in the inhibition efficiency, too. The best process conditions to obtain the lower weight loss (186 mg) and the higher inhibition efficiency (95.21%) were 0.7 g/l peel extract plus KI at 303 K. In addition, different halide ions resulted in a different inhibition efficiency (order of synergy: KI>KCl). The best temperature was 303 K, as the increase in temperature caused an increase in the corrosion rate. Corrosion processes typically occur at high temperatures, favored by the increased hot movement of inhibitor molecules, decreasing their adsorption capacity on the steel surface. As regards the thermodynamic parameters, the data indicated that the inhibition process turns out to be spontaneous ($\Delta G = -18.64$ kJ/mol at 303 K). In addition, the presence of pumpkin peel plus halide ions resulted in an increase of the activation energy (from 10.72 kJ/mol to 50.81 kJ/mol) of the corrosive process, which in turn reduced the possibility that a corrosive process of the metal will be established (Umoru et al. 2023). Table 7 shows the previously mentioned examples of PBPs corrosion inhibitors, considering the preparation method and the percentage of inhibition efficiency.

5.2 *Wastewater Agent*

Chemical dyes used in textiles, leather, paper, printing, and cosmetics applications pollute landfills, leading to catastrophic problems for nature and humans. Pumpkin seeds could be transformed into water purification agents to fight this environmental problem. Their use could be considered an emerging strategy because their seeds

Table 7 PBPs corrosion inhibitors

By-product	Metal	Method	% Inhibition efficiency	Reference
Seeds	Mild steel	HCl 1 M	95.00	Paiva et al. (2024)
	Mild steel	HCl 1 M	93.23	Okai et al. (2021)
	Aluminum alloy 7075-T6	3.5% NaCl	92.70	Radi et al. (2021)
Peel	Mild steel	HCl 1 N	93.00	Anbarasi and Vasudha (2014)
	Mild steel	H_2SO_4 1 N	98.00	Anbarasi (2016)
	Grade 304 austenitic stainless steel	H_2SO_4 1 M	95.21	Umoru et al. (2023)

are rich in organic carbon, a material capable of reducing dyes in water. Pumpkin seeds were dried, milled in a saw-milling apparatus, carbonized at 250 °C for 2 h under atmospheric conditions to obtain carbon microparticles with various sizes (from 100 to 1000 μm) and tested in isothermal conditions in their adsorption efficiency toward curcumin, taken as a model of ideal dye. FTIR analysis was performed to confirm the successful conversion of the waste material to carbonaceous material. The results were compared using the kinetic approach based on different standard adsorption isotherm models (Langmuir, Freundlich, Temkin, and Dubinin-Radushkevich); in addition, the effect of particle size on the adsorption phenomena was also studied. The experimental results showed that the adsorption on the surface of carbon microparticles occurred in monolayer with a physical phenomenon because the active areas are located only on the outer surface of the particles, and no surface structure in the carbon is available. In addition, the smaller particles directly impacted the increase of adsorption capacity, which can be explained by the fact that a larger surface area allows more cooperative adsorption interaction. The maximum adsorption efficiency was 76–78% (Nandiyanto 2020; Nandiyanto et al. 2021).

In 2019, Rashid et al. (2019) demonstrated the possibility of using pumpkin by-products as a naturally activated carbon-based adsorbent to decolorize water from methylene blue (MB). Pumpkin peels after carbonization procedure at high temperatures (250, 350, 450, and 550 °C) for 1 h have been added to beetroot extract and functionalized with different acid treatments (citric acid, nitric acid, and oxalic acid) to obtain active carbon. The adsorption performance was evaluated by placing a constant mass of prepared carbon materials in contact with the MB solution. The dye-absorbent capacity has been monitored at 665 nm after a stirring step for 180 min. The results indicated a higher adsorption capacity than other materials, with a 198.15 mg/g MB value. The ability of pumpkin peel to be a helpful biosorbent has been supported by the study of Omari et al. (2022). The aim was to use the red peel to prepare a carbon adsorbent material functionalized with $ZnCl_2$ to remove MB and Congo Red from water. As in the previously mentioned study, activation was a necessary pre-treatment to achieve a high impregnation rate yield; in particular, using $ZnCl_2$ enhanced the yield from 30 to 70.79%. Using sulfuric acid as a pre-treatment, the effective application of pumpkin peel as a green and easy biochar was also investigated. In this case, only chemically treated peels presented an adsorption capacity toward MB (207.5 mg/g) with an isotherm value higher than most of the other adsorbents (Bal et al. 2021).

Water can be contaminated by various agents, including the above-mentioned chemical dyes derived mainly from cotton and silk farms, as well as by other agents, such as antibiotics. Their presence in water leads to increased antibiotic resistance, an issue that is no longer emerging. Pumpkin shells can effectively remove ciprofloxacin (CIP), one of the most widely administered antibiotics in the world, from aqueous systems. In fact, after optimizing the adsorption conditions, carbonized pumpkin shells activated by KOH could adsorb more than 99% CIP, with an adsorption capacity of 884.9 mg/g (Alacabey 2022). Biochar prepared by impregnating pumpkin peels with H_3PO_4 in equal mass ratio at 60 °C for 1 h and subsequent heating at 700 °C for 1 h and washing with distilled water for reaching neutrality showed a CIP adsorption capacity of 153.9 mg/g (from the Langmuir equation), with the maximum efficiency at pH = 8. A good desorption efficiency was observed in 0.1 M NaOH (Özer and İmamoğlu 2024).

Table 8 provides a summary of PBPs applications as wastewater purifying agents.

5.3 *Natural Textile Dye*

Previously, PBPs were considered as adsorbent agents of toxic chemical dyes. However, it is also known that using pumpkin peel as a natural dye could avoid the previously mentioned problem. The interest in using natural dyes from waste plant sources has grown in recent years. Because of the problems generated by synthetic and chemical dyes, the ability to obtain dyes from plant sources appears to be an essential goal, particularly for the textile sector. Pumpkin peel, rich in different carotenoid pigments, could be used as vegetable colorants and repurposed for manufacturing bio-based clothes (Che et al. 2023). Recently, natural dye production from pumpkin peel and its applicability versus a chemical dye has been investigated. In addition, to consider the environmental benefits of using natural dyes, treated waters have been used for cultivating the chickpea plant crop. The dyeing

Table 8 Summary of PBPs applications as wastewater purifying agents

Method	Absorption capacity (mg/g)	Reference
Carbonization with beetroot extract Activation by acid treatments	198.15	Rashid et al. (2019)
Carbonization Activation by $ZnCl_2$	100.60	Omari et al. (2022)
Carbonization Activation by H_2SO_4	207.50	Bal et al. (2021)
Carbonization Activation by KOH	884.90	Alacabey (2022)

process was analyzed on white cotton cloth with the help of natural mordant (vinegar) through color measurements, and the degradation and discoloration profiles were also considered. White cotton cloth has taken more colors from pumpkin peel, and applying the mordant resulted in appreciable washing, light fastness properties, and good rubbing fastness.

In terms of degradation and discoloration, natural and chemical dyes were treated with *Bacillus licheniformis* and *Aspergillus niger*. Using *B. licheniformis*, the maximum dye discoloration was achieved on the third day at 37 °C, pH 7, and the degradation value was 89.2% and 80.9%, respectively. Differently, using *A. niger*, the maximum discoloration was reached in the second day at 30 °C with different pH conditions and degradation values (76.8% at pH 4 for natural dye and 70.2% at pH 5 for chemical dye). Reusing treated water for irrigating chickpea seeds, considering soil fertility, plant growth rate, and the number of leaves in cultivated plants as approximate parameters, was also investigated. The efficiency of chickpea growth demonstrated that the reuse of natural effluents is an excellent strategy for growing plants without harming soil fertility and plant yield and avoiding issues generated by chemical effluents that not only pollute the soil but also damage plant yield (Karthikeyan and Vidya 2020).

5.4 Energy

Supercapacitors are one of the electrical energy storage technologies undergoing extensive developments in recent years, delivering higher power densities (10–20 kW/kg) than batteries and higher energy densities (typically 4–5 Wh/kg) than dielectric capacitors. Supercapacitors can mainly be viewed as pulse power devices with a typical discharge time ranging from a few seconds to less than 1 s, the ability for rapid charge, and, concerning batteries, lower energy densities but a wider working temperature range, better safety, higher reliability, environmental friendliness, and maintenance-free operation (Goikolea and Mysyk 2017). Electrical double-layer capacitors (EDLCs) are the most common type of supercapacitors. They are widely used as advanced electrochemical power sources thanks to their simple operation via electrosorption of ions from an electrolyte onto high-surface-area electrodes, guaranteeing rapid charge propagation dynamics and long cycle life performance (>100,000 cycles) and making them suitable for applications in electric vehicles, consumer electronics, electric pulse devices, as well as industrial power and energy management (Wang et al. 2012). Carbon-based materials such as nanotubes, nanofibers, and graphene have been tested as electrodes in EDLCs thanks to their high specific surface area and tunable pore sizes (Arumugam et al. 2023; Frackowiak et al. 2000; Wei et al. 2017; Yoo et al. 2011). Anyway, their relatively high costs and some intrinsic operability limits (for example, the graphene tendency to aggregation during cycling) make it necessary to find more efficient alternatives to these materials. Recently, activated carbon-based materials obtained by biomass of both vegetal and animal origins have attracted increasing attention as electrode materials in supercapacitors, with the dual aims of environmental protection and "clean" energy storage, thanks to their very high specific surface area (up to 3000 m^2/g) (Pontiroli et al. 2019; Leng et al. 2021; Liu et al. 2016), interconnected porous structures, good electrical conductivity, and excellent chemical stability coupled with and relatively easy processing methods and low costs (Bai et al. 2016).

The biomass undergoes two different preparation steps: a thermal treatment (pyrolysis) under controlled temperature and atmosphere conditions (air-free) to favor the formation of the solid residue called biochar concerning the liquid (bio-oil) and gaseous (bio-gas) fraction, followed by a chemical activation generally performed with strong bases and a second thermal treatment to increase the specific surface area of the carbon surface creating micropores (Pontiroli 2019).

Concerning pumpkin wastes, Gopiraman et al. (2017) prepared biochar and subsequently interconnected meso/microporous activated carbons from seeds and tested them as electrodes for supercapacitors. The fresh seeds were washed with distilled water, oven-dried at 60 °C for 24 h, and subsequently heated in a muffle furnace under air from room temperature to 300 °C with a heating rate of 1 °C/min and isothermally carbonized for 1 h. The obtained biochar was ground with NaOH in a 1:2 weight ratio, heated up to different temperatures (600, 700, 800, and 900 °C) under N_2 flow for 1 h to increase the porosity, washed with 1 M HCl solution for obtaining neutrality, and dried at 60 °C for 24 h. The BET analysis showed an increase of 3 times in the specific surface area and 4.8 times in the pores volume with the activation temperature, reaching, respectively, 2029 m^2/g and 1.30 cm^3/g for the 900 °C annealed sample. These values are comparable with those obtained for other promising activated carbon materials prepared from biomass, such as soy flour (Sevilla et al. 2017) and rice husk (Liu et al. 2016), although they are lower than those obtained for pumpkin pulp and peel biochar (up to 2968 m^2/g, Bai et al. 2016). The resultant porous carbons, containing heteroatoms (O and N) as confirmed by X-ray photoemission spectroscopy, were tested as working electrodes in an electrochemical cell (with platinum wire and Ag/AgCl as the counter and reference electrodes, respectively) giving a maximum specific capacitance (C_s) of 303 F/g in 1.0 M H_2SO_4 at a scan rate of 5 mV and stability up to 6000 cycles. This promising value is comparable with those reported for other waste-derived activated carbon electrode materials (229 F/g for poultry litter (Pontiroli et al. 2019), 220 F/g for corn fibers (Gopiraman et al. 2017) even if lower than the performance of pumpkin pulp biochar (up to 419 F/g) and of graphene-based electrodes (up to 550 F/g) (Liu 2016).

The biochar, bio-oil, and bio-gas products can also be obtained from biomass through microwave pyrolysis, which offers several advantages over conventional thermal treatments: microwave-assisted pyrolysis transfers heat energy through molecular interactions within the biomass rather than through heat transfer from external sources, resulting in faster and more uniform heating, lower processing times, and higher energy yields, making it a greener and more sustainable option (Allende et al. 2023; Ethaib et al. 2020). This technique was applied to pumpkin peel at three operating power level conditions (0.9, 1.2, and 1.5 kW). The lowest tested power generated the lowest bio-oil amount (20.3 wt%) and the highest biochar yield (11 wt%). However, biochar with high carbon content (70.33%), low oxygen amount (23%), and significant porosity, suitable for energetic issues, was obtained by treating the biomass at 1.2 kW for 40 min. In the same conditions, bio-oil exhibited high aromatic functional groups and low oxygen content, indicating superior quality. The energy conversion efficiency of biochar and bio-oil production was estimated at 6.99%, and it could also be improved by considering the production of methane and hydrogen as bio-gasses. The energy balance analysis demonstrated that the food waste feedstock generated an output energy of 0.07 kWh, primarily attributed to the high energy generation from bio-oil. By considering the biochar yield and its fixed C content, the pumpkin peel demonstrated potential for CO_2 sequestration with a rate of 14.29 g CO_2 eq/kg.

Moreover, the starting pumpkin peel had a 25 wt% moisture content, and the elemental composition analysis revealed 48.79% carbon, 7.52% hydrogen, 3.97% nitrogen, and 39.72% oxygen, estimating a higher heating value (HHV; the amount of heat released by a unit mass or volume of fuel, initially at 25 °C, once it is combusted and the products have returned to a temperature of 25 °C) of 23.2 MJ/kg (Demiral and Samdan 2016), in line with those reported for other raw biomasses, generally lying in the range of 15–20 MJ/kg (Chen 2015) suggesting the possibility to use this biomass as a solid fuel for bioenergy production (as a reference, coal HHVs range from 25 to 35 MJ/kg).

The non-edible pumpkin seed oil (*Cucurbita pepo L.*) was also employed for biodiesel production through the transesterification process, thanks to its high oil content, ranging from 42 to 45%. GC-MS profile of pumpkin seed oil methyl ester showed the presence of linoleic acid (43.72% wt), oleic acid (37.07% wt), palmitic acid (12.51% wt), and stearic acid (5.43% wt) (Schinas et al. 2009) and a calorific value of 38.08 MJ/kg, within the limits of biodiesel standard EN 14214. The CO and hydrocarbon emissions during the operation of a direct injection diesel engine were lower for this biodiesel concerning diesel fuel (Ramakrishnan et al. 2018).

6 Conclusions and Future Perspectives

Pumpkin (*Cucurbita maxima*) produces quite a large number of by-products with very versatile applications at different industrial levels. These promising reuses are essential in developing a circular economy with economic and environmental advantages.

Peels and seeds represent the main PBPs used, as they are extremely rich in bioactive compounds. Seeds and seed oil comprise fats, proteins, fiber, minerals, and fatty acids (oleic and linoleic acids). Peels are rich in pectins characterized by GalA.

Therefore, these PBPs can represent essential functional ingredients in food and food supplements with a promising possibility to obtain nutritional claims concerning, for example, their effect on fat metabolism or as a "source of fiber." In addition, pectin peel has many exploitable functional properties (water solubility, viscosity, and emulsifying properties), while seed and peel oils exhibit antioxidant activity due to the content of tocopherols and carotenoids, respectively.

Seeds are the main PBPs used in non-food/health applications as corrosion inhibitors and wastewater agents and in producing biochar, bio-oil, and bio-gas. Conversely, pumpkin peel can be applied as a natural textile dye for its carotenoid content.

A complementary key for the reuse of PBPs is represented by the ongoing progress of innovative green technologies, which lead to cost minimization (time and reagents consuming save).

References

Abd-Elnoor EV (2019) Hypoglycemic and hypolipidemic effects of pumpkin seeds powder and oil on alloxan-induced diabetic in rats. Egypt J Food Sci 47:255–269

Abed HM, Luma Khairy H (2023) The effect of adding pumpkin seed oil on physicochemical and sensory properties of the mayonnaise. IOP Conf Ser Earth Environ Sci 1214(1):012037

Abou-Elella F, Mourad R (2015) Anticancer and antioxidant potentials of ethanolic extracts of *Phoenix dactylifera*, *Musa acuminata* and *Cucurbita maxima*. Res J Pharm Biol Chem Sci 6:710

Achilonu MC, Nwafor IC, Umesiobi DO, Sedibe MM (2018) Biochemical proximates of pumpkin (*Cucurbitaeae* spp.) and their beneficial effects on the general well-being of poultry species. J Anim Physiol Anim Nutr 102:5–16

Agarkova EY, Kruchinin AG, Glazunova OA, Fedorova TV (2019) Whey protein hydrolysate and pumpkin pectin as nutraceutical and prebiotic components in a functional mousse with antihypertensive and bifidogenic properties. Nutrients 11:2930

Alacabey İ (2022) Antibiotic removal from the aquatic environment with activated carbon produced from pumpkin seeds. Molecules 27:1380

Aljahani AH (2022) Wheat-yellow pumpkin composite flour: Physico-functional, rheological, antioxidant potential and quality properties of pan and flat bread. Saudi J Biol Sci 29:3432–3439

Allende S, Brodie G, Jacob MV (2023) Breakdown of biomass for energy applications using microwave pyrolysis: A technological review. Environ Res 226:115619

Al-Okbi SY, Mohamed D, Kandil E, Ahmed EK, Mohammed S (2016) Antioxidant and anticancer effect of Egyptian and European pumpkin seed oil. Res J Pharm Biol Chem Sci 7:574

Al-Turky HM, Abo Chameh G, Ibrahim B (2022) Effect of defatting and extracting solvent on the antioxidant activities in seed extracts of two species of Syrian pumpkin. J Food Meas Charact 16:4813–4821

Amin MZ, Rity TI, Uddin MRM, Rahman MM, Uddin MJ (2020) A comparative assessment of anti-inflammatory, anti-oxidant and anti-bacterial activities of hybrid and indigenous varieties of pumpkin (*Cucurbita maxima* Linn.) seed oil. Biocatal Agric Biotechnol 28:101767

Amoo IA, Eleyinmi AF, Ilelaboye N, Akoja SS (2004) Characterisation of oil extracted from gourd (*Cucurbita maxima*) seed. Food Agric Environ 2:38–39

Anbarasi K (2016) Electrochemical and corrosion inhibition studies of *Cucurbita maxima*. Orient J Chem 32:2139–2145

Anbarasi K, Vasudha VG (2014) Chemical science review and letters corrosion inhibition potential of *Cucurbita maxima* plant extract on mild steel in acid media. Chem Sci Rev Lett 3:45–51

Antunović Z, Klir Ž, Šperanda M, Sičaja V, Čolović D, Mioč B, Novoselec J (2018) Partial replacement of soybean meal with pumpkin seed cake in lamb diets: Effects on carcass traits, haematochemical parameters and fatty acids in meat. S Afr J Anim Sci 48:695–704

Arumugam B, Mayakrishnan G, Subburayan Manickavasagam SK, Kim SC, Vanaraj R (2023) An overview of active electrode materials for the efficient high-performance supercapacitor application. Crystals 13:1118

Aziz A, Noreen S, Khalid W, Ejaz A, Faiz ul Rasool I, Maham, Munir A, Miral Javed F, Ercisli S, Okcu Z, Marc RA, Nayik GA, Ramniwas S, Uddin J (2023) Pumpkin and pumpkin byproducts: phytochemical constitutes, food application and health benefits. ACS Omega 8:23346–23357

Bai S, Tan G, Li X, Zhao Q, Meng Y, Wang Y, Zhang Y, Xiao D (2016) Pumpkin-derived porous carbon for supercapacitors with high performance. Chem Asian J 11:1828–1836

Baississe S, Fahloul D (2018) Rheological behavior and electrokinetic properties of pectin extracted from pumpkin (*Cucurbita maxima*) pulp and peel using hydrochloric acid solution. Chem Pap 72:2647–2658

Baississe S, Dridri S, Fahloul D, Mezdour S (2022) Structure of pumpkin pectin and its effect on its technological properties. Appl Rheol 32:34–55

Bal D, Özer Ç, İmamoğlu M (2021) Green and ecofriendly biochar preparation from pumpkin peel and its usage as an adsorbent for methylene blue removal from aqueous solutions. Water Air Soil Pollut 232:1–16

Balbino S, Dorić M, Vidaković S, Kraljić K, Škevin D, Drakula S, Voučko B, Čukelj N, Obranović M, Ćurić D (2019) Application of cryogenic grinding pretreatment to enhance extractability of bioactive molecules from pumpkin seed cake. J Food Process Eng 42:e13300

Bardaa S, Ben Halima N, Aloui F, Ben Mansour R, Jabeur H, Bouaziz M, Sahnoun Z (2016) Oil from pumpkin (*Cucurbita pepo* L.) seeds: evaluation of its functional properties on wound healing in rats. Lipids Health Dis 15:73

Batool M, Ali Nawaz Ranjha MM, Roobab U, Manzoor MF, Farooq U, Nadeem HR, Nadeem M, Kanwal R, AbdElgawad H, Al Jaouni SK, Selim S, Ibrahim SA (2022) Nutritional value, phytochemical potential, and therapeutic benefits of pumpkin (*Cucurbita* sp.). Plants 11:1394

Benítez LPT, Castellar PJM, Percy EDA, Bravo MJB (2014) Plant extracts used as corrosion inhibitors. Informador Técnico 78:154–163

Białek M, Rutkowska J, Adamska A, Bajdalow E (2016) Partial replacement of wheat with pumpkin seed flour in muffins offered to children. CYTA J Food 14:391–398

Boujemaa I, El Bernoussi S, Harhar H, Tabyaoui M (2020) The influence of the species on the quality, chemical composition and antioxidant activity of pumpkin seed oil. Oilseeds Fats Crops Lipids 27:40

Čakarević J, Torbica A, Belović M, Tomić J, Sedlar T, Popović L (2021) Pumpkin oil cake protein as a new carrier for encapsulation incorporated in food matrix: Effect of processing, storage and in vitro digestion on bioactivity. Int J Food Sci Technol 56:3400–3408

Che J, Teng X, Ji J (2023) Extraction and application of pumpkin peel colorants for natural textile dyeing. AATCC J Res 10:153–162

Chen W-H (2015) Chapter 10. Torrefaction. In: Pandey A, Negi S, Binod P, Larroche C (eds) Pretreatment of biomass processes technologies book. Elsevier, London, pp 173–192

Colombo R, Pellicorio V, Barberis M, Frosi I, Papetti A (2024) Recent advances in the valorization of seed wastes as source of bioactive peptides with multifunctional properties. Trends Food Sci Technol 144:104322

Demiral I, Samdan CA (2016) Preparation and characterisation of activated carbon from pumpkin seed shell using H_3PO_4. Anadolu Univ J Sci Technol A 17:125–138

Ebert S, Jungblut F, Herrmann K, Maier B, Terjung N, Gibis M, Weiss J (2022) Influence of wet extrudates from pumpkin seed proteins on drying, texture, and appearance of dry-cured hybrid sausages. Eur Food Res Technol 248:1469–1484

Egan AM, Dinneen S (2018) What is diabetes? Medicine 47:1

Erdoğan B, Özdestan-Ocak Ö (2022) Determination the effects of pumpkin and rosehip seed oils on heterocyclic aromatic amine formation in meatballs by high-performance liquid chromatography. J Food Process Preserv 46:e16299

Ethaib S, Omar R, Kamal SMM, Biak DRA, Zubaidi SL (2020) Microwave-assisted pyrolysis of biomass waste: a mini-review. Processes 8:1190

FAO Food and Agriculture Organization of the United Nation FAOSTAT (2019) Available from http://www.fao.org/faostat/en/#data/QC. Accessed 29 Mar 2024

Farooqi MA, Kesharwani A, Mishra MK, Kesharwani A (2022) Antihyperlipidemic and hepatoprotective effect of *Cucurbita maxima* (Duch) seed in paracetamol induced hepatotoxicity and high fat diet-induced hypercholesterolemic rats. Neuroquantology 20:6247–6255

Frackowiak E, Metenier K, Bertagna V, Beguin F (2000) Supercapacitor electrodes from multiwalled carbon nanotubes. Appl Phys Lett 77:2421

Frosi I, Balduzzi A, Moretto G, Colombo R, Papetti A (2023) Towards valorization of food-waste-derived pectin: recent advances on their characterization and application. Molecules 28:6390

Gao D, Helikh A, Duan Z, Shang F, Liu Y (2022) Development of pumpkin seed meal biscuits. East Eur J Enterprise Technol 2:36–42

Gaweł-Bęben K, Czech K, Strzępek-Gomółka M, Czop M, Szczepanik M, Lichtarska A, Kukula-Koch W (2022) Assessment of *Cucurbita* spp. peel extracts as potential sources of active substances for skin care and dermatology. Molecules 27:761

Goikolea E, Mysyk R (2017) Chapter four—Nanotechnology in electrochemical capacitors. In: Rodriguez-Martinez LM, Omar N (eds) Micro and nano technologies, emerging nanotechnologies in rechargeable energy storage systems. Elsevier, Boston, pp 131–169

Gopiraman M, Saravanamoorthy S, Kim S-H, Chung I-M (2017) Interconnected meso/microporous carbon derived from pumpkin seeds as an efficient electrode material for supercapacitors. Carbon Lett 24:73–81

Hagos M, Chandravanshi BS, Redi-AbShiro M, Yaya EE (2023) Determination of total phenolic, total flavonoid, ascorbic acid contents and antioxidant activity of pumpkin flesh, peel and seeds. Bull Chem Soc Ethiop 5:1093–1108

Hartmann GL, Marconato AM, Marques M, Santos R, Aparecida Do Amaral L, Freitas Dos Santos E, Novello D (2020) Addition of pumpkin peel flour affect physicochemical and sensory characteristics of bovine burger. Int J Res Granthaalayah 8:254–263

Hussain A, Kausar T, Din A, Murtaza A, Jamil MA, Noreen S, Iqbal MA (2021) Antioxidant and antimicrobial properties of pumpkin (*Cucurbita maxima*) peel, flesh and seeds powders. J Biol Agric Healthc 11:6

Hussain A, Kausar T, Sehar S, Sarwar A, Ashraf AH, Jamil MA, Noreen S, Rafique A, Iftikhar K, Quddoos MY (2022a) Determination of total phenolics, flavonoids, carotenoids, β-carotene and DPPH free radical scavenging activity of biscuits developed with different replacement levels of pumpkin (*Cucurbita maxima*) peel, flesh and seeds powders. Turk J Agric Food Sci Technol 10:1506–1514

Hussain A, Kausar T, Sehar S, Sarwar A, Ashraf AH, Jamil MA, Noreen S, Rafique A, Iftikhar K, Quddoos MY, Aslam J, Majeed MA (2022b) A comprehensive review of functional ingredients, especially bioactive compounds present in pumpkin peel, flesh and seeds, and their health benefits. Food Chem Adv 1:100067

Indrianingsih AW, Rosyida VT, Apriyana W, Hayati SN (2019) Comparisons of antioxidant activities of two varieties of pumpkin (*Cucurbita moschata* and *Cucurbita maxima*) extracts. IOP Conf Ser Earth Environ Sci 251:012021

Isutsa DK, Mallowa SO (2013) Increasing leaf harvest intensity enhances edible leaf vegetable yields and decreases mature fruit yields in multi-purpose pumpkin. J Agric Biol Sci 8:610–615

Jahan F, Islam BM, Moulick SP, Al Bashera M, Hasan SM, Tasnim N, Saha T, Boby F, Waliullah M, Saha KA, Hossain A, Ferdousi L, Rahman MM, Saha BK, Bhuiyan MNK (2023) Nutritional characterization and antioxidant properties of various edible portions of *Cucurbita maxima*: A potential source of nutraceuticals. Heliyon 9:e16628

Kalyna VS, Kondratiuk NV, Uvarova NV, Rozghon OG, Bohuslavskyi VV, Lytsenko MV (2021) Characteristics and composition of fractional products from pumpkin seeds. J Chem Technol 29:660–670

Kanareikina S, Kanareikin V, Ganieva E, Burakovskaya N, Shadrin M, Khalepo O, Babaeva M, Nikolaeva N, Voskanyan O (2019) The structure development of yogurt with vegetable ingredients. Int J Recent Technol Eng 8:1587–1592

Karthikeyan G, Vidya A (2020) Production and application of natural dye from skin of yellow pumpkin vegetable. Int J Recent Sci Res 11:37828–37839

Kaur S, Panghal A, Garg MK, Mann S, Khatkar SK, Sharma P, Chhikara N (2020) Functional and nutraceutical properties of pumpkin—a review. Nutr Food Sci 50:384–401

Klir Z, Castro-Montoya JM, Novoselec J, Molkentin J, Domacinovic M, Mioc B, Dickhoefer U, Antunovic Z (2017) Influence of pumpkin seed cake and extruded linseed on milk production and milk fatty acid profile in Alpine goats. Animal 11:1772–1778

Konopacka D, Seroczyńska A, Korzeniewska A, Jesionkowska K, Niemirowicz-Szczytt K, Płocharski W (2010) Studies on the usefulness of Cucurbita maxima for the production of ready-to-eat dried vegetable snacks with a high carotenoid content. LWT Food Sci Technol 43:302–309

Kushawaha DK, Yadav M, Chattergji S, Srivastava AM, Watal G (2017) Evidence based study of antidiabetic potential of *C. maxima* seeds in vivo. J Tradit Complement Med 7:466–470

Laaraj S, Ouahidi I, Aarab L (2022) Photoabsorbent, antioxidative, and anti-inflammatory properties of three traditional cosmetic seed oils. Trop J Nat Prod Res 6:1790–1793

Lacatusu I, Arsenie LV, Badea G, Popa O, Oprea O, Badea N (2018) New cosmetic formulations with broad photoprotective and antioxidative activities designed by amaranth and pumpkin seed oils nanocarriers. Ind Crop Prod 123:424–433

Leichtweis MG, Molina AK, Pires TCS, Dias MI, Calhela R, Bachari K, Ziani BEC, Oliveira MBPP, Pereira C, Barros L (2022) Biological activity of pumpkin byproducts: antimicrobial and antioxidant properties. Molecules 27:8366

Leichtweis MG, Molina AK, Petropoulus SA, Carocho M, Pires TCSP, Dias MI, Calhelha R, Oliveira MBPP, Pereira C, Barros L (2023) Valorization of pumpkin peel as a source of bioactive compounds: optimization of heat- and ultrasound-assisted extraction. Molecules 28:3168

Lemus-Mondaca R, Marin J, Rivas J, Sanhueza LC (2019) Pumpkin seeds (*Cucurbita maxima*). A review of functional attributes and by-products. Revista Chilena de Nutrición 6:783–791

Leng L, Xiong Q, Yang L, Li H, Zhou Y, Zhang W, Jiang S, Li H, Huang H (2021) An overview on engineering the surface area and porosity of biochar. Sci Total Environ 763:144204

Li Y, Gao J, Lv J, Lambo MT, Wang Y, Wang L, Zhang Y (2022) Replacing soybean meal with high-oil pumpkin seed cake in the diet of lactating Holstein dairy cows modulated rumen bacteria and milk fatty acid profile. J Dairy Sci 106:1803–1814

Liu D, Zhang W, Lin H, Li Y, Lu H, Wang Y (2016) A green technology for the preparation of high capacitance rice husk-based activated carbon. J Clean Prod 112:1190–1198

Lotfi S, Fakhraei J, Mansoori Yarahmadi H (2021) Dietary supplementation of pumpkin seed oil and sunflower oil along with vitamin E improves sperm characteristics and reproductive hormones in roosters. Poult Sci 100:101289

Mahmoodpoor A, Medghalchi M, Nazemiyeh H, Asgharian P, Shadvar K, Hamishehkar H (2018) Effect of *Cucurbita maxima* on control of blood glucose in diabetic critically ill patients. Adv Pharm Bull 8:347–351

Maldonade IR, Amaro GB, Luengo RFA, Ribeiro RLV, Lozada MIO, Oliveira LL, Machado ER (2020) Phytochemical characterization of pumpkin seed with antiparasitic action. In: International symposium on advances in production and processing of medicinal and aromatic plants, 1287

Marcel MR, Chacha JS, Ofoedu CE (2022) Nutritional evaluation of complementary porridge formulated from orange-fleshed sweet potato, amaranth grain, pumpkin seed, and soybean flours. Food Sci Nutr 10:536–553

Matsuda M, Huh Y, Ji R (2019) Roles of inflammation, neurogenic inflammation, and neuroinflammation in pain. J Anesth 33:131–139

Mdluli H, Kidane SW, Nkambule TP, Shelembe JS (2024) Valorisation of pumpkin seeds through wheat-sorghum-pumpkin seed composite flour muffins: optimization using D-optimal mixture design. Cogent Food Agric 10:2303833

Monica SJ, John S, Madhanagopal R, Sivaraj C, Khusro A, Arumugam P, Gajdács M, Lydia DE, Sahibzada MUK, Alghamdi S, Almehmadi M, Emran TB (2022) Chemical composition of pumpkin (*Cucurbita maxima*) seeds and its supplemental effect on Indian women with metabolic syndrome. Arab J Chem 15:103985

Moniruzzaman M, Jinnah MM, Islam S, Biswas J, Al-Imran, Pramanik MJ, Uddin MS, Saleh MA, Zaman S (2022) Biological activity of *Cucurbita maxima* and *Momordica charantia* seed extracts against the biofilm-associated protein of *Staphylococcus aureus*: An in vitro and in silico studies. Inform Med Unlocked 33:101089

Nandiyanto ABD (2020) Isotherm adsorption of carbon microparticles prepared from pumpkin (*Cucurbita maxima*) seeds using two-parameter monolayer adsorption models and equations. Moroccan J Chem 8:745–761

Nandiyanto ABD, Hofifah SN, Inayah HT, Yani IF, Putri SR, Apriliani SS, Anggraeni S, Usdiyana D, Rahmat A (2021) Adsorption isotherm of carbon microparticles prepared from pumpkin (*Cucurbita maxima*) seeds for dye removal. Iraqi J Sci 62:1404–1414

Nishimura M, Ohkawara T, Sato H, Takeda H, Nishihira J (2014) Pumpkin seed oil extracted from *Cucurbita maxima* improves urinary disorder in human overactive bladder. J Tradit Complement Med 4:72–74

Okai V, Chahul HF, Wuana RA, Barnabas IA, Tolufashe GF (2021) Corrosion inhibition potentials of *Cucurbita* polyesteramide urethane on mild steel in hydrochloric acid medium: Experimental and computational studies. Sci Afr 12:e00776

Omari S, Nedjhioui M, Hamidi N, Benkortbi O (2022) Synthesis and characterization of activated carbon from red pumpkin skin for the removal of ionic dyes. Water Pract Technol 17:1197–1217

Ordoñez Lozada MI, Maldonade Rodrigues I, Bobrowski Rodrigues D, Santos DS, Ortega Sanchez BA, de Souza PEN, Longo JP, Amaro GB, de Oiliveira LDL (2021) Physicochemical characterization and nano-emulsification of three species of pumpkin seed oils with focus on their physical stability. Food Chem 343:128512

Özer Ç, İmamoğlu M (2024) Removal of ciprofloxacin from aqueous solutions by pumpkin peel biochar prepared using phosphoric acid. Biomass Convers Biorefin 14:6521–6531

Paiva MV, Massafra de Oliveira S, da Silva M, de Almeida C, Rodrigues de Araujo J, Soares Archanjo B, D'Elia E (2024) Pumpkin (*Cucurbita maxima*) seed-derived nitrogen, phosphorus, and sulfur carbon quantum dot as an inhibitor of corrosion for mild steel in HCl solution. J Mater Res Technol 28:2504–2515

Papargyropoulou E, Lozano R, Steinberger JK, Wright N, Ujang ZB (2014) The food waste hierarchy as a framework for the management of food surplus and food waste. J Clean Prod 76:106–115

Pazinato Cuco R, Cardozo-Filho L, da Silva C (2019) Simultaneous extraction of seed oil and active compounds from peel of pumpkin (*Cucurbita maxima*) using pressurized carbon dioxide as solvent. J Supercrit Fluids 143:8–15

Peter EA, Hudson N, Alice ON, Stanley O, William T, Ijani AS, Anne S (2013) Evaluation of micronutrients in seeds of pumpkin varieties grown by smallholder farmers in the Lake Victoria basin. Afr J Food Sci Technol 4:221–228

Pizzino G, Irrera N, Cucinotta M, Pallio G, Mannino F, Arcoraci V, Squadrito F, Altavilla D, Bitto A (2017) Oxidative stress: harms and benefits for human health. Oxid Med Cell Longev 2017:8416763

Pontiroli D, Scaravonati S, Magnani G, Fornasini L, Bersani D, Bertoni G, Milanese C, Girella A, Ridi F, Verucchi R, Mantovani L, Malcevschi A, Riccò M (2019) Super-activated biochar from poultry litter for high-performance supercapacitors. Microporous Mesoporous Mater 285:161–169

Promsakha na Sakon Nakhon P, Jangchud K, Jangchud A, Prinyawiwatkul W (2017) Comparisons of physicochemical properties and antioxidant activities among pumpkin (*Cucurbita moschata* L.) flour and isolated starches from fresh pumpkin or flour. Int J Food Sci Technol 52:2436–2444

Provesi JG, Dias CO, de Mello D, Castanho Amboni R, Amante ER (2012) Characterisation and stability of quality indices on storage of pumpkin (*Cucurbita moschata* and *Cucurbita maxima*) purees: Stability on storage of pumpkin purees. Int J Food Sci Technol 47:67–74

Quintana SE, Marsiglia RM, Machacon D, Torregroza E, García-Zapateiro LA (2018) Chemical composition and physicochemical properties of squash (*Cucurbita moschata*) cultivated in Bolivar Department (Colombia). Contemp Eng Sci 11:1003–1012

Radi M, Melian R, Galai M, Dkhirche N, Makha M, Verma C, Fernandez C, EbnTouhami M (2021) Pumpkin seeds as an eco-friendly corrosion inhibitor for 7075-T6 alloy in 3.5% NaCl solution: Electrochemical, surface and computational studies. J Mol Liq 337:116547

Ramakrishnan M, Rathinam TM, Viswanathan K (2018) Biodiesel from plant seed oils as an alternate fuel for compression ignition engines—a review. Environ Sci Pollut Res 25:4621–4631

Rashid J, Tehreem F, Rehman A, Kumar R (2019) Synthesis using natural functionalization of activated carbon from pumpkin peels for decolourization of aqueous methylene blue. Sci Total Environ 671:369–376

Rezig L, Chouaibi M, Msaada K, Hamdi S (2012) Chemical composition and profile characterisation of pumpkin (*Cucurbita maxima*) seed oil. Ind Crop Prod 37:82–87

Rezig L, Chibani F, Chouaibi M, Dalgalarrondo M, Hessini K, Guéguen J, Hamdi S (2013) Pumpkin (*Cucurbita maxima*) seed proteins: sequential extraction processing and fraction characterization. J Agric Food Chem 61:7715–7721

Rezig L, Chouaibi M, Ojeda-Amador RM, Gomez-Alonso S, Salvador MD, Fregapane G, Hamdi S (2018) *Cucurbita maxima* pumpkin seed oil: from the chemical properties to the different extracting techniques. Notulae Botanicae Horti Agrobotanici 46:663–669

Rezig L, Harzalli Z, Gharsallah K, Mahfoudhi N, Chouaibi M, Majdoub H, Oueslati I (2022) Microwave and roasting impact on pumpkin seed oil and its application in full-fat mayonnaise formula. Foods 11:2732

Saavedra MJ, Aires A, Dias C, Almeida JA, De Vasconcelos MCBM, Santos PEA (2015) Rosa Evaluation of the potential of squash pumpkin by-products (seeds and shell) as sources of antioxidant and bioactive compounds. J Food Sci Technol 52:1008–1015

Salami A, Asefi N, Kenari RE, Gharekhani M (2020) Addition of pumpkin peel extract obtained by supercritical fluid and subcritical water as an effective strategy to retard canola oil oxidation. J Food Meas Charact 14:2433–2442

Schinas P, Karavalakis GÃ, Davaris C, Anastopolous G, Karonis D, Zannikos FE, Stournas S, Lois E (2009) Pumpkin (*Cucurbita pepo* L.) seed oil as an alternative feedstock for the production of biodiesel in Greece. Biomass Bioenergy 33:44–49

Setsiripakdee A, Lourith N, Kanlayavattanakul M (2019) In vitro and in vivo removal efficacies of a formulated pumpkin seed oil makeup remover. J Surfactant Deterg 22:1461–1467

Sevilla M, Ferrero GA, Fuertes AB (2017) One-pot synthesis of biomass-based hierarchical porous carbons with a large porosity development. Chem Mater 29:6900–6907

Sharma M, Bhat R (2021) Extraction of carotenoids from pumpkin peel and pulp: comparison between innovative green extraction technologies (ultrasonic and microwave-assisted extractions using corn oil). Foods 10:787

Sobczak P, Zawislak K, Starek A, Zukiewicz-Sobczak W, Sagan A, Zdybel B, Andrejko D (2020) Compaction process as a concept of press-cake production from organic waste. Sustainability 12:1567

Sojak M, Głowacki S (2010) Analysis of giant pumpkin (*Cucurbita maxima*) drying kinetics in various technologies of convective drying. J Food Eng 99:323–329

Stahn T, Storandt R, Grebenteuch S, Rohn S, May D, Dolsdorf C, Pleissner D (2023) Utilization of brewer's spent grains and agricultural residues in pig feed formation. Sustainability 15:13774

Steiner Z, Prakatur I, Novoselec J, Samac D, Klir Ž, Antunović B, Benak S, Gantner V, Ronta M (2020) Influence of pumpkin seed cake on production and slaughter indicators for broiler chickens. Agric Forestry 66:25–31

Susmi TF, Khan MR, Hasan N, Aktar A, Amin MZ (2023) Bioactivity profiling of native and hybrid varieties of pumpkin peel (*Cucurbita maxima* Linn.). J Agric Food Res 14:100813

Švarc-Gajić J, Rodrigues F, Moreira MM, Delerue-Matos C, Morais S, Dorosh O, Silva AM, Bassani A, Dzedik V, Spigno G (2022) Chemical composition and bioactivity of oilseed cake extracts obtained by subcritical and modified subcritical water. Bioresour Bioprocess 9:114

Tabrez S, Khan AU, Hoque M, Suhail M, Khan MI, Zughaibi TA (2022) Biosynthesis of ZnO NPs from pumpkin seeds' extract and elucidation of its anticancer potential against breast cancer. Nanotechnol Rev 11:2714–2725

Tadee P, Chukiatsiri K, Amornlerdpisan D, Paserakung A, Kittiwan N (2020) Antimicrobial effect of Japanese pumpkin (*Cucurbita maxima*) extract on local bovine mastitis pathogen. Vet Integr Sci 18:141–152

Tekin-Cakmak ZH, Atik I, Karasu S (2021) The potential use of cold-pressed pumpkin seed oil by-products in a low-fat salad dressing: The effect on rheological, microstructural, recoverable properties, and emulsion and oxidative stability. Foods 10:2759

Tomar PPS, Nikhil K, Singh A, Selvakumar P, Roy P, Sharma AK (2014) Characterization of anticancer, DNase and antifungal activity of pumpkin 2S albumin. Biochem Biophys Res Commun 448:349–354

Tsaknis J, Lalas S, Lazos ES (1997) Characterization of crude and purified pumpkin seed oil. Grasas Aceites 48:267–272

Umoru EP, Emmanuel Awe F, Ifeanyi Uche J, Abiodun Babatunde O, Aliyu Salaha I, Abiodun Babatunde O (2023) Investigation of the adsorptive and inhibitive properties of *Cucurbita maxima* peel extract and halide ions as inhibitors for stainless steel in 1M H_2SO_4 solution. Commun Phys Sci 10:17–28

Uzlaşir T, Aktaş N, Gerçekaslan KE (2020) Pumpkin seed oil as a partial animal fat replacer in bologna-type sausages. Food Sci Anim Resour 40:551–562

Vahlensieck W, Heim S, Patz B, Sahin K (2022) Beneficial effects of pumpkin seed soft extract on lower urinary tract symptoms and quality of life in men with benign prostatic hyperplasia: a meta-analysis of two randomized, placebo-controlled trials over 12 months. Clin Phytosci 8:13

Valdez-Arjona LP, Ramírez-Mella M (2019) Pumpkin waste as livestock feed: Impact on nutrition and animal health and on quality of meat, milk, and egg. Animals 9:769

Vargas MA, Bernal C, Martínez R (2021) Protease-assisted process for tryptophan release from pumpkin (*Cucurbita maxima*) seed protein extracts. J Food Process Preserv 46:e16290

Voučko B, Novotni D, Balbino S, Mustač NČ, Drakula S, Dujmić F, Habuš M, Jarni K, Ćurić D (2022) Utilization of pumpkin seed oil cake and proso millet flour in enhancing gluten-free bread quality. J Food Process Preserv 46:e17070

Wahid S, Alqahtani A, Khan RA (2021) Analgesic and anti-inflammatory effects and safety profile of *Cucurbita maxima* and *Cucumis sativus* seeds. Saudi J Biol Sci 28:4334–4341

Wang G, Zhang L, Zhang J (2012) A review of electrode materials for electrochemical supercapacitors. Chem Soc Rev 41:797

Wei K, Kim KO, Song KH, Kang CY, Lee JS, Gopiraman M, Kim IS (2017) Nitrogen- and oxygen-containing porous ultrafine carbon nanofiber: a highly flexible electrode material for supercapacitor. J Mater Sci Technol 33:424

Yoo JJ, Balakrishnan K, Huang J, Meunier V, Sumpter BG, Srivastava A, Conway M, Reddy ALM, Yu J, Vajtai R, Ajayan PM (2011) Ultrathin planar graphene supercapacitors. Nano Lett 11:1423

Yoo S-H, Lee B-H, Lee H, Lee S, Bae IY, Lee HG, Fishman LM, Chau HK, Savary BJ, Hotchkiss AT (2012) Structural characteristics of pumpkin pectin extracted by microwave heating. J Food Sci 77:11

Phytochemicals from Tomato (*Solanum lycopersicum* L.) By-Products

Maggy M. Casey Nairenge, Martha Ikechukwu, Natascha Cheikhyoussef, Ahmed A. Hussein, and Ahmad Cheikhyoussef

Abstract Tomatoes (*Solanum lycopersicum* L.) are among the most valuable vegetables worldwide. This appreciation is associated with their wide varieties and unique organoleptic characteristics. The industrial processing of tomatoes generates massive amounts of by-products, mainly used in the formulation of livestock feeding material or disposed of without further valorization. The by-products include peels, seeds, stems and cores, pomace, and juice residues, and they have been reported to be used in animal feed, fertilizers, dietary supplements, and food additives. Tomato by-products contain beneficial compounds such as carotenoids (lycopene, α-, β-, γ-, and ξ-carotenes, phytoene, phytofluene, and neurosporene), flavonoids (lutein), polyphenols, sterols, terpenes, and tocopherols, which have been reported to be used in the treatment of heart disease, inflammation, and certain cancers. By transforming tomato by-products into valuable materials, we can reduce the environmental impact of waste and contribute to developing a more sustainable tomato industry. This chapter will provide insights into the latest literature on tomatoes' by-products, phytochemical contents, biological properties, and potential applications.

M. M. Casey Nairenge
Department of Biochemistry, Microbiology & Biotechnology, Faculty of Agriculture, Engineering and Natural Sciences, University of Namibia, Windhoek, Namibia

M. Ikechukwu
Sam Nujoma Marine and Coastal Resources Research Centre, Sam Nujoma Campus University of Namibia, Henties Bay, Namibia

N. Cheikhyoussef
Ministry of Higher Education, Technology, and Innovation (MHETI), Windhoek, Namibia

A. A. Hussein
Chemistry Department, Cape Peninsula University of Technology, Cape Town, South Africa

A. Cheikhyoussef (✉)
Science and Technology Division, Multidisciplinary Research Services, Centre for Research Services, University of Namibia, Windhoek, Namibia
e-mail: acheikhyoussef@unam.na

M. F. Ramadan (ed.), *Bioactive Phytochemicals in By-products from Bulb, Flower and Fruit Vegetables*, https://doi.org/10.1007/978-3-031-77399-0_12

Keywords Carotenoids · Lycopene · Natural pigment · Tomato by-products · Tomato industry

Abbreviations

BDNF	Brain-derived neurotrophic factor
FAOSTAT	Food and Agriculture Organization Statistical Database
FFA	Free fatty acid
HDF	Human dermal fibroblasts
PUFA	Polyunsaturated fatty acid
TVB-N	Total volatile basic nitrogen
UVA	Ultraviolet-A
WPTC	World Processing Tomato Council

1 Introduction and Economic Values of Each Discussed By-Product

Tomatoes (*Solanum lycopersicum* L.) are one of the most cherished, cultivated, and processed crops worldwide due to their taste and various forms (Szabo et al. 2018; Tiwari et al. 2020). Tomato processing is a significant industry worldwide, generating large quantities of by-products. The current world production for processing and fresh consumption in 2021 amounted to just over 189.1 million metric tons, with China, Egypt, India, Turkey, and the United States being the world's five top tomato producers (FAOSTAT 2024). Of that total, over 39 million tons of tomatoes are processed to produce products such as tomato paste, sauces, puree, ketchup, soup, juice, canned tomatoes, and tomato powder (Eslami et al. 2022; FAOSTAT 2024; Sarkar and Kaul 2014; Shah et al. 2021; Silva et al. 2019; World Processing Tomato Council 2020). This global demand for tomatoes and their processed forms has led to an abundance of by-products, including peels, seeds, pulp, and other solids remaining after extraction of juice, typically referred to as pomace (Coelho et al. 2019; Eslami et al. 2022; Ibrahim and Labaki 2022; Kumar et al. 2021; Sarkar and Kaul 2014; Shah et al. 2021; Silva et al. 2019; Szabo et al. 2019). The composition of these by-products can vary depending on the specific processing methods employed and the variety and quality of the tomatoes used. According to several researchers, these by-products constitute approximately 1.5–5% of the total of processed tomatoes (Del Valle et al. 2006; Eslami et al. 2022; Kalogeropoulos et al. 2012; Knoblich et al. 2005; Palomo et al. 2019; Silva et al. 2019). Eslami et al. (2022) further illustrated that the by-product generally constituted 56% pulp and peels and 44% seeds; however, Palomo et al. (2019) stated that the by-product typically consists of 78% peel and 22% pulp and crushed seeds. In the study by Silva et al. (2019), the by-product comprised 61.5% of tomato skin, while Kumar et al. (2021) stated that approximately 5–10% of tomato pomace is seed waste.

Tomato pomace is often used as a component in animal feed and fertilizers (Benítez et al. 2018; Knoblich et al. 2005; Liadakis et al. 2022a; Strati and Oreopoulou 2014; Travieso et al. 2022), providing dietary fiber and nutrients, serving as a sustainable feed ingredient. Additionally, it has been commonly utilized in powder form as a nutritional additive to various foods such as wheat flour-based products, meat products, dairy and oil products, as well as sweet products (Choo and Saik 2021; Gebeyew et al. 2015; Kumar et al. 2021; Lu et al. 2019). Additionally, tomato processing generates significant by-products that are often underutilized or discarded, representing an environmental inconvenience for producers (Liadakis et al. 2022a; Travieso et al. 2022; Rossini et al. 2013). Approximately 8.5 million tons of waste is generated globally each year, and about 4 million tons of by-product organic matter is generated (Liadakis et al. 2022a; Szabo et al. 2018; World Processing Tomato Council 2020; Trombino et al. 2021), representing a considerable loss of potential resources. The utilization of tomato processing by-products has gained significant attention in recent years due to the increasing focus on sustainable food production and waste reduction. The growing awareness of sustainability has sparked interest in valorizing and transforming these by-products into value-added products and reducing the environmental impact generated in the tomato processing industry (Sarkar and Kaul 2014). This review aims to assess the bioactive compounds found in tomato processing by-products along with their biological and functional properties to evaluate their food and non-food applications and valorization for non-health products.

2 Chemical Composition and Bioactive Compounds of Extracts from By-Products

Previous studies have widely ascertained that tomato by-products are still rich in valuable bioactive compounds, which, once recovered, could be utilized in different industrial sectors (Coelho et al. 2019; Eslami et al. 2022; Ibrahim and Labaki 2022; Liadakis et al. 2022a; Sarkar and Kaul 2014; Kumar et al. 2021; Shah et al. 2021; Silva et al. 2019; Szabo et al. 2018). Understanding the composition of these by-products (Table 1) is essential for identifying their potential applications and determining the best methods for their utilization. These bioactive compounds include dietary fibers, proteins, carbohydrates, carotenoids, phenolic compounds, vitamins and minerals, and fatty acids and essential oils; however, a range of factors, including tomato species, soil and climate conditions, degree of ripening, post-harvest storage conditions, and industrial processing parameters, will affect the amounts of chemical and bioactive compounds found in the by-products (Ibrahim and Labaki 2022; Travieso et al. 2022). Applying green technologies in the extraction procedures of tomatoes' by-products can allow high to optimum recovery of these high-added-value compounds (Coelho et al. 2023).

Table 1 Nutritional characterization of tomato by-products according to several authors (mg/kg)

Ash	Protein	Fiber	Fat	Carbohydrates	Sugars	References
54,200	190,000	336,000	94,000	–	172,000	Travieso et al. (2022)
35,500	326,900	294,200	154,300	433,100	–	Özbek et al. (2020)
42.1	176.2	524.4	21.9	–	–	Nour et al. (2018)
6570	30,750	107,040	9020	–	48,450	Del Valle et al. (2006)
36,400	176,600	495,500	96,600	590,700	95,400	Alvarado et al. (2001)
–	230,700	250,400	120,300	–	–	Abaza et al. (1987)
50,700	220,300	280,500	130,700	–	–	Bhargava and Talapatra (1968)

–, values not given

2.1 Dietary Fibers

According to several studies, dietary fibers constitute the major component of tomato pomace on a dry matter basis, with reported contents in the literature being highly variable from 25 to 64.50% (Azabou et al. 2020a; Eslami et al. 2022; Del Valle et al. 2006; Liadakis et al. 2022a; Silva et al. 2019; Travieso et al. 2022). A study by Silva et al. (2019) detected that dietary fiber had the highest content, ranging from 48.62 to 53.97 g dry weight of 100 g from six batches of tomato pomace. The majority of the fiber comes from tomato peels. Then from the seeds, furthermore fiber can be classified as soluble (pectins, β-glucans, galactomannan gums, and a wide variety of non-digestible oligosaccharides) or insoluble (lignin, cellulose, and hemicelluloses) (Liadakis et al. 2022a; Herrera et al. 2010; Szabo et al. 2018; Travieso et al. 2022).

2.2 Proteins and Amino Acids

Tomato by-products contain a significant amount of protein. According to Eslami et al. (2022), proteins range from 15.08 to 24.67%. Several studies have noted that tomato seeds contribute most of the protein in tomato pomace ranging from 20 to 40% on a dry weight basis and consisting of a combination of globulin, albumin, prolamin, and glutamate components (Alvarado et al. 2001; Eslami et al. 2022; Liadakis et al. 1998; Mechmeche et al. 2017; Persia et al. 2003; Szabo et al. 2018). According to the literature, the primary amino acids in tomato seeds and peels are glutamic acids and aspartic acid and the most abundant essential amino acids include arginine, threonine, lysine, and leucine (Eslami et al. 2022; Kalogeropoulos et al. 2012; Knoblich et al. 2005; Liadakis et al. 2022a; Mechmeche et al. 2017; Silva et al. 2019; Szabo et al. 2018). Sarkar and Kaul (2014) observed total essential amino acids of 39.5% in tomato seed protein.

2.3 *Fatty Acids and Oils*

Tomato processing by-products, particularly tomato seeds, are rich in fatty acids and oil. The oils extracted from tomato seeds contain saturated and unsaturated fatty acids, with the predominant fatty acids including linoleic acid, oleic acid, palmitic acid, and stearic acid (Eslami et al. 2022; Liadakis et al. 2022a; Travieso et al. 2022). Camara et al. (2001) examined the fatty acid composition of tomato pomace and found that it contained between 75 and 80% total unsaturated fatty acids, 53% polyunsaturated fatty acids, and 23% monounsaturated fatty acids. Similarly, a study by Nour et al. (2018) detected that unsaturated fatty acids (77.04%) predominate saturated fatty acids (22.72%) in dried tomato waste. Elbadrawy and Sello (2011) investigated tomato peels, and with a proportion of 52.41% of all fatty acids, linoleic acid represented the majority of fatty acids, followed by oleic acid, while linolenic acid only accounted for 4.26% and palmitoleic acid for only 1.82%.

2.4 *Total Phenolics and Carotenoids*

Tomato waste is a significant source of carotenoids (Liadakis et al. 2022b). Carotenoids are lipid-soluble pigments that give various fruits and vegetables characteristic colors (Szabo et al. 2018). The main carotenoids of tomato pomace are lycopene, β-carotene, lutein, cis-β-carotene, and zeaxanthin (Eslami et al. 2022; Liadakis et al. 2022b). According to Ibrahim and Labaki (2022), carotenoids represent more than 40% of tomato processing waste, with lycopene predominant, followed by β-carotene and lutein. Liadakis et al. (2022a) observed carotenoids constituting about 272–554 mg/kg of tomato pomace. Lycopene is the primary pigment in tomatoes, giving the fruit its distinctive red color (Luengo et al. 2014; Szabo et al. 2018). With the skin containing five times more lycopene than the pulp and seeds (Ibrahim and Labaki 2022; Liadakis et al. 2022b), lycopene constitutes 80–90% of the total carotenoids (Eslami et al. 2022; Luengo et al. 2014; Szabo et al. 2018). According to Szabo et al. (2018), β-carotene is a vital precursor of vitamin A, the most eaten plant pigment and after lycopene, the second most prevalent carotenoid. However, González et al. (2011) identified β-carotene as having a higher concentration than lycopene in seeds, while Kalogeropoulos et al. (2012) found the amount of β-carotene in by-products to be higher (14.9 vs. 8.6 mg/100 g) than in unprocessed tomatoes.

According to Ibrahim and Labaki (2022), flavonoids (rutin, naringenin, naringenin chalcone, myricetin, and quercetin) and phenolic acids (ellagic, chlorogenic, salicylic, vanillic, gallic, p-coumaric, ferulic, and caffeic acids) are the main phenolic compounds found in tomato processing waste, the concentration of these phenolic compounds was higher in tomato skin than in seeds and pulp with rutin being the most abundant flavonoid followed by naringenin. González et al. (2011) detected ellagic (143.4 mg/kg) and chlorogenic (76.3 mg/kg) acids to be the most prevalent phenolic acids, while only rutin (29.2 mg/kg) and myricetin (63.7 mg/kg) were the only flavonoids detected in the dried tomato waste. However, Liadakis et al. (2022a, b) observed flavonoids in significant concentrations (415–688 mg QE/kg), especially flavonols like kaempferol and quercetin in tomato pomace. Nour et al. (2018) analyzed tomato waste samples and found an average total phenolic content of 1229.5 mg GAE/kg, with flavonoids representing 415.3 mg QE/kg of the phenolic compounds. Moreover, Kalogeropoulos et al. (2012) found that hydroxycinnamic acids accounted for 23.3% of the simple polyphenols, and flavonoids comprised 73% of the total tomato processing by-products.

2.5 *Minerals and Vitamins*

According to the data in the literature, the major minerals found in tomato pomace include magnesium, calcium, zinc, manganese, copper, iron, sodium, phosphorus, and potassium, with potassium presenting the highest concentrations (Azabou et al. 2020a; Elbadrawy and Sello 2011; González et al. 2011; Jeguirim 2022; Liadakis et al. 2022b). Nour et al. (2018) detected a concentration of 30.3 g/kg potassium in tomato pomace. At the same time, Elbadrawy and Sello (2011) analyzed tomato peels and observed a lower concentration of 0.001097 g/kg for potassium. Carmona-Cabello et al. (2022) state that the mineral content in tomato waste is represented in the ash, while Del Valle et al. (2006) observed the total mineral content of tomato pomace ranging between 5.31 and 13.25 g/kg. Additionally, Herrera et al. (2010) analyzed tomato fiber and found a slightly higher total mineral content of 39.2 g/kg.

Moreover, the seeds contain a significant amount of vitamin E, primarily in the form of γ-tocopherol, which makes up 97–98% of the content, and α-tocopherol, which makes up 2–3% (Ibrahim and Labaki 2022). In a Botinestean et al. (2013) study, γ-tocopherol was the dominant homolog, followed by α-tocopherol and δ-tocopherol. In contrast, Kalogeropoulos et al. (2012) observed α-tocopherol as the highest tocopherol with 155.7 mg/kg on a dry weight basis of the by-product. Some studies have also detected vitamin C in tomato pomace, such as Sandei et al. (2019), in agreement with Alvarado et al. (2001), observing 0.207 g/kg and 0.278 g/kg, respectively.

3 Biological and Functional Properties of Extracts and Bioactive Compounds from By-Products

Kaboré et al. (2022) reported on the richness of tomatoes' by-products in macronutrients and micronutrients and their potential capabilities for undernutrition prevention and reduction of overnutrition incidence. They reported average carbohydrate, protein, and lipid contents of 15.43%, 11.71%, and 5.4% (DM) in the peels and

58.75%, 15.4%, and 22.2% in the seeds, respectively. The interesting contents of minerals in tomato by-products may prevent mineral deficiency malnutrition due to their main role in functioning the immune system (Weyh et al. 2022).

Beta-carotene (Table 2) possesses a strong antioxidant capacity and has the potential to boost the immune function (Coelho et al. 2023; Jialal et al. 1991; Li et al. 2023), cardiovascular health, vasodilatory, coronary (Coelho et al. 2023; Gammone et al. 2015; Pawar et al. 2023; Viuda-Martos et al. 2014), chemopreventive properties (Coelho et al. 2023), and maintaining the mental health and cognitive performance (Abrego-Guandique et al. 2023). A study by Bakker et al. (2016) indicated that higher plasma β-carotene and α-carotene concentrations in women are less susceptible to breast cancer risk. Furthermore, supplementation rich in lycopene is effective against aging-related inflammatory and oxidative stress-induced neurodegeneration (Coelho et al. 2023; Jha et al. 2024; Zhao et al. 2017). Laranjeira et al. (2022) reported several valorization techniques for treating tomato by-products to obtain bioactive compounds and indicated their applications in inflammation and cancer management.

The most demonstrated function of beta-carotene is the production of provitamin A in humans. Therefore, industrial wastes of tomato's by-products represent a cheap source of bioactive compounds, such as α-, β-, γ-, and ξ-carotene, lutein, lycopene, neurosporene, phytoene, and phytofluene, that could be valorized to lower the number of challenges associated with deficiency of vitamin A in populations with limited nutritional resources (Szabo et al. 2018). Lycopene, α-tomatine, and tomatidine from fresh and processed tomatoes showed several health effects and biological activities, including anticarcinogenic and cardioprotective (Friedman 2013),

Table 2 Bioactive compounds found in tomato by-products and their functional properties

Compound	Functional properties	References
β-carotene	Antioxidant properties, inhibition of cardiovascular disease, breast cancer inhibition, improved mental cognitive performance, provitamin A synthesis	Jialal et al. (1991); Li et al. (2023); Gammone et al. (2015); Pawar et al. (2023); Abrego-Guandique et al. (2023); Bakker et al. (2016); Szabo et al. (2018); Valdez-Morales et al. (2014)
α-carotene	Breast cancer inhibition	Bakker et al. (2016)
Lycopene	Protects against aging-related inflammatory and oxidative stress-induced neurodegeneration. Anticarcinogenic, cardioprotective, and hepatoprotective effects	Jha et al. (2024); Zhao et al. (2017); Friedman (2013); Bhatia et al. (2018)
α-tocopherol	Photoprotection, antioxidation properties	Ascenso et al. (2013); Fernández-García (2014); Petruk et al. (2016); Zhang et al. (2023)
Exopolysaccharides	Antioxidant and anticytotoxic activity	Tommonaro et al. (2008)
Phenolics	Antioxidant and antimicrobial activities	Perea-Domínguez et al. (2018); Añibarro-Ortega et al. (2020)

Gokul and Muralidhara (2014) showed the potential of the aqueous extracts of tomato seeds to attenuate rotenone-induced oxidative stress and neurotoxicity in the Drosophila system and to alleviate the behavioral phenotype, mitochondrial dysfunction, and neurotoxicity as well as oxidative stress alleviation in mice.

Aspartic acid, beta-carotene, glutamic acid, and lycopene from tomato's by-products can be transformed as incorporated nutrients in several functional food formulations (Szabo et al. 2018). A study by Stahl and Sies (2012) reported that daily consumption of tomato-based products could reduce DNA damage significantly from treatment with $Fe2+$ and increase the protection from UV and metal ions transition. Similarly, several studies confirmed the photoprotection and the dermal delivery applications for α-tocopherol, naringenin, and lycopene-based formulations to protect against oxidative stress generated in human dermal fibroblasts (HDF) exposed to ultraviolet-A (UVA) light (Ascenso et al. 2013; Fernández-García 2014; Petruk et al. 2016; Zhang et al. 2023). Clinical trials for lycopene-containing tomato by-products reported positive effects on human health against reactive oxygen species and prevented some non-communicable diseases in humans (Grabowska et al. 2019; Kulczyński et al. 2017).

Recently, oral supplements from tomato by-products have been developed, such as Lumenato (Tarshish et al. 2022, 2023) and Lycomato (Tarshish and Hermoni 2023), with reported improving effects on the strength of the skin barrier and skin health and appearance respectively. Lumenato and Lycomato are considered a new generation of diet supplements (nutraceuticals) to be used as smart tissue carriers for innovative contemporary use in cosmeceuticals (Morganti et al. 2024). In a study by Hong et al. (2022), it was found that tomato extract, when combined with lemon, had a synergistic effect on anti-oxidative stress and hippocampal neurogenesis. Furthermore, Yoon et al. (2024) confirmed the synergistic effects of tomatoes and lemons impact on improving cognitive function via increasing brain-derived neurotrophic factor (BDNF) levels in aged mice animal models.

4 Food and Non-food Applications of Extracts and Bioactive Compounds from By-Products

Tomato by-products such as tomato pomace can be a source of beneficial phytochemicals such as carotenoids, phenolic compounds, vitamins and fiber, and other compounds with antioxidants (Stoica et al. 2018; Valdez-Morales et al. 2014). Such beneficial bioactive compounds make them ideal for developing high-value products with food and non-food applications, as illustrated in Fig. 1.

4.1 Food Applications of Extracts and Bioactive Compounds from By-Products

Phytochemicals such as carotenoids are beneficial antioxidants that protect against disease by enhancing the human immune system. Humans cannot synthesize these

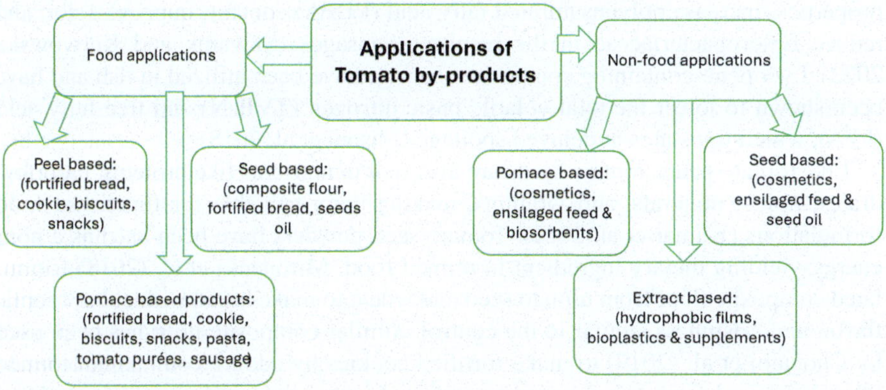

Fig. 1 Food and non-food applications of tomato by-products

carotenoids, and thus, they need to be consumed in diets such as tomato and tomato-based foods (Przybylska 2020). Lycopene, a carotenoid abundant in tomato peels, has been investigated for its potential application in food to enhance nutritional value and health benefits (Kurasova and Sarsenova 2023). Dried tomato pomace powder, high in lycopene, has been used as a functional food component in food items to improve numerous sensory and nutritional properties.

Lycopene-rich dry powders have extensively been included as a functional food ingredient in the flour of baked products such as cookies/biscuits and corn chips, leading to improved nutritional content (Paulino et al. 2020). Salem (2020), for instance, used tomato pomace powder to increase the protein, fiber, minerals, and phenolic content while decreasing carbohydrates and enhancing the sensory characteristics of biscuits. Tomato pomace powder has been utilized as an ingredient in multigrain cookies, improving their nutritional value with increased lycopene, beta-carotene, and phenolic content while reducing moisture, fat, and carbohydrates (Kadam and Mane 2022). Tomato pomace powder, up to 20% of the lycopene-rich dry powders, has been incorporated in gluten-free snack dough, and it was found to reduce oil usage cooking time, improve antioxidant activity, nutritional profile, and even extend the shelf life of the product (El-Araby 2022; Rehal et al. 2022).

Similarly, López-Bermúdez et al. (2022) used tomato peel powder to replace cookie dough fat partially. Up to 30% of replacements scored consumer preference like the controls. Wirkijowska et al. (2023) used tomato pomace to develop fortified bread by adding ground tomato pomace and the bread resulted in increased protein, lipid, and dietary fiber content and reduced carbohydrate content compared to the control. Powder of tomato pomace has also been utilized to develop fortified pasta by incorporating it in pasta dough, resulting in improved nutritional profile, sensory quality, and starch digestibility (Teterycz and Sobota 2023; Padalino et al. 2015). Tomato pomace powder has also been used in an instant soup mix, with a 50% addition showing maximum acceptability (Rehal and Kaur 2022). Lycopene-rich dry powders are a thickening ingredient in commercial tomato purées (Donegà et al. 2015). Tomato pomace powder has been used to develop functional and healthier meat products (Domínguez et al. 2020). For instance, pomace powder has been used as an ingredient sausage formulation and has been shown to enhance antioxidant properties, increase polyunsaturated fatty acid (PUFA) content, improve color, and reduce Enterobacteriaceae in the resulting sausages (Skwarek and Karwowska 2022). Lycopene-containing composite coatings have been utilized in fish and have been shown to lower the total volatile basic nitrogen (TVB-N) and free fatty acid (FFA), which translates to delayed spoilage (Ehsani et al. 2018).

Like tomato peels, tomato seeds are also rich in nutrients like proteins, carbohydrates, lipids, minerals, and vitamins, making them valuable for functional food formulations (Kumar et al. 2022). Tomato seed powders have been used as a non-energy-yielding dietary ingredient in refined food. Mironeasa et al. (2018) formulated composite flour from tomato seed and wheat to make bread with color, aroma, flavor, and crispiness similar to the control. Similar composite flour has been used by Chouaibi et al. (2019) to make fortified cookies by adding commercial tomato fiber developed from tomato seed to cookie dough, which enhanced the cookies' antioxidant activity and increased breaking strength. According to Choe et al. (2021), tomato seed-based composite flour contains compounds with radical scavenging anti-inflammatory properties and modulates gut microbiota, suggesting its potential as a value-added food ingredient.

Besides its use in composite flour formula, tomato seeds can be dried and then processed into seed oil. Tomato seed oil has been shown to have low acidity, high phenol content, and radical scavenging activity (Capocasale and Zappia 2017), thus can be used to enhance antioxidant properties in condiments and confectioneries (Sangeetha et al. 2023). Lycopene extracts have been investigated for their potential utilization as edible coatings for fruits, and it is an effective anti-browning agent for fresh-cut apples (Azam et al. 2024). A recent study by Lima et al. (2022) reviewed the impact of different agronomic practices including postharvest management, and processing methods on the functional compounds of tomatoes and their byproducts. The study highlighted how secondary metabolites such as biogenic amines, carotenoids, and phenolic compounds are produced as a response to environmental harsh conditions. Another application of tomato by-products was explored to improve the quality aspect of olive oil. The ethanolic extract of tomato by-products was added to the refined olive and olive-pomace oils and successfully enhanced their oxidative stability (Azabou et al. 2020b). A recent study by Chabi et al. (2024) summarized the current knowledge on the recent food applications of tomato pomace, focusing on its use as a functional ingredient for improving the functional properties and shelf life of foods.

4.2 Non-food Applications of Extracts and Bioactive Compounds from By-Products

Including tomato by-products in animal diets has been explored due to their nutritional value and health benefits, with variable effects on different animals and their respective products. Tomato pomace has been investigated as a potential ingredient in various animal feeds. Including tomato by-products for goat feeds (Ventura et al.

2009) and especially tomato pomace in the diet of lactating goats has been shown to increase milk production and fat composition (Mizael et al. 2020). Adding tomato pomace to the conventional feed growing lambs reduces the consumption of the costly conventional feeds without compromising the meat quality (Valenti et al. 2018). Incorporation of tomato pomace powder in the diet of laying hens has been shown to increase concentrations of serum and egg yolk lycopene, β-carotene, lutein, vitamin A, and egg production persistence with increasing supplemental tomato powder concentration (Akdemir et al. 2012). Dietary replacement of maize silage with up to 10% enhances milk vitamin contents, antioxidant, and immune functions in lactating dairy cows (Tuoxunjiang et al. 2020). Tomato pomace has been used to produce nutraceuticals and pharmaceuticals, tapping into their rich polyphenols and carotenoid content (López-Yerena et al. 2023). Other non-food applications of tomato pomace include its utilization in producing low-cost biosorbents for dye and heavy metals removal from wastewater (Rajan et al. 2022; Azabou et al. 2020a). Utilization of whole tomato by-products, such as tomato pomace, may affect the end products' physicochemical properties, making them unsuitable for intended purposes.

Consequently, bioactive compounds extracted from the by-product improve their applicability and quality. For instance, lycopene-enriched extracts from tomato pomace have been used in cosmetic emulsions, showing antioxidant effects and influencing the products' odor, color, and acceptability (Costa et al. 2021). Tomato pomace extract has been shown to have film-forming properties when used as an ingredient in cosmetic formulas, thus recommended for and potential for eco-designed formulations (Jamaleddine et al. 2022). Cutin has been extracted from tomato pomace and then used to manufacture hydrophobic films that can potentially replace the petroleum-based polymers in food packaging, making tomato processing more sustainable and profitable (Simões et al. 2023). Cutin-derived oligomers from tomato pomace have been shown to exert antimicrobial properties, making them ideal for applications in materials like biodegradable plastics and supplements (Escórcio et al. 2022). Tomato seeds have been reported as a rich source of antimicrobial agents (Taveira et al. 2010), and several antimicrobial activities have been reported (Szabo et al. 2019). Cutin-based polyesters have been fabricated using a mixture of unsaturated and polyhydroxylated fatty acids isolated from tomato pomace fractionation using alkaline hydrolysis (Benítez et al. 2018). Active films were formulated from polyvinyl alcohol (PVA) mixed with itaconic acid and chitosan and enriched with tomato by-product extract to develop new bioactive food packaging (Szabo et al. 2020).

5 Valorization of By-Products for Non-health Purposes, i.e., Source of Bioenergy

Tomato paste by-product is a good carbon source for the FucoPol bioprocess production (Antunes et al. 2017). FucoPol is a fucose-rich extracellular polysaccharide with film-forming capacity. It is synthesized by *Enterobacter* A47 and is considered a rich source of fucose, a precursor of human milk oligosaccharides (Szabo et al. 2018). Bioethanol production has been reported from the tomato pomace materials, including skins, seeds, and vascular tissues and the tomato by-products of the lycopene supercritical carbon dioxide extraction (tomato serum and exhausted matrix) (Lenucci et al. 2013). Tomato waste is considered a cheap source of feedstocks and bioethanol production, with further optimization needed as a potential alternative for bioenergy (Szabo et al. 2018).

The biorefinery cascade processing concept of tomato pomace yielded carotenoids, lycopene-rich oleoresin, seed oil, and protein, which are considered a valuable source of lignocellulosic substrates for bioethanol production, unrevealing the cascade fractionation of the compounds mentioned above from tomato industrial by-products as alternatives for bioenergy production (Kehili et al. 2016). Tomato peel waste is a promising source of pectin with high inhibition efficiency (73%) for pure tin corrosion (Grassino et al. 2016). Protein isolates from tomato seeds were reported to have a good emulsion stabilization effect, showing their potential to be used as emulsifiers in the food and beverage industry at neutral conditions (Sarkar et al. 2016).

References

Abaza MA, Nour AN, Borhami BE, El-Shazly K (1987) Nutritive value of vegetable and fruit waste as animal feeds. Alexandria J Agric Sci 32:75–83

Abrego-Guandique DM, Bonet ML, Caroleo MC, Cannataro R, Tucci P, Ribot J, Cione E (2023) The effect of beta-carotene on cognitive function: a systematic review. Brain Sci 13(10):1468

Akdemir F, Orhan C, Sahin N, Sahin K, Hayirli A (2012) Tomato powder in laying hen diets: effects on concentrations of yolk carotenoids and lipid peroxidation. Br Poult Sci 53(5):675–680

Alvarado A, Pacheco-Delahaye E, Hevia P (2001) Value of a tomato byproduct as a source of dietary fibre in rats. Plant Foods Hum Nutr 56(4):335–348

Añibarro-Ortega M, Pinela J, Ćirić A, Martins V, Rocha F, Soković MD et al (2020) Valorisation of table tomato crop by-products: Phenolic profiles and in vitro antioxidant and antimicrobial activities. Food Bioprod Process 124:307–319

Antunes S, Freitas F, Sevrin C, Grandfils C, Reis MA (2017) Production of FucoPol by *Enterobacter* A47 using waste tomato paste by-product as sole carbon source. Bioresour Technol 227:66–73

Ascenso A, Pinho S, Eleutério C, Praça FG, Bentley MV, Oliveira H, Santos C, Silva O, Simões S (2013) Lycopene from tomatoes: vesicular nanocarrier formulations for dermal delivery. J Agric Food Chem 61(30):7284–7293

Azabou S, Louati I, Ben Taheur F, Nasri M, Mechichi T (2020a) Towards sustainable management of tomato pomace through the recovery of valuable compounds and sequential production of low-cost biosorbent. Environ Sci Pollut Res 27:39402–39412

Azabou S, Sebii H, Taheur FB, Abid Y, Jridi M, Nasri M (2020b) Phytochemical profile and antioxidant properties of tomato by-products as affected by extraction solvents and potential application in refined olive oils. Food Biosci 36:100664

Azam I, ul Ain Q, Sidiq M, Muzzaffar S (2024) Lycopene extract as an antibrowning agent for edible coating of fresh cut apples. J Food Sci Technol 1–9

Bakker MF, Peeters PH, Klaasen VM, Bueno-de-Mesquita HB, Jansen EH, Ros MM, Travier N, Olsen A, Tjønneland A, Overvad K, Rinaldi S, Romieu I, Brennan P, Boutron-Ruault MC, Perquier F, Cadeau C, Boeing H, Aleksandrova K, Kaaks R, Kühn T et al (2016) Plasma carotenoids, vitamin C, tocopherols, and retinol and the risk of breast cancer in the European Prospective Investigation into Cancer and Nutrition cohort. Am J Clin Nutr 103(2):454–464

Benítez JJ, Castillo PM, del Río JC, León-Camacho M, Domínguez E, Heredia A, Guzmán-Puyol S, Athanassiou A, Heredia-Guerrero JA (2018) Valorization of tomato processing by-products: fatty acid extraction and production of bio-based materials. Materials 11(11)

Bhargava B, Talapatra SK (1968) Studies on fruit and vegetable factory wastes. Indian Vet J 45:948–952

Bhatia N, Singh B, Koul A (2018) Lycopene treatment stalls the onset of experimentally induced hepatocellular carcinoma: a radioisotopic, physiological and biochemical analysis. Hepatoma Res 4(9):1–16

Botinestean C, Schreiner M, Jianu I (2013) Influence of solvent used for extraction on tocopherols content of tomato seed oil. Bull Univ Agric Sci Vet Med Cluj-Napoca Agric 70(2):431–432

Camara M, Del Valle M, Torija E, Castilho C (2001) Fatty acid composition of tomato pomace. Acta Hortic 542:175–180

Capocasale M, Zappia C (2017) Tomato seed oil for edible use: Cold break, hot break, and harvest year effects. J Food Process Preserv 41(6)

Carmona-Cabello M, Zorpas AA, Dorado MP (2022) Biorefinery concept for the industrial valorization of tomato processing by-products. In: Tomato processing by-products sustainable applications. Academic, Cambridge, pp 371–420

Chabi IB, Zannou O, Dedehou ESCA, Ayegnon BP, Oscar Odouaro OB, Maqsood S, Galanakis CM, Pierre Polycarpe Kayodé A (2024) Tomato pomace as a source of valuable functional ingredients for improving physicochemical and sensory properties and extending the shelf life of foods: A review. Heliyon 10(3):e25261

Choe U, Sun J, Bailoni E, Chen P, Li Y, Gao B et al (2021) Chemical composition of tomato seed flours, and their radical scavenging, anti-inflammatory and gut microbiota modulating properties. Molecules 26(5):1478

Choo WS, Saik WH (2021) Valorization of fruit and vegetable waste for bioactive pigments: extraction and utilisation. In: Valorization of agri-food wastes and by-products. Academic, Cambridge, pp 61–81

Chouaibi M, Rezig L, Boussaid A, Hamdi S (2019) Insoluble tomato-fiber effect on wheat dough rheology and cookies' quality. Ital J Food Sci 31(1)

Coelho M, Pereira R, Rodrigues AS, Teixeira JA, Pintado ME (2019) Extraction of tomato by-products' bioactive compounds using ohmic technology. Food Bioprod Process 117:329–339

Coelho MC, Rodrigues AS, Teixeira JA, Pintado ME (2023) Integral valorisation of tomato by-products towards bioactive compounds recovery: Human health benefits. Food Chem 410:135319

Costa A, Marques M, Congiu F, Paiva A, Simoes P, Ferreira A et al (2021) Evaluating the presence of lycopene-enriched extracts from tomato on topical emulsions: Physico-chemical characterization and sensory analysis. Appl Sci 11(11):5120

Del Valle M, Cámara M, Torija ME (2006) Chemical characterization of tomato pomace. J Sci Food Agric 86(8):1232–1236

Domínguez R, Gullón P, Pateiro M, Munekata PE, Zhang W, Lorenzo JM (2020) Tomato as potential source of natural additives for meat industry. A review. Antioxidants 9(1):73

Donegà V, Marchetti MG, Pedrini P, Costa S, Tamburini E (2015) Valorization of tomato dried peels powder as thickening agent in tomato purees. J Food Process Technol 6(11):1000511

Ehsani A, Jasour MS, Agh N, Hashemi M, Khodadadi M (2018) Rancidity development of refrigerated rainbow trout (Oncorhynchus mykiss) fillets: Comparative effects of in vivo and in vitro lycopene. J Sci Food Agric 98(2):559–565

El-Araby GM (2022) Processing untraditional formula from snacks using apple and tomato pomace powder as a source of dietary fiber and antioxidants. J Food Dairy Sci 13(4):59–64

Elbadrawy E, Sello A (2011) Evaluation of nutritional value and antioxidant activity of tomato peel extracts. Arabian J Chem Suppl 2(9):1010–1018

Escórcio R, Bento A, Tome AS, Correia VG, Rodrigues R, Moreira CJ et al (2022) Finding a needle in a haystack: Producing antimicrobial cutin-derived oligomers from tomato pomace. ACS Sustain Chem Eng 10(34):11415–11427

Eslami E, Carpentieri S, Pataro G, Ferrari G (2022) A comprehensive overview of tomato processing by-product valorization by conventional methods versus emerging technologies. Foods 12(166)

FAOSTAT Statistical Database, Food and Agriculture Organization of the United Nations (2024). Available at http://www.fao.org/faostat/en/#home

Fernández-García E (2014) Photoprotection of human dermal fibroblasts against ultraviolet light by antioxidant combinations present in tomato. Food Funct 5(2):285–290

Friedman M (2013) Anticarcinogenic, cardioprotective, and other health benefits of tomato compounds lycopene, α-tomatine, and tomatidine in pure form and in fresh and processed tomatoes. J Agric Food Chem 61(40):9534–9550

Gammone MA, Riccioni G, D'Orazio N (2015) Carotenoids: potential allies of cardiovascular health? Food Nutr Res 59:26762

Gebeyew K, Animut G, Urge M, Feyera T (2015) The effect of feeding dried tomato pomace and concentrate feed on body weight change, carcass parameter and economic feasibility on Hararghe highland sheep, eastern Ethiopia. J Vet Sci Technol 6(2)

Gokul K, Muralidhara (2014) Oral supplements of aqueous extract of tomato seeds alleviate motor abnormality, oxidative impairments and neurotoxicity induced by rotenone in mice: relevance to Parkinson's disease. Neurochem Res 39(7):1382–1394

González M, Cid M, Lobo G (2011) Usage of tomato (*Lycopersicum esculentum* Mill.) seeds in health. In: Preedy VR, Watson RR, Patel VB (eds) Nuts and seeds in health and disease prevention. Elsevier, San Diego

Grabowska M, Wawrzyniak D, Rolle K, Chomczyński P, Oziewicz S, Jurga S, Barciszewski J (2019) Let food be your medicine: nutraceutical properties of lycopene. Food Funct 10(6):3090–3102

Grassino AN, Halambek J, Djaković S, Rimac Brnčić S, Dent M, Grabarić Z (2016) Utilization of tomato peel waste from canning factory as a potential source for pectin production and application as tin corrosion inhibitor. Food Hydrocoll 52:265–274

Herrera PG, Sánchez-Mata MC, Cámara M (2010) Nutritional characterization of tomato fibre as a useful ingredient for food industry. Innov Food Sci Emerg Technol 11(4):707–711

Hong JY, Ban JJ, Quan QL, Eom JE, Shin HS, Chung JH (2022) Mixture of tomato and lemon extracts synergistically prevents PC12 cell death from oxidative stress and improves hippocampal neurogenesis in aged mice. Foods 11(21):3418

Ibrahim M, Labaki M (2022) Extraction and formulation of valuable components from tomato processing by-products. In: Tomato processing by-products. Academic, Cambridge, pp 77–116

Jamaleddine, A., De Caro, P., Bouajila, J., Evon, P., Haddad, J. G., El-Kalamouni, C., ... & Merah, O. (2022). In vitro bioactivities of extracts from tomato pomace. Frontiers in Bioscience-Landmark, 27(9), 259

Jeguirim SE (2022) Tomato wastes valorization for bio-based materials production. In: Tomato processing by-products. Academic, Cambridge, pp 149–170

Jha NK, Kumar Nelson V, Nuli MV, Kanna S, Gahtani RM, Hani U et al (2024) Unveiling the impact of Aging on BBB and Alzheimer's disease: Factors and Therapeutic implications. Ageing Res Rev 98:102224

Jialal I, Norkus EP, Cristol L, Grundy SM (1991) Beta-carotene inhibits the oxidative modification of low-density lipoprotein. Biochim Biophys Acta 1086(1):134–138

Kaboré K, Konaté K, Sanou A, Dakuyo R, Sama H, Santara B, Compaoré EWR, Dicko MH (2022) Tomato by-products, a source of nutrients for the prevention and reduction of malnutrition. Nutrients 14(14):2871

Kadam AN, Mane KA (2022) Enrichment of multigrain cookies with tomato pomace powder. Int J Adv Eng Res Sci 9:5

Kalogeropoulos N, Chiou A, Pyriochou V, Peristeraki A, Karathanos VT (2012) Bioactive phytochemicals in industrial tomatoes and their processing byproducts. LWT Food Sci Technol 49:213–216

Kehili M, Schmidt LM, Reynolds W, Zammel A, Zetzl C, Smirnova I, Allouche N, Sayadi S (2016) Biorefinery cascade processing for creating added value on tomato industrial by-products from Tunisia. Biotechnol Biofuels 9:261

Knoblich M, Anderson B, Latshaw D (2005) Analyses of tomato peel and seed byproducts and their use as a source of carotenoids. J Sci Food Agric 85:1166–1170

Kulczyński B, Gramza-Michałowska A, Kobus-Cisowska J, Kmiecik D (2017) The role of carotenoids in the prevention and treatment of cardiovascular disease—current state of knowledge. J Funct Foods 38:45–65

Kumar M, Tomar M, Bhuyan DJ, Punia S, Grasso S, Sá AGA, Carciofi BAM, Arrutia F, Changan S, Radha, Singh S, Dhumal S, Senapathy M, Satankar V, Anitha T, Sharma A, Pandiselvam R, Amarowicz R, Mekhemar M (2021) Tomato (*Solanum lycopersicum* L.) seed: A review on bioactives and biomedical activities. Biomed Pharmacother 142

Kumar M, Chandran D, Tomar M, Bhuyan DJ, Grasso S, Sá AGA et al (2022) Valorization potential of tomato (*Solanum lycopersicum* L.) seed: nutraceutical quality, food properties, safety aspects, and application as a health-promoting ingredient in foods. Horticulturae 8(3):265

Kurasova, L. A., & Sarsenova, A. Z. (2023). Development of technology for obtaining lycopene-containing dry powder from tomato pomace. The Journal of Almaty Technological University, 2, 91–98

Laranjeira T, Costa A, Faria-Silva C, Ribeiro D, de Oliveira JMPF, Simões S, Ascenso A (2022) Sustainable valorization of tomato by-products to obtain bioactive compounds: their potential in inflammation and cancer management. Molecules 27(5):1701

Lenucci MS, Durante M, Anna M, Dalessandro G, Piro G (2013) Possible use of the carbohydrates present in tomato pomace and in byproducts of the supercritical carbon dioxide lycopene extraction process as biomass for bioethanol production. J Agric Food Chem 61(15):3683–3692

Li XY, Meng L, Shen L, Ji HF (2023) Regulation of gut microbiota by vitamin C, vitamin E and β-carotene. Food Res Int 169:112749

Liadakis GN, Tzia C, Oreopoulou V, Thomopoulos CD (1998) Isolation of tomato seed meal proteins with salt solutions. J Food Sci 63:450–453

Liadakis G, Katsouli M, Chanioti S, Giannou V, Tzia C (2022a) Identification, quantification, and characterization of tomato processing by-products. In: Tomato processing by-products. Academic, Cambridge, pp 1–32

Liadakis G, Kekes T, Frakolaki G, Giannou V, Tzia G (2022b) Ingredients for food products. In: Tomato processing by-products. Academic, Cambridge, pp 117–148

Lima GPP, Gómez HAG, Seabra Junior S, Maraschin M, Tecchio MA, Borges CV (2022) Functional and nutraceutical compounds of tomatoes as affected by agronomic practices, postharvest management, and processing methods: A mini-review. Front Nutr 9:868492

López-Bermúdez YN, Aldana-Heredia JF, Sánchez-Camargo ADP, Hernández-Carrión M (2022) Valorization strategies for a by-product of organic tomato processing as potential ingredient in functional food formulations. Front Food Sci Technol 2:893795

López-Yerena A, Domínguez-López I, Abuhabib MM, Lamuela-Raventós RM, Vallverdú-Queralt A, Pérez M (2023) Tomato wastes and by-products: upcoming sources of polyphenols and carotenoids for food, nutraceutical, and pharma applications. Crit Rev Food Sci Nutr 1–18

Lu Z, Wang J, Gao R, Ye F, Zhao G (2019) Sustainable valorisation of tomato pomace:a comprehensive review. Trends Food Sci Technol 86:172–187

Luengo E, Álvarez I, Raso J (2014) Improving carotenoid extraction from tomato waste by pulsed electric fields. Front Nutr 1(12)

Mechmeche M, Kachouri F, Chouabi M, Ksontini H, Setti K, Hamdi M (2017) Optimization of extraction parameters of protein isolate from tomato seed using response surface methodology. Food Anal Methods 10:809–819

Mironeasa S, Codină GG, Mironeasa C (2018) Effect of composite flour made from tomato seed and wheat of 650 type of a strong quality for bread making on bread quality and alveograph rheological properties. Int J Food Eng 4:22–26

Mizael WC, Costa RG, Rodrigo Beltrão Cruz G, Ramos de Carvalho FF, Ribeiro NL, Lima A et al (2020) Effect of the use of tomato pomace on feeding and performance of lactating goats. Animals 10(9):1574

Morganti P, Morganti G, Chen H-D, Coltelli M-B, Gagliardini A (2024) Smart tissue carriers for innovative cosmeceuticals and nutraceuticals. Cosmetics 11:20

Nour V, Panaite TD, Ropota M, Turcu R, Trandafir I, Corbu AR (2018) Nutritional and bioactive compounds in dried tomato processing waste. CyTA J Food 16(1):222–229

Özbek ZA, Çelik K, Ergönül PG, Hepçimen AZ (2020) A promising food waste for food fortification: characterization of dried tomato pomace and its cold pressed oil. J Food Chem Nanotechnol 6:9–17

Padalino L, Conte A, Lecce A, Likyova D, Sicari V, Pellicanò TM et al (2015) Durum wheat whole-meal spaghetti with tomato peels: how by-product particles size can affect final quality of pasta. J Food Process Technol 6:10

Palomo I, Concha-Meye A, Lutz M, Said M, Sáez B, Vásquez A, Fuentes E (2019) Chemical characterization and antiplatelet potential of bioactive extract from tomato pomace (byproduct of tomato paste). Nutrients 11(456)

Paulino SLJ, Adrián ÁTG, Gabriela EAL, Maribel VM, Sergio MG (2020) Nutraceutical potential of flours from tomato by-product and tomato field waste. J Food Sci Technol 57:3525–3531

Pawar VA, Srivastava S, Tyagi A, Tayal R, Shukla SK, Kumar V (2023) Efficacy of bioactive compounds in the regulation of metabolism and pathophysiology in cardiovascular diseases. Curr Cardiol Rep 25(9):1041–1052

Perea-Domínguez XP, Hernández-Gastelum LZ, Olivas-Olguin HR, Espinosa-Alonso LG, Valdez-Morales M, Medina-Godoy S (2018) Phenolic composition of tomato varieties and an industrial tomato by-product: free, conjugated and bound phenolics and antioxidant activity. J Food Sci Technol 55:3453–3461

Persia M, Parsons C, Schang M, Azcona J (2003) Nutritional evaluation of dried tomato seeds. Poult Sci 82:141–146

Petruk G, Raiola A, Del Giudice R, Barone A, Frusciante L, Rigano MM, Monti DM (2016) An ascorbic acid-enriched tomato genotype to fight UVA-induced oxidative stress in normal human keratinocytes. J Photochem Photobiol B Biol 163:284–289

Przybylska, S. (2020). Lycopene–a bioactive carotenoid offering multiple health benefits: a review. International Journal of Food Science & Technology, 55(1), 11–32

Rajan A, Kumar S, Sunil CK, Radhakrishnan M, Rawson A (2022) Recent advances in the utilization of industrial byproducts and wastes generated at different stages of tomato processing: Status report. J Food Process Preserv 46(11):e17063

Rehal JK, Kaur I (2022) Development and evaluation of soup mix utilizing tomato pomace powder. J Krishi Vigyan 11(1):59–62

Rehal JK, Aggarwal P, Dhaliwal I, Sharma M, Kaushik P (2022) A tomato pomace enriched gluten-free ready-to-cook snack's nutritional profile, quality, and shelf life evaluation. Horticulturae 8(5):403

Rossini G, Toscano G, Duca D, Corinaldesi F, Pedretti F, Riva G (2013) Analysis of the characteristics of the tomato manufacturing residues finalised to the energy recovery. Biomass Bioenergy 51:177–182

Salem BR (2020) Use of tomato pomace, mango seeds kernel and pomegranate peels powders for the production of functional biscuits. Zagazig J Agric Res 47(4):1011–1023

Sandei L, Stingone C, Vitelli R, Cocconi E, Zanotti A, Zoni C (2019) Processing tomato by-products re-use, secondary raw material for tomato products with new functionality. Acta Hortic 1233

Sangeetha K, Ramyaa RB, Khaneghah AM, Radhakrishnan M (2023) Extraction, characterization, and application of tomato seed oil in the food industry: An updated review. J Agric Food Res 11:100529

Sarkar A, Kaul P (2014) Evaluation of tomato processing by-products: a comparative study in a pilot scale setup. J Food Process Eng 3:299–307

Sarkar A, Kamaruddin H, Bentley A, Wang S (2016) Emulsion stabilization by tomato seed protein isolate: Influence of pH, ionic strength and thermal treatment. Food Hydrocoll 57:160–168

Shah KK, Modi B, Lamsal B, Shrestha J, Aryal SP (2021) Bioactive compounds in tomato and their roles in disease prevention. Fundam Appl Agric 6(2):210–222

Silva YPA, Borbaa BC, Pereiraa VA, Reisa MG, Caliarib M, Brooks MS, Ferreira TAPC (2019) Characterization of tomato processing by-product for use as a potential functional food ingredient: nutritional composition, antioxidant activity and bioactive compounds. Int J Food Sci Nutr 70(2):150–160

Simões A, Coelhoso IM, Alves VD, Brazinha C (2023) Recovery and purification of cutin from tomato by-products for application in hydrophobic films. Membranes 13(3):261

Skwarek P, Karwowska M (2022) Fatty acids profile and antioxidant properties of raw fermented sausages with the addition of tomato pomace. Biomolecules 12(11):1695

Stahl W, Sies H (2012) Photoprotection by dietary carotenoids: concept, mechanisms, evidence and future development. Mol Nutr Food Res 56(2):287–295

Stoica RM, Tomulescu C, Cășărică A (2018) Tomato by-products as a source of natural antioxidants for pharmaceutical and food industries—a mini-review. Sci Bull Ser F Biotechnol 22

Strati IF, Oreopoulou V (2014) Effect of extraction parameters on the carotenoid recovery from tomato waste. Food Res Int 65:311–321

Szabo K, Cătoi AF, Vodnar DC (2018) Bioactive compounds extracted from tomato processing by-products as a source of valuable nutrients. Plant Foods Hum Nutr 73:268–277

Szabo K, Diaconeasa Z, Cătoi AF, Vodnar DC (2019) Screening of ten tomato varieties processing waste for bioactive components and their related antioxidant and antimicrobial activities. Antioxidants 8(8):292

Szabo K, Teleky BE, Mitrea L, Călinoiu LF, Martău GA, Simon E et al (2020) Active packaging-poly (vinyl alcohol) films enriched with tomato by-products extract. Coatings 10(2):141

Tarshish E, Hermoni K (2023) Beauty from within: Improvement of skin health and appearance with Lycomato a tomato-derived oral supplement. J Cosmet Dermatol 22(6):1786–1798

Tarshish E, Hermoni K, Sharoni Y, Muizzuddin N (2022) Effect of Lumenato oral supplementation on plasma carotenoid levels and improvement of visual and experiential skin attributes. J Cosmet Dermatol 21(9):4042–4052

Tarshish E, Hermoni K, Muizzuddin N (2023) Effect of Lumenato a Tomato derived oral supplement on improving skin barrier strength. Skin Res Technol 29(11):e13504

Taveira M, Silva LR, Vale-Silva LA, Pinto E, Valentao P, Ferreres F et al (2010) Lycopersicon esculentum seeds: an industrial byproduct as an antimicrobial agent. J Agric Food Chem 58(17):9529–9536

Teterycz D, Sobota A (2023) Use of high-protein and high-dietary-fibre vegetable processing waste from bell pepper and tomato for pasta fortification. Foods 12(13):2567

Tiwari I, Shah KK, Tripathi S, Modi B, Shrestha J, Pandey HP, Bhattarai BP, Rajbhandari BP (2020) Post-harvest practices and loss assessment in tomato (*Solanum lycopersicum* L.) in Kathmandu. Nepal J Agric Nat Resour 3:335–352

Tommonaro G, Poli A, De Rosa S, Nicolaus B (2008) Tomato derived polysaccharides for biotechnological applications: chemical and biological approaches. Molecules 13(6):1384–1398

Travieso MD, de Evan T, Marcos CN, Molina-Alcaide E (2022) Tomato by-products as animal feed. In: Tomato processing by-products. Academic, Cambridge, pp 33–76

Trombino S, Cassano R, Procopio D, Di Gioia ML, Barone E (2021) Valorization of tomato waste as a source of carotenoids. Molecules 26(5062)

Tuoxunjiang H, Yimamu A, Li XQ, Maimaiti R, Wang YL (2020) Effect of ensiled tomato pomace on performance and antioxidant status in the peripartum dairy cow. J Anim Feed Sci 29(2):105–114

Valdez-Morales M, Espinosa-Alonso LG, Espinoza-Torres LC, Delgado-Vargas F, Medina-Godoy S (2014) Phenolic content and antioxidant and antimutagenic activities in tomato peel, seeds, and byproducts. J Agric Food Chem 62(23):5281–5289

Valenti B, Luciano G, Pauselli M, Mattioli S, Biondi L, Priolo A et al (2018) Dried tomato pomace supplementation to reduce lamb concentrate intake: Effects on growth performance and meat quality. Meat Sci 145:63–70

Ventura MR, Pieltain MC, Castanon JIR (2009) Evaluation of tomato crop by-products as feed for goats. Anim Feed Sci Technol 154(3–4):271–275

Viuda-Martos M, Sanchez-Zapata E, Sayas-Barberá E, Sendra E, Pérez-Álvarez JA, Fernández-López J (2014) Tomato and tomato byproducts. Human health benefits of lycopene and its application to meat products: a review. Crit Rev Food Sci Nutr 54(8):1032–1049

Weyh C, Krüger K, Peeling P, Castell L (2022) The role of minerals in the optimal functioning of the immune system. Nutrients 14(3):644

Wirkijowska A, Zarzycki P, Teterycz D, Nawrocka A, Blicharz-Kania A, Łysakowska P (2023) The influence of tomato and pepper processing waste on bread quality. Appl Sci 13(16):9312

World Processing Tomato Council (2020) The 2020 processed tomato yearbook 2020. Available at http://www.tomatonews.com/pdf/yearbook/2020/index.html#48

Yoon KN, Cui Y, Quan QL, Lee DH, Oh JH, Chung JH (2024) Tomato and lemon extracts synergistically improve cognitive function by increasing brain-derived neurotrophic factor levels in aged mice. Br J Nutr 131(7):1105–1114

Zhang X, Zhou Q, Qi Y, Chen X, Deng J, Zhang Y, Li R, Fan J (2023) The effect of tomato and lycopene on clinical characteristics and molecular markers of UV-induced skin deterioration: A systematic review and meta-analysis of intervention trials. Crit Rev Food Sci Nutr 63:1–20

Zhao B, Ren B, Guo R, Zhang W, Ma S, Yao Y, Yuan T, Liu Z, Liu X (2017) Supplementation of lycopene attenuates oxidative stress induced neuroinflammation and cognitive impairment via Nrf2/NF-κB transcriptional pathway. Food Chem Toxicol 109:505–516

Part V
Future Trends and Conclusion

Future Trends and Conclusion for Bioactive Phytochemicals in By-Products from Bulb, Flower, and Fruit Vegetables

Mohamed Fawzy Ramadan

Abstract Vegetable production and processing biowaste as well as vegetable post-harvest by-products are disposed of. These wastes contain a wealth of beneficial phytochemicals. Consequently, it stimulates scientific curiosity to investigate the phytochemical composition of these by-products and attempt to determine their possible uses. A common deficiency in knowledge is the precise phytochemical makeup of each kind of by-product derived from vegetables. The most crucial elements for conducting consecutive in vitro and in vivo investigations in the field are not entirely agreed upon or commonly accepted. The variety of bioactive chemicals obtained by various authors frequently complicates comparisons. Botanicals' uncertain safety profile may complicate their application, and more toxicological data are required to support this. This work highlights the current situation and valorization potential of bulb, flower, and fruity vegetable by-products.

Keywords Vegetable · Circle economy · Environment · Waste · Phytochemistry · Processing · Production · Challenge

1 Introduction

Modern vegetables come from a long history of selecting plants with enhanced edibility, such as bigger leaves, larger bulbs, tastier flesh, and fancier colors. The main edible parts of common vegetables include leaves, roots, bulbs, flowers, stems, and fruits. In the market, inedible residues from vegetable production, tissue culture transplants, and postharvest culls are usually discarded, buried, or incinerated. This waste is a treasure trove of plenty of bioactive phytochemicals. Thus, it evokes

M. F. Ramadan (✉)
Department of Clinical Nutrition, Faculty of Applied Medical Sciences, Umm Al-Qura University, Makkah, Saudi Arabia
e-mail: mfhassanien@uqu.edu.sa

M. F. Ramadan (ed.), *Bioactive Phytochemicals in By-products from Bulb, Flower and Fruit Vegetables*, https://doi.org/10.1007/978-3-031-77399-0_13

scientific interest to explore the phytochemical content in these by-products and tries to figure out their potential use (Espinosa-Alonso et al. 2020; Wang et al. 2023; Cano-Lamadrid and Artes-Hernandez 2021; Jalal et al. 2023; Tlais et al. 2020; Santos et al. 2022; Sorrenti et al. 2023; Reguengo et al. 2022; Teshome et al. 2023).

Using waste and residuals to extract bioactive compounds may make them "zero-profit." It means that for each product sale (zero-profit), it may return a used product. Consequently, this is a recurring incentive for food and nutrition scientists to investigate bioactive compounds in vegetable by-products. Meanwhile, the relevance and coherence of vegetable bioactives are of value to the investigation. The interest in vegetable phytochemicals has been expanding to taste and aroma. For example, one known ethanol extract of garlic with 5.0 ppm of s-allyl cysteine and 0.6 ppm of total thiosulfinates has been claimed to use the aroma in pharmacy products (Taghian Dinani and van der Goot 2023; Rodriguez Garcia and Raghavan 2022; Jiménez-Moreno et al. 2020; Kumar et al. 2021; Kainat et al. 2022; Lemes et al. 2022; Rifna et al. 2023; Majumder and Annegowda 2021; Espinosa-Alonso et al. 2020; Reguengo et al. 2022).

2 The Objective of the Work

The global demands of vegetable consumption have increased by-products from both the production and processing of vegetables. This work illustrates and screens the current situation and valorization potential of bulb, flower, and fruity vegetable by-products. Research on identifying and quantifying both primary and secondary metabolites is mandatory in screening the bioactive phytochemicals from the selected by-products despite recognizing their variation within the different extraction segments. The objectives for this work were, namely: (a) guided isolation of pure phytochemicals from each vegetable by-product, which was subjected to in-depth phytochemical and biological profiling, and (b) quantification of the bioactive compounds containing phenolics, carotenoids, and other bioactive compounds (Lau et al. 2021; Taghian Dinani and van der Goot 2023; Sharma et al. 2021; Ueda et al. 2022; Trigo et al. 2022; Ganesh et al. 2022; Jalal et al. 2023; Zhu et al. 2023; Malenica et al. 2022).

This work has established the following strategy as a set of working hypotheses: (a) extraction; (b) purification techniques using guided isolation methods; (c) in-depth phytochemical profiling of the purified extracts; (d) in vitro biological profiling of the extra- and intra-cellular activities and, if possible, toxicity testing. Optimizing extraction techniques for improved extraction and separation of components within various plant parts and tissues is a critical parameter in the study of vegetables, given their morphological complexity. A fruitful strategy would be combining different organic solvents and environmentally friendly extracts, such as supercritical fluids and green solvents like ionic liquids, with the classical and/or solvent-assisted microwave- and ultrasound-assisted extraction to guarantee higher quality in terms of bioactivity and safety profile (Manzoor et al. 2021; Poletto et al.

2020; Hernández-Montesinos et al. 2024; Sultanbawa and Sivakumar 2022; Trifan et al. 2022; Halder and Jha 2023; Domínguez et al. 2021; Cucu et al. 2022).

3 Bioactive Phytochemicals

Phytochemicals are compounds in plants that are involved in biological activities. They are not essential for the human body, but these compounds promote optimal health. For example, it can support normal physiological functions like enzyme functions, hormonal balance, DNA repair, and anti-inflammatory, antioxidant, and antimicrobial impacts. Various plant bioactive compounds, including carbohydrates, lipids, proteins, vitamins, and phenolic compounds, have been discovered. According to their specific chemical structures, the phenolic compounds usually consist of several hydroxyls (-OH) and conjugated unsaturation. Many bioactive compounds are contained in edible plants and fruits, and their by-products have high chemical diversity worldwide, such as flavonoids, alkaloids, saponins, and terpenoids. It is worth noting that some of these bioactive compounds also possess radical scavenging ability in our body and contribute to reducing cardiovascular disease (CVD) risk factors (Alì et al. 2021; Santhiravel et al. 2022; Cote et al. 2022; Makhaik et al. 2021; Howes et al. 2020; George et al. 2021; Nistor et al. 2022).

Considering the high health risk of cardiovascular or cerebrovascular diseases, abdominal obesity, metabolic syndrome, diabetes mellitus, and related diseases, exploring dietary components or formulas containing antioxidants and bioactive phenolic compounds in plant resources has become a research focus. The existing studies suggest that vegetable by-products are a great source of these bioactive phytochemicals. Many organic bioactive compounds could be released or produced during various processing methods, such as drying, fermentation, or extraction with different solvents, which have attracted interest and attention worldwide. Moreover, extraction of organic compounds from the vegetable by-products of the corresponding frozen products with various solvents has become a feasible method practiced for some recent investigations (Coman et al. 2020; Reguengo et al. 2022; Cano-Lamadrid and Artes-Hernandez 2021; Majumder and Annegowda 2021; Sorrenti et al. 2023; Galali et al. 2020; Gomez-Garcia et al. 2020).

4 Importance of Phytochemicals for Human Health

The interest in natural compounds has increased due to their use in disease prevention, health care, and wellness, as they can be used with the same efficiency and fewer side effects than their synthetic counterparts. The emphasis on diet and the consumption of vegetables, as carriers of bioactive compounds, prevents diseases, particularly chronic ones, and contributes to well-being. Many epidemiological studies have supplemented the diet with bioactive components that contribute to human health maintenance and can enhance the efficacy of drug therapies in many cases. These bioactive compounds can help reduce oxidative stress, improve brain function and immune function, optimize glucose metabolism, reduce the risk of cardiovascular disease, and have cholesterol-lowering and anti-inflammatory effects. Moreover, because of their antioxidant properties, they can protect against cancer and act as anti-aging products. Many of these compounds enter the plant in the last 2 weeks of maturation. For this reason, by-products obtained from these vegetables can represent a good source of bio-compounds of potential interest from a nutraceutical and health perspective (Samtiya et al. 2021; Banwo et al. 2021; Favela-González et al. 2020; Dhalaria et al. 2020; Zhao et al. 2021; Guan et al. 2021; Thiruvengadam et al. 2021).

Hence, this study focuses on by-products from bulb, flower, and fleshy vegetables, highlighting some phytochemicals and radical scavenging actives. Considering the proliferation of properties surrounding diet and using "bio" products, especially when referring to by-products, these are biologically active compounds that promote a healthier lifestyle. In terms of phytochemicals, by-products contain a high content of taste and color-associated compounds, including flavonoids, phenolic acids, and pro-vitamin A, as well as citric acid, a principal organic acid and important antioxidants (Salem et al. 2023; Annegowda and Majumder 2021; Coman et al. 2020; Kaur and Ghoshal 2024; Stoica et al. 2023; Kaur et al. 2024; Gomez-Molina et al. 2024; Lalramhlimi et al. 2022).

5 Challenges and Future Directions

It is essential to reveal the bioactive phytochemicals in by-products from bulb, flower, and fruity vegetables. These bioactive phytochemicals within by-products have emerged as promising functionalities with health benefits. However, it is essential to note that several challenges exist in identifying the bioactive phytochemicals from the above categories of by-products. Therefore, it is critical to consider the future directions that can be employed to undertake this kind of research while exploring the bioactive phytochemicals in the by-products of different vegetables.

Challenges There is often a lack of detailed information on the phytochemical composition of each type of by-products obtained from fresh vegetables. There is no complete or widely accepted consensus on the most important characteristics to carry out successively in vitro and in vivo studies in the field. The diversity of obtained bioactive compounds by different authors often makes comparisons intricate. Application of botanicals might be hindered because of the unpredictable safety profile, and more toxicological data are needed for this purpose.

Future Directions The unexploited by-products of bulb, flower, and fruity vegetables have the potential to be the source of numerous new bioactive agents with healthy properties. Despite the growing amount of scientific literature concerning the phytochemical composition of fresh vegetables, in general, because of the complexity of mixtures of phytochemicals present in them, much less is known about which of these compounds may be bioactive (Annegowda and Majumder 2021; Coman et al. 2020; Marcillo-Parra et al. 2021; Shabir et al. 2022; Kaur et al. 2024; Tsegay and Mulaw 2024; Salem et al. 2023; Mohammadnezhad et al. 2023; Nguyen et al. 2023).

6 Conclusion

Some countries import crop by-products as animal feed due to a lack of water or labor for onsite usage in foreign countries. Some countries also export crops, mainly peeled, and sometimes import those peeled crops. It is time to explore the practical bioactive phytochemicals in these by-products. Bulb white and red (purple) vegetables, flower vegetables, and some crops with edible flowers and fruits (such as corn, squash, tomato, and eggplant) were chosen as observation materials in Kagoshima, Kyushu. The respective by-product edible parts and peel of imported ones had more phytochemical quantity and activity than the original nutrients of bulbs, flowers, or fruits. Phytochemicals localize in cropped tissues other than peel and outer green leaves. There are differences in phytochemicals, even in the vegetables of the same families. The subjects of the following studies include models of induced or reduced phytochemicals in human cancer. Additionally, the potential health effects of by-products are explored, with the hypothesis that methods like those targeting human cancer cell reduction may show broader applicability in promoting healthier diets among more people.

Treating the by-products onsite to prevent epidemics and reduce processing and waste treatment costs is essential. Exploring the bioactive phytochemicals in the bulb, flower, and fruity vegetable by-products and evaluating them is increasingly required worldwide (Richards and Yabar 2023; Shurson et al. 2023; Pueppke et al. 2020; Gocht et al. 2021).

References

Alì S, Davinelli S, Accardi G, Aiello A, Caruso C, Duro G, Ligotti ME, Pojero F, Scapagnini G, Candore G (2021) Healthy ageing and Mediterranean diet: A focus on hormetic phytochemicals. Mech Ageing Dev 200:111592

Annegowda HV, Majumder P (2021) Valuable bioactives from vegetable wastes. In: Valorization of agri-food wastes and by-products. Academic, pp 83–109. https://www.sciencedirect.com/article/pii/B9780128240441000039

Banwo K, Olojede AO, Adesulu-Dahunsi AT, Verma DK, Thakur M, Tripathy S, Singh S, Patel AR, Gupta AK, Aguilar CN, Utama GL (2021) Functional importance of bioactive compounds of foods with potential health benefits: A review on recent trends. Food Biosci 43:101320

Cano-Lamadrid M, Artes-Hernandez F (2021) By-products revalorization with non-thermal treatments to enhance phytochemical compounds of fruit and vegetables derived products: A review. Foods 11(1):59

Coman V, Teleky BE, Mitrea L, Martău GA, Szabo K, Călinoiu LF, Vodnar DC (2020) Bioactive potential of fruit and vegetable wastes. Adv Food Nutr Res 91:157–225

Cote B, Elbarbry F, Bui F, Su JW, Seo K, Nguyen A, Lee M, Rao DA (2022) Mechanistic basis for the role of phytochemicals in inflammation-associated chronic diseases. Molecules 27(3):781

Cucu AA, Baci GM, Cucu AB, Dezsi Ş, Lujerdean C, Hegeduş IC, Bobiş O, Moise AR, Dezmirean DS (2022) Calluna vulgaris as a valuable source of bioactive compounds: exploring its phytochemical profile, biological activities and apitherapeutic potential. Plants 11(15):1993

Dhalaria R, Verma R, Kumar D, Puri S, Tapwal A, Kumar V, Nepovimova E, Kuca K (2020) Bioactive compounds of edible fruits with their anti-aging properties: A comprehensive review to prolong human life. Antioxidants 9(11):1123

Domínguez R, Pateiro M, Munekata PE, McClements DJ, Lorenzo JM (2021) Encapsulation of bioactive phytochemicals in plant-based matrices and application as additives in meat and meat products. Molecules 26(13):3984

Espinosa-Alonso LG, Valdez-Morales M, Aparicio-Fernandez X, Medina-Godoy S, Guevara-Lara F (2020) Vegetable by-products. Food wastes and by-products: Nutraceutical and health potential. Wiley-Blackwell, Hoboken, NJ, pp 223–266

Favela-González KM, Hernández-Almanza AY, De la Fuente-Salcido NM (2020) The value of bioactive compounds of cruciferous vegetables (Brassica) as antimicrobials and antioxidants: A review. J Food Biochem 44(10):e13414

Galali Y, Omar ZA, Sajadi SM (2020) Biologically active components in by-products of food processing. Food Sci Nutr 8:3004–3022

Ganesh KS, Sridhar A, Vishali S (2022) Utilization of fruit and vegetable waste to produce value-added products: Conventional utilization and emerging opportunities—a review. Chemosphere 287:132221

George BP, Chandran R, Abrahamse H (2021) Role of phytochemicals in cancer chemoprevention: insights. Antioxidants 10(9):1455

Gocht A, Consmüller N, Thom F, Grethe H (2021) Economic and environmental consequences of the ECJ genome editing judgment in agriculture. Agronomy 11(6):1212

Gomez-Garcia R, Campos DA, Aguilar CN, Madureira AR, Pintado M (2020) Valorization of melon fruit (*Cucumis melo* L.) by-products: phytochemical and biofunctional properties with emphasis on recent trends and advances. Trends Food Sci Technol 99:507–519

Gomez-Molina M, Albaladejo-Marico L, Yepes-Molina L, Nicolas-Espinosa J, Navarro-León E, Garcia-Ibañez P, Carvajal M (2024) Exploring phenolic compounds in crop by-products for cosmetic efficacy. Int J Mol Sci 25(11):5884

Guan R, Van Le Q, Yang H, Zhang D, Gu H, Yang Y, Sonne C, Lam SS, Zhong J, Jianguang Z, Liu R (2021) A review of dietary phytochemicals and their relation to oxidative stress and human diseases. Chemosphere 271:129499

Halder M, Jha S (2023) Medicinal plants and bioactive phytochemical diversity: a fountainhead of potential drugs against human diseases. In: Medicinal plants: biodiversity, biotechnology and conservation. Springer Nature, Singapore, pp 39–93

Hernández-Montesinos IY, Carreón-Delgado DF, Lazo-Zamalloa O, Tapia-López L, Rosas-Morales M, Ochoa-Velasco CE, Hernández-Carranza P, Cruz-Narváez Y, Ramírez-López C (2024) Exploring agro-industrial by-products: phenolic content, antioxidant capacity, and phytochemical profiling via FI-ESI-FTICR-MS untargeted analysis. Antioxidants 13(8):925

Howes MJR, Perry NS, Vásquez-Londoño C, Perry EK (2020) Role of phytochemicals as nutraceuticals for cognitive functions affected in ageing. Br J Pharmacol 177(6):1294–1315

Jalal H, Giammarco M, Lanzoni L, Akram MZ, Mammi LM, Vignola G, Chincarini M, Formigoni A, Fusaro I (2023) Potential of fruits and vegetable by-products as an alternative feed source for sustainable ruminant nutrition and production: a review. Agriculture 13(2):286

Jiménez-Moreno N, Esparza I, Bimbela F, Gandía LM, Ancín-Azpilicueta C (2020) Valorization of selected fruit and vegetable wastes as bioactive compounds: Opportunities and challenges. Crit Rev Environ Sci Technol 50(20):2061–2108

Kainat S, Arshad MS, Khalid W, Zubair Khalid M, Koraqi H, Afzal MF, Noreen S, Aziz Z, Al-Farga A (2022) Sustainable novel extraction of bioactive compounds from fruits and vegetables waste for functional foods: a review. Int J Food Prop 25(1):2457–2476

Kaur A, Ghoshal G (2024) Valorization of carrot and turnip processing wastes and by-products. In: Ray RC (ed) Roots, tubers, and bulb crop wastes: management by biorefinery approaches. Springer Nature, Singapore, pp 183–214

Kaur B, Das D, Panesar G, Panesar PS, Riar CS, Zakaria ZA (2024) Technological and analytical approaches for extraction of bioactive compounds and nutraceuticals from fruit and vegetable processing byproducts. In: Bioactive compounds and nutraceuticals from plant sources. Apple Academic, pp 157–201

Kumar K, Srivastav S, Sharanagat VS (2021) Ultrasound assisted extraction (UAE) of bioactive compounds from fruit and vegetable processing by-products: A review. Ultrason Sonochem 70:105325

Lalramhlimi B, Mukherjee D, Chakraborty I, Ghosh N, Chattopadhyay A, Dey RC (2022) Fruit and vegetable wastes as livestock feeds. In: Ray RC (ed) Fruits and vegetable wastes: valorization to bioproducts and platform chemicals. Springer Nature, Singapore, pp 139–168

Lau KQ, Sabran MR, Shafie SR (2021) Utilization of vegetable and fruit by-products as functional ingredient and food. Front Nutr 8:661693

Lemes AC, Egea MB, Oliveira Filho JGD, Gautério GV, Ribeiro BD, Coelho MAZ (2022) Biological approaches for extraction of bioactive compounds from agro-industrial by-products: a review. Front Bioeng Biotechnol 9:802543

Majumder P, Annegowda HV (2021) Fruit and vegetable by-products: Novel ingredients for a sustainable society. In: Valorization of agri-food wastes and by-products. Academic, pp 133–156

Makhaik MS, Shakya AK, Kale R (2021) Dietary phytochemicals: As a natural source of antioxidants. Antioxidants-benefits, sources, mechanisms of action, p 646

Malenica D, Kass M, Bhat R (2022) Sustainable management and valorization of agri-food industrial wastes and by-products as animal feed: for ruminants, non-ruminants and as poultry feed. Sustainability 15(1):1–23

Manzoor M, Singh J, Gani A, Noor N (2021) Valorization of natural colors as health-promoting bioactive compounds: Phytochemical profile, extraction techniques, and pharmacological perspectives. Food Chem 362:130141

Marcillo-Parra V, Tupuna-Yerovi DS, González Z, Ruales J (2021) Encapsulation of bioactive compounds from fruit and vegetable by-products for food application-A review. Trends Food Sci Technol 116:11–23

Mohammadnezhad P, Valdés A, Álvarez-Rivera G (2023) Bioactivity of food by-products: An updated insight. Curr Opin Food Sci 52:101065

Nguyen TL, Ora A, Häkkinen ST, Ritala A, Räisänen R, Kallioinen-Mänttäri M, Melin K (2023) Innovative extraction technologies of bioactive compounds from plant by-products for textile colorants and antimicrobial agents. Biomass Convers Biorefin 1–30

Nistor M, Pop R, Daescu A, Pintea A, Socaciu C, Rugina D (2022) Anthocyanins as key phytochemicals acting for the prevention of metabolic diseases: an overview. Molecules 27(13):4254

Poletto P, Alvarez-Rivera G, Torres TM, Mendiola JA, Ibañez E, Cifuentes A (2020) Compressed fluids and phytochemical profiling tools to obtain and characterize antiviral and anti-inflammatory compounds from natural sources. Trends Anal Chem 129:115942

Pueppke SG, Nurtazin S, Ou W (2020) Water and land as shared resources for agriculture and aquaculture: Insights from Asia. Water 12(10):2787

Reguengo LM, Salgaço MK, Sivieri K, Júnior MRM (2022) Agro-industrial by-products: Valuable sources of bioactive compounds. Food Res Int 152:110871

Richards D, Yabar H (2023) Promoting energy and resource recovery from livestock waste: Case study Yuge Farm, Japan. Case Stud Chem Environ Eng 7:100299

Rifna EJ, Misra NN, Dwivedi M (2023) Recent advances in extraction technologies for recovery of bioactive compounds derived from fruit and vegetable waste peels: A review. Crit Rev Food Sci Nutr 63(6):719–752

Rodriguez Garcia SL, Raghavan V (2022) Green extraction techniques from fruit and vegetable waste to obtain bioactive compounds-A review. Crit Rev Food Sci Nutr 62(23):6446–6466

Salem MA, Mansour HEA, Mosalam EM, El-Shiekh RA, Ezzat SM, Zayed A (2023) Valorization of by-products derived from onions and potato: extraction optimization, metabolic profile, outstanding bioactivities, and industrial applications. Waste Biomass Valorization 14(6):1823–1858

Samtiya M, Aluko RE, Dhewa T, Moreno-Rojas JM (2021) Potential health benefits of plant food-derived bioactive components: An overview. Foods 10(4):839

Santhiravel S, Bekhit AEDA, Mendis E, Jacobs JL, Dunshea FR, Rajapakse N, Ponnampalam EN (2022) The impact of plant phytochemicals on the gut microbiota of humans for a balanced life. Int J Mol Sci 23(15):8124

Santos D, da Silva JAL, Pintado M (2022) Fruit and vegetable by-products' flours as ingredients: A review on production process, health benefits and technological functionalities. LWT Food Sci Technol 154:112707

Shabir I, Pandey VK, Dar AH, Pandiselvam R, Manzoor S, Mir SA, Shams R, Dash KK, Fayaz U, Khan SA, Jeevarathinam G (2022) Nutritional profile, phytochemical compounds, biological activities, and utilisation of onion peel for food applications: a review. Sustainability 14(19):11958

Sharma S, Katoch V, Kumar S, Chatterjee S (2021) Functional relationship of vegetable colors and bioactive compounds: Implications in human health. J Nutr Biochem 92:108615

Shurson GC, Dierenfeld ES, Dou Z (2023) Rules are meant to be broken—rethinking the regulations on the use of food waste as animal feed. Resour Conserv Recycl 199:107273

Sorrenti V, Burò I, Consoli V, Vanella L (2023) Recent advances in health benefits of bioactive compounds from food wastes and by-products: Biochemical aspects. Int J Mol Sci 24(3):2019

Stoica F, Rațu RN, Veleșcu ID, Stănciuc N, Râpeanu G (2023) A comprehensive review on bioactive compounds, health benefits, and potential food applications of onion (*Allium cepa* L.) skin waste. Trends Food Sci Technol 141:104173

Sultanbawa Y, Sivakumar D (2022) Enhanced nutritional and phytochemical profiles of selected underutilized fruits, vegetables, and legumes. Curr Opin Food Sci 46:100853

Taghian Dinani S, van der Goot AJ (2023) Challenges and solutions of extracting value-added ingredients from fruit and vegetable by-products: a review. Crit Rev Food Sci Nutr 63(25):7749–7771

Teshome E, Teka TA, Nandasiri R, Rout JR, Harouna DV, Astatkie T, Urugo MM (2023) Fruit by-products and their industrial applications for nutritional benefits and health promotion: a comprehensive review. Sustainability 15(10):7840

Thiruvengadam M, Venkidasamy B, Subramanian U, Samynathan R, Ali Shariati M, Rebezov M, Girish S, Thangavel S, Dhanapal AR, Fedoseeva N, Lee J (2021) Bioactive compounds in oxidative stress-mediated diseases: targeting the NRF2/ARE signaling pathway and epigenetic regulation. Antioxidants 10(12):1859

Tlais AZA, Fiorino GM, Polo A, Filannino P, Di Cagno R (2020) High-value compounds in fruit, vegetable and cereal byproducts: an overview of potential sustainable reuse and exploitation. Molecules 25(13):2987

Trifan A, Zengin G, Sinan KI, Sieniawska E, Sawicki R, Maciejewska-Turska M, Skalikca-Woźniak K, Luca SV (2022) Unveiling the phytochemical profile and biological potential of five Artemisia species. Antioxidants 11(5):1017

Trigo JP, Alexandre EM, Saraiva JA, Pintado ME (2022) High value-added compounds from fruit and vegetable by-products—characterization, bioactivities, and application in the development of novel food products. Crit Rev Food Sci Nutr 60(8):1388–1416

Tsegay ZT, Mulaw G (2024) Antimicrobial activities and mode of action of bioactive substances from vegetable and fruit byproducts as a current option for valorization. Waste Biomass Valorization 1–28

Ueda JM, Pedrosa MC, Heleno SA, Carocho M, Ferreira IC, Barros L (2022) Food additives from fruit and vegetable by-products and bio-residues: a comprehensive review focused on sustainability. Sustainability 14(9):5212

Wang C, Li M, Duan X, Abu-Izneid T, Rauf A, Khan Z, Mitra S, Emran TB, Aljohani AS, Alhumaydhi FA, Thiruvengadam M (2023) Phytochemical and nutritional profiling of tomatoes; impact of processing on bioavailability—a comprehensive review. Food Rev Int 39(8):5986–6010

Zhao XX, Lin FJ, Li H, Li HB, Wu DT, Geng F, Ma W, Wang Y, Miao BH, Gan RY (2021) Recent advances in bioactive compounds, health functions, and safety concerns of onion (*Allium cepa* L.). Front Nutr 8:669805

Zhu Y, Luan Y, Zhao Y, Liu J, Duan Z, Ruan R (2023) Current technologies and uses for fruit and vegetable wastes in a sustainable system: A review. Foods. https://doi.org/10.3390/foods12101949

Correction to: Phytochemicals from Cucumber (*Cucumis sativus*) By-products

M. Z. Bauleth, J. Sheehama, Natascha Cheikhyoussef, Ahmad Cheikhyoussef, M. Makhaba, R. Sharma, C. N. Cupido, and Ahmed A. Hussein

Correction to:
Chapter 9 in: M. F. Ramadan (ed.), *Bioactive Phytochemicals in By-products from Bulb, Flower and Fruit Vegetables*, https://doi.org/10.1007/978-3-031-77399-0_9

The chapter "Phytochemicals from Cucumber (*Cucumis sativus*) By-products" was inadvertently published with the incorrect author name "M. Mkhaba". The correct author name is M. Makhaba.

The updated version of this chapter can be found at
https://doi.org/10.1007/978-3-031-77399-0_9

M. F. Ramadan (ed.), *Bioactive Phytochemicals in By-products from Bulb, Flower and Fruit Vegetables*, https://doi.org/10.1007/978-3-031-77399-0_14

The manufacturer's authorised representative in the EU is Springer Nature Customer Service Centre GmbH, Europaplatz 3, 69115 Heidelberg, Germany. If you have any concerns regarding our products, please contact ProductSafety@springernature.com

Printed and bound by CPI Group (UK) Ltd, Croydon, CR0 4YY
30/09/2026
02225563-0010